Lecture Notes in Mathematics 1541

D.A. Dawson B. Maisonneuve J. Spencer

Ecole d'Eté de Probabilités de Saint-Flour XXI - 1991

Editor: P.L. Hennequin

Springer-Verlag
Berlin Heidelberg New York
London Paris Tokyo
Hong Kong Barcelona
Budapest

Authors

Donald A. Dawson
Carleton University
Department of Mathematics
Ottawa, Ontario, Canada K 15586

Bernard Maisonneuve
Université de Grenoble I
Institut Fourier
B.P. 74
F-38402 Saint Martin d'Hères Cedex, France

Joel Spencer
New York University
Courant Institute of Mathematical Sciences
251 Mercer Street
New York, NY 10012, USA

Editor

Paul-Louis Hennequin
Université Blaise Pascal
Clermont-Ferrand
Mathématiques Appliquées
63177 Aubière Cedex, France

Mathematics Subject Classification (1991): 60-02, 05C80, 35R60, 60G48, 60G55, 60J25, 60J35, 60J80

ISBN 3-540-56622-8Springer-Verlag Berlin Heidelberg New York
ISBN 0-387-56622-8Springer-Verlag New York Berlin Heidelberg

Typesetting: Camera-ready by author/editor
46/3140-543210 - Printed on acid-free paper

INTRODUCTION

Ce volume contient les cours donnés à l'Ecole d'Eté de Calcul des Probabilités de Saint-Flour du 18 Août au 4 Septembre 1991.

Nous remercions les auteurs qui ont effectué un gros travail de rédaction définitive qui fait de leurs cours un texte de référence.

L'Ecole a rassemblé soixante cinq participants dont 33 ont présenté, dans un exposé, leur travail de recherche.

On trouvera ci-dessous la liste des participants et de ces exposés dont un résumé pourra être obtenu sur demande.

Afin de faciliter les recherches concernant les écoles antérieures, nous redonnons ici le numéro du volume des "Lecture Notes" qui leur est consacré :

Lecture Notes in Mathematics

1971 : n° 307 - 1973 : n° 390 - 1974 : n° 480 - 1975 : n° 539 -
1976 : n° 598 - 1977 : n° 678 - 1978 : n° 774 - 1979 : n° 876 -
1980 : n° 929 - 1981 : n° 976 - 1982 : n° 1097 - 1983 : n° 1117 -
1984 : n° 1180 - 1985 - 1986 et 1987 : n° 1362 - 1988 : n° 1427 -
1989 : n° 1464 - 1990 : n° 1527

Lecture Notes in Statistics

1986 : n° 50

TABLE DES MATIERES

Donald D. DAWSON : "MEASURE-VALUED MARKOV PROCESSES"

Bernard MAISONNEUVE : "PROCESSUS DE MARKOV : NAISSANCE, RETOURNEMENT, REGENERATION"

Joël SPENCER : "NINE LECTURES ON RANDOM GRAPHS"

MEASURE-VALUED MARKOV PROCESSES

Donald A. DAWSON

MEASURE-VALUED MARKOV PROCESSES

Donald Dawson
Department of Mathematics and Statistics
Carleton University
Ottawa, Canada

CONTENTS

1. INTRODUCTION

The central theme of these lectures is the construction and study of measure-valued Markov processes. This subject is in the midst of rapid development and has been stimulated from several different directions including branching processes, population genetics models, interacting particle systems and stochastic partial differential equations. The objective of these notes is to provide an introduction to these different aspects of the subject with some emphasis on their interrelations and also to outline some aspects currently under development. Chapters 1-9 provide an introduction to some of the main ideas and tools in the theory of measure-valued processes. Chapters 10-12 cover topics currently under active development and are primarily intended as an introduction to the growing literature devoted to these aspects of measure-valued processes. Throughout the emphasis is given to outlining the main lines of development rather than attempting a systematic detailed exposition. In Section 1.5 we describe in more detail the structure of these notes (including interrelations between the chapters) as well as mention the principal methods. In the remaining sections of this introduction we outline the roots of the theory of measure-valued Markov processes and the major topics to be discussed in these notes.

By a *measure-valued Markov process* we will always mean a Markov process whose state space is $M(E)$, the space of Radon measures on $(E,\mathcal{E})$, where E is a Polish space and $\mathcal{E} = \mathcal{B}(E)$ is the σ-algebra of Borel subsets of E.

1.1. Particle Systems and their Empirical Measures

Consider a system of N E-valued processes $\{Z_i(t): i=1,\dots,N,\ t\ge 0\}$. The associated (*normalized*) *empirical measure process* is defined by

$$(1.1.1)\qquad X(t) = N^{-1}\sum_{i=1}^{N} \delta_{Z_i(t)}.$$

We will show in the next chapter that empirical measure processes of type (1.1.1) in which the $\{Z_i\}$ form an exchangeable Markov system are in fact measure-valued Markov processes. Exchangeable particle systems arise naturally in many fields including statistical physics, population biology and genetic algorithms.

In Section 2.10 and Chapter 4 we will proceed to study a related class of measure-valued processes on $\mathbb{R}^d$ which arise from spatially distributed population models in which the number of particles, $N(t)$, at time t, is no longer conserved but in which particles undergo birth and death. In this case we consider atomic measures of the form

$$(1.1.2) \qquad X(t) = m(N(0)) \sum_{i=1}^{N(t)} \delta_{Z_i(t)},$$

where m(N(0)) is the mass of each particle.

1.2. Limits of Particle Systems and Stochastic Partial Differential Equations

Non-atomic measure-valued processes arise naturally in the limit, $N \to \infty$, of systems of the form (1.1.1) or in the corresponding *high density* limit of (1.1.2). In many such cases a law of large numbers phenomenon (*propagation of chaos)* occurs and the limiting process is deterministic. For example this occurs in the usual *mean-field* or *McKean-Vlasov limit* (cf. Gärtner (1986), Léonard (1986), Sznitman (1989)). The high density limit $N(0) \longrightarrow \infty$, $m(N(0))\cdot N(0) \longrightarrow m$, of systems of the form (1.1.2) can also give rise to (deterministic) linear or nonlinear partial differential equations including reaction diffusion equations, hydrodynamic equations, etc. (cf. e.g. Oelschläger (1989), De Masi and Presutti (1991), Sznitman (1989)).

The main emphasis in these lectures will be to study such limits when the limiting process is itself random. However we do not consider the frequently studied case of fluctuations around a law of large numbers limit in which normalized and centered sequences are studied (cf. Holley and Stroock (1979)), but rather study the non-centered and therefore *non-negative* limits. For example in the context of symmetrically interacting diffusions *random* McKean-Vlasov limits can arise (cf. Sect. 5.8.1). In the next chapter we will consider sequences of finite exchangeable systems which arise in the study of population genetics and genetic algorithms which converge to a random measure-valued limit.

It should be emphasized that in many applications it is the finite particle systems themselves which are of primary interest. However qualitative properties of the limiting process can often provide insight into the collective behavior of the former. In addition these limiting processes possess rich mathematical structures which are of interest in their own right.

If we begin with the empirical measure of a system of particles in $\mathbb{R}^d$ one possibility is that the limiting measure-valued process has the form $X(t,dx) = \tilde{X}(t,x)dx$, where $\tilde{X}(t,x)$ denotes the *density process* and that $\tilde{X}(t,.)$ belongs to an appropriate linear space, V, of non-negative measurable functions. It would also be reasonable to expect that $\tilde{X}(t,x)$ is described as the solution of a *stochastic partial differential equation.* There are a number of different ways to formulate a stochastic partial differential equation (see e.g. Walsh, (1986)). For example the *integral form* of such an equation is given in terms of an $\mathcal{S}'(\mathbb{R}^d)$-valued Wiener process $\{W(t):t\geq 0\}$ with covariance operator Q_0 as follows: for each $\phi\in V^*$, a linear space of test functions in duality with V, with canonical bilinear form $\langle .,.\rangle$ on $V\otimes V^*$,

$$(1.2.1)\qquad \langle X(t),\phi\rangle-\langle X(0),\phi\rangle = \int_0^t \langle A(s,X(s)),\phi\rangle ds + \int_0^t \langle B(s,X(s))W(ds),\phi\rangle$$

where for each s $A(s,.):V \to V$, and for each $v\in V$ $B(s,v)$ is a linear operator such that the Itô integral $\int_0^t \langle B(s,X(s))W(ds),\phi\rangle$ is well-defined yielding a martingale with increasing process $\int_0^t Q(s,X(s))(\phi,\phi)ds$ where $Q(s,v) := B(s,v)Q_0(s,v)B^*(s,v)$. Under certain natural conditions on $A(s,.)$, $B(s,.)$ and $Q(s,.)$ the solution to equation (1.2.1) is non-negative and hence measure-valued. An important class of stochastic partial differential equations of this form were first developed by Pardoux (1975) and generalized by Krylov and Rozovskii (1981). Equations of this type do occur in some cases including the study of stochastic flows in $\mathbb{R}^d$ (cf. Kunita (1986), Rozovskii (1990)), turbulent flows (cf. Chow (1978)) and random McKean-Vlasov limits. However it turns out that we must also consider more general measure-valued diffusions, that is, *singular* measure-valued as well as *density-valued* processes.

1.3 Some Basic Classes of Measure-valued Processes

To introduce these notions let us consider at a purely formal level the measure-valued analogue, $\{X_t:t\geq 0\}$, of a finite dimensional diffusion process associated with a second order elliptic operator. For functions, F, in an appropriate domain $D(G) \subset bC(M(E))$ (the bounded continuous functions on $M(E)$), a second order infinite dimensional differential operator would have the form

$$(1.3.1)\qquad G(t)F(\mu) = \int_E A(t,\mu,dx)(\delta F(\mu)/\delta\mu(x)) + 1/2\int_E\int_E (\delta^2F(\mu)/\delta\mu(x)\delta\mu(y))\ Q(t,\mu;dx,dy)$$

where $\dfrac{\delta F(\mu)}{\delta\mu(x)} := \lim_{\varepsilon\downarrow 0} (F(\mu+\varepsilon\delta_x)-F(\mu))/\varepsilon$, $A(t,.)$ generates a deterministic evolution on $M(E)$, and $Q(t,\mu;dx,dy)$ is a symmetric signed measure on $E\times E$ such that

$\iint \phi(x)\phi(y)Q(t,\mu;dx,dy) \geq 0,\ \forall\ \phi\in bC(E)$. Formally,

$$Q(t,\mu;dx,dy)$$
$$= \frac{d}{ds}\left[E[X_{t+s}(dx)X_{t+s}(dy)\,|\,X_t=\mu]-E[X_{t+s}(dx)\,|\,X_t=\mu]\cdot E[X_{t+s}(dy)\,|\,X_t=\mu]\right]\Big|_{s=0}.$$

Some additional structure is required in order to guarantee that X_t is *non-negative*. If $E = \{e_1,\ldots,e_d\}$, then $M(E) = (\mathbb{R}_+)^d$ and

$$G = \frac{1}{2}\sum_{i,j=1}^{d} a_{ij}(x)\frac{\partial}{\partial x_i}\frac{\partial}{\partial x_j} + \sum_{i=1}^{d} b_i(x)\frac{\partial}{\partial x_i}.$$

In this finite dimensional case Stroock and Varadhan (1979) established the existence and uniqueness to the martingale problem in the non-degenerate case $\inf_x \left(\sum_{i,j}\xi_i\xi_j a_{ij}(x)\right) > 0\ \forall\ (\xi_1,\ldots,\xi_d) \neq 0$. The degenerate case is more difficult. Results assuming that there is a Lipschitz square root are found in [EK, p. 374]. In the case $E = \{e_1\}$, then $M(E) = \mathbb{R}_+$ and this reduces to the generator of a one-dimensional diffusion given by

$$Gf(x) = b(x)f'(x) + \frac{1}{2}a(x)f''(x).$$

If $a(.)$ and $b(.)$ are Lipschitz and $b(0)\geq 0$, $a(0)=0$, then the resulting process is non-negative. An analogous condition for the measure-valued case would be that $A(t)$ generates a positivity-preserving evolution and that $Q(t,\mu;A\times A) = 0$ if $\mu(A) = 0$. Hence by their very nature measure-valued processes correspond to the highly degenerate case and this is one reason for the difficulty in developing a general theory. In the subsequent sections we will establish the existence of measure-valued diffusions with generators of form (1.3.1) in a number of cases. For example we will consider the case $A(t,v) = Av + R(t,v)$, $v\in D(A) \subset C(\mathbb{R}^d)$, in which A is the generator of a Feller semigroup on $C_0(\mathbb{R}^d)$ (the *spatial motion* or *mutation semigroup*), the *interaction term* $R(.,.)$ satisfies certain regularity conditions and in which Q has one of the following two basic forms:

(i) sampling-replacement: $Q_S(\mu{:}dx,dy) = \gamma(\delta_x(dy)\mu(dx) - \mu(dx)\mu(dy))$, $\gamma>0$,

(ii) branching: $Q_B(\mu{:}dx,dy) = c(x)\delta_x(dy)\mu(dx)$, $c(x)\geq 0$.

We will refer to these two classes of processes (with $R \equiv 0$) as *Fleming-Viot processes* and *continuous superprocesses*, respectively, and will study them in considerable detail in Chapters 2-9. In the case of branching we will also consider more generalized branching mechanisms involving jumps. In both cases special tools are available: infinite divisibility and Laplace functionals in the case of branching, and exchangeability and duality in the case of sampling-replacement. Not only are these special classes of interest in their own right but they are natural building blocks for developing and studying more general classes of measure-valued processes.

A rapidly developing area of research in measure-valued processes is the study of Fleming-Viot and continuous supeprocesses with interaction term, R, given by a nonlinear function. The effect of interactions is to destroy much of the simplicity (e.g. infinite divisibility, closed moment hierarchy, etc.) of these processes and this makes their study much more complicated. In Chapter 10 we give a brief introduction to some approaches and recent literature on this subject.

1.4. Structure and Behavior of Measure-valued Processes

Having introduced some families of measure-valued processes we now consider their local spatial structures and sample path behavior.

By *local spatial structure* is meant the determination as to whether the mass distribution at fixed times is atomic, absolutely continuous or singular and in the last case to investigate the nature of the support. In terms of *sample path behavior* we will consider questions of sample path continuity as well as determine which sets can be "hit" by the process.

Both branching and sampling systems share another important common feature - namely the central role played by *family structures.* In both cases tracing back through the family history of a typical individual turns out to be a powerful tool in the analysis of both the local spatial structure and long-time behavior of the system. This approach is currently an active area of research. In Chapters 11-12 we will give an introduction to the main features of this development.

Although in these notes we will only briefly mention the long-time behavior of measure-valued Markov processes we would like to mention here the importance and scope of this subject. The study of the *long-time behavior* of stochastic systems includes notions of transience, recurrence, invariant measure, ergodicity, large deviations from the invariant measure, and phase transitions. For example in the spatial setting transience is related to local extinction and spatial clumping - two closely related phenomena (cf. Dawson and Perkins (1991), Gorostiza and Wakolbinger (1991)). The long-time behavior of measure-valued processes is also closely related to that for infinite particle systems and many of the same ideas and difficulties arise (cf. Liggett (1985)).

1.5. Structure of these Notes

Now, let us describe in more detail the structure of these notes. First, a significant part is devoted to the systematic treatment of the two classes of measure-valued processes introduced above, namely *superprocesses* and the *Fleming-Viot processes*. Both classes are of importance in view of various applications as well as due to the central role they play in the development of the whole theory of measure-valued processes. Thus, superprocesses arise as the measure-valued generalization of branching processes (cf. also sections 1.3 - 1.4 above), whereas the Fleming-Viot processes are the measure-valued analogues of various processes arising in population genetics (in particular they describe the genetic structure of a population). These two classes are in fact closely related; the main difference is the fact that the Fleming-Viot processes are *probability measure*-valued processes, and their total mass does not change in time.

Although these two classes of measure-valued processes possess many analogous properties, the methods required to deal with them are often different. The comparison of these classes and attempts to provide a unified approach to them can serve as stimulus for further developments in the theory of measure-valued processes. Processes of both classes can be constructed as weak limits of properly scaled particle systems (i.e. branching particle systems in the case of superprocesses and the Moran particle processes in the case of Fleming-Viot processes). The approach based on particle system approximations is developed in Chapters 2, 4, and 12 on various levels of generality. In order to introduce some basic notions, in Chapter 2 this approach is carried out at a very intuitive level, for the simplest representatives of both classes.

Chapter 3 contains a summary of some important results in the general theory of random measures adapted to the purposes of these notes. We include this chapter just to make the notes to some extent self-contained; the reader familiar with the theory of random measures can omit this chapter in the first reading.

Certain properties of measure-valued processes can be studied by use of the appropriate analytical tools. In particular, path properties of certain superprocesses can be studied in terms of properties of the corresponding log-Laplace (non-linear evolution) equations. The methodology of log-Laplace functionals is developed in Chapter 4. In addition in this chapter we also present in more detail the inter-relations between branching particle systems and superprocesses (cf. sections 4.4 and 4.6).

The analytical methods developed in Chapter 4 are also applied in Chapter 9 to the investigation of the sample path properties of the class of superprocesses known as *super-Brownian motion*. The log-Laplace equation can be used to derive certain path properties of super-Brownian motion which are analogous to those of the classical Brownian motion (e.g. support dimension, carrying dimension, Hausdorff measure, modulus of continuity, range and multiple points, etc.).

A powerful tool systematically used in these notes which provides a unified approach to both classes and which is a basic tool for the development of the theory of measure-valued processes is the martingale problems characterization. Various martingale problems are considered in Chapter 5 (for the Fleming-Viot processes) and Chapter 6 (for superprocesses). In Chapter 5 we also illustrate the usefulness of the martingale problem technique to formulate wider classes of the measure-valued processes including time inhomogeneous versions of Fleming-Viot processes and some random flow models.

The structure of conditional distributions of measure-valued processes is also of interest. In Chapter 8 it is shown that the continuous superprocess conditioned to have constant mass and normalized is a Fleming-Viot process and in general the continuous superprocess conditioned on the total mass processes is a time inhomogeneous Fleming-Viot process. This is carried out using the martingale problem characterizations in Chapters 5 and 6 for both classes of measure-valued processes as well as the *stochastic calculus of measure-valued process* developed in Chapter 7. The latter includes the consideration of martingale measures, stochastic integrals and a Cameron-Martin-Girsanov formula. Applications are also given to continuity of sample paths, the weighted occupation times and a more general class of functionals.

The general theory of random measures, the martingale problem technique, andand the stochastic calculus developed in Chapters 3 - 7 are all used in Chapter 8 to determine conditions under which the main classes of measure-valued processes are either pure atomic or absolutely continuous (cf. sections 8.2 and 8.3).

In the same way as the application of log-Laplace functionals provides a tool for the analysis of superprocesses in Chapters 4 and 7, the *method of duality* provides a tool for the analysis of Fleming-Viot processes. Moreover, this method enables us to reveal the interrelations between the Fleming-Viot process and the Moran particle process in more depth. The method of duality is developed in sections 2.8, 5.5 - 5.6, 5.8 and 10.2 on various levels of generality (which correspond to the consideration of wider and wider classes of Fleming-Viot processes).

In Chapter 10 we build more complex models, namely superprocesses with interactions and Fleming-Viot processes with selection. A number of specific models (e.g. noncritical branching, branching with nonlinear death rates, multitype branching, the Fleming-Viot processes with recombination and selection, stepping stone models etc.) are also discussed in this chapter. The results of Chapter 10 illustrate the power of the combination of the martingale problem technique and either stochastic calculus or duality in the context of measure-valued processes.

In Chapter 11 we collect some powerful techniques for the study of random measures adapted to the purposes of the development of Chapter 12. In particular, in Chapter 11 we obtain *an infinite particle representation of the Fleming-Viot process* (cf. section 11.3) and *a cluster representation for superprocess* (cf. section 11.5).

In Chapter 12 we consider *family* structures of superprocesses and Fleming-Viot processes. Note that in order to distinguish them we refer to the dynamical family structure of superprocesses as *historical processes* and to the dynamical family structure of Fleming-Viot processes as *genealogical processes*. Let us point out that when studying family structures various tools developed in previous chapters have to be adapted to this more complicated historical/genealogical setting (e.g. martingale problem (cf. sections 12.3.3 and 12.5.1), branching particle system approximation (cf. section 12.4), etc.). Chapter 12 is intended as an introduction to the rapidly developing theory of historical (genealogical) processes.

The consideration of family structures provides a setting and level of generality which has already led to the solution of several earlier unsolved problems of the theory of measure-valued processes including some of the results described in Chapter 9. The culmination of Chapter 12 is a complete probabilistic description of the genealogical structure of a population at a fixed time in terms of an infinite random tree for both the continuous superprocess and Fleming-Viot process as well as a proof that the laws of the two resulting infinite random trees are mutually singular.

1.6. Acknowledgements

I would like to thank Professor P.L. Hennequin for the invitation to participate in the 1991 session of the Saint Flour Ecole d'Eté de Calcul des Probabilités and the participants for their encouragement. I would like to thank in particular E.B. Dynkin, E.A. Perkins, and T.G. Kurtz for providing preprints of their recent work on which much of these notes is based. I would also like to thank L.G. Gorostiza, Hao Wang, and V. Vinogradov for their advice concerning the preparation of these notes.

I would also like to thank the Natural Sciences and Engineering Research Council of Canada for financial support which has made possible much of my collaborative work over the years. Above all I thank my family for their continuing moral support.

1.7. Index of Some Frequently Used Notation and References

$\mathbb{R}^d$, $\mathbb{Z}^d$ d-dimensional euclidean space, lattice

$\mathbb{N}$ natural numbers

$\binom{n}{k} = n!/(k!(n-k)!)$

E denotes a Polish space (sometimes a compact metric space)

$D_E = D([0,\infty),E)$ denotes the set of càdlàg functions from $[0,\infty)$ into E

Note that $D_{M_F(E)}$ or $D_{M_1(E)}$ is often denoted by **D** for the sake of brevity

$\sigma\{f_1,\ldots,f_n\}$ = σ-algebra generated by functions $f_1,\ldots,f_n$

$\mathcal{F}_1 \otimes \mathcal{F}_2 = \sigma\{F_1 \times F_2 : F_1 \in \mathcal{F}_1, F_2 \in \mathcal{F}_2\}$

$\mathcal{F}_1 \vee \mathcal{F}_2$ = smallest σ-algebra containing $\mathcal{F}_1$ and $\mathcal{F}_2$

$\mathcal{F}_{t+} = \bigcap_{\varepsilon>0} \mathcal{F}_{t+\varepsilon}$ (if $\mathcal{F}_t$ is an increasing family of σ-algebras)

$\mathcal{F}^{uc}$ = universal completion of σ-algebra $\mathcal{F}$

(the universal completion is assumed where appropriate)

ϕ is $\mathcal{F}_1/\mathcal{F}_2$-measurable $\Leftrightarrow$ $\phi^{-1}(B) \in \mathcal{F}_1$ $\forall$ $B \in \mathcal{F}_2$

$\mathcal{E} = \mathcal{B}(E)$ = the σ-algebra of Borel subsets of E

$\mathcal{E}$ also denotes the bounded $\mathcal{E}$-measurable functions

$p(\mathcal{E})$ = non-negative functions in class $\mathcal{E}$

$pp(\mathcal{E})$ = functions f in class $\mathcal{E}$ with $\inf f > 0$

$b(\mathcal{E})$ = bounded functions in class $\mathcal{E}$

C(E) continuous functions on E

$bC_b(E)$ bounded continuous functions with bounded supports

$C_c(E)$ continuous functions with compact support

$C_0(\mathbb{R}^d)$, $C_0(E)$ continuous functions tending to zero at infinity (if E is locally compact)

$C^k(\mathbb{R}^d)$ functions whose partial derivatives of order k or less exist and are continuous

$C^\infty(\mathbb{R}^d)$ functions whose partial derivatives of all orders exist and are continuous

$M_1(E)$ probability measures on $\mathcal{E}$

$M_F(E)$ finite measures on E

$M_{1,N}(E)$ probability measures consisting of N atoms of mass 1/N.

$f \circ g$ = composition of functions

f', f" first and second derivatives of the function f

$\langle\mu,f\rangle := \int f\,d\mu$ if $f \in b\mathcal{E}$, $\mu \in M_F(E)$

μ^n = n-fold product measure $\mu(dx_1)\ldots\mu(dx_n)$

$\mu_1 * \mu_2$ = convolution of μ_1 and μ_2

supp(μ) = topological support of the measure μ

$\Sigma_{N-1} = \{(s_1,\ldots,s_N): s_i \ge 0,\ \Sigma\, s_i = 1\}$

$\Xi_N(x_1,\ldots,x_N) := N^{-1} \sum_{i=1}^{N} \delta_{x_i}$

B(x,r) denotes the open ball with centre x and radius r in a metric space

$\xrightarrow{bp}$ bounded pointwise convergence

$\Rightarrow$ weak (narrow) convergence of probability measures

f.d.d. in sense of finite dimensional distributions

$\overset{\mathcal{D}}{=}$ equality in distribution

δ_z with $z \in E$ denotes the unit mass at z

1 = constant function $1(x) = 1\ \forall x$

1_A or $1(A)$ = indicator function of the set A

$a \wedge b = \min(a,b)$, $a \vee b = \max(a,b)$

A^c = complement of A

Δ = Laplacian operator

$\Delta_\alpha = -(-\Delta)^{\alpha/2}$

Id = Identity operator

S^* = Adjoint of operator S

E(X) expectation of the random variable X (without reference to law)

P(X) = expectation of X under probability measure P, $P(A) = P(1_A)$

$P \ll Q$ P is absolutely continuous with respect to Q

$\mathcal{L}aw(X)$ = probability law of X

$\mathcal{P}ois(\lambda)$ = Poisson distribution with mean λ

$\mathcal{E}xp(a)$ = Exponential distribution with mean a

Frequent References:

[D] Dawson (1992)

[DP] Dawson and Perkins (1991)

[EK] Ethier and Kurtz (1985)

[LS] Liptser and Shiryayev (1989)

2. FROM PARTICLE SYSTEMS TO MEASURE-VALUED PROCESSES

In this chapter we introduce two classes of measure-valued processes using only rather elementary methods and in particular the method of moment measures. Our starting point is a class of exchangeable finite particle systems and we study the limits of their normalized empirical measures as the number of initial particles tends to infinity. The resulting limit process is the Fleming-Viot probability-measure-valued process. Before this we introduce the setting of measure-valued Feller processes. The last two sections of this chapter provide a preliminary introduction to weak convergence of processes, martingale problems and branching particle systems, all of which will be developed in some generality in subsequent chapters.

2.1 Measure-Valued Feller Processes.

Let (E,d) be a compact metric space, $C(E)$ the space of continuous functions, $\mathcal{E} = \mathcal{B}(E)$ the σ-algebra of Borel subsets of E, and $M_1(E)$ the space of probability measures on E. We denote by $b\mathcal{E}$ (resp. $pb\mathcal{E}$) the bounded (resp. non-negative bounded) $\mathcal{E}$-measurable functions on E. If $\mu\in M_1(E)$ and $f\in b\mathcal{E}$, we define $\langle\mu,f\rangle := \int_E f d\mu$. Note that $M_1(E)$ furnished with the topology of weak convergence is a compact metric space (recall that $\mu_n \overset{w}{\Longrightarrow} \mu$ if and only if $\langle\mu_n,f\rangle \longrightarrow \langle\mu,f\rangle$ $\forall$ $f\in bC(E)$ cf. [EK, Chap. 3].

Let $\mathbf{D} = D([0,\infty),M_1(E))$ be furnished with the usual Skorohod topology (cf. Section 3.6) and $X_t:\mathbf{D}\longrightarrow M_1(E)$, $X_t(\omega) := \omega(t)$ for $\omega\in\mathbf{D}$. Let $\mathcal{D}^o_t = \sigma\{X_s: 0\le s\le t\}$, $\mathcal{D} = \vee\mathcal{D}^o_t = \mathcal{B}(\mathbf{D})$, $\mathcal{D}_t = \mathcal{D}^o_{t+} := \bigcap_{\varepsilon>0} \mathcal{D}^o_{t+\varepsilon}$. For any $\mathcal{D}_t$-stopping time τ, $\mathcal{D}_\tau := \{A\in\mathcal{D}: A\cap\{\tau\le t\} \in \mathcal{D}_t$ $\forall t\}$. Then $(\mathbf{D},(\mathcal{D}_t)_{t\ge0},\mathcal{D},(X_t)_{t\ge0})$ defines the *canonical probability-measure-valued process*.

Recall that $\mathbf{D}$ and $M_1(\mathbf{D})$ are both Polish spaces. If $P\in M_1(\mathbf{D})$, and $F\in b\mathcal{D}$, we let $P(F) := \int F dP$. (We sometimes also use the notation $E[X]$ to denote the expectation of a random variable X.)

For $t\ge0$, define $\Pi_t:M_1(\mathbf{D})\longrightarrow M_1(M_1(E))$ by $\Pi_t P := P\circ X_t^{-1}$. Then for fixed $P\in M_1(\mathbf{D})$, the mapping $t\longrightarrow \Pi_t P \in D([0,\infty), M_1(M_1(E)))$. (N.B. However the mapping $P \longrightarrow \Pi_{\bullet} P$ is not continuous from $M_1(\mathbf{D})$ to $D([0,\infty),M_1(M_1(E)))$.)

By an $M_1(E)$-valued *stochastic process* we will mean a family of probability measures $\{P_\mu:\mu\in M_1(E)\}$ on $(\mathbf{D},\mathcal{D},(\mathcal{D}_t)_{t\ge0})$ such that

(i) $P_\mu(X(0)=\mu) = 1$, that is $\Pi_0 P_\mu = \delta_\mu$,

(ii) the mapping $\mu \longrightarrow P_\mu$ from $M_1(E)$ to $M_1(\mathbf{D})$ is measurable.

It is said to be *time homogeneous strong Markov* if for every $\{\mathcal{D}_t\}$-stopping time τ, $\mu\in M_1(E)$, with $P_\mu(\tau<\infty)=1$,

(iii) $P_\mu[F(X(\tau+t))|\mathcal{D}_\tau] = T_tF(X(\tau))$, P_μ-a.s.

for all $F\in b\mathcal{B}(M_1(E))$, $t\geq 0$, where

$$T_tF(\mu) := P_\mu F(X(t)) = \int F(\nu)\mathcal{P}(t,\mu,d\nu).$$

The *transition function* is defined by $\mathcal{P}(t,\mu;.) := \Pi_tP_\mu(.)$. Let $(C(M_1(E)),\|.\|)$ denote the Banach space of continuous functions on $M_1(E)$ with $\|F\| := \sup_\mu |F(\mu)|$. The process is a *Feller process* if in addition $T_t:C(M_1(E)) \longrightarrow C(M_1(E))$ $\forall t>0$ and $\|T_tF-F\| \longrightarrow 0$ as $t\to 0$. Then $\{T_t:t\geq 0\}$ forms a *strongly continuous semigroup of positive contraction operators* on $C(M_1(E))$, that is, $T_tF \geq 0$ if $F\geq 0$ and $\|T_tF\| \leq \|F\|$ for $F\in C(M_1(E))$. Given a Feller semigroup the *strong infinitesimal generator* is defined by

$$\mathfrak{G}F := \lim_{t\downarrow 0} \frac{T_tF-F}{t} \quad \text{(where the limit is taken in the norm topology).}$$

The *domain* $\tilde{\mathfrak{D}}(\mathfrak{G})$ of $\mathfrak{G}$ is the subspace of $C(M_1(E))$ for which this limit exists. Since $\int_0^\infty e^{-\lambda t}T_tFdt \in \tilde{\mathfrak{D}}(\mathfrak{G})$ if $\lambda>0$ and $F\in C(M_1(E))$, it follows that $\tilde{\mathfrak{D}}(\mathfrak{G})$ is dense in $C(M_1(E))$. A subspace $\mathfrak{D}_0 \subset \tilde{\mathfrak{D}}(\mathfrak{G})$ is a *core* for $\mathfrak{G}$ if the closure of the restriction of $\mathfrak{G}$ to $\mathfrak{D}_0$ is equal to $\mathfrak{G}$.

<u>Lemma 2.1.1.</u> Let $\mathfrak{G}$ be the generator of a strongly continuous contraction semigroup $\{T_t\}$ on $C(M_1(E))$. Let $\mathfrak{D}_0$ be a dense subspace of $C(M_1(E))$ and $\mathfrak{D}_0 \subset \mathfrak{D}(\tilde{\mathfrak{G}})$. If $T_t:\mathfrak{D}_0 \longrightarrow \mathfrak{D}_0$, then it is a core for $\mathfrak{G}$.

(A similar statement is true for a semigroup $\{S_t\}$ defined on $C(E)$, with generator A and domain $D(A)$.)

<u>Proof.</u> [EK, Ch. 1, Prop. 3.3]

In order to formulate an $M_1(E)$-valued Feller process we must first introduce some appropriate subspaces of $C(M_1(E))$ which can serve as a core for the generator.

The *algebra of polynomials*, $C_P(M_1(E))$, is defined to be the linear span of *monomials* of the form

$$F_{f,n}(\mu) = \int f(\mathbf{x})\mu^n(d\mathbf{x})$$
$$= \int_E \cdots \int_E f(x_1,\ldots,x_n)\mu(dx_1)\ldots\mu(dx_n)$$

where $f\in C(E^n)$.

The function $F\in C(M_1(E))$ is said to be *differentiable* if the limit

$$F^{(1)}(\mu;x) := \frac{\delta F(\mu)}{\delta\mu(x)} := \lim_{\varepsilon\downarrow 0} (F(\mu+\varepsilon\delta_x)-F(\mu))/\varepsilon = \frac{\partial}{\partial\varepsilon} F(\mu+\varepsilon\delta_x)\Big|_{\varepsilon=0}$$

exists for each $x\in E$ and belongs to $C(E)$ $\forall\mu\in M_1(E)$. The set of functions for which $F^{(1)}(\mu;x)$ is jointly continuous in μ and x is denoted by $C^{(1)}(M(E))$.

The *second derivative* is defined by

$$F^{(2)}(\mu;x,y) := \frac{\delta^2 F(\mu)}{\delta\mu(x)\delta\mu(y)} = \frac{\partial^2}{\partial\varepsilon_1\partial\varepsilon_2} F(\mu+\varepsilon_1\delta_x+\varepsilon_2\delta_y)\Big|_{\varepsilon_1=\varepsilon_2=0}$$

if it exists for each x and y and belongs to $C(E\times E)$ $\forall$ $\mu\in M_1(E)$.

Let $C^{(k)}(M_1(E))$ denote the set of functions for which $F^{(k)}(\mu;x_1,\dots,x_k)$ exists and is continuous on $M_1(E)\times E^k$.

Lemma 2.1.2 (a) $C_P(M_1(E))$ is dense in $C(M_1(E))$ and convergence determining in $M_1(M_1(E))$.

(b) Functions in $C_P(M_1(E))$ are infinitely differentiable, and the first and second derivatives are given by

$$\frac{\delta F_{f,n}(\mu)}{\delta\mu(x)} = \sum_{j=1}^{n} \int_E\cdots\int_E f(x_1,\dots,x_{j-1},x,x_{j+1},\dots,x_n) \prod_{i\neq j} \mu(dx_i)$$

$$\frac{\delta^2 F_{f,n}(\mu)}{\delta\mu(x)\delta\mu(y)} =$$

$$\sum_{j=1}^{n}\sum_{\substack{k=1\\ j\neq k}}^{n} \int_E\cdots\int_E f(x_1,\dots,x_{j-1},x,x_{j+1},\dots,x_{k-1},y,x_{k+1},\dots,x_n) \prod_{i\neq j,k} \mu(dx_i)$$

Proof. (a) The linear span of the space in question is an algebra of functions on the compact metric space $M_1(E)$. In order to verify that $C_P(M_1(E))$ separates points it suffices to note that $\mu\in M_1(E)$ is uniquely determined by $\{\langle\mu,\phi\rangle: \phi\in C(E)\}$. The first part of the result is then an immediate consequence of the Stone-Weierstrass theorem.

If $\int F(\mu)P_n(d\mu)\to\int F(\mu)P(d\mu)$ as $n\to\infty$ for all F belonging to a dense subset of $C(M_1(E))$, then it is true for all $F\in C(M_1(E))$. This proves that $C_P(M_1(E))$ is convergence determining in $M_1(M_1(E))$.

(b) follows by a simple calculation. □

2.2 Independent Particle Systems: Dynamical Law of Large Numbers

Let $\{S_t:t\geq 0\}$ be a Feller semigroup on the Banach space $(C(E),\|.\|)$ where $\|.\|$ is the supremum norm, with E compact. Then the domain D(A) of the infinitesimal generator A is a dense subspace of C(E). We assume that there exists a separating algebra of functions

$$D_0 \subset D(A), \quad S_t D_0 \subset D_0.$$

Consequently D_0 is a core for A (cf. Lemma 2.1.1).

Let $P(t,x,dy)$ denote the transition function of $\{S_t\}$, that is,

$$S_t f(x) = \int_E f(y)P(t,x,dy), \quad f \in C(E).$$

It will be convenient to work with a canonical version of the Feller process which will be described in the following result. Let $D_E = D([0,\infty),E)$ denote the space of càdlàg functions from $[0,\infty)$ into E. Then D_E is a Polish space if it is furnished with the Skorohod topology (see Section 3.6 for these definitions).

Proposition 2.2.1. Let $\{S_t\}$ be a Feller semigroup on C(E) with E compact. Then

(a) For each $x \in E$ there exists a probability measure P_x on $\mathcal{B}(D_E)$ satisfying

(2.2.1) $P_x(\omega(0)=x) = 1$, and for $s \le t$,

(2.2.2) $P_x(f(\omega(t)) | \sigma(\omega(u):u \le s) = (S_{t-s}f)(\omega(s))$, P_x-a.s. $\forall f \in C(E)$.

(b) There exists a standard probability space $(\Omega,\mathcal{F},Q_A)$ and a measurable mapping ζ: $(E\times\Omega,\mathcal{E}\otimes\mathcal{F}) \longrightarrow (D_E,\mathcal{B}(D_E))$, such that for each $x \in E$

(2.2.3) $Q_A(\{\omega:\zeta(x,\omega)\in B\}) = P_x(B) \quad \forall B \in \mathcal{B}(D_E)$.

Furthermore, $\zeta(.,\omega)$ is continuous at x for Q_A-a.e. ω, for each $x \in E$.

The resulting measurable random function is denoted by $(\Omega,\mathcal{F},Q_A,\{\zeta(x)\}_{x\in E})$.

(A *standard probability space* is one which is isomorphic to [0,1] with Lebesgue measure.)

Proof. (a) Given $x \in E$, the existence of P_x satisfying (2.2.2) is a standard result on the existence of a càdlàg version of the Feller process (e.g. [EK Chapt. 4, Thm 2.7]).

(b) It can also be shown that the mapping $x \longrightarrow P_x$ from E to $M_1(D_E)$ is continuous when the latter is given the weak topology (see Section 3.1). Since the map $x \longrightarrow P_x$ is continuous, the existence of a representation $(\Omega,\mathcal{F},Q_A,\{\zeta(x)\}_{x\in E})$, ξ: $\Omega\times E \longrightarrow D_E$, such that

(i) for each $x \in E$, $\xi(.,x)$ is measurable and has law P_x,

and

(ii) $\xi(\omega,.)$ is continuous at x for Q_A-a.e. ω for each $x \in E$,

follows from the extension of Skorohod's almost sure representation theorem due to Blackwell and Dubins (1983). From this the existence of a jointly measurable version of ζ follows by a standard argument (cf. Neveu (1964, III-4)). □

A system of N independent particles $\{Z(t):t\ge 0\} := \{Z_1(.),\dots,Z_N(.)\}$ each undergoing an A-motion in E and with initial value $Z_i(0)$ having law $\mu \in M_1(E)$ is then realized on $((E\times\Omega)^N,(\mu\otimes Q_A)^N)$ by

$$Z_i(((e_1,\omega_1),\ldots,(e_N,\omega_N)),t) = \zeta(e_i,\omega_i)(t),\ i=1,\ldots,N,$$
$$((e_1,\omega_1),\ldots,(e_N,\omega_N)) \in (E\times\Omega)^N.$$

Then $\mathbf{Z}(t)$ is an E^N-valued Markov process with semigroup

$$S_t^N f(x_1,\ldots,x_N) = \int_E \cdots \int_E f(y_1,\ldots,y_N)P(t,x_1,dy_1)\ldots P(t,x_N,dy_N),\ f\in C(E^N).$$

The semigroup $\{S_t^N:t\geq 0\}$ is strongly continuous on the closure of D_0^N (:= algebra generated by $\{f_1(x_1)\ldots f_N(x_N):\ f_i\in D_0,\ i=1,\ldots,N\}$), which is $C(E^N)$, and hence S_t^N is a Feller semigroup associated with a process with values in E^N. The corresponding generator is

$$A^{(N)} := \sum_{i=1}^{N} A_i \quad \text{on} \quad D(A^{(N)}) \subset C(E^N).$$

where A_i denotes the action of A on the ith variable. Furthermore it easily follows that $S_t^N:D_0^N \longrightarrow D_0^N$ and therefore D_0^N is a core for $A^{(N)}$.

The associated *empirical measure process* is given by

$$X^N(t) = \Xi_N(Z_1(t),\ldots,Z_N(t)) := N^{-1} \sum_{i=1}^{N} \delta_{Z_i(t)} \in M_1(E).$$

It will follow from Proposition 2.3.3 that $X^N(\cdot)$ is also a Feller process with state space $M_{1,N}(E)$, the space of measures consisting of atoms whose masses are multiples of 1/N and contained in E. We will denote its generator by $\mathfrak{G}_N^A$.

Let $\mathfrak{D}_0(\mathfrak{G}_N^A) := \{F_{f,n}(\mu) = \langle\mu^n,f\rangle:\ f\in D_0^n,\ n\leq N\}$.

For $F_{f,n}\in\mathfrak{D}_0(\mathfrak{G}_N^A)$ and $\mu_N = \frac{1}{N}\sum_{i=1}^{N}\delta_{z_i}$,

$$F_{f,n}(\mu_N) = N^{-n}\sum_{i_1=1}^{N}\cdots\cdots\sum_{i_n=1}^{N} f(z_{i_1},\ldots,z_{i_n})$$

$$= N^{-n}\sum_{k=1}^{n}\ \sum_{p\in\rho_k^n}\ \sum_{j_1=1}^{N}\cdots\sum_{j_k=1}^{N} f(z_{j_{p1}},\ldots,z_{j_{pn}})$$

where for $1\leq k<n$, ρ_k^n denotes the set of partitions $p:\{1,\ldots,n\}\longrightarrow\{1,\ldots,k\}$.

$$\mathfrak{G}_N^A F_{f,n}(\mu_N) = \langle\mu_N^n,A^{(n)}f\rangle$$

$$+ N^{-n}\sum_{k=1}^{n-1}\sum_{p\in\rho_k^n}\sum_{j_{p1}=1}^{N}\cdots\sum_{j_{pk}=1}^{N}(A^{(k)}f^{(p)} - A^{(n)}f)(z_{j_{p1}},\dots,z_{j_{pn}})$$

$$= F_{A^{(n)}f,n}(\mu_N) + R(N,n,f)(\mu_N)$$

and for $p\in\rho_k^n$, $f^{(p)}(z_{j_1},\dots,z_{j_n}) = f(z_{j_{p1}},\dots,z_{j_{pn}})$.

Since

$$R(N,n,f)(\mu_N) = N^{-n}\sum_{k=1}^{n-1}\sum_{p\in\rho_k^n}\sum_{j_{p1}=1}^{N}\cdots\sum_{j_{pk}=1}^{N}(A^{(k)}f^{(p)} - A^{(n)}f)(z_{j_{p1}},\dots,z_{j_{pn}})$$

it follows that $|R(N,n,f)(\mu_N)| \le c(n)\|f\|_{A,n}/N$ where for $f\in D_0^n$, $\|f\|_{A,n} := \|f\| + \max_k \max_{p\in\rho_k^n} \|A^{(k)}f^{(p)}\|$. Note that $\|S_t^N f\|_{A,n} \le \|f\|_{A,n}$ and therefore $S_t^N : D_0^n \to D_0^n$.

By the law of large numbers $X^N(0) \Rightarrow X(0) := \mu$. Using the above expression for the generator we can then show (as a special case of Theorem 2.7.1) that $X^N(t) \Rightarrow X(t)$, which is a deterministic $M_1(E)$-valued process characterized as the unique solution of the *weak equation*

$$\langle X(t),f\rangle = \langle X(0),f\rangle + \int_0^t \langle X(s),Af\rangle ds \quad \forall\ f\in D(A),$$

that is, formally, $\frac{\partial X}{\partial t} = A^*X$ where A^* denotes the adjoint of A.

This implies that

$$\langle X(t),f\rangle = \langle X(0),S_t f\rangle.$$

This is the simplest example of a *dynamical law of large numbers* and is a degenerate case of the McKean-Vlasov limit of exchangeably interacting particle systems. For detailed developments on the McKean Vlasov (or mean-field) limit of interacting particle systems and the related phenomenon of propagation of chaos the reader is referred to Gärtner (1988), Léonard (1986) and Sznitman (1989).

2.3 Exchangeable Particle Systems

Let $\mathcal{Per}(N)$ denote the set of permutations of $\{1,\dots,N\}$. A continuous function $f:E^N\to\mathbb{R}$ is said to be *symmetric*, $f\in C_{sym}(E^N)$, if $f = \tilde{\pi}f$ $\forall\ \pi\in\mathcal{Per}(N)$, where $(\tilde{\pi}f)(z_1,\dots,z_N) := f(z_{\pi1},\dots,z_{\pi N})$.

Given $z_1,\dots,z_N \in E$ (not necessarily distinct) the associated *empirical measure* is defined by

$$\Xi_N(z_1,\ldots,z_N) := N^{-1}\sum_{i=1}^{N}\delta_{z_i} \in M_1(E).$$

The mapping $\Xi_N: E^N \longrightarrow M_1(E)$ is clearly $\sigma(C_{sym}(E^N))$-measurable. On the other hand, given a measure $\mu = \sum_{i=1}^{M} a_i\delta_{z_i} + \nu \in M_1(E)$, with $z_1,\ldots,z_M$ distinct, and ν nonatomic, let $\Sigma(\mu) := \{(z_1;a_1)\ldots,(z_M;a_M)\} \in (E\times\mathbb{R}_+)^M$, mod($\mathcal{Per}(M)$) (i.e. unordered M-tuples). The mapping $\mu \longrightarrow \Sigma(\mu)$ is measurable from $(M_1(E),\mathcal{B}(M_1(E)))$ to $\cup_{M=1}^{\infty}(E\times\mathbb{R}_+)^M$ where the latter is furnished with the smallest σ-algebra containing $\sigma(C_{sym}(E\times\mathbb{R}_+)^M)$ for each M (cf. Theorem 3.4.1.1(d)). Consequently, if $\mu\in M_{1,N}(E)$, then the mapping $\mu \longrightarrow ((z'_1,n_1),\ldots,(z'_k,n_k))$ where the $z'_1,\ldots,z'_k$ are the distinct locations of the atoms and the n_k are their multiplicities is $(M(E),\mathcal{B}(M(E))$-measurable. Then the unordered n-tuple $(z_1,\ldots,z_n)$ is given by listing the distinct $z'_1,\ldots,z'_k$ with the appropriate multiplicities. Thus we obtain the following.

Lemma 2.3.1 The sub-σ-algebras $\sigma\{C_{sym}(E^N)\}$ and $\sigma\{\Xi_N\}$ of $\mathcal{B}(E^N)$ coincide. In particular, if $f\in C_{sym}(E^N)$, then $f(z_1,\ldots,z_N)$ is $\sigma(\Xi_N)$-measurable.

Proof. If $\Sigma\left(\Xi_N(z_1,\ldots,z_N)\right) = ((z'_1,n_1),\ldots,(z'_k,n_k))$, and $f\in C_{sym}(E^N)$ then $f(z_1,\ldots,z_N) = f(z'_1,\ldots,z'_1,\ldots,z'_k,\ldots,z'_k)$ (with z'_i repeated n_i times for each $i = 1,\ldots,k$). □

The E-valued random variables $Z_1,\ldots,Z_N$ are *exchangeable* if the joint distributions of $Z_1,\ldots,Z_N$ and $Z_{\pi 1},\ldots,Z_{\pi N}$ are identical for any $\pi\in\mathcal{Per}(N)$. The probability law P on $\mathcal{B}(E^N)$ of the exchangeable random variables $Z_1,\ldots,Z_N$ is uniquely determined by its restriction to the sub-σ-algebra $\sigma(C_{sym}(E^N))$. Let $M_{1,ex}(E^N)$ denote the family of exchangeable probability laws on E^N. Then $C_{sym}(E^N)$ is $M_{1,ex}(E^N)$-determining, that is, if $\mu_1,\mu_2 \in M_{1,ex}(E^N)$ and $\int_{E^N} f(x)\mu_1(dx) = \int_{E^N} f(x)\mu_2(dx)$ $\forall$ $f\in C_{sym}(E^N)$, then $\mu_1=\mu_2$. Moreover if $\mu\in M_{1,ex}(E^N)$, $g \in pC_{sym}(E^N)$, and $\langle\mu,g\rangle < \infty$, then $\mu_g(A) := \langle\mu,g1_A\rangle/\langle\mu,g\rangle \in M_{1,ex}(E^N)$.

Given a Polish space S let $D_S = D([0,\infty);S)$ denote the space of càdlàg functions from $[0,\infty)$ to S furnished with the Skorohod topology (cf. Ch. 3, Sect. 6). Given $\pi\in\mathcal{Per}(N)$, let $\tilde{\pi}:E^N \longrightarrow E^N$ be defined by $(\tilde{\pi}x)_i = x_{\pi i}$ for $x = (x_1,\ldots,x_N) \in E^N$ and $\tilde{\pi}:D_{E^N} \longrightarrow D_{E^N}$ be defined by $(\tilde{\pi}x)_i(t) = x_{\pi i}(t)$.

An *exchangeable system of* N *particles* is defined by an exchangeable probability law P on D_{E^N}, or equivalently,

(i) an exchangeable initial distribution $\Pi_0 P$ on E^N, and

(ii) a family $\{P_y : y \in E^N\}$ of conditional distriubtions on D_{E^N} which satisfies $P_{\tilde{\pi} y} = P_y \circ \tilde{\pi}^{-1}$ or $P_{\tilde{\pi} y}(\tilde{\pi} A) = P_y(A)$ for every $y \in E^N$, $A \in \mathcal{D}$, and $\pi \in \mathcal{Perm}(N)$.

We next give a simple criterion which implies that an E^N-valued Markov process is exchangeable.

<u>Lemma 2.3.2</u> Let $\mathbf{Z} = (Z_1, \ldots, Z_N)$ be an E^N-valued càdlàg Markov process with transition function $p(s,x;t,dy)$. Then Z is an exchangeable system provided that the marginal distributions $P(\mathbf{Z}(t) \in .)$, $t \in \mathbb{R}_+$ are exchangeable and $p(s,y;t,B) = p(s,\tilde{\pi} y;t,\tilde{\pi} B)$ for every $\pi \in \mathcal{Perm}(N)$, $y \in E^N$, $B \in \mathcal{B}(E^N)$, or equivalently,

(2.3.1) $\quad (S_t f(\tilde{\pi}.))(\tilde{\pi}^{-1} x) = (S_t f)(x), \quad f \in C(E^N).$

Note that in the case of a time homogeneous Feller semigroup $\{S_t\}$ and with generator A and core $D_0(A)$ the above criterion is implied by

(2.3.2) $\quad (Af(\tilde{\pi}.))(\tilde{\pi}^{-1} y) = (Af(.))(y), \quad f \in D_0(A).$

<u>Proof.</u> Let $m \in \mathbb{Z}_+$, $t_1, \ldots, t_m \in \mathbb{R}_+$ and $\pi \in \mathcal{Perm}(N)$. Then for $B_i^j \in \mathcal{B}(E)$,

$$P_{y_0}\left(\mathbf{Z}(t_i) \in \prod_{j=1}^N B_i^{\pi^{-1} j}, \ i=1,\ldots,m\right)$$

$$= \int_{\prod_j B_1^{\pi^{-1} j}} \cdots \int_{\prod_j B_m^{\pi^{-1} j}} p(0,y_0;t_1,dy_1) \prod_{i=1}^{m-1} p(t_i,y_i;t_{i+1},dy_{i+1})$$

$$= \int_{\prod_j B_1^{\pi^{-1} j}} \cdots \int_{\prod_j B_m^{\pi^{-1} j}} p(0,\tilde{\pi} y_0;t_1,\tilde{\pi} dy_1) \prod_{i=1}^{m-1} p(t_i,\tilde{\pi} y_i;t_{i+1},\tilde{\pi} dy_{i+1})$$

$$= \int_{\prod_j B_1^j} \cdots \int_{\prod_j B_m^j} p(0,\tilde{\pi} y_0;t_1,dy_1) \prod_{i=1}^{m-1} p(t_i,y_i;t_{i+1},dy_{i+1})$$

$$= P_{\tilde{\pi} y_0}\left(\mathbf{Z}(t_i) \in \prod_{j=1}^N B_i^j, \ i=1,\ldots,m\right),$$

since by assumption $p(t_i,\tilde{\pi} y_i;t_{i+1},\tilde{\pi} dy_{i+1}) = p(t_i,y_i;t_{i+1},dy_{i+1})$.

Thus the finite dimensional distributions of $P_{\tilde{\pi} y_0}$ and $P_{y_0} \circ \tilde{\pi}^{-1}$ coincide which yields the result. □

Proposition 2.3.3 Let $Z = (Z_1,\ldots,Z_N)$ be an E^N-valued càdlàg exchangeable Feller process. Then the *empirical measure process* $X(t) := \Xi_N(Z(t))$ is a càdlàg $M_1(E)$-valued Feller Markov process.

Proof. For each $\phi \in C(E)$, $\int \phi(x)X(t,dx) = N^{-1} \sum_{i=1}^{N} \phi(Z_i(t))$ is càdlàg and hence $X(t) \in D([0,\infty);M_1(E))$, a.s.

Let $\mathcal{F}_t^Z := \sigma\{Z(s):0\le s\le t\}$. Then in order to prove the Markov property for $X(.)$ it suffices to show that

$$P(X(t+s)\in .\,|\sigma(X(t))\vee\mathcal{F}_t^Z) = P(X(t+s)\in .\,|\sigma(X(t)), \text{ a.s.}$$

It follows from the Markov property of Z and the inclusion $\mathcal{F}_t^Z \supset \sigma(X(t))$, that the left hand side equals $P(X(t+s)\in .\,|\sigma(Z(t))$ a.s. Hence it suffices to show that $P(\Xi_N(Z(t+s))\in .\,|\sigma(Z(t)) = P(\Xi_N(Z(t+s))\in .\,|\sigma(X(t))$. Since $(Z(t),Z(t+s))$ forms N $(E\times E)$-valued exchangeable random variables by hypothesis, $P(\Xi_N(Z(t+s))\in .\,|\sigma(Z(t)) = P(\Xi_N(\tilde{\pi}Z(t+s))\in .\,|\sigma(\tilde{\pi}Z(t))) = P(\Xi_N(Z(t+s))\in .\,|\sigma(\tilde{\pi}Z(t)))$ a.s. Thus $P(\Xi_N(Z(t+s))\in .\,|\sigma(Z(t))$ is a symmetric function of $Z(t)$ and by Lemma 2.3.1 we conclude that there is a $\sigma(\Xi_N(Z(t))$-measurable version of $P(\Xi_N(Z(t+s))\in .\,|\sigma(Z(t))$, and this yields the Markov property. Finally, note that the assumption that Z is càdlàg and Feller implies that Ξ_N is also càdlàg. □

2.4 Random Probability Measures, Moment Measures and Exchangeable Sequences.

Let X be a random probability measure on E, E Polish. Then the *nth moment measure* is a probability measure defined on E^n as follows:
$M_n(dx_1,\ldots,dx_n) = E(X(dx_1)\ldots X(dx_n))$. It is the probability law of n-exchangeable E-valued random variables, $\{Z_1,\ldots,Z_n\}$. Noting that this is a consistent family and using Kolmogorov's extension theorem we can associate with every random probability measure on E an exchangeable sequence of E-valued random variables, $\{Z_n:n\in\mathbb{N}\}$. The converse result is related to de Finetti's theorem and is formulated in Chapter 11.

Lemma 2.4.1 (a) A random probability measure X on E is uniquely determined by its moment measures of all orders.

(b) The sequence $\{X_n\}$ of random probability measures with moment measures $\{M_{n,m},n,m\in\mathbb{N}\}$ converges weakly to a random probability measure X with moment measures $\{M_m\}$ if and only if $M_{n,m} \Rightarrow M_m$ as $n\to\infty$ for each $m\in\mathbb{N}$.

Proof. This follows from Lemma 2.1.2(a), Corollary 3.2.7 and Lemma 3.2.8. □

2.5 The Moran Particle Process

Let $(A,D(A))$ be the generator of a Feller semigroup $\{S(t)\}$ on $C(E)$ with E compact and consider the N-particle system introduced in Section 2.2.

We now add a pair interaction, called *replacement sampling*, to the independent motions considered above. In this model each particle jumps at rate $(N-1)\gamma/2$, $\gamma>0$, to the location of one of the remaining (N-1) particles, that is, the times between replacements are i.i.d. exponential with mean $2/\gamma N(N-1)$. At the time of a jump a pair (i,j) is chosen at random and the jth particle jumps to the location of the ith particle.

Let $\{Z^N(t)\} = \{Z^N_1(t),\ldots,Z^N_N(t)\} \in E^N$ where $Z^N_j(t)$ denotes the site of the jth individual at time t, where again $\{Z^N_1(0),\ldots,Z^N_N(0)\}$ are i.i.d. with law μ. This process is obtained by a *direct construction* on a probability space on which a countable independent collection of exponential random variables and copies of $(\Omega,\mathcal{F},Q_A,\{\zeta(x)\}_{x\in E})$ of Proposition 2.2.1 are defined. To make this precise first consider the probability space

$$\left(\mathbb{R}^{N(N-1)},\mathcal{B}(\mathbb{R}^{N(N-1)}),\nu;\{\eta^{i,j}\}_{\substack{i,j=1,..,N\\ i\neq j}}\right)$$

where $\nu := \prod_{i=1}^{N(N-1)}\left(\frac{\gamma}{2}\cdot e^{-\gamma u_i/2}du_i\right)$ and the $\eta^{i,j}$ are simply the coordinate functions.

The basic probability space on which the process is defined is

$$\left\{(E,\mu)^N\otimes(\mathbb{R}^{N(N-1)}\times\Omega,\nu\otimes Q_A)^{\mathbb{N}},\{e_k:k=1,\ldots,N\},\{\eta_k,\zeta_k\}_{k\in\mathbb{N}}\right\}.$$

Then $\{\eta_k\}$ is an i.i.d. sequence of copies of $\{\eta^{i,j}\}$ and $\{\zeta_k\}$ is an i.i.d. sequence of copies of the process $(\Omega,\mathcal{F},Q_A,\{\zeta(x)\}_{x\in E})$. The process $\{\mathbf{Z}(t):t\geq 0\} = \{Z_1(.),\ldots,Z_N(.)\}$ is defined as follows:

$\tau_0:=0$, $\tau_k := \tau_{k-1} + \min\limits_{i,j}\eta_k^{i,j}$, and without loss of generality we can assume that the pair (i_k,j_k) is uniquely defined by $\min\limits_{i,j}\eta_k^{i,j} = \eta_k^{i_k,j_k}$.

For $0\leq t<\tau_1$ set

$$Z_i(t) := \zeta_i(e_i;t)$$

for $\tau_k\leq t<\tau_{k+1}$, $k\geq 1$, set

$$Z_{j_k}(t) := \zeta_{N+k}(Z_{i_k}(\tau_k);(t-\tau_k)), \quad a(N+k) = i_k,$$

$$Z_i(t) := \zeta_m(Z_{a(m)}(s(m));(t-s(m)) \text{ for } i\neq j_k,$$

$$\text{if} \quad Z_i(t) = \zeta_m(Z_{a(m)}(s(m));(t-s(m)), \text{ for } \tau_{k-1}\leq t<\tau_k$$

(where $(Z_{a(m)}, s(m))$ denotes the starting place and time of the mth particle motion; and a(m) denotes the immediate *ancestor* of the mth particle motion).

Remark. This probability space representation also provides information on the *family structure* of the population. For example, consider the ℓth particle at time t. If we define

$$k(\ell,t) := \max\,\{k:\tau_k\le t,\ \ell\in(i_k,j_k)\},$$

then i_k (or j_k) is the sibling of ℓ. This family structure will be studied in Ch. 12.

Since the sojourn times are independent random variables and the jumps have a Markov chain structure, it is easy to verify that the process $\{Z^N(t)\}$ is a stationary Markov process with state space E^N. We define the semigroup

$$(V^N_t f)(x_1,\dots,x_N) := E[f(Z^N_1(t),\dots,Z^N_N(t))\,|\,Z^N_1(0)=x_1,\dots,Z^N_N(0)=x_N],$$

for $f\in C(E^N)$.

Theorem 2.5.1. (a) $\{Z^N(t)\}$ is a Feller process with semigroup $\{V^N_t\}$, where $V^N_t f$, $f\in$ given by the unique solution of the evolution equation

(2.5.1)

$$(V^N_t f)(x_1,\dots,x_N)$$

$$= e^{-N(N-1)\gamma t/2}(S^N_t f)(x_1,\dots,x_N)$$

$$+\frac{1}{2}\gamma\sum_{\substack{i,j\\ i\ne j}}\int_0^t e^{-N(N-1)\gamma s/2}\, S^N_s(\Theta_{ij}(V^N_{t-s}f))(x_1,\dots,x_N)ds$$

where $\{S^N_t\}$, D_0, $\|.\|_{A,N}$ are the same as those defined in Sect. 2.2, and $\Theta_{ij}:C(E^N)\to C(E^N)$ is defined by

$$(\Theta_{ij}f)(x_1,\dots,x_N) := f(x_1,\dots,x_i,\dots,x_{j-1},x_i,x_{j+1},\dots,x_N)$$

that is, $(\Theta_{ij}f)(x_1,\dots,x_N)$ depends only on $\{x_k:k\ne j\}$ and is obtained by evaluating f at $\tilde{x}_1,\dots,\tilde{x}_N$ with

$$\tilde{x}_k = x_k \quad \text{if } k\ne j$$
$$\tilde{x}_j = x_i.$$

(b) $V^N_t:D^N_0\to D^N_0$ and $\|V^N_t f\|_{A,N}\le\|f\|_{A,N}$ and consequently D^N_0 is a core for $\{V^N_t\}$.

The process Z^N is called the *N-particle Moran process,* and the corresponding *Moran generator* has the form

(2.5.2)

$$L_N f(x_1,\dots,x_N) = A^{(N)}f(x_1,\dots,x_N) + \frac{1}{2}\gamma\Big(\sum_{i\neq j}\Theta_{ij}f-f\Big)(x_1,\dots,x_N)$$

where $A^{(N)}f(x_1,\dots,x_N) = \sum_{j=1}^{N} A_j f(x_1,\dots,x_N)$ and A_j denotes the action of A on the jth variable.

(Note also that Equation (2.5.1) is formally equivalent to the differential equation $\frac{\partial}{\partial t}V_t^N f = L_N V_t^N f$, $V_0^N f = f$.)

Remarks 2.5.2

(a) Note that $L_2 f(x_1,x_2) = A_1 f(x_1,x_2)+A_2 f(x_1,x_2)$

$$+ \tfrac{1}{2}\gamma(f(x_1,x_1)-f(x_1,x_2)) + \tfrac{1}{2}\gamma(f(x_2,x_2)-f(x_1,x_2)).$$

(b) We also define a function of (N-1)-variables $\tilde{\Theta}_{ij}f$ as follows:

$$(\tilde{\Theta}_{ij}f)(y_1,\dots,y_{N-1}) := f(x_1,\dots,x_N)$$

with $x_k = y_k$ for $k < i\vee j$, $k\neq i\wedge j$

$x_{i\vee j}=x_{i\wedge j} = y_{i\wedge j}$

$x_k = y_{k-1}$ for $k > i\vee j$.

Note that $\Theta_{ij}f$ can be written in terms of $\tilde{\Theta}_{ij}f$ as follows:

$$(\Theta_{ij}f)(x_1,\dots,x_N) = (\tilde{\Theta}_{ij}f)(x_1,\dots,x_{j-1},x_{j+1},\dots,x_N),$$

and if μ is a probability measure, then $\int(\Theta_{ij}f)d\mu^N = \int(\tilde{\Theta}_{ij}f)d\mu^{N-1}$.

Proof of Theorem 2.5.1. Conditioning on the time and position of the first interaction jump, τ_1, we obtain the following equation for the semigroup V_t^N of this process:

(2.5.3) $(V_t^N f)(x_1,\dots,x_N)$

$$= E[E[f(Z_1^N(t),\dots,Z_N^N(t))\,|\,\mathcal{F}_\tau^Z]\,|\,Z_1^N(0)=x_1,\dots,Z_N^N(0)=x_N]$$

$$= E[E[1(\tau_1>t)f(Z_1^N(t),\dots,Z_N^N(t))]\,|\,Z_1^N(0)=x_1,\dots,Z_N^N(0)=x_N]$$

$$+ \tfrac{1}{2}\gamma\sum_{i\neq j}\int_0^t E\Big[1(\tau_1\in ds)\Theta_{ij}\Big(E[f(Z_1^N(t),\dots,Z_N^N(t))\,|\,Z_1^N(s)=y_1,\dots,Z_N^N(s)=y_N]\Big)$$

$$\left. |Z_1^N(0)=x_1,\dots,Z_N^N(0)=x_N] \right\}$$

where $\Theta_{ij}:C(E^N)\to C(E^N)$ is defined as above. Substituting the definition of $V^N_{t-s}f$ and the exponential distribution of the first jump time τ_1 on the right hand side of (2.5.3) we obtain (2.5.1).

If $f(x_1,\dots,x_N) = \sum_{i=1}^N g_i(x_i)$, then the solution can be written explicitly. Then noting that $\Theta_{ij}\sum_k\sum_\ell g_{k\ell}(x_k,x_\ell)$ can be written in terms of functions of the form $\sum_k\sum_\ell g_{k\ell}(x_k,x_\ell)$ and $\sum_{i=1}^N g_i(x_i)$, we can then solve this system of equations for functions of the form $f(x_1,\dots,x_N) = \sum_k\sum_\ell g_{k\ell}(x_k,x_\ell)$. Continuing in this way we can prove that this system of equations has a unique solution for any initial $f\in C(E^N)$. It is easy to verify that the solution $\{V^N_t f\}$ defines a strongly continuous semigroup on $C(E^N)$. Moreover, the facts that $\|S^N_t f\|_{A,N} \le \|f\|_{A,N}$ and $\|\Theta_{ij}f\|_{A,N} \le \|f\|_{A,N}$, together with equation (2.5.1) imply that $\|V^N_t f\|_{A,N} \le \|f\|_{A,N}$. Therefore $V^N_t:D^N_0\to D^N_0$ and D^N_0 is a core by Lemma 2.1.1. The required expression for the generator L_N (cf. (2.5.2)) is then obtained by differentiation. □

Proposition 2.5.3. L_N satisfies the exchangeability condition (2.3.2).

Proof. In order to emphasize the main point, for the moment, we assume that the generator of the jth particle motion is $A^{\{j\}}$ and the interaction involving the ith and jth particles is given by $\Theta^{(i,j)}$.

Then $A^{\{j\}}$ denotes the action of operator $A^{\{j\}}$ on the kth variable x_k and $\Theta^{(i,j)}_{\ell k}$ denote the action of the operator $\Theta^{(i,j)}$ on the pair (x_ℓ,x_k). Then

$$L_N f(x_1,\dots,x_N) = \sum_{j=1}^N (A_j^{\{j\}}f)(x_1,\dots,x_N) + \tfrac{1}{2}\gamma\cdot\Big(\sum_{i\ne j}\Theta_{ij}^{(i,j)}f-f\Big)(x_1,\dots,x_N).$$

We then have

$$L_N[\tilde{\pi}f](\tilde{\pi}^{-1}\mathbf{x}) = \sum_{j=1}^N (A_j^{\{\pi^{-1}j\}}f)(\mathbf{x}) + \tfrac{1}{2}\gamma\sum_{i\ne j}\Big(\Theta_{ij}^{(\pi^{-1}i,\pi^{-1}j)}f-f\Big)(\mathbf{x})$$

$$= (Lf)(\mathbf{x})$$

since $A_j^{(\pi^{-1}j)} f = A_j f$ and $\Theta_{ij}^{(\pi^{-1}i,\pi^{-1}j)} f = \Theta_{ij} f$ in the case under consideration. □

2.6 The Measure-valued Moran Process

Now consider the empirical process $X^N(t)$;

$$X^N(t;B) := \Xi_N(\mathbf{Z}(t))(B) = \frac{1}{N}\sum_{i=1}^{N} 1_B(Z_i(t)) \quad B \in \mathcal{E},$$

with state space $M_{1,N}(E)$ (with E compact). Using Propositions 2.3.3, 2.5.3 and Theorem 2.5.1 we conclude that X^N is a càdlàg Feller process with state space $M_{1,N}(E)$.

The corresponding semigroup is defined by

$$T_t^N F(\mu) := EF(X^N(t)\,|\,X^N(0)=\mu), \quad F\in C(M_1(E)),\ \mu\in M_{1,N}(E).$$

If $F_{f,n}$ is a monomial of degree $n \le N$ (cf. Sect. 2.1), and $\mu_N = \sum_{i=1}^{N} \delta_{x_i}$, then

$$F_{f,n}(\mu_N) = N^{-n} \sum_{x_{i_1}=1}^{N} \cdots \sum_{x_{i_n}=1}^{N} f(x_{i_1},\ldots,x_{i_n}).$$

$$= N^{-n} \sum_{k=1}^{n} \sum_{p\in\rho_k^n} \sum_{j_{p1}=1}^{N} \cdots \sum_{j_{pk}=1}^{N} f(x_{j_{p1}},\ldots,x_{j_{pn}})$$

where the ρ_k^n are as defined in section 2.2.

Then

(2.6.1)

$$T_t^N F_{f,n}(\mu_N) = N^{-n} \sum_{k=1}^{n} \sum_{p\in\rho_k^n} \sum_{j_{p1}=1}^{N} \cdots \sum_{j_{pk}=1}^{N} (V_t^N f^{(p)})(x_{j_1},\ldots,x_{j_n})$$

where V_t^N is as in (2.5.1).

Let $\mathfrak{G}_N^A$ denote the generator of the measure-valued process coming from the empirical measure of N independent A-processes as in Section 2.2. Then the generator of the measure-valued process X^N is given by

(2.6.2) $$\mathfrak{G}_N F(\mu) = \mathfrak{G}_N^A F(\mu) + \tfrac{1}{2}\gamma N(N-1)\int_E\int_E [F(\mu - \delta_x/N + \delta_y/N) - F(\mu)]\mu(dx)\mu(dy).$$

Note that we can also interpret the transitions which determine the second term in $\mathfrak{G}_N$ as the death of an individual and the birth of a new individual whose site is that of a second individual chosen at random from the remaining population - this is the *sampling replacement* mechanism of population genetics.

Let $\mathfrak{D}_0(\mathfrak{G})= C_{P,D_0}(M_1(E)) \ := \{F_{f,n}(\cdot): f \in D_0^n,\ n\in\mathbb{N}\}$.

where $F_{f,n}(\mu) := \int\ldots\int f(x_1,\ldots,x_n)\mu^n(dx)$.
If $F_{f,n} \in \mathfrak{D}_0(\mathfrak{G})$, then

$$\mathfrak{G}^A_N F_{f,n}(\mu_N)$$

$$= \langle\mu^n_N, A^{(n)}f\rangle + N^{-n}\sum_{k=1}^{n-1}\sum_{p\in\rho^n_k}\sum_{j_{p1}=1}^{N}\cdots\sum_{j_{pk}=1}^{N}(A^{(k)}f^{(p)} - A^{(n)}f)(z_{j_1},\ldots,z_{j_n})$$

$$= F_{A^{(n)}f,n}(\mu_N) + R_1(N,n,f)(\mu_N)$$

and the remainder term satisfies $|R_1(N,n,f)(\mu_N)| \le c_1(n)\|f\|_{A,n}/N$.
Using Taylor's formula again , we obtain that for $F = F_{f,n}\in \mathfrak{D}_0(\mathfrak{G})$,

(2.6.3)

$$\mathfrak{G}_N F(\mu) = \int(A\,\frac{\delta F(\mu)}{\delta\mu(x)})\mu(dx) + \frac{\gamma}{4}\iint\left[\frac{\delta^2 F(\mu)}{\delta\mu(x)^2} + \frac{\delta^2 F(\mu)}{\delta\mu(y)^2} - 2\,\frac{\delta^2 F(\mu)}{\delta\mu(x)\delta\mu(y)}\right]\mu(dx)\mu(dy)$$
$$+ R(N,n,f)(\mu)$$

where $|R(N,n,f)(\mu_N)| \le c_2(n)\|f\|_{A,n}/N$.

The limiting generator is called the *Fleming-Viot generator:*

$$\text{(2.6.4)}\qquad \mathfrak{G}F(\mu) = \int\left(A\,\frac{\delta F(\mu)}{\delta\mu(x)}\right)\mu(dx) + \frac{\gamma}{2}\iint\frac{\delta^2 F(\mu)}{\delta\mu(x)\delta\mu(y)}\,Q_S(\mu;dx,dy)$$

where $Q_S(\mu;dx,dy) := \mu(dx)\delta_x(dy) - \mu(dx)\mu(dy)$.
For $F_{f,n}\in \mathfrak{D}_0(\mathfrak{G})$,

$$\mathfrak{G}F_{f,n}(\mu) = \langle\mu^n, A^{(n)}f\rangle + \frac{1}{2}\,\gamma\sum_{\substack{i,j\\ i\ne j}}\left(\langle\tilde{\Theta}_{ij}f,\mu^{(n-1)}\rangle - \langle f,\mu^{(n)}\rangle\right)$$

(2.6.5)

$$= F_{A^{(n)}f,n}(\mu) + \frac{1}{2}\,\gamma\sum_{\substack{i,j\\ i\ne j}}\left(F_{\Theta_{ij}f,n}(\mu) - F_{f,n}(\mu)\right)$$

$$= F_{L_n f}(\mu),$$

where $\tilde{\Theta}_{ij}: C(E^n) \to C(E^{n-1})$ is defined as in Remark 2.5.2(b) and L_n is Moran particle generator defined by (2.5.2).

2.7 The Fleming-Viot Limit

In this section we show that as $N \to \infty$, the processes X^N converge (in the sense of weak convergence of finite dimensional distributions (f.d.d.) to a measure-valued process introduced by Fleming and Viot (1979).

Let $\mu_N = N^{-1} \sum_{j=1}^{N} \delta_{Z_j}$, with $\{Z_j\}$ exchangeable E-valued random variables and assume that the random measures $\mu_N \Rightarrow \mu \in M_1(E)$ (for example, $\{Z_j\}$ are i.i.d. μ).

Theorem 2.7.1. Let $\mathcal{D}_0(\mathfrak{G}) := \{F_{f,n}(\cdot): f \in D_0^n, n \in \mathbb{N}\}$. Assume that $\mu_N \Rightarrow \mu$ as $N \to \infty$. Then as $N \to \infty$ the processes $\{X_N\}$ with generators $\mathfrak{G}_N$ (defined by (2.6.5)) converge (f.d.d.) to a Feller process taking values in $M_1(E)$, with initial value μ, semigroup $\{T_t\}$ defined on $\mathcal{D}_0(\mathfrak{G})$ by

$$T_t F_{f,n}(\mu) := F_{V_t^n f}(\mu) = \int \cdots \int (V_t^n f)(\mathbf{x}) \mu^n(d\mathbf{x}), \quad f \in D_0^n,$$

where V_t^n is as in Theorem 2.5.1. Moreover $\mathcal{D}_0(\mathfrak{G})$ is a core for the generator $\mathfrak{G}$. *This process is called the Fleming-Viot process with mutation operator A and sampling rate γ and will be referred to as the S(A,γ)-Fleming-Viot process.*

Proof. We first verify that $\{T_t\}$ defines a Feller semigroup on $C(M_1(E))$. By the same argument as in Lemma 2.1.2 we can verify that $\mathcal{D}_0(\mathfrak{G})$ is dense in $C(M_1(E))$. For $F_{f,n} \in \mathcal{D}_0(\mathfrak{G})$,

$$T_t F_{f,n}(\mu) := F_{V_t^n f}(\mu) = \int \cdots \int (V_t^n f)(\mathbf{x}) \mu^n(d\mathbf{x})$$

is strongly continuous on $\mathcal{D}_0(\mathfrak{G})$ by Theorem 2.5.1(a) and $V_t^n: D_0^n \to D_0^n$ by Theorem 2.5.1(b). Therefore $\{T_t\}$ is a Feller semigroup on $C(M_1(E))$ and $T_t: \mathcal{D}_0(\mathfrak{G}) \to \mathcal{D}_0(\mathfrak{G})$. Thus $\mathcal{D}_0(\mathfrak{G})$ is a core for the generator $\mathfrak{G}$ (by Lemma 2.1.1).

In order to complete the proof of the weak convergence of the finite dimensional distributions of X^N to this limiting Feller process note that by (2.6.3)

$$\mathfrak{G}_N F_{f,n}(\mu_N) = \mathfrak{G} F_{f,n}(\mu_N) + R(N,n,f)(\mu_N)$$

where $|R(N,n,f)(\mu)| \le c(n)\|f\|_{A,n}/N$ and $\|V_t^n f\|_{A,n} \le \|f\|_{A,n}$.

Then recalling the definition of $\{T_t^N\}$, from the beginning of Section 2.6 we have

$$T^N_t F_{f,n}(\mu_N) = T_t F_{f,n}(\mu_N) + \int_0^t T_{t-s} R(N,n,V^N_s f)(\mu_N) ds$$

and as above $|R(N,n,V^N_s f)(\mu_N)| \le c(n)\|f\|_{A,n}/N$.

Therefore if $\mu_N \to \mu$, $T_t F_{f,n}(\mu_N) = \int (V^n_t f)(\mathbf{x})\mu^n_N(d\mathbf{x}) \to \int (V^n_t f)(\mathbf{x})\mu^n(d\mathbf{x})$

and $\int_0^t T_{t-s} R_N(N,n,V^N_s f)(\mu_N) ds \to 0$.

This proves the convergence $T^N_t F_{f,n}(\mu_N) \to T_t F_{f,n}(\mu)$.

Thus we have proved that $E(F_{f,n}(X_N(t))|X_N(0)=\mu_N) \to T_t F_{f,n}(\mu)$ for all polynomials $F_{f,n}$. Since $M_1(E)$ is compact, and $\{X_N(t):N\in\mathbb{N}\}$ are tight random measures for each t, then this proves (since polynomials are convergence determining) that

$$T_t F_{f,n}(\mu) = \int F_{f,n}(\nu) P_{t,\mu}(d\nu)$$

for some probability measure $P_{t,\mu}$ on $M_1(E)$. Thus we have proved that the law of measure-valued Moran process $X_N(t)$ converges to $P_{t,\mu}(d\nu)$. Since $\{V^n_t f\} \in C(E^n)$, the mappings $\mu \to T_t F_{f,n}(\mu)$ and $\mu \to P_{t,\mu}$ are continuous. Using the Markov property the convergence of the finite dimensional distributions follows by a standard argument. □

2.8. The Moment Measures of Fleming-Viot.

Since $\mathfrak{G} F_{t,n}(\mu) = F_{L_n f}(\mu)$, the limiting moment measures, M_n, of the Fleming-Viot process, X(t), satisfy the weak form of the system of equations dual to (2.5.1), that is,

(2.8.1)

$$\partial M_n(t;dx_1,\dots,dx_n)/\partial t$$
$$= \sum_{i=1}^n (A^i)^* M_n(t;dx_1,\dots,dx_n) - \tfrac{1}{2}\gamma n(n-1) M_n(t;dx_1,\dots,dx_n)$$
$$+ \tfrac{1}{2}\gamma \sum_i \sum_{j\ne i} M_{n-1}(t;dx_1,\dots,dx_{i-1},dx_{i+1},\dots,dx_n)\,\delta_{x_j}(dx_i)$$

$$M_n(0;dx_1,\dots,dx_n) = \mu(dx_1)\dots\mu(dx_n).$$

Proposition 2.8.1 The solution of the moment equations is given as follows:

for $f \in C_{sym}(E^k)$,

$$T_t F_{f,k}(\mu) = \int\dots\int f(x_1,\dots,x_k) M_k(t,\mu;dx_1,\dots,dx_k)$$

$$= \langle \mu^k (S^k_t)^*, f\rangle + \sum_{j=1}^{k-1} \int_0^t ds \langle \mu^j (S^j_s)^*, U^k_j(t-s) f\rangle$$

where for $f \in C_{sym}(E^k)$

$$(S_t^k f)(x_1,\dots,x_k)$$

$$:= \quad e^{-\gamma k(k-1)t/2} \int \dots \int p(t,x_1,dy_1)\dots p(t,x_k,dy_k) f(y_1,\dots,y_k)$$

for $\nu \in M_1(E^k)$

$$\nu(S_t^k)^*(dy_1,\dots,dy_k)$$

$$:= \quad e^{-\gamma k(k-1)t/2} \int \nu(dx_1,\dots,dx_k) p(t,x_1;dy_1)\dots p(t,x_k;dy_k),$$

where $p(t,x;dy)$ denotes the transition probability function of the A-process, * denotes the adjoint of an operator and $U_j^k(t)$ is inductively defined by

$$U_{k-1}^k(t) := \tfrac{1}{2}\gamma \Psi_k^{k-1} S_t^k, \quad t \geq 0,$$

$$U_j^k(t) := \int_0^t U_j^{k-1}(s) U_{k-1}^k(t-s) ds \text{ , for } k-j > 1,$$

$$(\Psi_k^{k-1} f)(x_1,\dots,x_{k-1}) := \sum_{\substack{i,j \\ i \neq j}} (\tilde{\Theta}_{ij} f)(x_1,\dots,x_{k-1}).$$

Proof. This obtained by induction from the above equations (2.8.1). (cf. Dynkin (1989c), Ethier and Griffiths (1992)). □

Corollary 2.8.2 For $f \in C(E^n)$, $t \geq 0$,

$$\int \dots \int f(y_1,\dots,y_n) M_n(t,\mu;dy_1,\dots,dy_n) = \int \dots \int V_t^n f(y_1,\dots,y_n) \mu(dy_1)\dots\mu(dy_n)$$

when $\{V_t^n\}$ is the n-particle Moran semigroup (Theorem 2.5.1) i.e. the moment measure $M_n(t,\mu;.)$ is the probability law at time t of the n particle Moran process with initial distribution i.i.d. μ.

Proof. This follows directly from the proof of Theorem 2.7.1. □

Remarks 2.8.3.

(1) For the moment let us begin with the system of moment equations (2.8.1) whose solution is given by Proposition 2.8.1. For $f \in C(E^n)$ let us define

$$T_t F_{f,n}(\mu) := \int \dots \int f(x_1,\dots,x_n) M_n(t,\mu;dx_1,\dots,dx_n).$$

We can then verify directly that

(i) $\{M_n(t,\mu;.) \in M_1(E^n): n=1,2,\dots\}$ forms a projective system of probability measures which define a sequence of E-valued exchangeable random variables and consequently by de Finetti's theorem (cf. Section 11.2) define a random probability measure on E,

(ii) $\{T_t\}$ defines a Feller semigroup on $C_p(M_1(E))$.

This gives an alternative approach to the construction of the Fleming-Viot process which does not require the consideration any approximating particle system.

(2) In this chapter we have gone full circle - starting with the Moran process we constructed the Fleming-Viot process as a weak limit and then considering the moment measures of the Fleming-Viot we have recovered the Moran process (Corollary 2.8.2). In Chapter 11 we will see that there is a deeper connection here and in fact a consistent family of n-particle Moran processes, $n \in \mathbb{N}$, is embedded in the Fleming-Viot process.

(3) Duality between the Moran and Fleming-Viot processes.

The relation between the Moran processes and the Fleming-Viot process expressed in Corollary 2.8.2 also illustrates a simple example of the type of *duality relation* which will be developed in Chapter 5. To describe this in a more general setting let us be given the following:

(i) $\{T_1(t,s): 0 \le s \le t\}$ an evolution family on E_1 with generator A^1_s, and domain $D(A^1)$, that is,

Domain of A^1_s contains $D(A^1)$ for each s,

$T_1(t,s): D(A_1) \to D(A_1)$, $\quad T_1(t,t) = Id$,

$$\frac{\partial T_1(t,s)}{\partial s} = -T_1(t,s)A^1_s, \quad \text{and}$$

$$\frac{\partial T_1(t,s)}{\partial t} = A^1_t T_1(t,s), \quad \text{for} \quad 0 < s < t.$$

Similarly, let $T_2(t,s)$ be an evolution family on E_2 with generator A^2_s.

<u>Proposition</u> <u>2.8.4.</u> Assume that $f \in C(E_1 \times E_2)$ and $g: [0,\infty) \to C(E_1 \times E_2)$ be such that $f(.,y) \in D(A^1)$ for each $y \in E_2$, and $f(x,.) \in D(A^2)$ for each $x \in E_1$, and

$$g(s,x,y) = A^1_s f(x,y) = A^2_s f(x,y)$$

where A^1_s acts on the first variable and A^2_s acts on the second variable. Then

$$T_1(t,0)f(x,y) = T_2(t,0)f(x,y).$$

Proof. From the assumptions we obtain that

$$\frac{d}{ds} T_2(t,s)T_1(s,0)f$$

$$= -T_2(t,s)A^2_s T_1(s,0)f + T_2(t,s)A^1_s T_1(s,0)f \equiv 0$$

and therefore $\quad T_1(t,0)f(x,y) = T_2(t,0)f(x,y)$.

<u>Remark</u> <u>2.8.5.</u> Let $E_1 = M_1(E)$ and $E_2 = \bigcup_{n=1}^{\infty} C(E^n) \cap D^n_0$, T^1_t be the Fleming-Viot

semigroup on $C_p(M_1(E))$, and T^2_t be the Moran semigroup on $C\left(\bigcup_{n=1}^{\infty} C(E^n)\right)$ (i.e. $T^2_t F(f) := F(V^n_t f)$ if $f \in C(E^n)$. Let $F(\mu,f) := \int f d\mu^n$. Then Proposition 2.8.4 can be applied to show that

$T_t F_{f,n}(\mu) = T^1_t F(\mu,f) = T^2_t F(\mu,f) = F(\mu, V^n_t f) = F_{V^n_t f}(\mu)$ if $\mu \in M_1(E)$, $f \in C(E^n)$.

This notion of duality will be systematically developed in Section 5.5.

2.9 Weak Convergence of the Approximating Processes and the Martingale Problem

In subsequent sections we will systematically develop the notions of weak convergence of measure-valued processes and measure-valued martingale problems. In this section we briefly introduce this approach by applying it to the Fleming-Viot process.

In particular we will show that in addition to weak convergence of finite dimensional distributions the laws of the measure-valued Moran processes X_N, their distributions $P^N_{\mu_N}$ are tight in the space of probability measures on $D([0,\infty),M_1(E))$ and consequently weak convergence of processes follows. This implies that the Fleming-Viot process can be realized as a càdlàg process (we will later show that it is actually continuous).

We will now show (by using some results to be proved in the next chapter) that the Fleming-Viot process can also be characterized as the unique solution of the martingale problem for $(\mathfrak{G},\mathfrak{D}_0(\mathfrak{G}))$ defined in Sect. 2.7.

Since $\{X_N(t)\}$ is a Feller process with generator $\mathfrak{G}_N$ and core $\mathfrak{D}_0(\mathfrak{G})$ (cf. Sect. 2.6), it follows that

$$M_N(t) := F(X_N(t)) - \int_0^t \mathfrak{G}_N F(X_N(s))ds, \quad F \in \mathfrak{D}_0(\mathfrak{G}),$$

is a bounded martingale under $P^N_{\mu_N}$.

Therefore for $t \in [0,T]$,

$$\text{(2.9.1)} \quad F_{f,n}(X_N(t)) - \int_0^t \mathfrak{G} F_{f,n}(X_N(s))ds = M_N(t) + \int_0^t R(N,F,s)(X_N(s))ds$$

and $\quad \sup_{0 \le s \le T} |R(N,F,s)| \le c(F)/N.$

In order to prove the tightness of the $P^N_{\mu_N}$ on $\mathbf{D} := D([0,\infty),M_1(E))$ we will use Theorem 3.6.4. Noting that $M_1(E)$ is compact (so that condition (i) of Theorem 3.6.4 is automatic) it suffices to show that for $\phi \in D_0(A)$, $\langle X_N(t),\phi\rangle$ are tight in $D([0,\infty),\mathbb{R})$. Applying (2.9.1) to $F_1(\mu) = \langle\mu,\phi\rangle$ and $F_2(\mu) = \langle\mu,\phi\rangle^2$, and then using Corollary 3.6.3 it follows that the laws of $\langle X_N(t),\phi\rangle$ are tight in $D([0,\infty),\mathbb{R})$. Since we already proved the convergence of the finite dimensional distributions in Theorem 2.7.1, this yields the weak convergence of the probability measures $P^N_{\mu_N}$ on **D**.

Now let $F\in \mathfrak{D}_0(\mathfrak{G})$. In this case the real-valued functional

$$F(x(t))-F(x(s)) - \int_s^t \mathfrak{G}F(x(u))du$$

is continuous on $D([0,\infty),M_1(E))$. Now let $H\in b\mathcal{D}_s\cap C(\mathbf{D})$. Therefore letting X denote the canonical process we get that

$$P_\mu\left[\left(F(X(t))-F(X(s)) - \int_s^t \mathfrak{G}F(X(u))du\right)H(X)\right]$$

$$= \lim_{N\to\infty} P^N_{\mu_N}\left[\left(F(X(t))-F(X(s)) - \int_s^t \mathfrak{G}F(X(u))du\right)H(X)\right]$$

$$= \lim_{N\to\infty} P^N_{\mu_N}\left[\left(F(X(t))-F(X(s)) - \int_s^t \mathfrak{G}_N F(X(u))du + \int_s^t R(N,F,u)du\right)H(X)\right]$$

$$= \lim_{N\to\infty} P^N_{\mu_N}\left[\left(M_N(t) - M_N(s) + \int_s^t R(N,F,u)du\right)H(X)\right]$$

$$= \lim_{N\to\infty} P^N_{\mu_N}\left[\left(\int_s^t R(N,F,u)du\right)H(X)\right] = 0.$$

This implies that

$$M_F(t) := F(X(t))-F(X(s)) - \int_s^t \mathfrak{G}F(X(u))du$$

is also a martingale for each $F\in\mathfrak{D}_0(\mathfrak{G})$ under P_μ. Therefore $\{P_\mu:\mu\in M_1(E)\}$ is *a solution to the martingale problem for* $(\mathfrak{G},\mathfrak{D}_0(\mathfrak{G}))$. In fact the family $\{P_\mu:\mu\in M_1(E)\}$ is uniquely characterized in this way since any solution to the martingale problem must have the <u>same</u> moment measures as the Fleming-Viot process. This can be verified by applying the martingale problem to polynomials which yield the moment equations solved above. The details of this argument will be given in greater generality below.

2.10 Branching Particle Systems

Let us for the moment continue in the same spirit and consider a simple branching particle system on a compact metric space E. The main difference from the Moran model is that the total number of particles is no longer constant in time. For this reason the basic state space is now M(E) the space of finite Borel measures on E. We will again follow the elementary approach based on moment measures to characterize the transition function for the limiting measure-valued process.

We consider a system of particles in the space E which move, die and produce offspring. We begin by assuming that during its lifetime each particle performs an

A-motion independently of the other particles.

In the case of critical branching when particles die they produce k particles with probability p_k, $k=0,1,2,\dots$, $\sum_k kp_k = 1$. We will also assume in this section that

$$m_2 := \sum k^2 p_k, \text{ and } \sum k^3 p_k < \infty.$$

After branching the resulting set of particles evolve in the same way and independently of each other starting off from the parent particle's branching site. Let N(t) denote the total number of particles at time t. We will denote their locations by $\{x_i(t): 1\le i\le N(t)\}$.

In order to obtain a measure-valued process by use of an appropriate scaling we assume that particles have mass ε and branch at rate c/ε.

For $B \in \mathcal{E}$, define

$$X_\varepsilon(t,B) := \varepsilon\left(\sum_{i=1}^{N(t)} 1_B(x_i(t))\right).$$

Let C_F denote the class of functions on M(E) of the form $F_f(\mu) = f(\langle\mu,\phi\rangle)$ with $f\in C_b(\mathbb{R})$, $\phi\in C(E)$. Let $\mathfrak{D}(\mathfrak{G}) := \{F_f(\mu): F_f(\mu)=f(\langle\mu,\phi\rangle);\ f\in C_b^\infty(\mathbb{R}),\ \phi\in D_0(A)\}$ where $D_0(A)$ is as in Sect. 2.2. Then $X_\varepsilon(.)$ is an M(E)-valued Feller process. The generator of $X_\varepsilon(\cdot)$ is defined on $\mathfrak{D}(\mathfrak{G})$, by

$$\mathfrak{G}_\varepsilon F(\mu) := G^A F_f(\mu) + c\varepsilon^{-2}\int\left\{\sum_k p_k[f(\langle\mu,\phi\rangle+\varepsilon(k-1)\phi(x)) - f(\langle\mu,\phi\rangle)]\right\}\mu(dx)$$

where G^A denotes the generator of the empirical process associated to particles performing independent A-motions in E.

Then for $F\in \mathfrak{D}(\mathfrak{G})$,

$$\mathfrak{G}_\varepsilon F(\mu) = \mathfrak{G}F(\mu) + O(\varepsilon)$$

where $\mathfrak{G}F(\mu) = f'(\langle\mu,\phi\rangle)\langle\mu,A\phi\rangle + \frac{1}{2}c(m_2-1)f''(\langle\mu,\phi\rangle)\langle\mu,\phi^2\rangle$.

Letting $\varepsilon\to 0$ we obtain a measure-valued process with generator defined on $\mathfrak{D}(\mathfrak{G})$ by

$$\mathfrak{G}F(\mu) := f'(\langle\mu,\phi\rangle)\langle\mu,A\phi\rangle + \tfrac{1}{2}c(m_2-1)f''(\langle\mu,\phi\rangle)\langle\mu,\phi^2\rangle$$

$$= \int A(\delta F(\mu)/\delta\mu(x))\mu(dx) + \tfrac{1}{2}c(m_2-1)\iint(\delta^2 F(\mu)/\delta\mu(x)\delta\mu(y))\delta_x(dy)\mu(dx).$$

To carry out the proof two changes are required to the arguments of section 2.7. First, since the total mass of this process is not uniformly bounded, a martingale inequality is required to verify that for each $\eta,T > 0$ there exists a compact subset K_η of M(E) such that $P(X(t)\in K_\eta\ \forall t\in[0,T]) > 1-\eta$. (This calculation is a special case of the proof of tightness which is given in Ch. 4.) Secondly, to characterize the limiting finite dimensional distributions we can again use moment measures. To do this we must check that the moment measures uniquely determine the random measure $X(t)$ conditioned on $X(s)$ with $s<t$. The condition of Corollary 3.2.10 can be verified for the limiting process $\{X(t)\}$ with generator $\mathfrak{G}$ using the fact that $E(e^{\theta X(t,E)}) < \infty$ for small $\theta > 0$ (see Ch. 4).

In the next chapter we will develop the technical tools necessary to carry out the proofs of limit theorems of this type in a more general setting and then we will return to the study of a general class of measure-valued branching processes in Chapter 4.

3. RANDOM MEASURES AND CANONICAL MEASURE-VALUED PROCESSES

The objective of this chapter is to collect some basic material on measures on Polish spaces, random measures, canonical measure-valued processes, homogeneous and inhomogeneous Markov processes and weak convergence of random measures and measure-valued processes which will be needed in subsequent chapters.

3.1. State Spaces for Measure-valued Processes

Let E be a Polish space with metric d, and $\mathcal{E}$ be the Borel σ-algebra.

Let $C(E)$ ($pC(E)$, $bC(E)$, $pbC(E)$, $bC_b(E)$, $C_c(E)$, respectively) denote the spaces of all continuous functions on E (non-negative, bounded, non-negative bounded, with bounded support, with compact support, continuous functions respectively).
Let $\mathcal{E}$ ($p\mathcal{E}$, $b\mathcal{E}$, $bp\mathcal{E}$, $b\mathcal{E}_b$, $\mathcal{E}_c$, etc.) denote the corresponding spaces of Borel measurable functions on E.

Let $M(E)$ denote the space of Radon measures on $(E,\mathcal{E})$, that is the Borel measures, μ, such that $\mu(K) < \infty$ for all compact sets $K \subset E$.

For $\mu \in M(E)$ and $f \in b\mathcal{E}_c$ let $\langle\mu,f\rangle := \int f d\mu$.

Let $M_F(E)$ and $M_1(E)$ denote the finite and probability measures on E, respectively.

3.1.1. Finite measures on a Polish space

Let $(M_F(E),\tau_w)$ denote the finite measures with the *weak topology* defined by

$\mu_n \Rightarrow \mu$ iff $\langle\mu_n,f\rangle \to \langle\mu,f\rangle \ \forall f \in bC(E)$.

Then $(M_F(E),\tau_w)$ is a Polish space.

3.1.2. Measures on Compact Metric Spaces

If E is compact then $M_F(E)$ may be compactified $\bar{M}_F(E) = M_F(E)\cup\{\infty_w\}$, where $\{\infty_w\}$ is a compactifying point and the topology τ is defined by

$\mu_n \Rightarrow \mu \in M_F(E)$ iff $\langle\mu_n,f\rangle \to \langle\mu,f\rangle \ \forall f \in C(E)$,

$\mu_n \Rightarrow \infty_w$ iff $\langle\mu_n,1\rangle \to \infty$.

Then $(\bar{M}_F(E),\tau)$ is also a compact metrizable space called the *Watanabe compactification* of $M_F(E)$.

3.1.3. Radon measures on a locally compact metric space

Let E be a locally compact metric space. Then the *vague topology* is defined by

$$\mu_n \xrightarrow{\tau_v} \mu \quad \text{iff } \langle\mu_n,f\rangle \longrightarrow \langle\mu,f\rangle \quad \forall f \in C_c(E).)$$

Then $(M(E),\tau_v)$ is a Polish space.

The locally compact space E can be compactified: set $\bar{E}= E\cup\{\infty_a\}$ such that $x_n \longrightarrow \infty_p$ if and only if $x_n \in K$ for at most finitely many n for each compact set $K\subset E$. Then $\{\mu\in M_F(\bar{E}):\mu(\{\infty_a\})=0\}$ can be identified with the finite Radon measures on E.

3.1.4. Locally finite measures on a Polish space

Let $\mathcal{E}_b$ be the ring of bounded sets in $\mathcal{E}$. Let $(M_{LF}(E),\tau_v)$ denote the collection of *locally finite* Borel measures, μ, that is, measures such that $\mu(A) < \infty$ $\forall$ $A\in \mathcal{E}_b$. We define the *vague topology*

$$\mu_n \xrightarrow{\tau_v} \mu \quad \text{iff } \langle\mu_n,f\rangle \longrightarrow \langle\mu,f\rangle \quad \forall\ f\in bC_b(E).$$

Then $(M_{LF}(E),\tau_v)$ is a Polish space.

3.1.5. p-Tempered Measures on $\mathbb{R}^d$

Let p>0 and $\phi_p(x) := (1+|x|^2)^{-p}$; $M_p(\mathbb{R}^d) = \{\mu:\langle\mu,\phi_p\rangle < \infty\}$ with the topology τ_p defined by

$$\mu_n \Longrightarrow \mu \quad \text{iff } \langle\mu_n,f\rangle \longrightarrow \langle\mu,f\rangle\ \forall\ f\in K_p(\mathbb{R}^d),$$

where $K_p(\mathbb{R}^d) = \{f:f = g + \alpha\cdot\phi_p, \alpha\in\mathbb{R},\ g\in C_c(\mathbb{R}^d)\}$. $(M_p(\mathbb{R}^d),\tau_p)$ is also a Polish space. Let $\{f_n:n\geq 1\}$ be a dense set in $C_c(\mathbb{R}^d)$ and $f_0 = \phi_p$. We define a metric on $M_p(\mathbb{R}^d)$ by

$$d_p(\mu,\nu) := \sum_{m=0}^{\infty} 2^{-m}\left(1 \wedge \left|\int f_m d\mu - \int f_m d\nu\right|\right). \tag{3.1.1}$$

Let $\dot{\mathbb{R}}^d := \mathbb{R}^d\cup\{\infty_p\}$, where ∞_p is an isolated point and extend ϕ_p to $\dot{\mathbb{R}}^d$ by defining $\dot{\phi}_p(\infty_p) = 1$. Let $M_p(\dot{\mathbb{R}}^d) = \{\mu:\langle\mu,\dot{\phi}_p\rangle < \infty\}$ with the p-vague topology $\mu_n \Longrightarrow \mu$ iff $\langle\mu_n,f\rangle \longrightarrow \langle\mu,f\rangle$ $\forall$ $f\in K_p(\dot{\mathbb{R}}^d)$ (defined as above). In this case $M_p(\dot{\mathbb{R}}^d)$ is locally compact and sets of the form $\{\mu:\langle\mu,\dot{\phi}_p\rangle \leq k\}$ are compact (cf. Iscoe (1986a)).

3.2 Random Measures and Laplace Functionals.

Let E be Polish, $(M_F(E),\tau_w)$ be as above and $\mathcal{M} = \mathcal{B}(M_F(E))$ denote the Borel subsets of $M_F(E)$. A *finite random measure on* E is given by a probability measure P on $(M_F(E),\mathcal{M})$.

A sequence $\{f_n\}$ in $pb\mathcal{E}$ is said to *converge bp* to f if $f_n(x) \longrightarrow f(x)$ $\forall x$, and $\exists$ $M < \infty$ such that $\sup_{n,x} f_n(x) \leq M$. Given $H \subset bp\mathcal{E}$, the bounded pointwise closure is the smallest collection of functions containing H which is closed under bp convergence.

Lemma 3.2.1. There exists a countable set $V = \{f_n\} \subseteq bpC(E)$ such that $1\in V$, and V is closed with respect to addition operation, and the bp closure of V is an element of $bp\mathcal{E}$. Then the set V is also convergence determining in $M_F(E)$.

Proof. [EK pages 111-112].

We can then define a metric on $M_F(E)$ in terms of $V = \{f_n\}$ (as in Lemma 3.2.1) as follows

$$(3.2.1) \qquad d(\mu,\nu) := \sum_{m=0}^{\infty} 2^{-m}\left(1 \wedge |\langle\mu,f_m\rangle-\langle\nu,f_m\rangle|\right).$$

Since V is convergence determining it follows that μ_n converges to μ in the topology τ_w if and only if $d(\mu_n,\mu) \to 0$ as $n\to\infty$.

Remark 3.2.2 If E is compact then $\langle\mu_n,1\rangle \to m < \infty$ implies that $\{\mu_n\}$ is relatively compact. From this we can verify the fact that d is a complete metric if E is compact. To obtain a complete metric in the general case we begin by recalling that every Polish space E is Borel isomorphic to a compact metric space (cf. Cohn (1980) Theorem 8.3.6). We can then put a metric on E such that it becomes a compact metric space and such that the Borel sets for this metric coincide with the Borel subsets of E. We then choose a countable subset $V' \subset bp\mathcal{E}$ which is a dense subset of the space of functions which are *continuous in the new metric*. We can and will assume that $1\in V'$ and that the functions in V' are strictly positive. Then a metric defined as above but using V' in place of V is complete.

Lemma 3.2.3. (i) If $\mathcal{A}$ is a class of Borel sets in E closed under finite intersections and containing a basis for the topology on E, then

$\mathcal{M} = \sigma\{f_A: A\in\mathcal{A}\}$ where $f_A(\mu) = \mu(A)$.

(ii) Let V be as in Lemma 3.2.1. Then $\mathcal{M} = \sigma\{\langle.,f\rangle: f\in V\}$.

Proof. (i) For $f\in bC(E)$, the mapping $\mu \to \langle\mu,f\rangle$ is continuous and hence $\mathcal{M}$-measurable. It is easy to check that $\{f: \mu \to \langle\mu,f\rangle$ is $\mathcal{M}$-measurable$\}$ is closed under bp limits. Since $bp\mathcal{E}$ is the bp closure of bC(E), this implies that $\mu \to \mu(A)$ is $\mathcal{M}$-measurable for any $A\in\mathcal{E}$ and hence $\mathcal{M} \supset \sigma\{f_A: A\in\mathcal{A}\}$. On the other hand for each f of the form $f = \sum_{i=1}^{n} a_i\chi_{A_i}$ with $A_i\in\mathcal{A}$, $a_i \in \mathbb{R}$, the mapping $\mu \to \langle\mu,f\rangle$ is $\sigma\{f_A: A\in\mathcal{A}\}$-measurable. By Dynkin's class theorem for functions ([EK p. 497]) the bp-closure of this class of functions contains bC(E). Hence for $f\in bC(E)$, $\mu \to \langle\mu,f\rangle$ is $\sigma\{f_A: A\in\mathcal{A}\}$-measurable which implies that $\mathcal{M} \subset \sigma\{f_A: A\in\mathcal{A}\}$.

(ii) The proof is similar to that of (i). □

Lemma 3.2.4. Let $\mathcal{A}$, V be as in Lemma 3.2.3. Then a probability measure, X, on $(M_F,\mathcal{M})$ is uniquely determined by the "finite dimensional distributions" of

$\{X(A_1),\ldots,X(A_n): A_i\in \mathcal{A}\}$ denoted by $\{P_{A_1\ldots A_n}:n\in\mathbb{N}\}$ or equivalently by $\{\langle X,f_1\rangle,\ldots,\langle X,f_n\rangle:f_i\in V,\ i=1,\ldots,n,\ n\in\mathbb{N}\}$ denoted by $\{P_{f_1,\ldots,f_n}:n\in\mathbb{N}\}$.

Proof. This follows immediately from Lemma 3.2.3. □

Let P be a probability measure on $(M_F(E),\mathcal{M})$. The associated *Laplace functional* is defined by

$$L(f) := \int_{M_F(E)} e^{-\langle\mu,f\rangle}P(d\mu), \quad f\in bp\mathcal{E}.$$

It is easy to check that L(.) is continuous under bp-limits.

Let $\mathcal{V} \subset \mathcal{B}(M_F(E))$ be the linear span of $\{F_f(.):f\in V\}$ where V is as in Lemma 3.2.1 and $F_f(\mu) := e^{-\langle\mu,f\rangle}$. Note that $\mathcal{V}$ is an algebra of functions since V is closed under addition.

Lemma 3.2.5.

(i) The space of $\mathcal{M}$-measurable functions is the bp closure of $\mathcal{V}$ and $\mathcal{M} = \sigma\{F_f:f\in V\}$.

(ii) P is uniquely determined by L(f), $f\in$ V.

Proof. (i) Sets of the form $\{\mu:\langle\mu,f_i\rangle \in O_i,\ i=1,\ldots,n\}$ where O_i are open in $\mathbb{R}$, $f_i\in V$ form a fundamental set of neighbourhoods in $M_F(E)$. Note that the bounded pointwise closure of finite linear combinations of the form $\sum a_j e^{-jx}$, a_j rational, contains sets of the form 1_O, O open. Hence the indicator, 1_{O_i}, of each fundamental open set O_i belongs to the bp-closure of $\mathcal{V}$. The result follows by Dynkin's class theorem for functions (cf. [EK, p. 497]).

(ii) then follows from (i). □

Theorem 3.2.6 (i) $\{F_f:f\in V\}$ is convergence determining in $M_1(M_F(E))$, that is, if $\int F_f(\mu)P_n(d\mu) \longrightarrow \int F_f(\mu)P(d\mu)$ as $n\to\infty$ $\forall$ $f\in V$, with $P_n,P \in M_1(M_F(E))$, then P_n converges weakly to P.

(ii) There exists a countable set of strictly positive $\mathcal{E}$-measurable functions V' such that if $\int F_f(\mu)P_n(d\mu) \longrightarrow L(f)$ as $n\to\infty$ $\forall$ $f\in V'$, for $\{P_n\}$ in $M_1(M_F(E))$, then there exists $P\in M_{\le 1}(M_F(E))$ (the subprobability measures) such that

$$L(f) = \int_{M_F(E)} F_f(\mu)\ P(d\mu)\ \forall\ f\in V'.$$

Proof. (i) The linear span of $\{F_f:f\in V\}$ is an algebra. We will show that it is *strongly separating*, that is, for every $\delta>0$ and μ there exists $\{F_{f_1},\ldots,F_{f_N}\}$ such that

$$\inf_{\nu\in(B(\mu,\delta))^c} \max_{1\le i\le N} |F_{f_i}(\mu)-F_{f_i}(\nu)| > 0$$

where $B(\mu,\delta)$ is a ball with center μ and radius δ (with respect to the metric d). Let $\mu\in M_F(E)$ and $\delta>0$. Let $2^{-m} < \delta/4$. If $d(\nu,\mu) > \delta$, then $\sum_{i=1}^{m} |\langle\nu,f_i\rangle-\langle\mu,f_i\rangle| \ge \delta/2$ and hence $\max_{1\le i\le m} |\langle\nu,f_i\rangle-\langle\mu,f_i\rangle| \ge \delta/2m$. But then $\max_{1\le i\le m} |F_{f_i}(\nu)-F_{f_i}(\mu)| \ge \text{const}\cdot\delta$, where the constant can be chosen independent of ν. Hence the $\{F_f\}$ are strongly separating. The result then follows since a strongly separating algebra of functions is convergence determining (cf. [EK p. 113]).

(ii) We can put a new metric on E making it a compact metric space and for which the Borel sets coincide with $\mathcal{E}$. We then choose (as in Remark 3.2.2) a countable set V' of strictly positive functions on E which are dense in the space of nonnegative functions continuous for the new metric and consequently convergence determining for $M_F(E)$. Recall that a sequence $\{P_n\} \in M_1(M_F(E))$, E compact, can be viewed as elements of $M_1(\overline{M_F(E)})$ where $\overline{M_F(E)}$ denotes the Watanabe compactification of $M_F(E)$. The result (ii) follows by noting that $M_1(\overline{M_F(E)})$ is compact, every element of $M_1(\overline{M_F(E)})$ gives a subprobability on $M_F(E)$, with the subsequent application of the continuity theorem for Laplace transforms and (i). The details are left to the reader.□

Corollary 3.2.7. Let $P_n\in M_1(M_F(E))$. To prove that $P_n \xrightarrow{w} P$ as $n\to\infty$ it suffices to show that

(i) $\{P_n\}$ are tight, that is, for every $\varepsilon>0$ there exists a compact subset $K_\varepsilon\subset M_F(E)$, such that $\sup_n P_n(K_\varepsilon^c) < \varepsilon$,

(ii) $L_n(f) = \int F_f(\mu)P_n(d\mu)$ converges as $n\to\infty$ for each $f\in V$.

Given $P\in M_1(M_F(E))$, we define the moment measures, when they exist, as follows: $M_n \in M_F(E^n)$,

$$M_n(A_1\times\ldots\times A_n) := \int_{M_F(E)} \mu(A_1)\ldots\mu(A_n)P(d\mu), \quad A_1,\ldots,A_n \in \mathcal{E}.$$

Subset A of $M_F(E)$ is *tight* if

(i) $\sup_{\mu\in A} \mu(E) < \infty$, and

(ii) for every $\varepsilon>0$ there exists a compact subset $K_\varepsilon \subset E$ such that

$$\sup_{\mu\in A} \mu(K_\varepsilon^c) < \varepsilon.$$

Lemma 3.2.8. Let $\{P_n\} \subset M_1(M_F(E))$. Consider the first moment measures $M_{1,n}(B) =$

$\int_{M_F(E)} \mu(B)P_n(d\mu),\ B\in \mathcal{E}.$

If the $\{M_{1,n}:n=1,2,\ldots\}$ exist and are tight in $M_F(E)$, then the $\{P_n\}$ are tight in $M_1(M_F(E))$.

Proof. Let $\ell := \sup_n M_{1,n}(E) < \infty$ and for $m\in \mathbb{N}$, K_m be a compact such that $\sup_n M_{1,n}(K_m^c) < 1/2^m$. Let

$$C_m := \{\mu:\mu(E) \le 2^{m+1}\ell,\ \forall k\ge m+2,\ \mu(K_{2k}^c) \le 2^{-k}\}.$$

Then C_m is compact by Prohorov's criterion and

$$P_n(C_m^c) \le \frac{\ell}{2^{m+1}\ell} + \sum_{k=m+2}^{\infty} P(\{\mu:\mu(K_{2k}^c) > 2^{-k}\})$$

$$\le 2^{-(m+1)} + \sum_{k=m+2}^{\infty} 2^k/2^{2k} = 2^{-m} \quad \text{for all } n.$$

Hence the $\{P_n\}$ are tight by Prohorov's criterion.□

Theorem 3.2.9 (a) Let $P \in M_1(M_F(E))$. Then in order that P is uniquely characterized by its moment measures $\{M_n:n=1,2,\ldots\}$ it suffices to show that

$$(3.2.2)\qquad \sum_{n=1}^{\infty} [M_n(E^n)]^{-1/2n} = +\infty.$$

(b) Let $\{P_m:m\in\mathbb{N}\} \subset M_1(M_F(E))$ and all moment measures exist for each m. If for each k, the kth moment measures $M_{k,m}$ of P_m converge weakly to a finite measure M_k on E^k, and $\sum_{k=1}^{\infty} [M_k(E^k)]^{-1/2k} = +\infty$, then there exists a unique $P\in M_1(M_F(E))$ such that P_m converges weakly to P and P has moment measures $\{M_k\}$.

Proof. (a) This is an extension of the classical result of Carleman to the case of random measures (cf. Zessin (1983)).

(b) follows as in Zessin (1983, Theorem 2.2). □

Corollary 3.2.10. A sufficient condition is that

$$\limsup_{n\to\infty} \left(\frac{[M_n(E^n)]^{1/n}}{n}\right) < \infty.$$

Proof. See for example Breiman (1968, Prop. 8.49). □

3.3 Poisson Cluster Random Measures and the Canonical Representation of Infinitely Divisible Random Measures

The *Poisson random measure* (with intensity measure $\Lambda(.) \in M_F(E)$) is given by the Laplace functional

$$(3.3.1)\qquad L_\Lambda(\phi) = \exp\left(-\int_E (1 - e^{-\phi(x)})\Lambda(dx)\right),\quad \phi \in bp\mathcal{E}.$$

Let E_1, E_2 be Polish spaces. The *Poisson cluster random measure* with finite intensity measure Λ on E_1 and cluster law $\{P_x\}_{x\in E_1}$ on $M_F(E_2)$ such that $\int_{E_1} P_x\Lambda(dx) \in M_F(E_2)$, is the random measure on E_2 defined by

(3.3.2) $$L_{\Lambda,\{P_x\}}(\phi) := \exp\left[-\int_{E_1} (1 - P_x e^{-\langle\cdot,\phi\rangle})\Lambda(dx)\right],\quad \phi\in bp\mathcal{E}(E_2).$$

A random measure X is said to be *infinitely divisible* if for every positive integer n, $X \overset{D}{=} X_1+\ldots+X_n$ where $X_1,\ldots,X_n$ are independent identically distributed random measures. The Poisson random measure and Poisson cluster random measures are infinitely divisible.

In fact the Poisson cluster random measure is closely related to the canonical representation of infinitely divisible random measures given in the following result.

Theorem 3.3.1. Let X be a random measure with values in $M_{LF}(E)$ with infinitely divisible law P. Then there exists a pair $(M,R) \in M_{LF}(E)\times M(M_{LF}(E))$ such that

(3.3.3a) $R(\{0\}) = 0$, and

(3.3.3b) $$\int_{M_{LF}(E)} (1-e^{-\mu(A)})R(d\mu) < \infty \quad \forall\ A\in \mathcal{E}_b$$

(3.3.3c) $$L(\phi) = E(e^{-\langle X,\phi\rangle}) = e^{-u(\phi)} \quad \forall\ \phi\in pb\mathcal{E}_b,$$

and the *log-Laplace functional*, $u(\cdot)$, is represented by

(3.3.3d) $$u(\phi) := \langle M,\phi\rangle + \int_{M_{LF}(E)} (1-e^{-\langle\nu,\phi\rangle})R(d\nu).$$

Conversely, any functional of this form is the Laplace functional of an infinitely divisible random measure.

Proof. See Matthes, Kerstan and Mecke (1978, Ch. 2), Kallenberg (1983) or Dawson (1991, Theorem 3.4.1).

The representation (3.3.3) is called the *canonical representation* and the measure R is called the *canonical measure* or the *KLM measure.*

To complete this section we introduce another generalization of the notion of Poisson cluster random measures. In this case the intensity Λ is itself a random measure on E_1, i.e. given by a probability measure P_I on $M_{LF}(E_1)$. Then the resulting random measure on E_2 has Laplace functional

(3.3.4) $$L_{I,\{P_x\}}(\phi) = \int\exp\left(-\int(1-P_x e^{-\langle\cdot,\phi\rangle})\Lambda(dx)\right)P_I(d\Lambda),\quad \phi\in bp\mathcal{E}_{2,b}.$$

and is called the *Cox cluster (or doubly stochastic) random measure*, X, on E_2 with random intensity I on E_1 and clustering law $\{P_x : x\in E_1\}$. We assume that $x \to P_x$ is a measurable mapping from E_1 to $M_1(M_F(E_2))$. Moreover I is assumed to be a random measure on E_1 and X be a random measure on $E_1\times E_2$. Assume that the pair (I,X) has joint probability law $P \in M_1(M_F(E_1) \times M_F(E_1 \times E_2))$ such that conditioned on I, X is a Poisson cluster random measure with intensity I on E_1 and cluster distribution $\delta_x \times P_x$ on $E_1 \times E_2$.

3.4. The Structure of Random Measures

3.4.1 Pure Atomic Random Measure

Every $\mu \in M_F(E)$ can be uniquely decomposed into $\mu = \mu_a + \mu_d$ where μ_a is a *pure atomic measure* and μ_d is a *diffuse* (nonatomic) measure, that is, $\mu_a = \sum a_j \delta_{x_j}$, and $\mu_d(\{x\}) = 0$ for every $x \in E$. We denote $\mu_a = \zeta_a(\mu)$ the mapping $\mu \to \mu_a$ defined on $M_F(E)$ and $M_a(E) := \{\mu \in M_F(E) : \mu = \zeta_a(\mu)\}$.

Let $\bar{\mathfrak{A}}_\infty = \{(a_1, a_2, \ldots) : a_1 \ge a_2 \ge \ldots \ge 0,\ \Sigma a_i \le 1\}$, and

$\mathfrak{A}_\infty = \{(a_1, a_2, \ldots) : a_1 \ge a_2 \ge \ldots \ge 0,\ \Sigma a_i = 1\}$.

Define $\vartheta : M_1(E) \longrightarrow \bar{\mathfrak{A}}_\infty$, by letting $\vartheta(\mu)$ be the vector of *descending order statistics* of the masses of the atoms of $\zeta_a(\mu)$ and let $|\vartheta(\mu)| \le \infty$ denote the number of non-zero atoms. We also define $\tilde{\vartheta}(\mu) := (\vartheta(\mu); x_1, \ldots, x_{|\vartheta(\mu)|})$ if $\mu_a = \sum_{i=1}^{|\vartheta(\mu)|} a_i \delta_{x_i}$, $a_1 \ge a_2 \ge \ldots$. Note that $\mu \in M_a(E) \cap M_1(E)$ if and only if $\vartheta(\zeta_a(\mu)) \in \mathfrak{A}_\infty$.

Let E be a separable metric space. A *partition family* for E is given by a collection $\{E_j^n, x_j^n, \rho_n ; n, j \in \mathbb{N}\}$ where

(a) each $E_j^n \in \mathcal{E}$, $x_j^n \in E_j^n$, $E_j^n \cap E_k^n = \emptyset$ if $j \ne k$,

(b) $\bigcup_j E_j^n = E$ for each $n \in \mathbb{N}$,

(c) $\{E_j^{n+1} : j \in \mathbb{N}\}$ is a refinement of $\{E_j^n : j \in \mathbb{N}\}$, that is, each $E_j^{n+1} \subset E_k^n$ for some k,

(d) $\rho_n := \sup_j \operatorname{diam}(E_j^n) \longrightarrow 0$ as $n \to \infty$.

Given $\eta > 0$, and the partition family $\{E_j^n, x_j^n\}$ we define a mapping $\xi_n : M_F(E) \longrightarrow M_F(E)$ as follows:

$$\xi_n(\mu) := \sum_j \mu(E_j^n) \delta_{x_j^n}.$$

Given $\eta > 1$, we define $\zeta_{\eta,n} : M_F(E) \longrightarrow M_F(E)$ by

$$\zeta_{\eta,n}(\mu) := \sum_j \left(\mu(E_j^n)\right)^\eta \delta_{x_j^n}$$

<u>Theorem 3.4.1.1</u> (a) $\zeta_a(\mu) = \lim_{\eta\downarrow 1}\lim_{n\to\infty} \zeta_{\eta,n}(\mu)$.

(b) The mapping $\mu \longrightarrow \mu_a$ is $\mathcal{B}(M_F(E))/\mathcal{B}(M_F(E))$ measurable.

(c) $M_a(E) := \{\mu:\mu(E) = \zeta_a(\mu)(E)\} \in \mathcal{B}(M_F(E))$.

(d) The mapping $\tilde{\vartheta}:M_1(E) \longrightarrow \bar{\mathfrak{A}}_\infty \times \bigcup_{i=0}^{\infty} E^i$ is measurable.

<u>Sketch of Proof.</u> (a) It is easy to verify that if $\mu = \sum a_i\delta_{y_i}$, then

$$\lim_{\eta\downarrow 1}\lim_{n\to\infty} \zeta_{\eta,n}(\mu) = \mu.$$

On the other hand if μ is diffuse then

$$\lim_{\eta\downarrow 1}\lim_{n\to\infty} \zeta_{\eta,n}(\mu) = 0.$$

(b) The mappings $\{\zeta_{\eta,n}\}$ are clearly measurable since the mappings $\mu \longrightarrow \mu(E^n_j)$ are measurable (cf. Lemma 3.2.3). This implies that ζ_a is measurable.

(c) follows immediately from the definition.

(d) For $n\in\mathbb{N}$ let π_n denote the permutation of $\{1,2,3,\ldots\}$ such that

$$\pi_n j < \pi_n i \text{ if } \mu(E^n_j) > \mu(E^n_i)$$

$$\pi_n j < \pi_n i \text{ if } \mu(E^n_j) = \mu(E^n_i) \text{ and } j < i.$$

Let $\tilde{\vartheta}_n(\mu) := (\{\mu(E^n_j)\},\{x^n_j\})$. Clearly $\mu(E^n_{\pi_n^{-1}j}) \longrightarrow a_j$ and x_j is one of the limit points of the sequence of finite sets $\{x^n_{\pi_n^{-1}k} : a_k=a_j\}$. □

We will now give another application of this technique used in the last proof to a class of Cox random measures. Let $x \longrightarrow P_x$ be a measurable mapping from E_1 to $M_1(M_F(E_2))$. Let I be a random measure on E_1 and X be a random measure on $E_1\times E_2$. Assume that the pair (I,X) has joint probability law $P \in M_1(M_F(E_1)\times M_F(E_1\times E_2))$ such that conditioned on I, X is a Poisson cluster random measure with intensity I on E_1 and cluster distribution $\delta_x\times P_x$ on $E_1\times E_2$. Intuitively we can think of the formation of X in two stages, first the choice of a set of points in E_1 and then associating to each of these points a cluster. The following conditioning result allows us to distintegrate the measure in this way.

Let $\mathcal{G}_0 := \sigma\{1_{(0,\infty)}(X(A\times E_2)):A\in\mathcal{E}_1\}$, $\mathcal{G}_1 = \sigma\{I(A):A\in\mathcal{E}_1\}$, and choose a version of the conditional expectation $E(X|\mathcal{G}_0\vee\mathcal{G}_1)$ which is almost surely a random measure on E. Recall that the set $\{I:I \text{ is nonatomic}\} \in \mathcal{G}_1$ (cf. Theorem 3.4.1.1(c)).

<u>Theorem 3.4.1.2.</u> Let X be a Cox cluster random measure on E_2 with random intensity I on E_1 and cluster law $\{P_x:x\in E_1\}$ and let $\{A^n_i,x^n_i,\rho_n:n,i\in\mathbb{N}\}$ be a partition family for E_1 with $\rho_n = 1/n$. Then

(a) P-a.s. on $\{I:I \text{ is nonatomic}\}$

$$E(X|\mathcal{G}_0 \vee \mathcal{G}_1) = \int \delta_x \times \left(\int \mu P_x(d\mu)\right) \tilde{X}(dx), \quad \text{and}$$

$$E\left(e^{-\langle X,\phi\rangle} | \mathcal{G}_0 \vee \mathcal{G}_1\right) = \exp\left(\int \log\left(\int e^{-\int \phi(x,y)\mu(dy)} P_x(d\mu)\right)\tilde{X}(dx)\right), \quad \phi \in pb\mathcal{E}_1 \times \mathcal{E}_2,$$

where $\tilde{X}$ is a random measure on E_1 satisfying

$$\tilde{X}(G) = \lim_{n\to\infty} \sum_{i=1}^{\infty} 1(A_i^n \subset G) 1_{(0,\infty)}(X(A_i^n \times E_2)) \quad \text{for } G \text{ open.}$$

(b) If I is a.s. nonatomic, then conditioned on I, $\tilde{X}$ is distributed as a Poisson random measure on E_1 with intensity I.

Proof. See [DP, Appendix, section 3].

3.4.2 Lebesgue Decomposition and Absolutely Continuous Random Measures.

Let X be a random measure defined on the probability space $(\Omega,\mathcal{F},P)$ and with values in $(M_F(E),\mathcal{B}(M_F(E))$ where E is a Polish space and $\nu \in M_F(E)$. For each $\omega\in\Omega$ denote the Lebesgue decomposition of $X(\omega)$ with respect to ν by $X = \zeta_{ac}^{\nu}(X)(\omega) + \zeta_s^{\nu}(X)(\omega)$ where for each $\omega\in\Omega$, $\zeta_{ac}^{\nu}(X)(\omega) \ll \nu$ (*absolutely continuous part*) and $\zeta_s^{\nu}(X)(\omega) \perp \nu$ (*singular part*).

Lemma 3.4.2.1. Then $\omega \longrightarrow \zeta_{ac}^{\nu}(X)(\omega)$ and $\omega \longrightarrow \zeta_s^{\nu}(X)(\omega)$ are both measurable maps of $(\Omega,\mathcal{F})$ into $(M_F(E),\mathcal{B}(M_F(E)))$, that is, $X_{ac}(\nu)$ and $X_s(\nu)$ are random measures.

Proof. (See Dai Yonglong (1982) or Cutler (1984)). We will simply sketch the proof that ζ_{ac}^{ν} is measurable where $\zeta_{ac}^{\nu}(\mu) + \zeta_s^{\nu}(\mu)$ denotes the Lebesgue decomposition of $\mu\in M_F(E)$. It suffices to show that ζ_{ac}^{ν} is $\mathcal{B}(M_F(E))/\mathcal{B}(M_F(E))$-measurable. The first step is to show (cf. Cutler (1984, Theorem 2.1.3)) that $N := \{(\mu_1,\mu_2):\mu_1\ll\nu,\mu_2\perp\nu\}$ is a measurable subset of $M_F(E)\times M_F(E)$. The mapping $\psi:M_F(E)\times M_F(E) \longrightarrow M_F(E)$ defined by $\psi(\mu,\gamma) := (\mu+\gamma)$ is continuous and the usual Lebesgue decomposition theorem implies that $\psi|N$ is one-to-one. Since $M_F(E)\times M_F(E)$ and $M_F(E)$ are Polish spaces we can conclude by Kuratowski's theorem (cf. Parthasarthy (1967, p. 15) that $\psi(B)$ is measurable for any measurable subset of N. For $B\in \mathcal{E}$, let $(\zeta_{ac}^{\nu})^{-1}(\{\mu:\mu(B)<\alpha\}) = \psi(N\cap\{(\mu_1,\mu_2):\mu_1(B)<\alpha\}) \in \mathcal{B}(M_F(E))$ which yields the result. □

Lemma 3.4.2.2. Let X be a random measure defined on $(\Omega,\mathcal{F},P)$ with values in $(M_F(\mathbb{R}^d),\mathcal{B}(M_F(\mathbb{R}^d)))$, Assume that

(i) there exists a Borel subset $N \subset \mathbb{R}^d$ of Lebesgue measure zero such that for each $z \in \mathbb{R}^d\backslash N$ there is a sequence $\varepsilon_n(z) \longrightarrow 0$ as $n \to \infty$, and as $n \to \infty$ $\dfrac{X(B(z,\varepsilon_n))}{\varepsilon_n^d}$ converges in distribution to a random variable $\eta(z)$ with $E(\eta(z)) < \infty$,

(ii)

(3.4.2.1) $$E(\langle X,\phi\rangle) := \int_{\mathbb{R}^d\setminus N} E(\eta(z))\phi(z)dz \qquad \forall\ \phi \in bC(\mathbb{R}^d).$$

Then X is almost surely an absolutely continuous measure on $\mathbb{R}^d$.

Proof. (cf. Dawson, Fleischmann and Roelly (1990)) From Lemma 3.4.2.1 the absolutely continuous component $X_{ac}(\omega) := \zeta_{ac}(X)(\omega)$ (with respect to Lebesgue measure) is measurable and the same is true for the singular component $X_s(\omega) := \zeta_\nu(X)$. Furthermore, for each ω by the Lebesgue density theorem the limit

(3.4.2.2) $$\lim_{n\to\infty} \frac{X(B(z,\varepsilon_n))}{\varepsilon_n^d} = \lim_{n\to\infty} \frac{X_{ac}(B(z,\varepsilon_n))}{\varepsilon_n^d} = \eta_{ac}(\omega,z)$$

exists for all $z\in \mathbb{R}^d\setminus N(\omega)$ where $N(\omega)$ is a Borel subset of $\mathbb{R}^d$ of Lebesgue measure zero and $\eta_{ac}(\omega,z)$ is a version of the Radon-Nikodym derivative of $X_{ac}(\omega)$ with respect to Lebesgue measure. It is easy to verify that $\eta_{ac}:\Omega\times\mathbb{R}^d \longrightarrow \mathbb{R}_+$ is $(\mathcal{F}\otimes\mathcal{B}(\mathbb{R}^d))$-measurable and (3.4.2.2) holds a.s. with respect to the product measure $P(d\omega)dz$. In particular, for almost all z, the relationships (3.4.2.2) are true with respect to convergence in distribution. Then by assumption (i), we conclude that $\eta_{ac}(.,z)$ coincides in distribution with $\eta(z)$, for almost all $z \in \mathbb{R}^d$. Therefore by (3.4.2.1)

$$\int_{M_F(\mathbb{R}^d)} P(d\omega)\int_{\mathbb{R}^d} \phi(z)X_{ac}(\omega,dz) = \int_{M_F(\mathbb{R}^d)} P(d\omega)\int_{\mathbb{R}^d} \phi(z)\eta_{ac}(\omega,z)dz =$$

$$= \int_{M_F(\mathbb{R}^d)} P(d\omega)\int_{\mathbb{R}^d} \phi(z)X(\omega,dz) \qquad \forall\ \phi\in pbC(\mathbb{R}^d).$$

Therefore since $X_{ac}(\omega) \le X(\omega)\ \forall\ \omega\in\Omega$, this implies that $X = X_{ac}$ P-a.s. and the proof is complete. □

3.5 Markov Transition Kernels and Laplace Functionals

Let E be a compact metric space and let $M_F(E)$ denote the space of finite Borel measures on E furnished with the weak topology. This space is a locally compact separable metric space and can be compactified as in section 3.1. Let ρ be a complete metric on $M_F(E)$, $C_0(M_F(E))$ denote the class of continuous functions that tend to zero at infinity.

We wish to consider possibly time inhomogeneous $M_F(E)$-valued Markov processes with transition probabilities $\mathcal{P}(s,t,\mu;d\nu)$. We can characterize a transition function on $M_F(E)$ in terms of a *Laplace transition functional*

$$L(s,t,\mu;\phi) = \int e^{-\langle\nu,\phi\rangle}\mathcal{P}(s,t,\mu;d\nu), \quad \phi \in bp\mathcal{E},$$

as follows:

(i) for s,t,μ fixed, $\phi \longrightarrow L(s,t,\mu;\phi)$ is the Laplace functional of a random measure on E,

(ii) for each $\phi \in bp\mathcal{E}$, $\mu \longrightarrow L(s,t,\mu;\phi)$ is $\mathcal{M}$-measurable,

(iii) $L(s,t,\mu;\phi) = \int \mathcal{P}(s,u,\mu;d\nu)L(u,t,\nu;\phi)$ for $s < u < t$.

In the time homogeneous case the transition function has the form

$$\mathcal{P}(t,\mu;d\nu) = \mathcal{P}(s,s+t,\mu;d\nu)$$

and we define the corresponding semigroup of contraction operators on $b\mathcal{B}(M_F(E))$ by

$$T_tF(\mu) := \int F(\nu)\mathcal{P}(t,\mu:d\nu).$$

The transition function is said to be *Feller* if

$$T_t:C_0(M_F(E))\longrightarrow C_0(M_F(E))$$

and $\|T_tF-F\| \longrightarrow 0 \ \forall \ F\in C_0(M_F(E))$.

Lemma 3.5.1 In order that a time homogeneous transition function on $M_F(E)$ with Laplace transition functional L correspond to a Feller semigroup on $C_0(M_F(E))$ it is necessary and sufficient that

(i) for fixed $t > 0$, the mapping $\mu \longrightarrow \mathcal{P}(t,\mu,.)$ from $M_F(E)$ to $M_1(M(E_F))$ be continuous, or equivalently,

$$\mu \longrightarrow L(t,\mu,\phi) := \int e^{-\langle\nu,\phi\rangle}\mathcal{P}(t,\mu;d\nu)$$

is continuous for each $\phi \in V$ (where V is as in Theorem 3.2.1),

(ii) $\lim_{\mu\to\Delta} \mathcal{P}(t,\mu;.) = \delta_\Delta$, i.e. $\lim_{\mu\to\Delta} L(t,\mu,1) = 0 \ \ \forall \ t\geq 0$,

(where Δ is as in the Watanabe compactification given in Sect. 3.1),

(iii) the mapping $t \longrightarrow \mathcal{P}(t,\mu;.)$ must be uniformly stochastically continuous at zero, i.e. for each $\varepsilon > 0$,

$$\lim_{t\downarrow 0} \sup_{\mu} [1 - \mathcal{P}(t,\mu;N_\varepsilon(\mu))] = 0,$$

where $N_\varepsilon(\mu) := \{\nu:\rho(\mu,\nu)<\varepsilon\}$.

Proof. See [Dynkin (1965 Vol. I, Chapt. 2, Sect. 5)].

Corollary 3.5.2 Under the other hypotheses condition (iii) follows from the weaker condition that $\mathcal{P}(t,.,.)$ be stochastically continuous, i.e.

$$\lim_{t\downarrow 0} \mathcal{P}(t,\mu;.) = \delta_\mu, \quad \text{or equivalently}$$

$$\lim_{t\downarrow 0} L(t,\mu,\phi) = e^{-\langle\mu,\phi\rangle} \text{for each } \mu \in M_F(E) \text{ and } \phi \in V.$$

Proof. [Dynkin (1965, Vol. I, Lemma 2.10)].

Given a Feller transition function there exists a càdlàg *canonical realization* of the measure-valued process (cf. [EK, Ch. 4, Th. 2.7]) which is denoted by $(\mathbf{D},\mathcal{D},(\mathcal{D}_t)_{t\geq 0},(X_t)_{t\geq 0},(P_\mu)_{\mu\in M_F(E)})$ where $\mathbf{D} = D([0,\infty),M_F(E))$, $X_t:\mathbf{D} \to M_F(E)$, $X_t(\omega) = \omega(t)$ for $\omega\in\mathbf{D}$, $\mathcal{D}_t^o = \sigma\{X_s: 0\leq s\leq t\}$, $\mathcal{D}_t = \mathcal{D}_{t+}^o$, $\mathcal{D} = \vee\mathcal{D}_t$, P_μ is a probability measure on $\mathcal{D}$, the Borel σ-algebra on $\mathbf{D}$, and the $\{P_\mu:\mu\in M_F(E)\}$ are strong Markov with transition function $\mathcal{P}(t,.;.)$.

We will also require a class of time inhomogeneous Markov process with Polish state space given by the following definition.

Definition. Let $(E,\mathcal{E})$ be a Polish space with its Borel σ-field. Let $\hat{E} \in \mathcal{B}([0,\infty))\times\mathcal{E}$ and set $E^t = \{x:(t,x)\in \hat{E}\}$. ($\hat{E}$ is called the *global state space.*) Let $(\Omega,\mathcal{F})$ be a measurable space and for $0\leq s\leq t<\infty$ let $\mathcal{F}^o_{[s,t]}$ be a sub-σ-algebra of $\mathcal{F}$ and if $[s,t] \subset [u,v]$ the $\mathcal{F}^o_{[s,t]} \subset \mathcal{F}^o_{[u,v]}$. Let $\mathcal{F}_{s,t} := \bigcap_{\varepsilon>0} \mathcal{F}^o_{[s,t+\varepsilon]}$ and $\mathcal{F}_{s,\infty} := \bigvee_{t\geq s} \mathcal{F}_{s,t}$. If $\tau\geq s$ is a $(\mathcal{F}_{s,t})_{t\geq s}$ stopping time,

$$\mathcal{F}_{s,\tau} := \{A\in\mathcal{F}_{s,\infty}: A\cap\{\tau\leq t\} \in \mathcal{F}_{s,t}\ \forall\ t\geq s\}.$$

Then $Z := (\Omega,(\mathcal{F}_{s,\infty}),(\mathcal{F}_{s,t})_{t\geq s},\{Z_t\},(P_{s,z})_{z\in E})_{s\geq 0}$ is called an *inhomogeneous Borel strong Markov process* (IBSMP) with càdlàg paths in $E^t \subset E$ iff:

(i) $\forall\ t \geq 0$, $Z_t:(\Omega,\mathcal{F}^o_{[t,t]}) \longrightarrow (E,\mathcal{E})$ is measurable and satisfies $\forall\ (s,z) \in \hat{E}$,

(ii) $\forall\ (s,z) \in \hat{E}$, $P_{s,z}$ is a probability measure on $(\Omega,\mathcal{F}_{s,\infty})$ such that

$$P_{s,z}(Z(s)=z,\ Z_t \in E^t\ \forall\ t\geq s \text{ and } Z_{\bullet} \text{ is càdlàg on } [t,\infty)) = 1,$$

and for all $A\in \mathcal{F}_{u,\infty}$,

$$(s,z) \longrightarrow P_{s,z}(A) \text{ is Borel measurable on } \hat{E}\cap([0,u]\times E),$$

(iii) if $(s,z) \in \hat{E}$, $\psi \in b\mathcal{B}([s,\infty)\times D([s,\infty),E))$ and $\tau \geq s$ is a stopping time with respect to $(\mathcal{F}_{s,t})_{t\geq s}$, then $P_{s,z}$-a.s. on $\{\tau<\infty\}$,

$$P_{s,z}(\psi(\tau,Z(\tau+\bullet))\,|\,\mathcal{F}_{s,\tau})(\omega) = P_{\tau(\omega),Z(\tau)(\omega)}(\psi(\tau(\omega),Z(\tau(\omega)+\bullet)))$$

(strong Markov property).

3.6 Weak Convergence of Processes

We have seen in Chapter 2 that measure-valued processes can arise as limits of particle systems. In order to carry out such limiting procedures it is necessary to develop tools to study the weak convergence of sequences of càdlàg canonical processes. In this section we review some of basic criteria for proving relative compactness for sequences of measure-valued processes. However since these criteria rely on general criteria for processes with values in Polish spaces we first consider the latter case.

Let (E,d) be a Polish space. Let Λ be the set of continuous, strictly increasing Lipschitz continuous functions from $[0,\infty)$ onto $[0,\infty)$ such that

$$\gamma(\lambda) := \sup_{s>t\geq 0} \left| \log \frac{\lambda(s)-\lambda(t)}{s-t} \right| < \infty.$$

For $\omega,\omega' \in D_E = D([0,\infty),E)$,

$$\rho(\omega,\omega') := \inf_{\lambda\in\Lambda} \left(\gamma(\lambda) + \int_0^\infty e^{-u}\left(1 \wedge \sup_{t\geq 0} d(\omega(t\wedge u),\omega'(\lambda(t)\wedge u))\right) du\right)$$

It is easy to verify that ρ defines a metric on D_E. The resulting topology is called the *Skorohod topology* (J_1-topology) on D_E. Let $\mathcal{D} := \mathcal{B}(D([0,\infty);E))$ denote the σ-algebra of Borel subsets with respect to this topology. Let $X_t(\omega) := \omega(t)$ for $\omega\in\mathbf{D}$ and $\mathcal{D}_t^0 := \sigma(X_s : 0\leq s\leq t)$, $\mathcal{D}_t := \bigcap_{\varepsilon>0} \mathcal{D}_{t+\varepsilon}^0 \subset \mathcal{D}$.

Then $(D_E,\mathcal{D},(\mathcal{D}_t)_{t\geq 0},(X_t)_{t\geq 0})$ denotes the *canonical stochastic process* on E. For $f\in C(E)$, let $\tilde{f}:D_E \longrightarrow D([0,\infty),\mathbb{R})$ be defined by $(\tilde{f}x)(t) := f(x(t))$.

<u>Theorem 3.6.1</u> Let (E,d) be a Polish space. Then the sequence $\{P_n\}$ of probability measures on D_E is relatively compact if and only if the following two conditions are satisfied:

(a) For every $\varepsilon>0$ and rational $t\geq 0$, there exists a compact set $K_{\varepsilon,t}$ such that

$$\sup_n P_n\{X(t)\in (K_{\varepsilon,t})^c\} < \varepsilon,$$

(b) For every $\varepsilon>0$ and $T>0$ there exists $\delta>0$ such that

$$\sup_n P_n\{w'(X,\delta,T)\geq\varepsilon\} \leq \varepsilon,$$

where for $x \in D_E$,

$$w'(x,\delta,T) := \inf_{\min|t_i-t_{i-1}|\geq\delta} \ \max_i \ \sup_{t_i\leq s<t\leq t_{i+1}} d(x(s),x(t))$$

where $0\leq t_i\leq T$.

<u>Proof.</u> See e.g. Billingsley (1968, Th. 15.2), [EK p. 128].

<u>Theorem 3.6.2</u> (Kurtz's Tightness Criterion for $D([0,\infty);E)$)

Let (E,d) be a Polish space. Let $\{P_n\}$ be a sequence of probability measures on D_E satisfying condition (a) of Theorem 3.6.1 Assume that for some $\beta>0$, each n and T, and $\delta>0$ $\exists$ a non-negative random variable $\gamma_n^T(\delta)\geq 0$ such that

$$E_n\left([1\wedge d(X(t+u),X(t))]^\beta \,|\, \mathcal{D}_t\right) \leq E_n[\gamma_n^T(\delta)\,|\,\mathcal{D}_t], \quad 0\leq t\leq T,\ 0\leq u\leq\delta,$$

$$\text{and} \quad \lim_{\delta\to 0} \sup_n E_n[\gamma_n^T(\delta)] = 0.$$

Then $\{P_n\}$ is relatively compact.

Proof. [EK, p. 138].

Corollary 3.6.3. Let (E,d) be a Polish space and $\{P_\alpha\}$ be a family of probability laws on $(D_E,\mathcal{D},(\mathcal{D}_t)_{t\geq0},(X_t)_{t\geq0})$. Let D_0 be a dense subset of a subalgebra $C_a(E) \subset bC(E)$.

Assume that for each $f\in D_0$ there exists a càdlàg adapted process Z_f such that

$$f(X(t)) - \int_0^t Z_f(s)ds \text{ is a } (\mathcal{D}_t,P_\alpha)\text{-martingale and that}$$

$$P_\alpha\left(\operatorname*{ess\,sup}_{s\leq t} |Z_f(s)|\right) < \infty, \text{ or}$$

$$P_\alpha\left(\left(\sup_\alpha E(\int_0^t |Z_f(s)|^p ds)\right)^{1/p}\right) < \infty \text{ for each } t < \infty \quad \text{for some } p\in(1,\infty).$$

Then the family $\{P_\alpha\circ(\tilde{f})^{-1}\}$ is tight in $D([0,\infty),\mathbb{R})$ for each $f\in C_a$.

Proof. See [EK, Ch. 3, Th. 9.4].

Theorem 3.6.4. (Jakubowski's criterion for tightness for $D([0,\infty),E)$)

Let (E,d) be a Polish space. Let $\mathbb{F}$ be a family of real continuous functions on E that separates points in E and is closed under addition, i.e. $f,g\in\mathbb{F} \Rightarrow f+g \in \mathbb{F}$. Given $f\in\mathbb{F}$, $\tilde{f}:D_E \longrightarrow D([0,\infty),\mathbb{R})$ is defined by $(\tilde{f}x)(t) := f(x(t))$. A sequence $\{P_n\}$ of probability measures on D_E is tight iff the following two conditions hold:

(i) for each $T>0$ and $\varepsilon>0$ there is a compact $K_{T,\varepsilon}\subset E$ such that

$$P_n(D([0,T],K_{T,\varepsilon})) > 1-\varepsilon,$$

(ii) the family $\{P_n\}$ is $\mathbb{F}$-weakly tight, i.e. for each $f\in\mathbb{F}$ the sequence $\{P_n\circ(\tilde{f})^{-1}\}$ of probability measures in $D([0,\infty),\mathbb{R})$ is tight.

Finally, if the sequence $\{P_n\}$ is tight, then it is relatively compact in the weak topology on $M_1(D_E)$.

Proof. Jakubowski (1986).

The criteria of Theorem 3.6.4 reduce the question of the relative compactness of $\{P_n\}$ on D_E to that for the real-valued case. A very useful criterion for the relative compactness in the latter case is given by the following well-known result of Aldous.

Theorem 3.6.5 (Aldous Conditions for tightness in $D([0,\infty),\mathbb{R})$)

Let $\{P_n\}$ be a sequence of probability measures on $D([0,\infty),\mathbb{R})$ and X be the canonical process. Assume that

(i) for each rational $t\geq0$, $P_n\circ X_t^{-1}$ is tight in $\mathbb{R}$,

(ii) given stopping times τ_n bounded by T and $\delta_n\downarrow0$ as $n\to\infty$, then

$$\lim_{n\to\infty} P_n(|X_{\tau_n+\delta_n} - X_{\tau_n}| > \varepsilon) = 0$$

or

(ii') $\forall\ \eta>0\ \exists\ \delta, n_0$ such that

$$\sup_{n\geq n_0} \sup_{\theta\in[0,\delta]} P_n(|X(\tau_n+\theta)-X(\tau_n)| > \varepsilon) \leq \eta.$$

Then the $\{P_n\}$ are tight.

Proof. Aldous (1978).

Based on this result Joffe and Métivier (1986) derived the following criterion for tightness of locally square integrable processes which we will now describe.

A càdlàg adapted process X, defined on $(\Omega,\mathcal{F},\mathcal{F}_t,P)$ with values in $\mathbb{R}$ is called a *D-semimartingale* if there exists an increasing càdlàg function A(t), a linear subspace $D(L)\subset C(\mathbb{R})$, and a mapping $L:(D(L)\times\mathbb{R}\times[0,\infty)\times\Omega) \longrightarrow \mathbb{R}$ with the following properties:

(ai) For every $(x,t,\omega)\in \mathbb{R}\times[0,\infty)\times\Omega$ the mapping $\phi \longrightarrow L(\phi,x,t,\omega)$ is a linear functional on D(L) and $L(\phi,.,t,\omega)\in D(L)$,

(aii) for every $\phi\in D(L)$, $(x,t,\omega) \longrightarrow L(\phi,x,t,\omega)$ is $\mathcal{B}(\mathbb{R})\times\mathcal{P}$-measurable, where $\mathcal{P}$ is the predictable σ-algebra on $[0,\infty)\times\Omega$, ($\mathcal{P}$ is generated by the sets of the form $(s,t]\times F$ where $F\in \mathcal{F}_s$ and s,t are arbitrary)

(bi) for every $\phi\in D(L)$ the process M^ϕ defined by

$$M^\phi(t,\omega) := \phi(X_t(\omega)) - \phi(X_0(\omega)) - \int_0^t L(\phi,X_{s-}(\omega),s,\omega)dA_s$$

is a locally square integrable martingale on $(\Omega,\mathcal{F},\mathcal{F}_t,P)$,

(bii) The functions $\psi(x) := x$ and ψ^2 belong to D(L).

The functions

$$\beta(x,t,\omega) := L(\psi,x,t,\omega)$$

$$\alpha(x,t,\omega) := L((\psi)^2,x,t,\omega) - 2x\beta(x,t,\omega)$$

are called the *local coefficients of first and second order.*

Theorem 3.6.6 (Joffe-Métivier Criterion for tightness of D-semimartingales)
Let $X^m = (\Omega^m,\mathcal{F}^m,\mathcal{F}^m_t,P^m)$ be a sequence of D-semimartingales with common D(L) and associated operators L^m, functions A^m, β_m, and α_m. Then the sequence $\{X^m:m\in\mathbb{N}\}$ is tight in $D([0,\infty),\mathbb{R})$ provided the following conditions hold:

(i) $\sup_m E|X^m_0|^2 < \infty$

(ii) there is a K>0 and a sequence of positive adapted processes $\{C^m_t:t\geq 0\}$ (on Ω^m for each m) such that for every $m\in \mathbb{N}$, $x\in \mathbb{R}$, $\omega\in \Omega^m$

(a) $|\beta_m(x,t,\omega)|^2 + \alpha_m(x,t,\omega) \leq K(C^m_t(\omega)+x^2)$

(b) for every T>0

$$\sup_m \sup_{t\in[0,T]} E[C_t^m] < \infty \text{ and } \lim_{k\to\infty} \sup_m P^m(\sup_{t\in[0,T]} C_t^m \geq k) = 0$$

(iii) there exists a positive function γ on $[0,\infty)$ and a decreasing sequence of numbers $\{\delta_m\}$ such that $\lim_{t\to 0} \gamma(t) = 0$, $\lim_{m\to\infty} \delta_m = 0$, and for all $0<s<t$ and all m,

$$(A^m(t)-A^m(s)) \leq \gamma(t-s) + \delta_m .$$

Further, if we set $M_t^m := X_t^m - X_0^m - \int_0^t \beta_m(X_{s-}^m,s,.)dA_s^m$, then for each $T>0$ there a constant K_T and m_0 such that for all $m\geq m_0$,

(iv) $E(\sup_{t\in[0,T]} |X_t^m|^2) \leq K_T(1+E|X_0^m|^2)$

and

(v) $E(\sup_{t\in[0,T]} |M_t^m|^2) \leq K_T(1+E|X_0^m|^2)$.

Proof. Joffe and Métivier (1986).

<u>Corollary 3.6.7.</u> Assume that for $T>0$ there is a constant K_T such that

$$\sup_m \sup_{\substack{t\leq T\\ x\in\mathbb{R}}} (|\alpha_m(t,x)|+|\beta_m(t,x)|) \leq K_T \text{ a.s.},$$

$$\sup_m (A^m(t)-A^m(s)) \leq K_T(t-s) \text{ if } 0\leq s\leq t\leq T,$$

$$\sup_m E|X_0^m|^2 < \infty,$$

and M_t^m is a square integrable martingale with $\sup_m E(|M_T^m|^2) \leq K_T$.

Then the $\{X^m : m\in\mathbb{N}\}$ are tight in $D([0,\infty),\mathbb{R})$.

<u>3.7 Weak Convergence and Continuity of Measure-valued Processes.</u>

Let $M_1[D(\mathbb{R}_+,M_F(E))]$, $M_1[D(\mathbb{R}_+,M_p(\mathbb{R}^d))]$ and $M_1[D(\mathbb{R}_+,\mathbb{R})]$ denote the spaces of probability measures on the respective Skorohod spaces with the topology of weak convergence. Note that different versions of the following theorem have been used by several authors (e.g. Roelly-Coppoletta (1986), Vaillancourt (1990)) and Gorostiza and Lopez-Mimbela (1990)).

<u>Theorem 3.7.1</u> (a) Assume that E is compact and $\mathbb{F}$ is a dense subset of C(E) closed under addition. A sequence $\{P_n\} \subset M_1[D([0,\infty),M_F(E)]$ is tight if and only if $\{P_n \circ \tilde{f}_\phi^{-1}\}$ is tight in $M_1[D([0,\infty),\mathbb{R})]$ for each $\phi\in \mathbb{F}$ where $f_\phi(\mu) = \langle\mu,\phi\rangle$, and $\tilde{f}_\phi : D([0,\infty),M_F(E)) \longrightarrow D([0,\infty),\mathbb{R})$ is defined by $(\tilde{f}_\phi x)(t) := f_\phi(x(t))$.

(b) A sequence $\{P_n\}$ in $M_1[D([0,\infty),M_p(\dot{\mathbb{R}}^d))]$ is tight if and only if $\{P_n \circ \tilde{f}_\phi^{-1}\}$ is tight in $M_1[D([0,\infty),\mathbb{R})]$ for each $\phi \in K_p(\dot{\mathbb{R}}^d)$ and $f_\phi(\mu) := \langle\mu,\phi\rangle$ (where $K_p(\dot{\mathbb{R}}^d)$ is defined as in Sect. 3.1).

Proof. The proof of (a) is similar but easier than that of (b) so that we will only prove the latter. The necessity follows from the continuity of the mapping $\tilde{f}_\phi$. By hypothesis $\{P_n \circ \tilde{f}_{\dot{\phi}_p}^{-1}\}$ is tight in $D([0,T],\mathbb{R})$ for each $T > 0$. Hence for each $\varepsilon>0$ there exists a compact $K_\varepsilon \subset D([0,T],\mathbb{R})$ such that $P_n \circ \tilde{f}_{\dot{\phi}_p}^{-1}(K_\varepsilon) \geq 1-\varepsilon$. But then by the characterization of compact sets in $D([0,T],\mathbb{R})$ (cf. Theorem 3.6.4(i)) there exists $k_\varepsilon > 0$ such that $K_\varepsilon \subset D([0,T],[-k_\varepsilon,k_\varepsilon])$. Let $\Gamma_{T,\varepsilon} := \{\mu : \mu \in M_p(\dot{\mathbb{R}}^d) : |\langle\mu,\dot{\phi}_p\rangle| \leq k_\varepsilon\}$ which is a compact subset of $M_p(\dot{\mathbb{R}}^d)$. But then $P_n(D([0,T];\Gamma_{T,\varepsilon})) = P_n \circ \tilde{f}_{\phi_p}^{-1}(D([0,T];[-k_\varepsilon,k_\varepsilon]) \geq 1-\varepsilon$ and hence condition (i) of Theorem 3.6.4 is satisfied. Now let $\mathbb{F}$ denote the class of functions on $M_p(\dot{\mathbb{R}}^d)$ of the form $f_\phi(\mu) := \langle\mu,\phi\rangle$ with $\phi \in K_p(\dot{\mathbb{R}}^d)$ and note that $\mathbb{F}$ separates points and is closed under addition. The result then follows by applying Theorem 3.6.4. □

It will also be useful to have a criterion which guarantees that a measure-valued process has continuous sample paths. We first recall the basic criterion for the path continuity which is due to Kolmogorov and Ibragimov.

Lemma 3.7.2. (a) Assume that a real-valued stochastic process $\{Y(t):t\geq 0\}$ satisfies

$$E|Y(t)-Y(s)|^r \leq C|t-s|^{1+\alpha}, \quad 0\leq s,t\leq T \tag{3.7.1}$$

for $r>0$, $\alpha>0$. Then (for fixed T) for any $\gamma \in (2,2+\alpha)$ and $\lambda>0$

$$P\left(\sup_{0\leq s<t\leq T} \frac{|Y(t)-Y(s)|}{|t-s|^\beta} \geq (8\gamma/(\gamma-2))(4\lambda)^{1/r}\right) \leq CA/\lambda \tag{3.7.2}$$

where $\beta = (\gamma-2)/r$ and A is a constant.

In particular Y has a β-Hölder continuous version for any $0 < \beta < \alpha/r$.

(b) Let $\{Z(t):t\in\mathbb{R}^d\}$ be a random field satisfying

$$E|Z(x)-Z(y)|^r \leq C|x-y|^{d+\alpha}$$

for $r>0$, $\alpha>0$. Then Z has a β-Hölder continuous version for any $0 < \beta < \alpha/r$.

Proof. See for example Stroock and Varadhan (1979, p. 49) or Walsh (1986, Chapt. 1, Cor. 1.2). □

Corollary 3.7.3. Let $\{X(t):t\geq 0\}$ be a process with values in $M_F(E)$, E compact, (resp. $M_p(\dot{\mathbb{R}}^d)$). Let $\{f_n\}$ be as in Lemma 3.2.1 (resp. 3.1.5) and assume that

(3.7.3) $\quad P\left(|\langle X(t),f_n\rangle-\langle X(s),f_n\rangle|^r\right) \le C|t-s|^{1+\alpha}$ for $r>1$, $\alpha>0$. Then $X \in C([0,\infty),M_F(E))$ (or $C([0,\infty),M_p(\dot{\mathbb{R}}^d))$, resp.), P-almost surely. In addition, for $\beta < \alpha/r$, there is a version of $\{X(t)t\ge 0\}$ that is β-Hölder continuous in the metric d defined by (3.2.1) (resp. (3.1.1)).

Proof. Let $Z_n := \sup_{0\le s<t\le T} \dfrac{|\langle X(t),f_n\rangle-\langle X(s),f_n\rangle|}{|t-s|^\beta}$ and $2 > \lambda > 1$. Using (3.7.2) we obtain

$$\sum_{n=0}^{\infty} P\left(Z_n \ge (8\gamma/(\gamma-2))(4)^{1/r}\lambda^{n/r}\right) \le \sum_{n=0}^{\infty} CA/\lambda^n < \infty.$$

Then by the Borel-Cantelli lemma,

$$\sup_{0\le s<t\le T} \frac{d(X(t),X(s))}{|t-s|^\beta} \le \sum Z_n/2^n < \infty, \quad \text{P-a.s.} \ \square$$

4. MEASURE-VALUED BRANCHING AND LOG-LAPLACE FUNCTIONALS

4.1 Some Introductory Remarks

Branching processes form one of the basic classes of stochastic processes and arise in a wide variety of applications. In the context of space-time stochastic models branching particle systems have served an important role in providing a rich class of mathematically tractable models. In this chapter we will investigate their measure-valued versions which first appeared in the work of Jirina (1964) and Watanabe (1968). The theory of measure-valued branching processes has developed into a mature subject. The reason for this is two-fold. In the first place there are three complementary mathematical structures which can be exploited - these are the log-Laplace nonlinear evolution equation to which one can apply analytical results, the family structure to which one can apply probabilistic tools and the martingale problem characterization to which one can apply the tools of stochastic calculus. In the second place measure-valued branching processes exhibit interesting dimension dependent local spatial structure and in addition when provided with spatially homogeneous initial conditions interesting long-time scale behavior.

In section 2.10 we introduced a class of branching particle systems and indicated how the corresponding measure-valued limit could be obtained using the methods described in Chapter 2. In this chapter we return to the study of branching systems but for a variety of reasons the setting of section 2.10 is too restrictive for our needs. We will generalize this in a number of directions as follows (together with our reasons for doing so):

(i) we will replace the space and time homogeneous motion and branching mechanism by space and time dependent mechanisms - this is needed for example for the study of branching when the offspring distribution is modulated by a random medium;

(ii) we will replace the "deterministic" clock which determines the branching times by allowing each particle to have its own clock which is given by an additive functional of its trajectory - this is used in the formulation of branching in fractal media,

(iii) we will replace the branching mechanism with finite third moment assumption by a general one in which the offspring distribution need not have even second moments;

(iv) we will replace the assumption that E is a compact metric space with the assumption that E is a Polish space - this is essential to our formulation of the historical process in Chapter 12;

(v) we will replace the space of finite measures by a space of σ-finite measures on E - this is essential to the study of general entrance laws and stationary distributions.

Having made all these extensions it is not surprising that the setting is now somewhat more complex and our methods are somewhat different. However even though the technical details are more complex, the simple heuristic ideas of Chapter 2 remain as our guiding principle.

4.2 Markov Transition Kernels and Log-Laplace Semigroups

Let E be a compact metric space and let $M_F(E)$ denote the space of finite Borel measures on E furnished with the weak topology.

An $M_F(E)$-valued Markov process $\{X_t\}$ with Laplace transition functional $L(s,t,\mu;\phi)$ is said to be a *measure-valued branching process* if it can be represented in the form

$$L(s,t,\mu;\phi) = E\left(e^{-\langle X_t,\phi\rangle} \,|X_s=\mu\right) = e^{-\int V_{s,t}\phi(x)\mu(dx)}, \quad \phi\in pb\mathcal{E},\ \mu\in M_F(E)$$

where the family of nonlinear operators $\{V_{s,t}: s\le t\}$ on $pb\mathcal{E}$ satisfies two basic structural properties:

(i) it forms a time inhomogeneous semigroup, that is, for $r\le s\le t$,

(4.2.1) $\qquad V_{r,s}(V_{s,t}) = V_{r,t},\quad V_{t,t}= I,$ and

(ii) for each s,t and $\mu\in M_F(E)$, the mapping $\phi \longrightarrow \int V_{s,t}\phi(x)\mu(dx)$ is the log-Laplace functional of a random measure on E.

A family of operators $\{V_{s,t}\}$ satisfying (i) and (ii) is called a *log-Laplace semigroup* (or *cumulant, Skorohod,* or *Ψ-semigroup*) and determines the finite dimensional distributions of a $M_F(E)$-valued Markov process (whose transition functional is given as above). The theory of Ψ-semigroups was developed in Watanabe (1968) and Silverstein (1969). In this section we establish the existence of measure-valued branching processes that includes most of the currently studied classes. For recent developments on the characterization a general class of measure-valued branching processes see Dynkin, Kuznetsov and Skorohod (1992).

4.3. A General Class of Branching Systems

The main result of this section is to establish the existence of a wide class of log-Laplace semigroups as solutions of a class of nonlinear evolution equations, which we will call "log-Laplace equations". For the moment we consider the homogeneous case and then this is a *mild* (or evolution) equation

(4.3.1) $$v(t,x) = S_t\phi(x) + \int_0^t S_{t-s}\Phi(v(s,.))(x)ds, \quad \text{for} \quad \phi\in pb\mathcal{E}$$

where A is the generator of a semigroup $\{S_t\}$ on $pb\mathcal{E}$, and

(4.3.2) $\Phi(\lambda) = -\frac{1}{2}c\lambda^2 + b\lambda + \int_0^\infty \left(1 - e^{-\lambda u} - \frac{\lambda u}{1+u} \right) n(du)$

and n(du) is a measure on $(0,\infty)$ satisfying $\int_0^1 u^2 n(du) < \infty$.

Although we do not require the solutions to be differentiable, it is convenient to carry out formal calculations using the following (formally equivalent) differential equation

(4.3.3) $\frac{\partial v(t,x)}{\partial t} = Av(t,x) + \Phi(v(t,x))$ *(log-Laplace Equation)*

$v(0,x) = \phi(x)$.

(Such formal calculations can then be justified using the mild equation.)

In this chapter we restrict our attention to the *critical case*:

$S_t 1 = 1,\ b(.) \equiv 0,\ 0 \le c < \infty$,

(4.3.4) $\left(\int_1^\infty u\ n(du) + \int_0^1 u^2 n(du) \right) < \infty$,

and thus

(4.3.5) $\Phi(\lambda) = -\frac{1}{2}c\lambda^2 + \int_0^\infty (1 - e^{-\lambda u} - \lambda u) n(du),\ \lambda \ge 0$.

The restriction $b \equiv 0$ is made for notational simplicity at this point. In Example 10.1.2.2 we will show that non-critical branching can be obtained from critical branching using a Cameron-Martin-Girsanov formula. However if the condition on n is removed new phenomena such as explosions can occur which we do not consider.

Note that for each x, $\Phi(x,.)$ is the log-Laplace function of an infinitely divisible random variable. If in addition we assume that $\int_0^1 u n(du) < \infty$, then $e^{-\Phi(\lambda)} = E\left(e^{-\lambda(Z_1 + Z_2 - E(Z_2))}\right)$ where Z_1 is a normal random variable and Z_2 is a non-negative infinitely divisible random variable. Heuristically, in this case the continuous state branching process which will be constructed below can be interpreted as the corresponding infinitely divisible process run with a clock speed proportional to the current mass.

We next describe the formulation of the general class of critical inhomogeneous superprocesses which we will consider. The basic ingredients are a motion process, branching rate, and branching mechanism.

Motion Process:

Let E be a Polish space, and $D_E = D([0,\infty),E)$ the space of càdlàg functions with the Skorohod topology as defined in Section 3.6. For $\omega \in D_E$ and $t \ge 0$, $W_t(\omega) := \omega(t)$.

For each $t\geq 0$, let there be given $E^t \subset E$ and define $\hat{E} := \{(s,x): x\in E^s\}$, and $\hat{D}_E := \{\omega\in D_E, E), \omega_t\in E^t \ \forall t\}$, $\mathcal{E}^t := \mathcal{E}\cap E^t$. We assume that $\hat{E}\in\mathcal{B}([0,\infty))\otimes\mathcal{E}$ and $\hat{\mathbf{D}}$ is a Borel subset of D_E. For $0\leq s\leq t$, $\mathcal{D}_{s,t} := \bigcap_{\varepsilon>0}\sigma\{W_u: s\leq u\leq t+\varepsilon\}$, $\mathcal{D} := \mathcal{D}_{0,\infty}$. For $s\geq 0$, let $D_{E,s} := D([s,\infty),E)$.

The process $(\hat{D}_E, \mathcal{D}, \mathcal{D}_{s,t}, \{W_t\}_{t\geq 0}, \{P_{s,x}\}_{s\geq 0, x\in E^s})$ is assumed to be a canonical *time-inhomogeneous Borel strong Markov process*, that is,

(i) $\forall (s,x)\in\hat{E}$, $P_{s,x}$ is a probability measure on $(\hat{D}_E, \mathcal{D}_{s,\infty})$ such that for all $A\in\mathcal{D}_{u,\infty}$, $(s,x)\longrightarrow P_{s,x}(A)$ is Borel measurable on $([0,u]\times E)\cap\hat{E}$ and $P_{s,x}(W_s=x) = 1$,

(ii) *(strong Markov property)* if $(s,x)\in\hat{E}$, $\psi\in b\mathcal{B}([s,\infty)\times D_{E,s})$ and $\tau\geq s$ is a $(\mathcal{D}_{s,t}: t\geq s)$-stopping time, then

$$P_{s,x}(\psi(\tau, W_{\tau+\cdot})\,|\,\mathcal{D}_{s,\tau})(\omega) = P_{\tau(\omega), W_{\tau(\omega)}}(\psi(\tau(\omega), W_{\tau(\omega)+\cdot}),$$

for $P_{s,x}$-a.e. ω on $\{\tau<\infty\}$.

The associated *inhomogeneous semigroup* $\{S_{s,t}\}$ is defined by:

$$S_{s,t}\phi(x) = P_{s,x}f(W_t),\quad 0\leq s\leq t,\ x\in E^s,\ \phi\in b\mathcal{E}.$$

Branching Rate:

Let $\kappa(dt)$ be a *positive continuous additive functional* of W, that is, $\kappa:\hat{D}_E\longrightarrow M_{LF}([0,\infty))$, such that $\kappa(r,t)$ is $\mathcal{D}_{r,t}$-measurable. In addition we assume that κ is a.s. absolutely continuous (with respect to Lebesgue measure) and that for $0\leq r\leq t$, $\sup_{x\in E^r} P_{r,x}(e^{\theta\kappa(r,t)}) < \infty \quad \forall\ \theta > 0.$

Branching Mechanism: For each (t,x), $x\in E^t$, $\Phi(t,x,.)$ is assumed to be of the form

$$\Phi(t,x,\lambda) = -\tfrac{1}{2}c(t,x)\lambda^2 + \int_0^\infty (1-e^{-\lambda u}-\lambda u)n(t,x,du),\quad \lambda\geq 0, \tag{4.3.6}$$

where $c\in pb(\mathcal{B}(\mathbb{R}_+)\otimes\mathcal{E})$, and n is a measurable mapping from $(\mathbb{R}_+, \mathcal{B}(\mathbb{R}_+))\times(E,\mathcal{E})$ to $M((0,\infty))$ satisfying

$$\sup_{t,x}\left(\int_1^\infty u\ n(t,x,du) + \int_0^1 u^2 n(t,x,du)\right) < \infty.$$

Note that

$$\Phi(t,x,0) = 0,\ \Phi(t,x,\lambda)\leq 0, \tag{4.3.7}$$

$$\Phi'(t,x,\lambda) = -c(t,x)\lambda - \int_0^\infty (1-e^{-\lambda u})u.n(t,x,du) \leq 0,$$

$$\Phi''(t,x,\lambda) = - c(t,x) - \int_0^\infty e^{-\lambda u} u^2 n(t,x,du) \le 0,$$

$$(-1)^{k+1}\Phi^{(k)}(t,x,0) \ge 0, \; k\ge 2.$$

(Here ', " denote the first and second derivatives with respect to λ and $\Phi^{(k)}$ denotes the kth derivative of Φ with respect to λ.)

The Log-Laplace Equation.

The basic *log-Laplace equation* is given by

$$(4.3.8) \qquad V_{s,t}\phi(x) = P_{s,x}\left[\phi(W_t) + \int_s^t \kappa(dr)\Phi(r,W_r,V_{r,t}\phi(W_r))\right].$$

where $\phi \in bp\mathcal{E}^t$. We will prove that this equation has a unique solution and yields a log-Laplace semigroup.

Strategy of the Proof: We will prove

(i) uniqueness of the solution to (4.3.8) by an analytical argument,

(ii) the existence of a solution to the log-Laplace equation is obtained by a probabilistic method based on a branching particle approximation which yields a family of operators $V^{\varepsilon}_{s,t}$ which converge to a solution of (4.3.8), that is,

$$V_{s,t}\phi = \lim_{\varepsilon\downarrow 0} V^{\varepsilon}_{s,t}\phi,$$

(iii) the probabilisitic argument simultaneously shows that $\exp\left(-\int V_{s,t}\phi(x)\mu(dx)\right)$ is the log-Laplace functional of a random measure,

(iv) the semigroup property of $\{V_{s,t}\}$ is proved using the uniqueness of solutions to (4.3.8).

We begin with four technical lemmas.

Lemma 4.3.1 (Dynkin's generalized Gronwall inequality)

Assume that $h:[0,\infty)\times D_E \longrightarrow \mathbb{R}$ is progressively measurable such that $\sup_{t,x} |h(t,x)| \le M$, and

$$(4.3.9) \qquad h(r,x) \le c_1 + c_2 P_{r,x}\int_r^t h(s,W)\kappa(ds) \quad \text{for} \quad r\in [t_0,t).$$

Then

$$(4.3.10) \quad h(r,x) \le c_1 P_{r,x}\, e^{c_2\kappa(r,t)} \qquad \text{for all} \quad r\in [t_0,t).$$

Proof. Using (4.3.9) and induction it is easy to verify that for each $n \ge 1$

$$h(r,x) \le c_1 \sum_{k=0}^{n} c_2^k\, P_{r,x}\int\!\!..\!\!\int 1(r<s_1<\ldots<s_k<t)\kappa(ds_1)\ldots\kappa(ds_k)$$

$$+ c_2^{n+1} P_{r,x} \int \cdots \int 1(r<s_1<\ldots<s_{n+1}<t)\kappa(ds_1)\ldots\kappa(ds_{n+1})h(s_{n+1},W_{s_{n+1}})$$

$$= c_1 \sum_{k=0}^{n} c_2^k P_{r,x}\kappa(r,t)^k/k! + R_{n+1},$$

where $|R_{n+1}| \le M\cdot c_2^{n+1} P_{r,x}\kappa(r,t)^{n+1}/(n+1)! \longrightarrow 0$ and the proof is complete.□

Lemma 4.3.2 (Uniform Lipschitz property)

Under the above assumptions on c and n, given $\lambda_0>0 \ \exists\ K(\lambda_0)>0$ such that

$|\Phi(t,x,\lambda_1)-\Phi(t,x,\lambda_2)| \le K(\lambda_0)|\lambda_1-\lambda_2| \quad \forall\ \lambda_1,\lambda_2 \in (0,\lambda_0)$.

Proof. If $\lambda_1 < \lambda_2$, then

$$|\Phi(t,x,\lambda_1)-\Phi(t,x,\lambda_2)| \le 2\lambda_0 \sup_{t,x} |\tfrac{1}{2}\, c(t,x)|\, |\lambda_1-\lambda_2|$$

$$+ \int_0^\infty |(e^{-\lambda_1 u}+\lambda_1 u)-(e^{-\lambda_2 u}+\lambda_2 u)|\, n(t,x,du).$$

Writing $|(e^{-\lambda_1 u}+\lambda_1 u)-(e^{-\lambda_2 u}+\lambda_2 u)| = \int_{\lambda_1}^{\lambda_2} u(1-e^{-\lambda u})d\lambda$, we verify that the second term

$$\le \int_{\lambda_1}^{\lambda_2} \left\{ \int_0^1 \lambda u^2 n(t,x,du) + \int_1^\infty u\ n(t,x,du) \right\} d\lambda$$

$$\le \sup_{t,x} \left(\int_0^\infty (u\wedge u^2) n(t,x,du) \right) |\lambda_1-\lambda_2|. \ \square$$

Lemma 4.3.3 Under the assumptions (4.3.6) the log-Laplace equation (4.3.8) has at most one solution.

Proof. Consider two solutions $V^1_{s,t}\phi(x)$ and $V^2_{s,t}\phi(x)$. Note that $\sup_x |V^i_{s,t}\phi(x)| \le \sup_x |\phi(x)| := \lambda_0$. Using Lemma 4.3.2 we obtain that

$$|V^2_{s,t}\phi(x) - V^1_{s,t}\phi(x)|$$

$$\le P_{s,x}\left[\int_s^t \kappa(dr) \left[|\Phi(r,W_r,V^2_{r,t}(W_r)) - \Phi(r,W_r,V^1_{r,t}(W_r))| \right] \right]$$

$$\le K(\lambda_0) P_{s,x}\left\{ \int_s^t \left[|V^2_{r,t}(W_r) - V^1_{r,t}(W_r)| \right] \kappa(dr) \right\}.$$

Then Lemma 4.3.1 implies that

$$|V^2_{s,t}\phi(x) - V^1_{s,t}\phi(x)| = 0 \quad \text{for all} \quad s < t.\ \square$$

Lemma 4.3.4 Let $\mathcal{G}$ be a measurable function, $\mathcal{G}:[0,\infty)\times E\times[0,1]:\longrightarrow[0,1]$.

Then the equations

$$(4.3.11)\quad w_{r,t}(x) = P_{r,x}\left[e^{-\phi(W_t)}e^{-\kappa(r,t)} + \int_r^t e^{-\kappa(r,s)}\kappa(ds)\mathcal{G}(s,W_s,w_{s,t}(W_s))\right]$$

and

$$(4.3.12)\quad w_{r,t}(x) = P_{r,x}\left[e^{-\phi(W_t)} + \int_r^t \kappa(ds)\ \mathcal{G}(s,W_s,w_{s,t}(W_s)) - \int_r^t \kappa(ds)w_{s,t}(W_s)\right]$$

are equivalent.

Proof. Assume (4.3.12) and now calculate

$$P_{r,x}\left\{\int_r^t e^{-\kappa(r,s)}\kappa(ds)\mathcal{G}(s,W_s,w_{s,t}(W_s))\right\}$$

$$= P_{r,x}\left\{-\int_r^t e^{-\kappa(r,s)}\kappa(ds)\int_s^t \mathcal{G}(u,W_u,w_{u,t}(W_u))\kappa(du)\right.$$

$$\left. + \int_r^t \kappa(ds)\mathcal{G}(s,W_s,w_{s,t}(W_s))\right\} \quad \textit{(integration by parts)}$$

$$= P_{r,x}\left\{-\int_r^t e^{-\kappa(r,s)}\kappa(ds)\left\{w_{s,t}(W_s) - e^{-\phi(W_t)} + \int_s^t \kappa(du)w_{u,t}(W_u)\right\}\right.$$

$$\left. + w_{r,t}(x) - e^{-\phi(W_t)} + \int_r^t \kappa(du)w_{u,t}(W_u)\right\}$$

((4.3.12) and Markov property)

$$= w_{r,t}(x) - P_{r,x}\left\{e^{-\kappa(r,t)}e^{-\phi(W_t)}\right\}$$

by simplification and another integration by parts. □

4.4 Branching Particle Approximations

In this section we will prove the existence of a solution to equation (4.3.8) and establish that it yields a log-Laplace semigroup.

Theorem 4.4.1 Given (W,κ,Φ) there exists a transition function on $M_F(E)$ with transition Laplace functional

$$(4.4.1)\qquad P_{s,\mu}\left(e^{-\langle X(t),\phi\rangle}\right) = \exp\left(-\int (V_{s,t}\phi)(x)\mu(dx)\right),\quad \phi\in bp\mathcal{E}^t,\ \mu\in M_F(E),$$

where $V_{s,t}\phi$ is the unique solution of the equation (4.3.8).
The associated $M_F(E)$-valued time inhomogeneous Markov process is called the (W,κ,Φ)-*superprocess*.

Proof. We first obtain the existence of a solution to a closely related equation via a probabilistic construction.

Lemma 4.4.2 Consider a branching particle system $\{Z_t\}$ with motion process W, branching rate $\kappa(dr)$ and offspring generating function $\mathcal{G}(t,x,.)$, i.e.

$$\mathcal{G}(t,x,z) = \sum_{n=0}^{\infty} p(t,x,n)z^n,$$

$$\sup_{t,x} \sum_{n=0}^{\infty} n.p(t,x,n) \le K,$$

where $p(t,x,n)\kappa(dt)$ is the probability that n offspring are produced by a branch at time t at position x.

(a) Then the branching particle system exists (i.e. no explosion) and the distribution of Z_t starting at time r with one particle at x has Laplace functional

$$w_{r,t}(x) := P^Z_{r,\delta_x} e^{-\langle Z_t,\phi\rangle}$$

which satisfies

$$(4.4.2) \qquad w_{r,t}(x) = P_{r,x}\left[e^{-\phi(W_t)} e^{-\kappa(r,t)} + \int_r^t e^{-\kappa(r,s)} \kappa(ds)\mathcal{G}(s,W_s,w_{s,t}(W_s))\right].$$

(b) $P^Z_{r,x}(\langle Z_t,1\rangle) \le P_{r,x}(e^{(K-1)\kappa(r,t)})$.

Proof. (a) The non-explosion property follows from Harris (1963, Ch. 5, Th. 9.1). To obtain equation (4.4.2) we condition on the existence and the time of the first branch in the interval [s,t] (cf. Dawson and Ivanoff (1978)).□

(b) This follows by taking $\phi \equiv \theta$, in (4.3.12) differentiating with respect to θ, evaluating at θ=0 and then using Lemma 4.3.1.

Proof of Theorem 4.4.1. The proof will be dividied into four steps.

Step 1: The branching particle approximation. Given Φ satisfying (4.3.6) it is easy to check that for any $0 < \varepsilon < 1$, $0 \le v \le 1$,

$$(4.4.3) \qquad \mathcal{G}^{\varepsilon}(t,x,v) := a(t,\varepsilon,x)^{-1}[a(t,\varepsilon,x)v - \varepsilon\Phi(t,x,(1-v)/\varepsilon)]$$

with $0 < a(t,\varepsilon,x) := -\Phi'(t,x,\varepsilon^{-1}) \le c_1/\varepsilon + c_2$

is the probability generating function of a non-negative integer-valued random variable with mean one. In fact using (4.3.6) and (4.3.7) it is easy to verify that $\mathcal{G}^{\varepsilon}(t,x,0) = -\varepsilon\Phi(t,x,1/\varepsilon) \ge 0$, $\mathcal{G}^{\varepsilon}(t,x,1) = 1$, $D\mathcal{G}^{\varepsilon}(t,x,v)|_{v=0} = 0$, $D\mathcal{G}^{\varepsilon}(t,x,v)|_{v=1} = 1$, and $D^k\mathcal{G}^{\varepsilon}(t,x,v)|_{v=0} \ge 0$ for all $k \ge 2$ (where D denotes differentiation with respect to v).

We now construct for each $\varepsilon > 0$ an approximating branching particle system $\{Z^{\varepsilon}_t\}$ to the (W,κ,Φ)-superprocess with initial measure μ at time r. We begin with an initial

Poisson random measure Z_r^ε with intensity μ/ε. Consider the resulting branching particle field, Z_t^ε with offspring generating function $\mathcal{G}^\varepsilon$, branching rate $\kappa^\varepsilon(ds) = a(s,\varepsilon,W_s)\kappa(ds)$, initial random measure Z_r^ε. and particle mass ε. Combining Lemma 4.4.2 with the Poisson cluster formula (3.3.2) we obtain the Laplace functional of the random measure $\varepsilon Z_t^\varepsilon$ as follows:

$$P_{r,\mathcal{P}ois(\mu/\varepsilon)}\, e^{-\langle Z_t^\varepsilon,\varepsilon\phi\rangle} = \exp\left(- \langle \mu, \frac{1-w_{r,t}^\varepsilon}{\varepsilon} \rangle \right)$$

where $w_{r,t}^\varepsilon$ is as in Lemma 4.4.2 with $\mathcal{G}$ replaced by $\mathcal{G}^\varepsilon$, and $w_{r,t}^\varepsilon$

$$w_{r,t}^\varepsilon(x) = P_{r,x}\Big[e^{-\varepsilon\phi(W_t)}\, e^{-\kappa^\varepsilon(r,t)} + \int_r^t e^{-\kappa^\varepsilon(r,s)}\kappa^\varepsilon(ds)\mathcal{G}^\varepsilon(s,W_s,w_{s,t}^\varepsilon(W_s))\Big].$$

Then by Lemma 4.3.4

$$w_{r,t}^\varepsilon(x) = P_{r,x}\Big\{e^{-\varepsilon\phi(W_t)} + \int_r^t \kappa^\varepsilon(ds)\mathcal{G}^\varepsilon(s,W_s,w_{s,t}^\varepsilon(W_s)) - \int_r^t \kappa^\varepsilon(ds)w_{s,t}^\varepsilon(W_s)\Big\}.$$

Putting $v_{r,t}^\varepsilon(x) = \dfrac{1-w_{r,t}^\varepsilon(x)}{\varepsilon}$ and substituting the expression for $\mathcal{G}^\varepsilon$ we get

$$(4.4.4)\quad v_{r,t}^\varepsilon(x) = P_{r,x}\Big[\Big(\frac{1-e^{-\varepsilon\phi(W_t)}}{\varepsilon}\Big) + \int_r^t \kappa(ds)\Phi(s,W_s,v_{s,t}^\varepsilon(W_s))\Big].$$

Then

$$(4.4.5)\quad 0 \le v_{r,t}^\varepsilon(x) \le P_{r,x}\Big(\frac{1-e^{-\varepsilon\phi(W_t)}}{\varepsilon}\Big) \le P_{r,x}\phi(W_t)$$

since $\Phi \le 0$.

<u>Step 2. The proof of convergence of log-Laplace functionals.</u>

<u>Lemma 4.4.3</u> $v_{r,t}^\varepsilon(x) \xrightarrow{\varepsilon\downarrow 0} v_{r,t}(x)$ exists (uniformly in x) and $v_{r,t}(x)$ is the unique solution to Equation (4.3.8).

<u>Proof.</u> The uniqueness follows from Lemma 4.3.3. It follows from (4.4.5) that $0 \le v_{r,t}^\varepsilon(.) \le \lambda_0$, provided that $0\le\phi\le\lambda_0$. Then using the Lipschitz property of Φ we obtain

$$|v^{\varepsilon_1}_{r,t} - v^{\varepsilon_2}_{r,t}|(x) \le P_{r,x} \left\| \frac{1-e^{-\varepsilon_1\phi(W_t)}}{\varepsilon_1} - \frac{1-e^{-\varepsilon_2\phi(W_t)}}{\varepsilon_2} \right\|_\infty$$

$$+ M(\lambda_0)P_{r,x}\int_r^t \kappa(ds)\|v^{\varepsilon_1}_{s,t}(W_s) - v^{\varepsilon_2}_{s,t}(W_s)\|_\infty.$$

Then by Lemma 4.3.1,

$$\|v^{\varepsilon_1}_{r,t} - v^{\varepsilon_2}_{r,t}\|_\infty \le \left\| \frac{1-e^{-\varepsilon_1\phi(.)}}{\varepsilon_1} - \frac{1-e^{-\varepsilon_2\phi(.)}}{\varepsilon_2} \right\|_\infty \sup_x P_{r,x} e^{M(\lambda_0)\kappa(r,t)}.$$

This yields the existence of the limit $v_{r,t}$. The fact that $v_{r,t}$ satisfies the equation (4.3.8) then follows from (4.4.3) by a bounded convergence argument. □

Step 3. Proof of the Semigroup Property of V.

We have $V_{r,t}\phi(y) = P_{r,y}[\phi(W_t) + \int_r^t \kappa(du)\Phi(u,W_u,V_{u,t}\phi(W_u))]$, $y \in E^r$,

$$V_{s,r}[V_{r,t}\phi](x)$$

$$= P_{s,x}[(V_{r,t}\phi)(W_r) + \int_s^r \kappa(du)\Phi(u,W_u,(V_{u,r}V_{r,t}\phi)(W_u))]$$

$$= P_{s,x}[P_{r,W_r}\phi(W_t) + P_{r,W_r}\int_r^t \kappa(du)\Phi(u,W_u,V_{u,t}\phi(W_u))]$$

$$+ P_{s,x}[\int_s^r \kappa(du)\Phi(u,W_u,(V_{u,r}V_{r,t})\phi(W_u))]$$

$$= P_{s,x}[\phi(W_t) + \int_s^r \kappa(du)\Phi(u,W_u,(V_{u,r}V_{r,t})\phi(W_u))$$

$$+ \int_r^t \kappa(du)\Phi(u,W_u,V_{u,t}\phi(W_u))] \quad \textit{(by Markov property of W).}$$

Hence $U_{u,t}\phi := V_{u,t}\phi$ for $u \ge r$

$$:= V_{u,r}V_{r,t}\phi \quad \text{for } u<r$$

satisfies the same log-Laplace equation as does $V_{u,t}\phi$, $\forall u$. Therefore applying uniqueness yields $V_{s,t} = V_{s,r}V_{r,t}$ for $s<r<t$.□

Step 4. Completion of the proof. We now complete the proof of Theorem 4.4.1. First note that the first moment measures, $P_{r,\mathcal{P}ois(\mu/\varepsilon)}(\varepsilon Z^\varepsilon_t(dx)) = S_{r,t}\mu \in M_F(E)$ due to the criticality of the branching and hence are tight by Lemma 3.2.8. Therefore the random measures $\{\varepsilon Z^\varepsilon_t : \varepsilon > 0\}$ are also tight. Together with Lemma 4.4.3 and Theorem 3.2.6 this implies that $X^\varepsilon_t := \varepsilon \cdot Z^\varepsilon_t$ converges weakly to a random measure with Laplace functional (4.4.1). Note that as $\varepsilon \downarrow 0$, $\varepsilon Z^\varepsilon_r$ converges weakly to μ and that $\mu \to \int (V_{s,t}\phi)(x)\mu(dx)$ is a $\mathcal{B}(M_F(E))$-measurable function. This completes the proof that (4.4.1) defines a Laplace transition functional. □

Corollary 4.4.6. As $\varepsilon \to 0$, the processes $\{X^{\varepsilon}_t\}$, $X^{\varepsilon}_t := \varepsilon Z^{\varepsilon}_t$ converge in the sense of finite dimensional distributions to the process $\{X(t)\}$ whose finite dimensional distributions are determined by the joint Laplace functional:

for $s \le t_1 \le \ldots \le t_n$, $\phi_1,\ldots,\phi_n \in pb\mathcal{E}$,

$$(4.4.6)\quad P_{s,\mu}\left(\exp\left(-[\langle X(t_1),\phi_1\rangle+\ldots+\langle X(t_n),\phi_n\rangle]\right)\right) = \exp\left(-\langle \mu, V_{t_1,\ldots,t_n}(\phi_1,\ldots,\phi_n)\rangle\right)$$

and V is recursively defined by $V_{t_1}(\phi_1) = V_{s,t_1}(\phi_1)$ with V_{s,t_1} given by the solution of (4.3.8), and

$$V_{t_1,\ldots,t_n}(\phi_1,\ldots,\phi_n) = V_{t_1,\ldots,t_{n-1}}(\phi_{n-1}+V_{t_{n-1},t_n}\phi_n).$$

Proof. (cf. Gorostiza and López-Mimbela (1992)) The proof is by induction on n. The proof in the case n=1 follows from the argument in the proof of Theorem 4.4.1. Letting $X^{\varepsilon} := \varepsilon Z^{\varepsilon}$, we have applying the Markov property of X^{ε}, and (4.4.3), that

$$P_{s,\mathcal{P}ois(\mu.\varepsilon)}\left(\exp\left(-[\langle X^{\varepsilon}(t_1),\phi_1\rangle+\ldots+\langle X^{\varepsilon}(t_n),\phi_n\rangle]\right)\right)$$

$$= P_{s,\mathcal{P}ois(\mu.\varepsilon)}\Big(\exp\Big(-[\langle X^{\varepsilon}(t_1),\phi_1\rangle+\ldots+\langle X^{\varepsilon}(t_{n-1}),\phi_{n-1}+V_{t_{n-1},t_n}\phi_n\rangle$$

$$+\langle X^{\varepsilon}(t_{n-1}),\varepsilon^{-1}\log(1-V_{t_{n-1},t_n}(1-e^{-\varepsilon\phi_n}))-V_{t_{n-1},t_n}\phi_n\rangle]\Big)\Big).$$

But by Lemma 4.4.3, $\|\varepsilon^{-1}\log(1-V_{t_{n-1},t_n}(1-e^{-\varepsilon\phi_n}))-V_{t_{n-1},t_n}\phi_n\| \longrightarrow 0$ as $\varepsilon\to 0$. Together with the induction hypothesis we conclude that

$$\lim_{\varepsilon\to\infty} P_{s,\mathcal{P}ois(\mu.\varepsilon)}\left(\exp\left(-[\langle X^{\varepsilon}(t_1),\phi_1\rangle+\ldots+\langle X^{\varepsilon}(t_n),\phi_n\rangle]\right)\right)$$

$$= \exp\left(-\langle \mu, V_{t_1,\ldots,t_n}(\phi_1,\ldots,\phi_n)\rangle\right)$$

and the proof is complete. □

This completes the construction of the finite dimensional distributions of the (W,κ,Φ)-superprocess. We can then construct a Markov process with these finite dimensional distributions on a probability space using Kolmogorov's extension theorem in the usual way. In the time homogeneous case in which $\kappa(ds) = ds$ and W is a right process Fitzsimmons (1988) proved the existence of a $M_F(E)$-valued right process with these finite dimensional distributions. An extension of this to the time inhomogeneous setting is given in [DP]. These results will not be described in detail in these notes. However the existence, in special cases, of versions with regular sample paths and the strong Markov property is discussed in the remainder of this chapter as well as in Chapter 6.

4.5 The (α,d,β)–Superprocess

In this special case the motion process W is a symmetric α-stable process in $E = \mathbb{R}^d$ with semigroup $\{S_t^\alpha\}$, generator $\Delta_\alpha = -(-\Delta)^{\alpha/2}$, $0<\alpha\le 2$, and with paths in $D_{\mathbb{R}^d} = D([0,\infty);\mathbb{R}^d)$. The branching mechanism in this special case is given by

$$\Phi(x,\lambda) = -\gamma\lambda^{1+\beta}, \quad 0 < \beta \le 1 \text{ (spatially homogeneous)}$$

$$\kappa(ds) = ds$$

$$n(du) = \frac{\beta(1+\beta)\gamma}{\Gamma(1-\beta)}\frac{1}{u^{\beta+2}}du, \quad c(.) \equiv 0, \text{ if } \beta < 1,$$

$$n(.) \equiv 0, \quad c(x) \equiv 1, \quad \text{if} \quad \beta = 1.$$

The approximating branching particle systems have offspring generating functions

$$g^\varepsilon(v) = v + (1+\beta)^{-1}(1-v)^{1+\beta}$$

which in this special case are independent of ε. In the case $\beta = 1$, this reduces to $g^\varepsilon(v) = \frac{1}{2} + \frac{1}{2}v^2$ *(binary branching)*. Also $\kappa^\varepsilon(ds) = \gamma(1+\beta)\varepsilon^{-\beta}ds$.
The resulting $M_F(\mathbb{R}^d)$-valued process is called the (α,d,β)-*superprocess* and is characterized by its log-Laplace functional $V_t\phi$ which is given by the unique solution of

$$(4.5.1)\quad V_t\phi(x) = S_t^\alpha\phi(x) - \gamma\int_0^t S_u^\alpha[(V_{t-u}\phi)^{1+\beta}]du$$

that is, $u(t,x) = (V_t\phi)(x)$ satisfies

$$\frac{\partial u}{\partial t} = \Delta_{d,\alpha}u - \gamma u^{1+\beta}, \quad u(0,x) = \phi(x) \in bp\mathcal{E}.$$

If $\phi(\cdot) \equiv \theta$, then

$$(4.5.2)\qquad u(\theta,t) = \frac{\theta}{(1+t\beta\gamma\theta^\beta)^{1/\beta}}.$$

Then

$$\lim_{\theta\uparrow\infty} u(\theta,t) = 1/(\beta\gamma t)^{1/\beta},$$

and consequently the *non-extinction* probability

$$P_\mu(\langle X_t,1\rangle = 0) = 1 - \exp\left(-\frac{\langle\mu,1\rangle}{(\beta\gamma t)^{1/\beta}}\right) \sim const\cdot t^{-1/\beta} \quad \text{as} \quad t \to \infty.$$

Note that this result can be derived from the results of Zolotorev (1957).

The following scaling property of the solution of the nonlinear evolution equation (4.5.1) plays an important role in the study of (α,d,β)-superprocesses.

Lemma 4.5.1 For $R > 0$,

(4.5.3) $$V_t[R^{-\alpha/\beta}\phi(./R)](y) = R^{-\alpha/\beta}(V_{t/R^\alpha}\phi)(y/R).$$

Proof. Using scaling relation $(S^\alpha_{t/R^\alpha}\phi)(y/R) = (S^\alpha_t\phi(./R))(y)$, we obtain

$$(R^{-\alpha/\beta}(V_{t/R^\alpha}\phi)(y/R))$$

$$= R^{-\alpha/\beta}S^\alpha_{t/R^\alpha}\phi(y/R) - \gamma\int_0^{t/R^\alpha} R^{-\alpha/\beta}\, S^\alpha_{(t/R^\alpha)-s}(V_s\phi)^{1+\beta}ds$$

$$= R^{-\alpha/\beta}\, S^\alpha_{t/R^\alpha}\phi(y/R)$$

$$- \gamma\int_0^t S^\alpha_{(t/R^\alpha)-(s'/R^\alpha)}[V_{(s'/R^\alpha)}\phi(./R)]^{1+\beta}\frac{R^{-\alpha/\beta}}{R^\alpha}ds'$$

$$= R^{-\alpha/\beta}(S^\alpha_{t/R^\alpha}\phi)(y/R) - \gamma\int_0^t S^\alpha_{t-s'}[[R^{-\alpha/\beta}V_{s'/R^\alpha}\phi]^{1+\beta}(./R)](y).$$

Hence it satisfies the same equation as $V_t(R^{-\alpha/\beta}\phi(./R))(y)$ and the result follows by uniqueness. □

4.6 Weak Convergence of Branching Particle Systems

In this section we will show that the (α,d,β)-superprocesses can be extended to the space $M_p(\dot{\mathbb{R}}^d)$ and realized as càdlàg strong Markov processes.

Let $p > d/2$ and if $\alpha < 2$, we make the additional assumption $p < (d+\alpha)/2$. Let $M_p(\dot{\mathbb{R}}^d)$ be defined as in Section 3.1.5 and let $C_p(\dot{\mathbb{R}}^d) := \{f\in C(\mathbb{R}^d\cup\{\infty\}): \lim_{|x|\to\infty} f(x)/\phi_p(x)$ exists$\}$ with norm $\|f\|_p := \sup_x |f(x)|/\phi_p(x)$ (recall that the added point $\{\infty\}$ is isolated). $M_p(\mathbb{R}^d)$ can be identified with $\{\mu:\mu\in M_p(\dot{\mathbb{R}}^d),\ \mu(\{\infty\}) = 0\}$ and will be furnished with the $C_p(\mathbb{R}^d)$ weak topology. The semigroup $\{S^\alpha_t\}$ can be extended to $C_p(\dot{\mathbb{R}}^d)$ (cf. Iscoe (1986a) in such a way that $S^\alpha_t f(\infty) = f(\infty)$, that is, $\{\infty\}$ is an absorbing point. Also $|\Delta_\alpha\phi_p(x)| \le \mathit{const}\ \phi_p(x)$ (cf. [D, Lemma 5.5.1]).

In order to construct the process we proceed as above but in addition we find $\mu_\varepsilon\ (\in M_F(\mathbb{R}^d)) \uparrow \mu \in M_p(\dot{\mathbb{R}}^d)$. We consider the approximating branching particle systems as above in which the initial measure is $\mathcal{P}ois(\mu_\varepsilon/\varepsilon)$. In this section we will show that the laws of the approximating branching particle systems are relatively compact in $M_1(D([0,\infty),M_p(\dot{\mathbb{R}}^d)))$.

Let $X^\varepsilon(t,du) = \varepsilon\, Z^\varepsilon(t,du)$ where $Z^\varepsilon(t,du)$ denotes the approximating branching particle system with $Z^\varepsilon(0)$ given by a Poisson random measure with intensity measure μ_ε, α-symmetric stable motions, $\kappa^\varepsilon(w,ds) = 1(w(s)\neq \infty)\gamma(1+\beta)\varepsilon^{-\beta}ds$, and offspring probability generating function

$$\mathcal{g}^\varepsilon(v) = v + (1+\beta)^{-1}(1-v)^{1+\beta}.$$

Let $p \in (d/2,(d+\alpha)/2)$ or $(d/2,\infty)$ if $\alpha = 2$.

We first need to state the appropriate maximal inequality.

Lemma 4.6.1 For $0 < \theta < \beta$, and $T > 0$

$$\text{(4.6.1)} \qquad E \sup_{t\le T} \left\{ \langle X^\varepsilon(t),\phi_p\rangle^{1+\theta} \right\} \le K\cdot(1+I(X_0^\varepsilon)) \le K' \quad \forall\ \varepsilon\in(0,1),$$

where $I(X_0^\varepsilon) := \sup_{t\le T} E\left(\langle X^\varepsilon(0),S_t^\alpha\phi_p\rangle^{1+\beta} + \langle X^\varepsilon(0),S_t^\alpha\phi_p\rangle \right)$ and K' is independent of ε. Further if $I(X_0^\varepsilon) < 1$, then

$$\text{(4.6.2)} \qquad E \sup_{t\le T} \left\{ \langle X^\varepsilon(t),\phi_p\rangle^{1+\theta} \right\} \le K''\cdot(I(X_0^\varepsilon))^{(1+\theta)/(1+\beta)}.$$

Proof. The proof is based on elementary inequalities involving some properties of the Laplace functional and Doob's maximal inequality. For a detailed proof see [D, Lemma 5.5.3].

Theorem 4.6.2. Let $X^\varepsilon(t,du) = \varepsilon\, Z^\varepsilon(t,du)$ where $Z^\varepsilon(t,du)$ denotes the approximating branching particle system with $Z^\varepsilon(0)$ given by a Poisson random measure with intensity measure μ_ε, α-symmetric stable motions, $\kappa^\varepsilon(w,ds) = 1(w(s)\neq \infty)\gamma(1+\beta)\varepsilon^{-\beta}ds$, and offspring probability generating function

$$\mathcal{g}^\varepsilon(v) = v + (1+\beta)^{-1}(1-v)^{1+\beta}.$$

Let $p \in (d/2,(d+\alpha)/2)$ if $0<\alpha<2$, or $(d/2,\infty)$ if $\alpha = 2$.

(a) Then the processes $\{X^\varepsilon(t,.)\}$ under $P^Z_{r,\mathcal{Pois}(\mu_\varepsilon/\varepsilon)}$ with $\mu_\varepsilon \longrightarrow \mu \in M_p(\mathbb{R}^d)$ converge weakly to a càdlàg $M_p(\dot{\mathbb{R}}^d)$-valued process whose Laplace functional is given by

$$P_\mu\left(e^{-\langle X(t),\phi\rangle} \right) = \exp\left(-\int v_t(x)\mu(dx)\right), \quad \mu\in M_p(\mathbb{R}^d),$$

where

$$\text{(4.6.3)} \qquad \partial v_t(x)/\partial t = \Delta_\alpha v_t - \gamma(v_t)^{1+\beta},$$
$$v_0(x) = \phi(x) \in C_p(\mathbb{R}^d).$$

(b) If $\mu(\{\infty\}) = 0$, then $\sup_t X(t,\{\infty\}) = 0$, P_μ-a.s.

(c) If $\mu \in M_p(\mathbb{R}^d)$, then X is a.s. càdlàg (for the $C_p(\mathbb{R}^d)$ topology).

Proof. The convergence of the Laplace functional of the finite dimensional distributions of X_t^ε to (4.4.6) when $\mu \in M_F(\mathbb{R}^d)$ was proved in Corollary 4.4.6. The convergence of the finite dimensional distributions when $\mu_\varepsilon \uparrow \mu \in M_p(\mathbb{R}^d)$ follows by verifying the tightness of the first moment measures - this is a simple consequence of the fact that S_t^α maps $C_p(\dot{\mathbb{R}}^d)$ into itself. It thus remains to prove the tightness of the processes X^ε in $D([0,\infty),M_p(\dot{\mathbb{R}}^d))$.

Proof of Tightness. (a) Let $\varepsilon_n \to 0$ as $n \to \infty$. By Theorem 3.7.1, in order to prove tightness of the processes $\{X^{\varepsilon_n}\}$ it suffices to show that for $\phi \in K_p(\mathbb{R}^d)$, the family $\{Z_n(t) := \langle X^{\varepsilon_n}(t),\phi\rangle;\ n \in \mathbb{Z}_+\}$ is tight in $D([0,\infty),\mathbb{R})$. Using the Aldous condition (Theorem 3.6.5) it then suffices to show that

$$Z_n(\tau_n+\delta_n)-Z_n(\tau_n) \longrightarrow 0 \quad \text{in distribution as } n \to \infty.$$

Here δ_n are positive constants converging to zero as $n \to \infty$ and τ_n is any stopping time of the process Z_n with respect to the canonical filtration, satisfying $\tau_n \le T$.

By the strong Markov property applied to the process X^{ε_n} we obtain that for $r,s \ge 0$,

$$L_n(\delta_n;s,r) = E\Big(\exp\{-sZ_n(\tau_n+\delta_n)-rZ_n(\tau_n)\}\Big)$$

$$= E\Big(\exp\{\langle -X^{\varepsilon_n}(\tau_n),v_{\delta_n}(s\phi)+v_0(r\phi)\rangle\}\Big)$$

where $\{v_t(s\phi):t\ge 0\}$ satisfies equation (4.6.3) (with ϕ replaced by $s\phi$).

Therefore

$$|L_n(0;s,r) - L_n(\delta_n;s,r)|$$

$$\le \|v_n(s\phi;\delta_n)-v_n(s\phi;0)\|_p \; E\{\sup_{t\le T} \langle X^{\varepsilon_n}(t),\phi_p\rangle\}$$

$$\le \text{const}\ \|v_n(s\phi;\delta_n)-v_n(s\phi;0)\|_p \qquad \textit{(by Lemma 4.6.1).}$$

A modification of the proof of Lemma 4.4.3 shows that $\|v_n(s\phi;\delta_n)-v_n(s\phi;0)\|_p \longrightarrow 0$ as $n \to \infty$ and therefore $|L_n(0)-L_n(\delta_n)| \longrightarrow 0$.

By Lemma 4.6.1 the sequence $\{Z_n(\tau_n+\delta_n),Z_n(\tau_n)\}$ is tight. Consider a subsequence $\{n_k\}$ such that $(Z_{n_k}(\tau_{n_k}+\delta_{n_k}),Z_{n_k}(\tau_{n_k}))$ converges in distribution. Since $|L_n(0;s,r) - L_n(\delta_n;s,r)| \longrightarrow 0$ we conclude that the limiting distribution has the form (Z_∞, Z_∞). This implies that $Z_n(\tau_n+\delta_n)-Z_n(\tau_n) \longrightarrow 0$ in distribution as $n \to \infty$ and the proof of tightness is complete.

(b) In order to prove that $\mu(\{\infty\}) = 0$ implies $\sup_t X(t,\{\infty\}) = 0$, P_μ-a.s. it suffices to show that for all $\delta > 0$, $T>0$

$$\lim_{R\to\infty} \sup_n E\left(\sup_{0\le t\le T}\left(\int_{|x|>R} \phi_p(x)X^{\varepsilon_n}(t,dx)\right)^{1+\theta}\right) < \delta \text{ for some } 0<\theta<\beta.$$

It follows from the properties of the approximating particle system that

$$P_{\mathcal{Pois}(\mu_\varepsilon/\varepsilon)} = P_{\mathcal{Pois}(\mu_{1,\varepsilon}/\varepsilon)} * P_{\mathcal{Pois}(\mu_{2,\varepsilon}/\varepsilon)} \quad (\text{* denotes convolution})$$

where $\mu_{1,\varepsilon} = \mu_\varepsilon(dx)1(|x|<K)$, $\mu_{2,\varepsilon} = \mu_\varepsilon(dx)1(|x|\ge K)$ (let $X^{1,\varepsilon}$, $X^{2,\varepsilon}$ denote the corresponding branching particle systems with initial conditions $\mathcal{Pois}(\mu_{1,\varepsilon}/\varepsilon)$, $\mathcal{Pois}(\mu_{2,\varepsilon}/\varepsilon)$, respectively).

Note that

$$\sup_{0<\varepsilon<1} E\left\{\sup_{t\le T} \langle X^{2,\varepsilon}(t),\phi_p 1(|x|>R)\rangle^{1+\theta}\right\}$$

$$\le \sup_{0<\varepsilon<1} E\left\{\sup_{t\le T} \langle X^{2,\varepsilon}(t),\phi_p\rangle^{1+\theta}\right\}$$

$$\le \mathit{const} \sup_{0<\varepsilon<1}\left\{\sup_{0\le t\le T} E\left(\langle X^{2,\varepsilon}(0),S_t^\alpha\phi_p\rangle^{1+\beta} + \langle X^{2,\varepsilon}(0),S_t^\alpha\phi_p\rangle\right)\right\}^{(1+\theta)/(1+\beta)}$$

by (4.6.1)

for $0<\theta<\beta$.

Now, since $\mu_\varepsilon \longrightarrow \mu$ and $\mu(\{\infty_p\}) = 0$, given $\delta > 0$ we can choose K in the definition of $\mu_{2,\varepsilon}$ sufficiently large so as to make this smaller than $\delta/2$ (uniformly for $0 < \varepsilon < 1$).

But for $0 < p' < p$,

$$\sup_{0<\varepsilon<1} E \sup_{t\le T}\left\{\langle X^{1,\varepsilon}(t),\phi_p 1(|x|>R)\rangle^{1+\theta}\right\}$$

$$\le (1+R^2)^{-(p-p')(1+\theta)} \cdot \sup_{0<\varepsilon<1} E \sup_{t\le T}\left\{\langle X^{1,\varepsilon}(t),\phi_{p'}\rangle^{1+\theta}\right\}$$

$$\le \mathit{const}\,(1+R^2)^{-(p-p')(1+\theta)} \cdot \left\{\sup_{0\le t\le T} E\left(\langle X^{1,\varepsilon}(0),S_t^\alpha\phi_{p'}\rangle^{1+\beta} + \langle X^{1,\varepsilon}(0),S_t^\alpha\phi_{p'}\rangle\right)\right\}^{(1+\theta)/(1+\beta)}.$$

We can then choose R to make this smaller than $\delta/2$ (uniformly in $0 < \varepsilon < 1$).

(c) The arguments in (b) prove the necessary Prohorov tightness condition for the $C_p(\mathbb{R}^d)$-topology. This completes the proof. □

Remark 4.6.3. Consider the (A,Φ)-superprocess when A is a Feller process on a compact metric space E, $\Phi(x,\lambda) = -\gamma\lambda^{1+\beta}$, $0 < \beta \le 1$ and $\mu\in M_F(E)$. A slightly simplified version of the proof of Theorem 4.6.2 yields tightness on $D([0,\infty),M_F(E))$ for the corresponding family of branching particle systems. In the case of Polish E, a new argument is required to verify the first condition in Jakubowski's criterion. The verification of the second condition can be carried out as in Theorem 4.6.2 provided that the domain of A contains a convergence determining family.

Let $D_0(\Delta_\alpha,p) \subset pC_p(\mathbb{R}^d)$ be closed under positive linear combinations and form a core for the semigroup $\{S_t^\alpha\}$ on $C_p(\mathbb{R}^d)$. Let us extend functions in $D_0(\Delta_\alpha,p)$ to $C_p(\dot{\mathbb{R}}^d)$ by defining

$$\dot{f}(\{\infty_p\}) := \lim_{|x|\to\infty} f(x)/\phi_p(x).$$

Also note that $S_t^\alpha\dot{f}(\infty_p) = \dot{f}(\infty_p)$ $\forall t$, $\forall$ $f\in C_p(\dot{\mathbb{R}}^d)$. Let $\{T_t:t\ge0\}$, the semigroup of operators on $C_0(M_p(\dot{\mathbb{R}}^d))$ associated with the (α,d,β)-superprocess, be defined as follows. Let $\mathcal{D}_0(G)$ denote the collection of functions on $M_p(\dot{\mathbb{R}}^d)$ of the form

$$F(\mu) = \exp\left\{-[\langle\mu,\dot{\phi}\rangle+\theta\langle\mu,\dot{\phi}_p\rangle)]\right\}$$

with $\phi \in D_0(\Delta_\alpha,p)$, $\theta>0$. It is easy to verify that $\mathcal{D}_0(G)$ is a dense subset of $C_0(M_p(\dot{\mathbb{R}}^d))$. For functions $F \in \mathcal{D}_0(G)$,

$$T_tF(\mu) := EF(X(t)|X(0)=\mu)$$

$$= \exp\left\{-[\langle\mu 1_{\mathbb{R}^d},V_t(\phi+\theta\phi_p)\rangle + \theta\mu(\{\infty_p\})]\right\}.$$

Note that this implies that if $F(\mu) = f(\{\mu(\infty_p)\})$ then $T_tF(\mu) = F(\mu)$ for all t (i.e. no branching occurs at $\{\infty_p\}$ - probabilistically this is a law of large numbers phenomenon).

Proposition 4.6.4. (a) The semigroup $\{T_t:t\ge0\}$ is Feller on $C_0(M_p(\dot{\mathbb{R}}^d))$.

(b) The process $\{X_t:t\ge0\}$ is strong Markov.

Proof.(a) (cf. Iscoe (1986a, Theorem 1.1)). It follows from Theorem 4.6.2 that $\{T_t\}$ is a contraction semigroup on $b\mathcal{B}(M_p(\dot{\mathbb{R}}^d))$. It remains to show that $T_t:C_0(M_p(\dot{\mathbb{R}}^d))\to C_0(M_p(\dot{\mathbb{R}}^d))$. To do this it suffices to show that for functions in $\mathcal{D}_0(G)$, $T_tF(\mu)$ is continuous on $M_p(\dot{\mathbb{R}}^d)$.

Thus it suffices to show that for such F, if $\mu_n\to\mu\ne\infty$ in $M_p(\dot{\mathbb{R}}^d)$, then

$$\lim_{n\to\infty} T_tF(\mu_n) = \lim_{n\to\infty} \exp\left\{-[<\mu_n 1_{\mathbb{R}^d}, V_t(\phi+\theta\phi_p)> + \theta\mu_n(\{\infty_p\})]\right\}$$

$$= \exp\left\{-[<\mu 1_{\mathbb{R}^d}, V_t(\phi+\theta\phi_p)> + \theta\mu(\{\infty_p\})]\right\}$$

and

$$\lim_{n\to\infty} T_tF(\mu_n) = 0 \text{ if } \phi=\phi_p \text{ and } \mu_n \longrightarrow \{\infty\}.$$

But this follows from the analytical result that $V_t:C_p(\mathbb{R}^d)\longrightarrow C_p(\mathbb{R}^d)$, $V_t\phi_p \geq const\cdot\phi_p$ and the fact that

$$\lim_{|x|\to\infty} V_t\phi_p(x)/\phi_p(x) = \lim_{|x|\to\infty} S_t^\alpha\phi_p(x)/\phi_p(x) = 1$$

(the second equality follows from an elementary argument and the first one follows from the Feynman-Kac formula).

(b) The strong Markov property follows directly from the Feller property and will also be a consequence of the results of Theorem 6.1.3. □

Remarks. 1. The fact that there is no branching at $\{\infty\}$ is a law of large numbers phenomenon and can be derived analytically from the fact that $\lim_{|x|\to\infty} V_t\phi_p(x)/\phi_p(x) = 1$. In order to get the former note that $\mu_n = a_n\delta_{x_n} \longrightarrow \delta_{\{\infty\}}$ in $M_p(\mathbb{R}^d)$ if and only if $|x_n|\to\infty$ and $a_n\phi_p(x_n)\longrightarrow 1$. But then X_t can be written as the normalized sum of $[\phi_p(x_n)^{-1}]$ i.i.d. random variables (plus a negligible remainder) and the result then follows from the weak law of large numbers). It also arises from the limit of the branching particle systems since we have assumed that $\kappa(w,ds) = 0$ if $w(s) = \infty$.

4.7 The Continuous B(A,c)-Superprocess

In this section we consider in greater detail the class of B(A,c)-superprocesses (which includes the $(\alpha,d,1)$-superprocesses of Sect. 4.5). First we give some notation. Let E be locally compact with one point compactification $\bar{E}$ and (A,D(A)) denote the generator of the A-Feller process on $\bar{E}$ and let $\beta = 1$. Then the B(A,c)-superprocess is characterized by the log-Laplace equation

$$(4.7.1) \qquad V_t\phi = S_t\phi - \tfrac{1}{2}c\int_0^t S_{t-s}(V_s\phi)^2 ds.$$

Lemma 4.7.1 Let $\phi\in b\mathcal{E}$. Then

(a) $P_\mu(<X_t,\phi>) = <\mu,S_t\phi>$,

(b) $P_\mu(\langle X_t,\phi\rangle^2) = \int_0^t \langle \mu S_{t-s},(S_s\phi)^2\rangle ds + \langle \mu,S_t\phi\rangle^2,$

(c) if $\phi\in bp\mathcal{E}$, then

$$P_\mu(\langle X_t,\phi\rangle^n) = \sum_{k=0}^{n-1}\binom{n-1}{k}\langle \mu,v_t^{(n-k)}\rangle P_\mu(\langle X_t,\phi\rangle^k)$$

where $v_t^{(1)} = S_t\phi$, and $v_t^{(n)} = \sum_{k=1}^{n-1}\binom{n-1}{k}\int_0^t S_{t-s}(v_s^{(k)}.v_s^{(n-k)})ds$, $n\geq 2$.

Proof. Let $u^{(n)}(t,\theta) := \frac{\partial^n}{\partial\theta^n} V_t(\theta\phi)$ for $\phi\in bp\mathcal{E}$. Then

(4.7.2) $\langle \mu,u^{(1)}(t,\theta)\rangle P_\mu(\exp(-\theta\langle X_t,\phi\rangle)) = P_\mu(\langle X_t,\phi\rangle\exp(-\theta\langle X_t,\phi\rangle))$.

Iteratively differentiating (4.7.2) with respect to θ and evaluating at $\theta=0$ yields (c). (a) and (b) follow as special cases. □

Let $\{f_n:n\in\mathbb{Z}_+\}$ be a convergence determining sequence in $pD(A)$ with $(\|Af_n\|+\|f_n\|) \leq 1$ and define

$$d(\mu,\nu) := \sum 2^{-n}\left(1 \wedge \left|\int f_n d\mu - \int f_n d\nu\right|\right).$$

Theorem 4.7.2. Let $\{X_t:t\geq 0\}$ be a $B(A,c)$-superprocess on the compact space $\bar{E}$, where $\bar{E}$ is the one-point compactification of E. Then for each $\mu\in M(\bar{E})$, $\{X_t:t\geq 0\}$ is P_μ-almost surely lies in $C([0,\infty),M(\bar{E}))$. In addition for $\beta < 1/4$ there exists a β-Hölder continuous version of $\{X_t:t\geq 0\}$ in the metric d.

Proof. Using Lemma 4.7.1 it can be verified that for $T<\infty$, $n\in \mathbb{Z}_+$,

(4.7.3) $P_\mu([\langle X_t,f_n\rangle-\langle X_s,f_n\rangle]^4) \leq C_T(t-s)^2, \quad 0\leq s<t\leq T.$

This implies that for each $n\in\mathbb{Z}_+$, $\langle X_t,f_n\rangle$ is P_μ-almost surely continuous. The β-Hölder continuity follows from (4.7.3) and Corollary 3.7.3. □

Remark. A stronger version of this result (β-Hölder continuity for $\beta < 1/2$) will be obtained in Prop. 7.3.1.

5. PROBABILITY MEASURE-VALUED PROCESSES AND MARTINGALE PROBLEMS

5.1. Martingale Problems and Markov processes

There is a number of possible approaches to the construction and characterization of measure-valued Markov processes. For example, in Chapters 2 and 4 we adopted semigroup methods to construct the Fleming-Viot and measure-valued branching processes. In this chapter we develop the martingale problem approach which serves as a unified setting in which to compare these two classes of processes and to use them as building blocks to construct more complex processes.

In this section we give a brief introduction to the martingale problem for E-valued processes with E a Polish space. (However note that we will later apply it with E replaced by $M_1(E)$ or $M_F(E)$.) We frequently refer to Ethier and Kurtz (1986, Chap.4) which should be consulted for a more complete exposition of the martingale problem method.

Let E be a Polish space and $D_{E,s} := D([s,\infty),E)$ with coordinate process $\{X_t\}$ and filtration $(\mathcal{D}_{s,t})_{t\geq s}$ where $\mathcal{D}_{s,t} := \bigcap_{\varepsilon>0} \mathcal{D}^o_{s,t+\varepsilon}$ where $\mathcal{D}^o_{s,t} := \sigma(\{X_u\}:s\leq u\leq t)$. Also let $\mathcal{D}_{s,\infty} := \bigvee_{n\in\mathbb{N}} \mathcal{D}_{s,s+n}$.

A *time inhomogeneous martingale problem* is given by a pair $(\mathfrak{D}(\mathfrak{G}),\mathfrak{G})$ satisfying:

(5.1.1) $\mathfrak{G}:[0,\infty)\times\mathfrak{D}(\mathfrak{G}) \longrightarrow b\mathcal{E}$, $\mathfrak{D}(\mathfrak{G}) \subseteq bC(E)$. and for each $F\in\mathfrak{D}(\mathfrak{G})$,
$(s,x) \longrightarrow (\mathfrak{G}F)(s,x) = (\mathfrak{G}(s)F)(x) \in b(\mathcal{B}(\mathbb{R}_+)\otimes\mathcal{E})$, and
for each $s\in[0,\infty)$, $\mathfrak{G}(s)$ is a linear mapping on $\mathfrak{D}(\mathfrak{G})$.

A probability measure P on $(D_{E,s},\mathcal{D}_{s,\infty},(\mathcal{D}_{s,t})_{t\geq s})$ is said to be a *solution to the s-initial value martingale problem for* $(\mathfrak{G},\mathfrak{D}(\mathfrak{G}))$ if for every $F\in\mathfrak{D}(\mathfrak{G})$,

(5.1.2) $$M_F(t,X) := F(X(t)) - F(X(s)) - \int_s^t (\mathfrak{G}F)(u,X(u))du$$

is a $((\mathcal{D}_{s,t})_{t\geq s},P)$-martingale.

Equivalently, P satisfies:

(5.1.3) $$H(X) \text{ is P-integrable and } \int_{D_{E,s}} H(X)P(dX) = 0 \quad \forall \ H\in \mathfrak{H}_s$$

where $\mathfrak{H}_s := \{H: H(u,t,F,G_u;X):s\leq u\leq t<\infty,\ F\in\mathfrak{D}(\mathfrak{G}),\ G_u\in b\mathcal{D}_{s,u}\}$, and

$$H(u,t,F,G_u;X) := [M_F(t,X)-M_F(u,X)]G_u(X).$$

The set of solutions to the initial value martingale problem at time 0 is a convex subset of $M_1(D_{E,0})$, denoted by $\mathfrak{P}(\mathfrak{G})$. Let $\mathfrak{P}_e(\mathfrak{G})$ denote the class of extreme points of $\mathfrak{P}(\mathfrak{G})$. For $P\in \mathfrak{P}(\mathfrak{G})$, let

$$\mathcal{H}^1(P) := \{M: M \text{ is a P-local martingale, } \|\sup_t |M_t|\|_1 < \infty\}$$

where $\|.\|_1$ is the $L^1(P)$-norm. Let $\mathcal{H}^1_{pr}(P)$ denote the smallest $\mathcal{H}^1(P)$-closed linear space containing $\{\int HdM_F \in \mathcal{H}^1(P)$:H is predictable, $F\in \mathfrak{D}(\mathfrak{G})\}$ and the constant process, where $\int HdM_F$ denotes the Itô stochastic integral.

Theorem 5.1.1 *(Jacod's Predictable Representation Theorem)*
If $P \in \mathfrak{P}(\mathfrak{G})$, then the following conditions are equivalent:
(i) $P\in \mathfrak{P}_e(\mathfrak{G})$,
(ii) $\mathcal{H}^1(P) = \mathcal{H}^1_{pr}(P)$ and $\mathcal{D}_0$ is P-trivial,
(iii) every bounded P-martingale, M, with $M_0=0$, which is orthogonal to M_F, $\forall F\in\mathfrak{D}(\mathfrak{G})$ is zero and $\mathcal{D}_0$ is P-trivial.
Proof. See Jacod (1979, Theorem 11.2).

Proposition 2.3.3 gives an example of a function of a Markov process that is Markov. In order to establish this type of result in the context of martingale problems we include the notion of *restricted martingale problem* which is formulated as follows.

Let $M_0(E) \subset M_1(E)$. Then $\mathcal{A} \subset pbC(E)$ is said to be $M_0(E)$-*determining* if
(i) $pb\sigma(\mathcal{A})$ is the bp closure of $\mathcal{A}$,
(ii) if $\mu_1,\mu_2 \in M_0(E)$ and $\mu_1 = \mu_2$ on $\sigma(\mathcal{A})$, then $\mu_1=\mu_2$, and
(iii) for each $\mu\in M_0(E)$ and $g\in pp\mathcal{A}$, $A \longrightarrow \mu_g(A) := \langle\mu,g1_A\rangle/\langle\mu,g\rangle \in M_0(E)$
(where $pp\mathcal{A} := \{g\in\mathcal{A}: \inf g > 0\}$).
Let $M_0(E) \subset M_1(E)$ and $\mathfrak{D}(\mathfrak{G}) \subset pbC(E)$ be $M_0(E)$-determining. Then a family of probability measures $\{P_{s,\mu}:\mu\in M_0(E)\}$, $M_0(E)\subset M_1(E)$ on $(D_{E,s},\mathcal{D}_{s,\infty},(\mathcal{D}_{s,t})_{t\geq s})$ is said to be a *solution to the restricted s-initial value martingale problem for* $(\mathfrak{G},\mathfrak{D}(\mathfrak{G}),M_0(E))$ if for every $\mu\in M_0(E)$, $\Pi_s P_{s,\mu}=\mu$, $\Pi_t P_{s,\mu} \in M_0(E)$ $\forall t\geq s$, and $P_{s,\mu}$ is a solution to the s-initial value martingale problem for $(\mathfrak{G},\mathfrak{D}(\mathfrak{G}))$. The initial value martingale problem is said to be *well-posed* if for every $s\geq 0$ and $\mu\in M_0(E)$, there is a *unique* solution to the martingale problem satisfying $\Pi_s P_{s,\mu}=\mu$.

The following result is a minor extension of a basic result of Stroock and Varadhan (1979, section 6.2) to include the case of restricted and time inhomogeneous martingale problems. (The extension to restricted martingale problems will be used to characterize exchangeable Markov systems and the time inhomogeneous martingale problem will be used in sections 5.5, 5.7 and Chapter 12.)

Theorem 5.1.2. Assume that
(i) $M_0(E)\subset M_1(E)$, $\mathfrak{D}(\mathfrak{G}) \subset ppbC(E)$ is $M_0(E)$-determining, $\mathfrak{G}:[0,\infty)\times\mathfrak{D}(\mathfrak{G})\longrightarrow$

$b(\sigma(\mathfrak{D}(\mathfrak{G})))$, and $(s,x) \longrightarrow (\mathfrak{G}F)(s,x) \in b(\mathcal{B}(\mathbb{R}_+)\otimes\sigma(\mathfrak{D}(\mathfrak{G})))\ \forall\ F\in \mathfrak{D}(\mathfrak{G})$,

(ii) for any $s\geq 0$ and any two solutions to the restricted s-initial value martingale problem for $(\mathfrak{G},\mathfrak{D}(\mathfrak{G}),M_0(E))$, denoted by $\{P^1_{s,\mu}:\mu\in M_0(E)\}$ and $\{P^2_{s,\mu}:\mu\in M_0(E)\}$, the marginal distributions $\Pi_t P^1_{s,\mu}$ and $\Pi_t P^2_{s,\mu}$ for $t\geq s$ agree on $\sigma\{\mathfrak{D}(\mathfrak{G})\}$ for all $t>s$ and $\mu\in M_0(E)$.

Then

(iii) $P_{s,\mu} := P^1_{s,\mu} = P^2_{s,\mu}$ on $(D_{E,s},\mathcal{D}^{\mathfrak{G}}_{s,\infty},(\mathcal{D}^{\mathfrak{G}}_{s,t})_{t\geq s})\ \forall\ \mu\in M_0(E)$ and $s\geq 0$, where $\mathcal{D}^{\mathfrak{G}}_{s,t} := \bigcap_{\varepsilon>0}\sigma\{F(X_u):F\in\mathfrak{D}(\mathfrak{G}), s\leq u\leq t+\varepsilon\}$ and $\mathcal{D}^{\mathfrak{G}}_{s,\infty} = \bigvee_{t\geq s}\mathcal{D}^{\mathfrak{G}}_{s,t}$, and

(iv) the process $(D_{E,s},\mathcal{D}^{\mathfrak{G}}_{s,t},(\mathcal{D}^{\mathfrak{G}}_{s,t})_{t\geq s},\{X_t\}_{t\geq s},\{P_{s,\mu}:\mu\in M_0(E)\})_{s\geq 0}$ is Markov, that is, if $B\in\mathcal{D}^{\mathfrak{G}}_{t_1,\infty}$, then

$$P_{s,\mu}(B|\mathcal{D}^{\mathfrak{G}}_{s,t_1}) = P_{s,\mu}(B|\sigma(F(X_{t_1}):F\in\mathfrak{D}(\mathfrak{G}))),\ P_{s,\mu}\text{-a.s.}$$

Sketch of Proof. To prove (iii) it suffices to show that the finite dimensional distributions agree. First consider $s\leq t_1<t_2$. In order to prove that the two distributions of $(X(t_1),X(t_2))$ agree on $\sigma(\mathfrak{D}(\mathfrak{G}))\otimes\sigma(\mathfrak{D}(\mathfrak{G}))$ it suffices to show that

$$(5.1.4)\quad P^1_{s,\mu}(g_1(X(t_1))g_2(X(t_2))) = P^2_{s,\mu}(g_1(X(t_1))g_2(X(t_2)))$$

for all $g_1,g_2\in\mathfrak{D}(\mathfrak{G})$ (because of assumption (i)). Since $g_1\in ppC(E)$, $P^1_{s,\mu}(g_1(X(t_1))) = P^2_{s,\mu}(g_1(X(t_1))) > 0$. Define probability measures on $\mathcal{D}^{\mathfrak{G}}_{t_1,\infty}$ by

$$Q^i_{t_1}(B) = P^i_{s,\mu}(g_1(X(t_1))1_B)/P^i_{s,\mu}(g_1(X(t_1))),\ i=1,2,\ B\in\mathcal{D}^{\mathfrak{G}}_{t_1,\infty}.$$

Using the assumption (i) it can be verified that $\{M_F(t):t\geq t_1\}$ is a $((\mathcal{D}^{\mathfrak{G}}_{t_1,t})_{t\geq t_1},Q^i_{t_1})$-martingale for each $F\in\mathfrak{D}(\mathfrak{G})$ and $i=1,2$. Thus $Q^1_{t_1}$ and $Q^2_{t_1}$ are solutions to the t_1-initial value martingale problem satisfying $\nu := \Pi_{t_1}Q^1_{t_1} = \Pi_{t_1}Q^2_{t_1}$ which by assumption (ii) belongs to $M_0(E)$. Then using assumption (ii), $Q^1_{t_1}(g_2(X(t_2))) = Q^2_{t_1}(g_2(X(t_2))) = P_{t_1,\nu}(g_2(X(t_2))$ which yields (5.1.4). Continuing in this way we can show that all the finite dimensional distributions of $P^1_{s,\mu}$ and $P^2_{s,\mu}$ agree and satisfy the Markov property (iv). □

Remark. Restricted and time inhomogeneous martingale problems will be needed at several places in these notes. However in order to simplify the exposition a little we will confine our attention to the time homogeneous ($\mathfrak{G}_s\equiv\mathfrak{G}\ \forall\ s\geq 0$) and unrestricted martingale problem in the remainder of this section. The appropriate modification for the time inhomogeneous and restricted case is then relatively straightforward.

<u>Theorem 5.1.3</u> Assume that there is a countable subset $\mathfrak{D}_0(\mathfrak{G})$ of $\mathfrak{D}(\mathfrak{G})$ such the assumptions of Theorem 5.1.2 (time homogeneous version) are satisfied with $\mathfrak{D}(\mathfrak{G})$ replaced by $\mathfrak{D}_0(\mathfrak{G})$ and with $M_0(E) = M_1(E)$. Then

(a) the mapping $\mu \longrightarrow P_\mu$ is measurable, $\mathfrak{P}_e = \{P_{\delta_x} : x \in E\}$ and $P_\mu = \int P_{\delta_x}\,\mu(dx)$,

(b) the process $(D_E, \mathcal{D}, (\mathcal{D}_t)_{t\geq 0}, \{P_\mu : \mu \in M_1(E)\})$ is strong Markov with transition measures $P_x := P_{\delta_x}$.

<u>Outline of Proof.</u> (See [EK, Ch. 4, Thms. 4.2 and 4.6] for details.)

(a) The set of solutions can be written in the form $\{P: \int H(X)P(dX)=0 \;\; \forall H \in \mathfrak{H}_0\}$ where $\mathfrak{H}_0 := \{H(s,t,F,G_s;X);\;\; s,t \text{ are rational, } F \in \mathfrak{D}_0(\mathfrak{G}),\ G_s \in (b\mathcal{D}_s)_0\}$ and $(b\mathcal{D}_s)_0$ is a countable set whose bounded pointwise closure coincides with $b\mathcal{D}_s$.

Since $\mathfrak{H}_0$ is countable, the set of solutions to the $(\mathfrak{G},\mathfrak{D}_0(\mathfrak{G}))$-martingale problem is a Borel subset, S, of $M_1(D_E)$. Note that the mapping $\Pi_0 : S \longrightarrow M_1(E)$ is continuous, one-to-one and onto since the martingale problem is well-posed. Thus by Kuratowski's theorem there is a measurable inverse, P_μ from $M_1(E)$ to S. Since the mapping $x \longrightarrow \delta_x$ is measurable this also yields the measurability of $P_x := P_{\delta_x}$. Given $\mu \in M_1(E)$, $P_\mu(H(X)) = \int P_{\delta_x}(H(X))\mu(dx) = 0 \;\forall H \in \mathfrak{H}$ and hence yields a solution P_μ to the martingale problem satisfying $\Pi_0 P_\mu = \mu$. The second statement in (a) follows by uniqueness.

(b) Let $\mu \in M_1(E)$, τ be a P_μ-a.s. finite stopping time and $g_1 \in pb\mathcal{D}_\tau$, such that $P_\mu(g_1) > 0$. Define probability measures on D_E by

$$P_1(g_2(X(\tau+.))) = P_\mu(g_1 P_\mu(g_2(X(\tau+.))\,|\,\mathcal{D}_\tau))/P_\mu(g_1), \text{ and}$$

$$P_2(g_2(X(\tau+.))) = P_\mu(g_1 P_{X(\tau)}(g_2(X(\tau+.))))/P_\mu(g_1),\;\; g_2 \in b\mathcal{D}.$$

In order to prove the strong Markov property it suffices to show that $P_1 = P_2$. Since $M_F(t)$ is a càdlàg martingale, the optional sampling theorem implies that $M_F(t+\tau) - M_F(\tau)$ is again a martingale. Hence both P_1 and P_2 are solutions to the martingale problem with the same initial condition. Thus they are almost surely equal by uniqueness. This yields the result. □

<u>Remark.</u> If the conditions of the above theorem are satisfied then $(\mathfrak{D}(\mathfrak{G}),\mathfrak{G})$ characterizes the process and will be called an *MP-generator* of the process.

In the remainder of this section we will assume that E is a compact metric space. We will then work with a martingale problem in which the basic state space is $M_1(E)$ (in place of E). In order to formulate the $M_1(E)$-valued martingale problem we will first introduce some appropriate subspaces of $C(M_1(E))$ which will serve as the domain of the generator.

Recall that the *algebra of polynomials*, $C_P(M_1(E))$, (resp. $B_P(M_1(E))$) is defined to be the linear span of *monomials* of the form

$$F_{f,n}(\mu) = \int f(x)\mu^n(dx)$$
$$= \int \ldots \int f(x_1,\ldots,x_n)\mu(dx_1)\ldots\mu(dx_n)$$

where $f \in \mathcal{B}(E^n)$, (resp. $C(E^n)$). We also consider the subspaces $\mathcal{B}_{sym}(M_1(E))$, (resp. $C_{sym}(M_1(E))$), and $\mathcal{B}_{pro}(M_1(E))$, (resp. $C_{pro}(M_1(E))$) in which f is symmetric and of the form $\Pi\phi_i(x_i)$, respectively.

For a positive integer m, $B_{F,m}(M_1(E))$, (resp. $C_{F,m}(M_1(E))$) denotes the algebra of all functions of the form

$$F_{f;\phi_1,\ldots,\phi_m}(\mu) = f(<\mu,\phi_1>,\ldots,<\mu,\phi_m>),\ f\in C^\infty(\mathbb{R}^m),$$

and $\phi_1,\ldots,\phi_m \in \mathcal{B}(E)$, (resp. $C(E)$).

The families $C_P(M_1(E))$, $C_{sym}(M_1(E))$, $C_{pro}(M_1(E))$ and $C_{F,m}(M_1(E))$ are all dense in $C(M_1(E))$ if E is compact (cf. Lemma 2.1.2 for the case $C_P(M_1(E))$). The first and second derivatives of functions in $C_P(M_1(E))$ were given in Lemma 2.1.2. For functions in $C_{F,m}(M_1(E))$ we have

$$\delta F_{f;\phi_1,\ldots,\phi_m}(\mu)/\delta\mu(x) = \sum_{j=1}^{m} F_{f_j;\phi_1,\ldots,\phi_m}(\mu)\phi_j(x)$$

$$\delta^2 F_{f;\phi_1,\ldots,\phi_m}(\mu)/\delta\mu(x)\delta\mu(y) = \sum_{j=1}^{m}\sum_{k=1}^{m} F_{f_{jk},\phi_1,\ldots,\phi_m}(\mu)\phi_j(x)\phi_k(y)$$

and $f_j(z_1,\ldots,z_m) = \partial f/\partial z_j$, $f_{jk}(z_1,\ldots,z_k) = \partial^2 f/\partial z_j\partial z_k$.

5.2. A Finite Type Version of the Moran Model

We will now revisit the Moran model (cf. section 2.5) from the viewpoint of martingale problems. However in this section we begin with a finite space of types.

Let $E_K = \{e_1,\ldots,e_K\}$ be a finite subset of E having K elements. E_K will serve as the *set of types*. We consider a population of N individuals in this space. We assume that the following two types of transition can occur:

(i) an individual changes from type x to type y with rate q^K_{xy} *(mutation)*;

(ii) an individual changes type at rate $\frac{1}{2}$ $(N-1)\gamma$ to a new type equal to that of a second individual chosen at random from the remaining population *(sampling-visitation)* or *(sampling-replacement)*, consequently the total jump rate due to this mechanism is $\frac{1}{2}\gamma N(N-1)$.

We can again consider this process from two points of view. First we can consider the exchangeable particle system $\mathbf{Z}(t) = \{Z_1(t),\ldots,Z_N(t)\} \in (E_K)^N$ where $Z_j(t)$ denotes the type of the jth individual at time t. This process has generator

$$L_N^K f(x_1,\ldots,x_N) = \sum_{i=1}^{N} A_i^K f(x_1,\ldots,x_N) + \frac{1}{2}\gamma \sum_{\substack{i,j\\ i\neq j}} [\Theta_{ji} f(x_1,\ldots,x_N) - f(x_1,\ldots,x_N)]$$

where $A^K g(x) := \sum_{y\in E_K} q_{x,y}^K [g(y)-g(x)]$ and A_i^K denotes the application of A^K to the variable x_i. Also let Θ_{ij} be as in Chapter 2.

Theorem 5.2.1. (a) The martingale problem with state space $E_K^N := (E_K)^N$ and with MP-generator $(L_N^K, C(E^N), M_1(E_K^N))$ is well-posed.

(b) The $(L_N^K,\ C_{sym}(E^N), M_{1,ex}(E_K^N))$-martingale problem is also well-posed and yields an exchangeable Markov system. (Here $M_{1,ex}(E_K^N)$ denotes the family of exchangeable laws on E_K^N.)

Proof. (a) The uniqueness of the one dimensional distributions follows since the martingale problem determines a finite system of N linear (Kolmogorov) equations for $P_\mu(\mathbf{Z}(t)=\mathbf{z})$, $\mathbf{z}\in(E_K)^N$, $\mu\in M_1((E_K)^N)$, which can be explicitly solved in terms of the matrix exponential.

(b) The solution of (a) is Markov by Theorem 5.1.2. Since $L_N^K(f(\tilde{\pi}\cdot))(\underline{x}) = (L_N^K f)(\tilde{\pi}\underline{x})$ it yields an exchangeable Markov jump process with state space $(E_K)^N$ (cf. Lemma 2.3.2) provided that the initial law $\mu\in M_{1,ex}(E_K^N)$. This proves the existence of a solution to the $(L_N^K,\ C_{sym}(E^N), M_{1,ex}(E_K^N))$-martingale problem. The uniqueness of the marginal distributions of the $(L_N^K, C_{sym}(E^N), M_{1,ex}(E_K^N))$-martingale problem follows as in (a). Since $C_{sym}(E^N)$ is $M_{1,ex}(E^N)$-determining (cf. Lemma 2.3.1) Theorem 5.1.2 then implies that the restricted martingale problem is well-posed. □

From the second viewpoint we consider the empirical process

$$X^N(t;B) := \Xi_N(\mathbf{Z}(t))(B) = \frac{\sum_{j=1}^{N} 1_B(Z_j(t))}{N}, \quad B \in \mathcal{E},$$

with state space $M_{1,N}(E_K)$.

Assume that $F \in C^3(M_1(E_K))$. Then the generator of the measure-valued process X^N is given by

$$\mathfrak{G}_N^K F(\mu) = N\int_{E_K} \sum_{y\in E_K} q_{x,y}^K [F(\mu - \delta_x/N + \delta_y/N) - F(\mu)]\mu(dx)$$

$$+ \frac{1}{2}\gamma N(N-1)\int_{E_K}\int_{E_K} [F(\mu - \delta_x/N + \delta_y/N) - F(\mu)]\mu(dx)\mu(dy)$$

$$= \int \{Q^K \frac{\delta F(\mu)}{\delta\mu(.)}\}(x)\mu(dx)$$

$$+ \frac{\gamma}{4}\iint \left[\frac{\delta^2 F(\mu)}{\delta\mu(x)^2} + \frac{\delta^2 F(\mu)}{\delta\mu(y)^2} - 2\frac{\delta^2 F(\mu)}{\delta\mu(x)\delta\mu(y)}\right]\mu(dx)\mu(dy)$$

$$+ R_N(F)/N.$$

Here we use Taylor's remainder formula, where $R_N(F)$ is bounded and $Q^K f(x) := \sum_{y\in E_K} q_{x,y}^K f(y)$. For functions of the form

$$F(\mu) = f(\langle\mu,\phi_1\rangle,\ldots,\langle\mu,\phi_K\rangle)$$

where $\phi_j(.) = 1_{\{e_j\}}(.)$ with $e_j \in E_K$, $f \in C^3(\mathbb{R}^K)$, and $\mu(\{e_i\}) = z_i$ this becomes $\mathfrak{G}_N^K F(\mu) = G_N^K f(\mu(\{e_1\}),\ldots,\mu(\{e_K\}))$, where

$$G_N^K f(z_1,\ldots,z_K) = \frac{\gamma}{2}\sum_{i,j=1}^K z_i(\delta_{ij} - z_j)\frac{\partial^2}{\partial x_i \partial x_j} f(z_1,\ldots,z_K)$$

$$+ \sum_{i=1}^K \left(\sum_{j\neq i} q_{ji}^K z_j - q_{ii}^K z_i\right)\frac{\partial}{\partial x_i} f(z_1,\ldots,z_K) + R_N(f)/N$$

where $R_N(f)$ is bounded in N.

<u>Theorem</u> <u>5.2.2.</u> The $(\mathfrak{G}_N^K, C_{pro}(M_1(E_K)), M_1(E_K))$-martingale problem is well-posed.
<u>Proof.</u> This follows from Lemma 2.3.1 and Theorem 5.2.1. □

5.3. The Finite Type Diffusion Limit

We now consider the limit of X^N as $N \to \infty$. We get

$\mathfrak{G}^K_\infty F(\mu) = G^K_\infty f(\mu(\{e_1\}),\ldots,\mu(\{e_K\}))$, where

$$G^K_\infty f(z_1,\ldots,z_K) = \frac{\gamma}{2} \sum_{i,j=1}^{K} z_i(\delta_{ij}-z_j) \frac{\partial^2}{\partial x_i \partial x_j} f(z_1,\ldots,z_K)$$

$$+ \sum_{i=1}^{K} \left(\sum_{j\neq i} q^K_{ji} z_j - q^K_{ii} z_i \right) \frac{\partial}{\partial x_i} f(z_1,\ldots,z_K)$$

Theorem 5.3.1. The $(\mathfrak{G}^K_\infty, C_{pro}(M_1(E_K)), M_1(E_K))$-martingale problem has a unique solution and this solution $\{P^K_\mu : \mu\in M_1(E_K)\}$ can be realized on $C([0,\infty),M_1(E_K))$.

Proof. (cf. Ethier (1976), Shiga (1981)). The existence is proved by a weak convergence argument. For each $F\in C_{F,m}(M_1(E_K))$ we have that $M_N(X^N(t))$ are uniformly bounded martingales where

$$M_N(\omega,t) := F(\omega(t)) - \int_0^t \mathfrak{G}^K_N F(\omega(s))ds,\ t\geq 0,\ \omega\in D([0,\infty),M_1(E_K)).$$

Hence $F(X^N(t))$ is a sequence of D-semimartingales which satisfies the tightness criteria of Joffe and Métivier (cf. Theorem 3.6.6 and Corollary 3.6.7). Thus the laws P^K_N of X^N are tight in $M_1(D([0,\infty),M_1(E_K)))$. Let P^K_∞ be a limit point. We next verify that P^K_∞ is a solution to the martingale problem for $\mathfrak{G}^K_\infty$. Note that $M_\infty : D([0,\infty),M_1(E_K)) \to D([0,\infty),\mathbb{R})$ is continuous and $\|M_N(.)(t)-M_\infty(.)(t)\| \leq N^{-1}t|R_N(F)|$. Then following the same argument as in Section 2.9 we can show that $M_\infty(\omega,t) := F(\omega(t)) - \int_0^t \mathfrak{G}^K_\infty F(\omega(s))ds$ is a martingale (with respect to P^K_∞).

We will defer the uniqueness argument since it will be a special case of the uniqueness argument in the next section. Similarly the proof that the process is a.s. continuous is a special case of Theorem 7.3.1. □

Remarks.

1. The limiting process can also be identified with a diffusion on the simplex $\{(x_1,\ldots,x_K):x_i\geq 0,\ i=1,\ldots,K;\ \Sigma x_i =1\}$. This is called the *Wright-Fisher diffusion* (with K alleles).

2. Note however that we cannot apply the basic uniqueness result of Stroock and Varadhan (1979) for finite dimensional diffusions since the coefficients are degenerate.

5.4 The Measure-Valued Diffusion Limit

Let $A:D(A)\ (\subset C(E)) \to C(E)$ generate a Feller semigroup $\{S_t\}$ on $C(E)$ given by a transition function $P(t,x,\Gamma)$, that is,

$$S_t f(x) = \int_E f(y)P(t,x,dy)$$

The operator A is called the *mutation operator* and $\{S_t\}$ is called the *mutation semigroup* for the process. As in Ch. 2 we extend $\{S_t\}$ to $\bigcup_m \mathcal{B}(E^m)$ by defining

$$S_t^m f(x_1,\ldots,x_m) := \int_E \cdots \int_E f(y_1,\ldots,y_m)P(t,x_1,dy_1)\ldots P(t,x_m,dy_m).$$

As in Section 2.2 let D_0 be a separating algebra of functions in D(A) satisfying $S_t:D_0 \to D_0$, $A^{(m)}$ denote the generator of $\{S_t^m\}$ and D_0^m also be defined as in Section 2.2.

We assume that there exists $\{E_K:K\in\mathbb{N}\}$ where $E_K = \{e_1,\ldots,e_K\} \subset E$ and $\{A_K:K\in\mathbb{N}\}$ which generate Feller semigroups on $C(E_K)$ such that $A_K \to A$ in the following sense:

$$Af(x)|_{E_K} = A_K f(x) + o(K)R_K(f), \quad x\in E_K,\ f\in D_0,$$

and $|R_K(f)|$ is bounded for each $f\in D_0$.

Now consider the martingale problem associated with Fleming-Viot generator

$$(5.4.1)\quad \mathfrak{G}F(\mu) = \int_E \left(A \frac{\delta F(\mu)}{\delta\mu(x)}\right)\mu(dx) + \frac{\gamma}{2}\int_E\int_E \frac{\delta^2 F(\mu)}{\delta\mu(x)\delta\mu(y)}\, Q_S(\mu;dx,dy),\ \mu\in M_1(E)$$

where $Q_S(\mu;dx,dy) := \mu(dx)\delta_x(dy) - \mu(dx)\mu(dy)$.

Let $\mathfrak{D}_0(\mathfrak{G})$ denote the class of functions $F_{f,n}(\mu)$ with $f\in D_0^n$, (cf. section 2.5 for the notation). Then

$$(5.4.2)\quad \mathfrak{G}F_{f,n}(\mu) = \langle\mu^n,A^{(n)}f\rangle + \frac{\gamma}{2}\sum_{\substack{i,j\\ i\neq j}}\left(\langle\mu^{n-1},\tilde{\Theta}_{ij}f\rangle - \langle\mu^n,f\rangle\right)$$

where $\tilde{\Theta}_{ij}:C(E^n)\to C(E^{n-1})$ is defined as in Remark 2.5.2(b). The operator $\mathfrak{G}_K$ will be defined as in (5.4.1) but with E replaced by E_K, A replaced by A_K and $M_1(E)$ replaced by $M_1(E_K)$.

<u>Theorem 5.4.1.</u> The processes X_K with laws $\{P_\mu^K\}$ (given in Theorem 5.3.1) converge weakly to a process with values in $M_1(E)$ which is the unique solution of the martingale problem for $(\mathfrak{G},\mathfrak{D}_0(\mathfrak{G}))$. (The limiting process coincides with the $S(A,\gamma)$-Fleming-Viot process on E.)

<u>Proof.</u> The proof of the weak convergence and the verification that each limit point is a solution for the martingale problem for $(\mathfrak{G},\mathfrak{D}(\mathfrak{G}))$ are essentially the same as that in Theorem 5.3.1. Let P_μ be a solution to the martingale problem. Then the uniqueness of the marginal distribution $\Pi_t P_\mu$ will be proved by establishing that the moment measures are uniquely determined (and in turn this implies that the mar-

tingale problem is well-posed by Theorem 5.1.2).

From the martingale problem applied to functions of class $\mathcal{B}_{pro}(M_1(E))$ we deduce the following equations for the moment measures $M_n(t;dx_1,\ldots,dx_n) := P_\mu(X_t(dx_1)\ldots X_t(dx_n))$:

$$\int\ldots\int \prod_{i=1}^{n} \phi_i(x_i)\; M_n(t;dx_1,\ldots,dx_n)$$

$$= e^{-\gamma n(n-1)t/2}\int\ldots\int\left(\prod_{i=1}^{n} S_t\phi_i(x_i)\right)\mu(dx_1)\ldots\mu(dx_n)$$

$$+ \frac{1}{2}\gamma \sum_{\substack{i,j\\ i\neq j}} \int_0^t ds\Big\{e^{-\gamma n(n-1)(t-s)/2}\int\ldots\int(S_{t-s}\phi_i S_{t-s}\phi_j)(x_j) \prod_{\ell\neq i,j}\left(S_{t-s}\phi_\ell(x_\ell)\right)$$

$$M_{n-1}(s;dx_1,..,dx_j,..,dx_{i-1},dx_{i+1}..,dx_{n-1})\Big\}$$

But this system of equations is just the evolution form of the system (2.8.1) thus identifying the finite dimensional distributions to be those given in Proposition 2.8.1. □

5.5. The Method of Duality

The method of duality has been widely applied in the study of infinite particle systems (cf. Liggett (1985)). In particular it was used to determine the ergodic behavior of the voter model by Holley and Liggett (1975). It was first applied to the diffusions arising in population genetics by Shiga (1980, 1981). Function-valued duals for measure-valued processes were introduced in Dawson and Hochberg (1982), Dawson and Kurtz (1982) and further developed in Ethier and Kurtz (1987, 1990).

In section 2.8 we introduced the idea of the duality between two evolution families and illustrated it by the relation between the Moran semigroup and the Fleming-Viot semigroup. In this section we give a more systematic development of this idea. In Chapter 10 we will apply it to construct the Fleming-Viot process with recombination and selection - a situation in which the moment measure equations are not closed and therefore the method of Ch. 2 cannot be used.

In this section we assume that E_1 is a Polish space, E_2 is a separable metric space and we consider time inhomogeneous martingale problems for E_1 and E_2-valued processes associated with $(\mathcal{D}(\mathfrak{G}_1),\mathfrak{G}_1)$ and $(\mathcal{D}(\mathfrak{G}_2),\mathfrak{G}_2)$, respectively.

Let $f\in b\mathcal{B}(E_1\times E_2)$. Let $\mathcal{D}(\mathfrak{G}_1) \subset b\mathcal{E}_1$, $\mathfrak{G}_1:[0,\infty)\times\mathcal{D}(\mathfrak{G}_1)\longrightarrow b\mathcal{E}_1$, $\mathcal{D}(\mathfrak{G}_2) \subset b\mathcal{E}_2$,

$\mathfrak{G}_2:[0,\infty)\times\mathfrak{D}(\mathfrak{G}_2)\longrightarrow\mathcal{E}_2$. Assume that $f(.,y)\in\mathfrak{D}(\mathfrak{G}_1)$ $\forall$ $y\in E_2$, $f(x,.)\in\mathfrak{D}(\mathfrak{G}_2)$ $\forall$ $x\in E_1$. For $s\geq 0$, let $g(s;x,y) := (\mathfrak{G}_1 f)(s;x,y)$ and $h(s;x,y) := (\mathfrak{G}_2 f)(s;x,y)$.

The t_1-initial value martingale problems for $(\mathfrak{G}_1,\{f(.,y),\ y\in E_2\})$ and $(\mathfrak{G}_2,\{f(x,.),x\in E_1\})$ are said to be *dual* for (μ_1,μ_2) with $\mu_i\in M_1(E_i)$, i=1,2 if for each solution $P_{\mu_1}\in M_1(D_{E_1,t_1})$ of the martingale problem for $\mathfrak{G}_1$ and each solution $Q_{\mu_2}\in M_1(D_{E_2,t_1})$ for $\mathfrak{G}_2$, and $t_2\geq t_1$,

$$\int_{E_2} P_{\mu_1}\Big[f(X(t_2),y)\Big]\mu_2(dy) = \int_{E_1} Q_{\mu_2}\Big[f(x,Y(t_2))\Big]\mu_1(dx)$$

where X and Y denote the respective coordinate processes in $D([0,\infty),E_1)$ and $D([0,\infty),E_2)$.

The martingale problem for $\mathfrak{G}_2$ is said to provide a *FK-dual representation* (FK is an abbreviation for Feynman-Kac) for the t_1-initial value martingale problem for $\mathfrak{G}_1$ for the pair (μ_1,μ_2) if there exists a function $\beta\in\mathcal{B}(\mathbb{R}_+)\otimes\mathcal{E}_2$ and a solution Q_{t_1,μ_2} for the t_1-initial value martingale problem for $\mathfrak{G}_2$ satisfying

$$\int_{E_1} Q_{t_1,\mu_2}\Big[|f(x,Y(t_2))|\exp\Big\{\int_{t_1}^{t_2}\beta(s,Y(s))ds\Big\}\Big]\mu_1(dx) < \infty,\quad t_2\geq t_1,$$

such that for <u>each</u> solution P_{t_1,μ_1} of the t_1-initial value martingale problem for $\mathfrak{G}_1$

$$\int_{E_2} P_{t_1,\mu_1}\Big[f(X(t_2),y)\Big]\mu_2(dy) = \int Q_{t_1,\mu_2}\Big[f(x,Y(t_2))\exp\Big\{\int_{t_1}^{t_2}\beta(s,Y(s))ds\Big\}\Big]\mu_1(dx).$$

The following result transforms the uniqueness problem for $\mathfrak{G}_1$ into the existence problem for $\mathfrak{G}_2$.

<u>Lemma 5.5.1.</u> Let $M_0(E_1)\subset M_1(E_1)$. Assume that the family $\{f(.,y):y\in E_2\}$ is $M_0(E_1)$-determining and $\forall y\in E_2$ $g(.,y)$ is $\sigma(\{f(.,y):y\in E_2\})$-measurable. Suppose that $\mathfrak{G}_2$ provides a FK-dual representation for the restricted t_1-initial value martingale problem for $(\{f(.,y):y\in E_2\},\mathfrak{G}_1,M_0(E))$ for every pair (μ,δ_y) with $\mu\in M_0(E_1)$, $y\in E_2$. Then for each $\mu\in M_0(E_1)$ there is a unique solution $P_{t_1,\mu}$ to the t_1-initial value martingale problem for $\mathfrak{G}_1$ satisfying $\Pi_{t_1}P_{t_1,\mu}=\mu$.

<u>Proof.</u> Let Q_{t_1,δ_y} be a solution of the t_1-initial value martingale problem for $(\mathfrak{G}_2,\delta_y)$. If $\mu\in M_0(E_1)$ and $P_{t_1,\mu}$ and $\tilde{P}_{t_1,\mu}$ are solutions of the retricted t_1-initial

value martingale problem for $(\mathfrak{G}_1,\mu)$, then by assumption

$$P_{t_1,\mu}[f(X(t_2),y)] = \int_{E_1} Q_{t_1,\delta_y}\left[f(x,Y(t_2))\exp\left\{\int_{t_1}^{t_2}\beta(s,Y(s))ds\right\}\right]\mu(dx)$$

$$= \tilde{P}_{t_1,\mu}[f(X(t_2),y)].$$

Since $\{f(.,y):y\in E_2\}$ is $M_0(E_1)$-determining, then $P_{t_1,\mu}(X(t_2)\in B)=\tilde{P}_{t_1,\mu}(X(t_2)\in B)$ $\forall B\in\mathcal{B}(E_1)$. The result now follows from Theorem 5.1.2. □

The following result provides a criterion for verifying the condition of Lemma 5.5.1.

Theorem 5.5.2. Let $\mu\in M_1(E_1)$, $P_{t_1,\mu} \in M_1((D([t_1,\infty),E_1),(\mathcal{D}^1_{t_1,t})_{t\ge t_1})$ and $\Pi_{t_1}P_{t_1,\mu}=\mu$. Let $\{Q_{t_1,\delta_y}:y\in E_2\}\subset M_1(D([t_1,\infty),E_2),(\mathcal{D}^2_{t_1,t})_{t\ge t_1})$ (with $\Pi_{t_1}Q_{t_1,\delta_y}=\delta_y$).

Let $\sigma\ge t_1$ be a $\{\mathcal{D}^2_{t_1,t}\}$-stopping time. Let $f\in \mathcal{E}_1\otimes\mathcal{E}_2$, $g,h\in \mathcal{B}(\mathbb{R}_+)\otimes\mathcal{E}_1\otimes\mathcal{E}_2$, $\beta\in\mathcal{B}(\mathbb{R}_+)\otimes\mathcal{E}_2$. Suppose that for each $T>t_1$ there exist a random variable Γ_T which is $P_{t_1,\mu}\otimes Q_{t_1,\delta_y}$-integrable for all $y\in E_2$, and a constant C_T such that

$$\sup_{\substack{t_1\le s\le T\\ t_1\le r\le\sigma\wedge T}} (\max|f(X(s),Y(r))|,|g(s;X(s),Y(r)|) \le \Gamma_T,\quad P_{t_1,\mu}\otimes Q_{t_1,\delta_y}\text{-a.s.}\forall y\in E_2,$$

$$\sup_{t_1\le u\le\sigma\wedge T}|\beta(u;Y(u))| \le C_T,\quad Q_{t_1,\delta_y}\text{-a.s., }\forall y\in E_2.$$

Assume that

(i) $f(X(t),y)-\int_{t_1}^{t} g(s;X(s),y)ds$ is a $(P_{t_1,\mu},\{\mathcal{D}^1_{t_1,t}\}_{t\ge t_1})$-martingale for each $y\in E_2$,

and

(ii) $\forall y\in E_2$, $f(x,Y(t\wedge\sigma)) - \int_{t_1}^{t\wedge\sigma} h(s;x,Y(s))ds$ is a $(Q_{t_1,\delta_y},\{\mathcal{D}^2_{t_1,t}\}_{t\ge t_1})$-martingale for each $x\in E_1$, and

(iii) $g(s;x,y) = h(t-s;x,y)+\beta(t-s;y)f(x,y)\quad \forall\ t_1\le s\le t,\ x\in E_1,\ y\in E_2.$

Then for all $t_1\le t\le T$, $\forall y\in E_2$,

$$P_{t_1,\mu}[f(X(t),y)] - \int_{E_1}\mu(dx)\left(Q_{t_1,\delta_y}\left[f(x,Y(t\wedge\sigma))\exp\left\{\int_{t_1}^{t\wedge\sigma}\beta(u;Y(u))du\right\}\right]\right)$$

$$= P_{t_1,\mu}\otimes Q_{t_1,\delta_y}\left[\int_{t_1}^{(t_1+t-\sigma)^+} g(s;X(s),Y(\sigma))\exp\left\{\int_{t_1}^{\sigma}\beta(u;Y(u))du\right\} ds\right].$$

Proof. If for $s,t \ge t_1$, $F(s,t) := P_{t_1,\mu} \otimes Q_{t_1,\delta_y}\left\{f(X(s),Y(t\wedge\sigma))\exp\left(\int_{t_1}^{t\wedge\sigma} \beta(v;Y(v))dv\right)\right\}$,

then from the hypotheses (i) and (ii), it can be shown that

$$F(s,t) = F(t_1,t) + P_{t_1,\mu} \otimes Q_{t_1,\delta_y}\left\{\int_{t_1}^{s}\left[g(u;X(u),Y(t\wedge\sigma))\exp\left(\int_{t_1}^{t\wedge\sigma} \beta(v;Y(v))dv\right)\right]du\right\}.$$

$$F(s,t) = F(s,t_1)$$

$$+ P_{t_1,\mu} \otimes Q_{t_1,\delta_y}\left\{\int_{t_1}^{t}\left[(h(u;X(s),Y(u))+\beta(u;Y(u)))1(u\le\sigma)\exp\left(\int_{t_1}^{u} \beta(v;Y(v))dv\right)\right]du\right\}.$$

Under the above hypotheses, it can verified that F is absolutely continuous in s for each t and absolutely continuous in t for each s. (Refer to Dawson and Kurtz (1982) or [EK p. 193] for the details.)

Let $F_1(s,t) := \partial F(s,t)/\partial s$, $F_2(s,t) := \partial F(s,t)/\partial t$. Then

$$F_1(s,t) = P_{t_1,\mu} \otimes Q_{t_1,\delta_y}\left[g(s;X(s),Y(t\wedge\sigma))\exp\left(\int_{t_1}^{t\wedge\sigma} \beta(v;Y(v))dv\right)\right], \text{ and}$$

$$F_2(s,t) = P_{t_1,\mu} \otimes Q_{t_1,\delta_y}\left[[h(t;X(s),Y(t\wedge\sigma))+\beta(t;Y(t))]1(t\le\sigma)\exp\left(\int_{t_1}^{t} \beta(v;Y(v))dv\right)\right].$$

Then it also follows from the above hypotheses that $\int_{t_1}^{T}\int_{t_1}^{T} |F_i(s,t)|dsdt < \infty$.

Let us next verify that

$$F(t,t_1)-F(t_1,t) = \int_{t_1}^{t} [F_1(s,t_1+t-s)-F_2(s,t_1+t-s)]ds.$$

Note that $\int_{t_1}^{\tau}\left(\int_{t_1}^{t} [F_1(s,t_1+t-s)-F_2(s,t_1+t-s)]ds\right)dt$

$$= \int_{t_1}^{\tau}\int_{s}^{\tau} F_1(t_1+t-s,s)dtds - \int_{t_1}^{\tau}\int_{s}^{\tau} F_2(s,t_1+t-s)dtds$$

$$= \int_{t_1}^{\tau} [F(t_1+\tau-s,s)-F(t_1,s)]ds - \int_{t_1}^{\tau} [F(s,t_1+\tau-s)-F(s,t_1)]ds$$

$$= \int_{t_1}^{\tau} [F(s,t_1)-F(t_1,s)]ds, \quad t_1\le\tau\le T.$$

Differentiating with respect to τ we get

$$F(\tau,t_1)-F(t_1,\tau) = \int_{t_1}^{\tau} [F_1(s,t_1+\tau-s)-F_2(s,t_1+\tau-s)]ds$$

for a.e. τ in $[t_1,T]$. The result for all $\tau\in[t_1,T]$ then follows by the continuity of $F(\tau,t_1)$ and $F(t_1,\tau)$.

Therefore

$$F(t,t_1)-F(t_1,t)$$

$$= P_{t_1,\mu}\Big\{f(X(t),y)\Big\} - \int_{E_1}\mu(dx)\Big(Q_{t_1,\delta_y}\Big\{f(x,Y(t\wedge\sigma))\exp\Big(\int_{t_1}^{t\wedge\sigma}\beta(v;Y(v))dv\Big)\Big\}\Big)$$

$$= \int_{t_1}^{t}[F_1(s,t_1+t-s)-F_2(s,t_1+t-s)]ds$$

$$= \int_{t_1}^{t}\Big\{P_{t_1,\mu}\otimes Q_{t_1,\delta_y}\Big[g(s;X(s),Y((t_1+t-s)\wedge\sigma))\exp\Big(\int_{t_1}^{(t_1+t-s)\wedge\sigma}\beta(v;Y(v))dv\Big)\Big]$$

$$- P_{t_1,\mu}\otimes Q_{t_1,\delta_y}\Big[[h(t_1+t-s;X(s),Y(t_1+t-s))+\beta(t_1+t-s;Y(t_1+t-s))]1(t_1+t-s\le\sigma)$$

$$\cdot\exp\Big(\int_{t_1}^{t_1+t-s}\beta(v;Y(v))dv\Big)\Big]\Big\}ds$$

$$= P_{t_1,\mu}\otimes Q_{t_1,\delta_y}\Big[\int_{t_1}^{(t_1+t-\sigma)^+} g(s;X(s),Y(\sigma))\exp\Big(\int_{t_1}^{\sigma}\beta(v;Y(v))dv\Big)ds\Big] \quad \text{(by Fubini).} \ \square$$

Corollary 5.5.3. Consider the time-homogeneous case $g(s;x,y) \equiv g(x,y)$, $h(s;x,y) \equiv h(x,y)$, $\beta(s;y) \equiv \beta(y)$ $\forall s$. In addition to the hypotheses of Theorem 5.5.2 assume that

i) $f \in bC(E_1\times E_2)$, and $\{f(.,y):y\in E_2\}$ is measure-determining on E_1,

ii) there exist stopping times $\sigma_K\uparrow t$ such that

$$\Big\{ (1 + \sup_x |g(x,Y(\sigma_K))|)\cdot\exp\Big(\int_0^{\sigma_K}|\beta(Y(u))|du\Big)\Big\}_K$$

are Q_{δ_y}-uniformly integrable $\forall$ $y\in E_2$, and

iii) $Q_{\delta_y}(Y(s-)\neq Y(s)) = 0$ for each $s\ge0$.

Then the $(\mathfrak{G}_1,\{f(.,y):y\in E_2\})$ initial value martingale problem is well-posed and

$$P_\mu[f(X(t),y)] = \int_{E_1}\mu(dx)\Big(Q_{\delta_y}\Big[f(x,Y(t))\exp\Big(\int_0^t\beta(Y(u))du\Big)\Big]\Big).$$

Proof. By Theorem 5.5.2

$$P_\mu[f(X(t),y)] - \int_{E_1} \mu(dx)\left(Q_{\delta_y}\left[f(x,Y(t\wedge\sigma_K))\exp\left\{\int_0^{t\wedge\sigma_K} \beta(Y(u))du\right\}\right]\right)$$

$$= P_\mu \otimes Q_{\delta_y}\left[\int_0^{(t-\sigma_K)^+} g(X(s),Y(\sigma_K))\exp\left\{\int_0^{\sigma_K} \beta(Y(u))du\right\} ds\right].$$

By (ii)

$$\lim_{K\to\infty} P_\mu \otimes Q_{\delta_y}\left[\int_0^{(t-\sigma_K)^+} |g(X(s),Y(\sigma_K))|\exp\left\{\int_0^{\sigma_K} \beta(Y(u))du\right\} ds\right]$$

$$\le \textit{const}\cdot \lim_{K\to\infty} Q_{\delta_y}\left[|t-\sigma_K| \sup_x |g(x,Y(\sigma_K))| \exp\left\{\int_0^{\sigma_K} |\beta(Y(u))|du\right\}\right]$$

$$= 0.$$

Moreover, by (iii)

$$f(x,Y(t\wedge\sigma_K))\exp\left\{\int_0^{t\wedge\sigma_K}\beta(Y(u))du\right\} \to f(x,Y(t))\exp\left\{\int_0^t \beta(Y(u))du\right\} \text{ as } K\to\infty,$$

Q_{δ_y}-a.s. and by (i),

$$\left|f(x,Y(t\wedge\sigma_K))\exp\left\{\int_0^{t\wedge\sigma_K}\beta(Y(u))du\right\}\right| \le \textit{const}\ \exp\left\{\int_0^t |\beta(Y(u))|du\right\}.$$

The result follows by dominated convergence. □

5.6. A Function-valued Dual to the Fleming-Viot Process.

Let E be a compact metric space. In this section we introduce a *function-valued dual* to the Fleming-Viot $M_1(E)$-valued process. The dual process will take values in the algebra of functions

$$C_{dir}(E^{\mathbb{N}}) = \sum_n C(E^n) \quad \text{(direct sum).}$$

Functions in $C_{dir}(E^{\mathbb{N}})$ are denoted by $f = (f_1, f_2, \ldots)$. A function $f \in C_{dir}(E^{\mathbb{N}})$ is said to be *simple* and $\#(f) = n$ if $f_n \ne 0$, and $f_m = 0$ for all $m\ne n$.

The set of simple functions is denoted by $C_{sim}(E^{\mathbb{N}})$.
Define $F: C_{dir}(E^{\mathbb{N}}) \to C(M_1(E))$ by

$$F(f,\mu) := \int_{E^{\mathbb{N}}} F(x)\mu^\infty(dx), \quad \mu^\infty = \prod_n \mu_n \text{ with } \mu_n = \mu\ \forall n$$

and note that $C_p(M_1(E))$ is the range of F.

Remark. We can also introduce appropriate normed spaces containing $C_{dir}(E^{\mathbb{N}})$, for example, we can consider the norm $\|f\| := \sum \|f_n\|_n < \infty$ where $\|f_n\|_n$ is the sup norm in $C(E^n)$. In addition certain Banach subspaces of $C(M_1(E))$ are required for the study of some processes related to the Fleming-Viot process (see Dawson and March (1992)).

Let $\{T_t\}$ be a Feller semigroup $\{T_t\}$ on $C(M_1(E))$ when $C(M_1(E))$ is given the supremum norm. If $T_tF(f,\mu)$ can be represented by $F(\hat{T}_tf,\mu)$, where $\{\hat{T}_t\}$ is a Markov semigroup on $C_{dir}(E^{\mathbb{N}})$, then the resulting $C_{dir}(E^{\mathbb{N}})$-valued process is called a *function-valued dual.* At least at the formal level a Markov semigroup on $C(E^{\mathbb{N}})$ can be identified with an infinite particle system in E. In Chapter 10, this idea will be developed for the Fleming-Viot process.

Now consider the Fleming-Viot process with generator $(C_P^A(M_1(E)),\mathfrak{G})$ where $C_P^A(M_1(E)) := \{F(f,\mu) \in C_P(M_1(E)),\ f\in D_0^n\}$ (cf. Sect. 5.4). If f is simple and $\#(f)=n$, we rewrite $F(f,\mu)$ as $F_{f,n}(\mu)$. For functions of this class the Fleming-Viot generator has the form:

$$\mathfrak{G}F_{f,n}(\mu) = F_{A^{(n)}f}(\mu) + \frac{\gamma}{2}\sum_{j=1}^{N}\sum_{k\neq j}[F_{\tilde{\Theta}_{jk}f,n-1}(\mu)-F_{f,n}(\mu)].$$

Now consider

$$KF(f) = A^{(\#(f))}f\cdot F'(f) + \frac{\gamma}{2}\sum_{\substack{j,k=1\\ j\neq k}}^{\#(f)}[F(\tilde{\Theta}_{jk}f)-F(f)]$$

where $\tilde{\Theta}_{jk}:D_0^n \to D_0^{n-1}$ is defined as in Remark 2.5.2b. K is the generator of a càdlàg Markov process $\{Y(t)\}$ with values in $C_{sim}(E^{\mathbb{N}})$ and law $\{Q_f:f \in C_{sim}(E^{\mathbb{N}})\}$ which evolves as follows:

(i) $Y(t)$ jumps from $C(E^n)$ to $C(E^{n-1})$ at rate $\frac{1}{2}\gamma n(n-1)$

(ii) at the time of a jump, f is replaced by $\tilde{\Theta}_{jk}f$

(iii) between jumps, $Y(t)$ is deterministic on $C(E^n)$ and evolves according to the semigroup $\{S_t^n\}$ with generator $A^{(n)}$.

Theorem 5.6.1 (a) Let $(\{X(t)\}_{t\geq 0},\{\tilde{P}_\nu:\nu\in M_1(M_1(E))\})$ be a solution to the Fleming-Viot martingale problem and the process $(\{Y(t)\}_{y\geq 0},\{Q_f:f\in C_{sim}(E^{\mathbb{N}})\})$ be defined as above with $Q_f = Q_{\delta_f}$.

Then

(a) these processes are dual, that is,

$$\tilde{P}_\nu(F(f,X(t))) = \int_{M_1(M_1(E))} Q_f\{F(Y(t),\mu)\}\nu(d\mu) \quad \forall \nu\in M_1(M_1(E)),\ f\in C_{sim}(E^{\mathbb{N}}),$$

and

(b) the Fleming-Viot martingale problem is well posed and $\{X(t)\}_{t\geq 0}$ is a strong Markov process with transition measures $P_\mu = \tilde{P}_{\delta_\mu}$, $\mu\in M_1(E)$.

Proof. In this case the function β is identically zero and for $f\in C_{sim}(E^{\mathbb{N}})$, $\#(f)=n$,

$$\mathfrak{G}F(f,\mu) = F(A^{(n)}f,\mu) + \frac{\gamma}{2}\sum_{j=1}^{n}\sum_{k\neq j}[F(\tilde{\Theta}_{jk}f,\mu)-F(f,\mu)]$$

$$= KF(f,\mu)$$

(a) then follows from Theorem 5.5.2. (b) then follows from Theorems 5.1.2 and 5.1.3. To apply Theorem 5.1.3 we note that we can choose a countable subset of $\tilde{C}\subset C_{sim}(E^{\mathbb{N}})$ such that $\{F(f,.):f\in\tilde{C}\}$ is $M_1(M_1(E))$-determining. □

Remarks. This argument can be modified to cover the case in which we do not have an algebra D_0 as assumed in Sect. 5.4 - see Ethier and Kurtz (1987).

The dual representation gives an alternative proof that the Fleming-Viot martingale problem is well-posed. However we have already done this using the fact that the moment equations are closed and relatively simple to solve. The real power of the dual process is exhibited when the moment equations are either much more complex or they are not closed. We will illustrate this in the next section and in Chap. 10.

In addition to proving that the martingale problem is well-posed the dual process has other useful applications. To illustrate one of these we will now use it to prove an ergodic theorem for the Fleming-Viot process with ergodic mutation semigroup.

Theorem 5.6.2 Let $\{X_t\}$ be a Fleming-Viot process with mutation semigroup $\{S_t\}$ on C(E) and Y(t) be the dual process defined above. Suppose there exists $\pi\in M_1(E)$ such that

$$\lim_{t\to\infty} S_t f(x) = \langle\pi,f\rangle \quad \forall x \text{ and } f\in C(E).$$

Then for $f\in C_{sim}(E^{\mathbb{N}})$, $X(0)=\mu$,

$$\lim_{t\to\infty} P_\mu[\langle X(t)^\infty,f\rangle] = Q_f[\langle\pi,Y(\tau_1)\rangle]$$

where $\tau_1 := \inf\{t:\#(Y(t)=1\}$.

Proof. If $\#(Y(0)) = n > 1$, then after $(n-1)$ jumps $\#(Y(t))$ absorbs at $Y(t)=1$.

Therefore

$$\lim_{t\to\infty} P_\mu[\langle X(t)^\infty,f\rangle] = \lim_{t\to\infty} E_f[\langle\mu,S(t-\tau_1)Y(\tau_1)\rangle]$$

$$= Q_f\langle\mu,\langle\pi,Y(\tau_1)\rangle\rangle = Q_f\langle\pi,Y(\tau_1)\rangle. \square$$

5.7. Examples and Generalizations of the Fleming-Viot Process

A number of variations of the basic Fleming-Viot model arise in the study of measure-valued processes and certain models in population genetics. In this section we will briefly review some of these.

5.7.1 Time Inhomogeneous Fleming-Viot.

Dynkin (1989c) introduced a Fleming-Viot process with time inhomogeneous mutation semigroup and Perkins (1991b) introduced a Fleming-Viot process in which the sampling rate varies in time and may be unbounded. We will briefly indicate the resulting Fleming-Viot process incorporating both types of time inhomogeneity.

Let $C_+ = \{\gamma:[0,\infty) \longrightarrow [0,\infty)$: γ cont., $\exists\ \tau_\gamma \in (0,\infty]$ such that $\gamma(t) > 0$ if $t \in [0,\tau_\gamma)$ and $\gamma(t) = 0$ if $t \ge \tau_\gamma\}$. C_+ is given the compact open topology. If $B \subset C_+$, let $B|_{T-} = \{\gamma|_{[0,T)}:\gamma\in B\}$ and $B|_T = \{\gamma(s)|_{[0,T]}:\gamma\in B\}$. Let $\{S_{r,t}\}$ be a strongly continuous evolution family on $C(E)$ with generator A_t and domain $D(A)$ (as in Prop. 2.8.4). (For example in Chapter 8 we will consider the case $A(t) = \sigma(t)A$ where $\sigma \in pC(\mathbb{R}_+)$.)

Theorem 5.7.1 Assume that $1/\gamma \in C_+$.

(a) There exists a unique solution $\{P_\mu^{S(A,\gamma)}:\mu\in M_1(E)\}$, to the initial value martingale problem: for each $\phi \in D(A)$,

$$M_t^{S,\gamma}(\phi) = \langle X_t,\phi\rangle - \langle\mu,\phi\rangle - \int_0^t \langle X_s,A_s\phi\rangle ds, \quad t < \tau_\gamma,$$

is an $(\mathcal{D}_t)$-martingale starting at 0 such that

$$\langle M^{S,\gamma}(\phi)\rangle_t = \int_0^t (\langle X_s,\phi^2\rangle - \langle X_s,\phi\rangle^2)\gamma(s)ds \quad \forall\ t < \tau_\gamma,$$

$$M_t^{S,\gamma}(\phi) = M_{\tau_\gamma -}^{S,\gamma}(\phi) \text{ for } t\ge\tau_\gamma,$$

and $X_0 = \mu$ and $X_t \equiv X_{\tau_\gamma}$ for all $t\ge\tau_\gamma$.

(b) Let $A_t=A$ be the generator of a Feller semigroup and $(\mu_n,\gamma_n|_{[0,T)}) \longrightarrow (\mu,\gamma|_{[0,T)})$ as $n\to\infty$ in $M_1(E)\times C_+|_{T-}$ $\forall\ T \le \tau_\gamma$. Then

$$P_{\mu_n}^{S(A,\gamma_n)}\Big|_{C([0,T],M_1(E))} \Longrightarrow P_\mu^{S(A,\gamma)}\Big|_{C([0,T],M_1(E))} \quad \text{as } n \to \infty \quad \forall\ T < \tau_\gamma.$$

(c) Let $\mathcal{P}_\gamma$ denote the transition function associated with the solution of the martingale problem in (a). Then

$$\int \mathcal{P}_\gamma(r,\mu;t,d\nu)\langle\nu^k,h\rangle = \langle\mu^k S_k(r,t),h\rangle + \sum_{j=1}^{k-1}\int_r^t ds\langle\mu^j S_j(r,s),U_j^k(s,t)h\rangle$$

where for $h\in \mathcal{B}_{sym}(E^k)$, $t<t_\gamma$

$$(S_k(r,t)h)(x_1,\dots,x_k)$$
$$:= \int e_k(r,t)p(r,x_1;t,dy_1)\dots p(r,x_k;t,dy_k)h(y_1,\dots,y_k)$$

and for $\nu \in M_1(E^k)$

$$(\nu S_k(r,t))(dy_1,\dots,dy_k) =$$
$$\int e_k(r,t)\nu(dx_1,\dots,dx_k)p(r,x_1;t,dy_1)\dots p(r,x_k;t,dy_k),$$

$$e_n(r,t) := \exp\left(-\tfrac{1}{2}\, n(n-1)\int_r^t \gamma(s)ds\right)$$

Ψ_k^{k-1} is as in Prop. 2.8.1

and
$$U_{k-1}^k(r,t) := \tfrac{1}{2}k\gamma(r)\Psi_k^{k-1}S_k(r,t) \text{ for } r<t$$
$$:= 0 \quad \text{for } r \ge t$$
$$U_j^k(r,t) = \int U_j^{k-1}(r,s)U_{k-1}^k(s,t)ds\ , \text{ for } k-j > 1.$$

<u>Outline of Proof.</u> (a) (cf. Perkins (1991b)) The proof of existence and uniqueness on the interval $[0,T]$ with $T < \tau_\gamma$ is similar to that for the time homogeneous case treated above. The main difference appears in the proof of uniqueness of the law of X_t for $0\le t\le T$. To accomplish this in this case we must use a time inhomogeneous function-valued dual process with generator $\{K_s: 0\le s\le t\}$ given by:

$$K_s F(f) = A_{t-s}^{(\#(f))}f\cdot F'(f) + \frac{\gamma(t-s)}{2}\sum_{\substack{j,k=1\\ j\ne k}}^{\#(f)} [F(\tilde{\Theta}_{jk}f)-F(f)].$$

(b) The tightness of $\left\{P_{\mu_n}^{S(A,\gamma_n)}\Big|_{C([0,T],M_1(E))} : n\in\mathbb{N}\right\}$ for $T<\tau_\gamma$ is proved by arguments similar to those used earlier and based on Theorem 3.6.6.

The weak convergence of the corresponding inhomogeneous function-valued dual processes can be proved using their direct contruction. The weak convergence of the finite dimensional distributions of $P^{S(A,\gamma_n)}_{\mu_n}$ is then obtained by using the Markov property and the dual representation.

(c) This follows by the same arguments as in the proof of Prop. 2.8.1. □

Remark. We will refer to the process considered in (a) as the $S(A, \gamma(.))$- Fleming-Viot process. This process will be used in Chapter 8.

5.7.2. The infinitely many neutral alleles model.

Let $E = [0,1]$ and $(Qf)(x) := \theta\int_0^1 (f(y)-f(x))dy,\ x\in E,\ \theta>0.$

This model satisfies the partition property, namely, given any finite partition $E = \bigcup_{j=1}^{K} E_j$, then $\{X(t,E_j): j=1,\dots,K\}$ is a finite dimensional Markov diffusion process as in Section 5.3. For a detailed discussion see Ethier and Kurtz (1990b, 1992b).

5.7.3. The Size Ordered Atom Process

Let $\mathfrak{A}_K := \{z=(z_1,\dots,z_K): z_1\ge 0,\dots,z_K\ge 0,\ \sum_{i=1}^{K} z_i \le 1\}$,

$\sigma_i : C([0,\infty),\mathfrak{A}_K)\longrightarrow [0,\infty],\quad \sigma_i(z) := \inf\{t\ge 0: z_i(t)=0\}.$

Define G_K on $\mathcal{D}(G_K) = C^2(\mathfrak{A}_K)$ by

$$G_K := \frac{\gamma}{2}\sum_{i,j=1}^{K} z_i(\delta_{ij}-z_j)\frac{\partial^2}{\partial z_i \partial z_j} + \frac{1}{2}\frac{K\theta}{(K-1)}\sum_{i=1}^{K}\Big\{\frac{1}{K} - z_i\Big\}\frac{\partial}{\partial z_i},\quad \theta>0,$$

$D_K = \{z\in C([0,\infty),\mathfrak{A}_K): \sigma_1(z)\ge\sigma_2(z)\ge\dots\ge\sigma_K(z)\}$.

For $K = 2$ the pair $\{z_1(t),z_2(t)\}$ can be identified with $(\max(z(t),1-z(t)), \min(z(t),1-z(t)))$ where $z(t)$ is the unique strong solution of the s.d.e.

$$dz(t) = \theta(1/2 - z(t))dt + \sqrt{\gamma z(t)(1-z(t))}\, dw(t).$$

This is called the *size-ordered atom process* for the *K-neutral-alleles diffusion*. In the limit of *infinitely many neutral alles* this becomes

$$G_\infty = \frac{1}{2}\sum_{i,j=1}^{\infty} z_i(\delta_{ij}-z_j)\frac{\partial^2}{\partial z_i \partial z_j} - \frac{1}{2}\theta\sum_{i=1}^{\infty} z_i\frac{\partial}{\partial z_i}.$$

(For a detailed discussion of this process see Ethier and Kurtz (1981).)

5.7.4. Continuous allele stepwise mutation of Ohta and Kimura

$E = \bar{\mathbb{R}}^d$, the one point compactification of $\mathbb{R}^d$ and

$$Af(x) = \frac{1}{2}\Delta f(x), \quad x \in \mathbb{R}^d$$
$$= 0, \quad x = \infty.$$

This is the model originally studied by Fleming and Viot (1979).

5.7.5. Shimizu's gene conversion model

Shimizu (1990) has studied the following *n-loci* model:

$$E = (E_0)^n,$$
$$Af(x_1,\ldots,x_n) = \sum_{j_1 \neq j_2} \{\beta_{j_1 j_2} f(x_1,\ldots,x_n) - f(x_1,\ldots,x_n)\}$$

where $\beta_{j_1 j_2}$ denotes the replacement of the variable x_{j_1} by x_{j_2}.

5.7.6. Infinitely many neutral alleles with ages

Let $E = [0,1]\times[0,\infty]$. The first component denotes the allele and the second the age of the allele, that is the time from the last mutation.

$$(Af)(x,a) = \frac{\partial}{\partial a} f(x,a) + \frac{1}{2}\theta\int (f(\xi,0)-f(x,a))P(x,d\xi)$$

$D(A) = C^{0,1}([0,1]\times[0,\infty])$, the space of functions continuous in the first variable and continuously differentiable in the second.

$P(x,d\xi)$ denotes the distribution of the allele which results from a mutation from an allele of type x. In the case $P(x,d\xi) = d\xi$ we obtain the infinitely many neutral alleles with ages model. For a detailed discussion see Ethier (1988).

5.7.7. Infinitely many sites model

For this model which was introduced by Ethier and Griffiths (1987) $E = [0,1]^{\mathbb{Z}_+}$ and the mutation operator is

$$Af(x_1,x_2,\ldots) := \frac{1}{2}\theta\int (f(y,x_1,x_2,\ldots)-f(x_1,x_2,\ldots))\nu_0(dy), \quad \theta>0.$$

Here the type of a gene is given by $x = (x_0,x_1,\ldots) \in E$ such that $x_0, x_1,\ldots$ is the sequence of sites at which mutations have occurred in the line of descent of the gene in question. (x_0 is the most recent mutation).

5.7.8. Weighted Sampling

The basic random sampling mechanism of the Fleming-Viot model can be modified to allow for *weighted sampling*. Here $\theta(x,y) \in C_{sym}(E^2)$ (and $\theta(x,x) = 1$), denotes the rate at which an individual of type x is replaced by one of type y. In the setting of the K-type N-particle approximation of section 5.2 with mutation operator A, we obtain

$$\mathcal{G}^K_N F(\mu) = N\int_{E_K} \sum_{y\in E_K} q^K_{x,y}[F(\mu - \delta_x/N + \delta_y/N)-F(\mu)]\mu(dx)$$

$$+\ \gamma N(N-1)\int_{E_K}\int_{E_K} [F(\mu - \delta_x/N + \delta_y/N)-F(\mu)]\theta(x,y)\mu(dx)\mu(dy)$$

$$= \int\{A^K \frac{\delta F(\mu)}{\delta\mu(.)}\}(x)\mu(dx)$$

$$+ \frac{\gamma}{4}\iint\left[\frac{\delta^2F(\mu)}{\delta\mu(x)^2} + \frac{\delta^2F(\mu)}{\delta\mu(y)^2} - 2\frac{\delta^2F(\mu)}{\delta\mu(x)\delta\mu(y)}\right]\theta(x,y)\mu(dx)\mu(dy)$$

$$+ R_N(F)/N$$

Letting $N \to \infty$ and then $K \to \infty$, we obtain

$$\mathcal{G}F(\mu) = \int\left(A\,\frac{\delta F(\mu)}{\delta\mu(.)}\right)(x)\mu(dx)$$

$$+ \frac{\gamma}{4}\iint\left[\frac{\delta^2F(\mu)}{\delta\mu(x)^2} + \frac{\delta^2F(\mu)}{\delta\mu(y)^2} - 2\frac{\delta^2F(\mu)}{\delta\mu(x)\delta\mu(y)}\right]\theta(x,y)\mu(dx)\mu(dy)$$

$$= \int\left(A\,\frac{\delta F(\mu)}{\delta\mu(.)}\right)(x)\mu(dx)$$

$$+ \frac{\gamma}{2}\iint\frac{\delta^2F(\mu)}{\delta\mu(x)\delta\mu(y)}[\theta(x,z)\delta_x(dy)\mu(dx)\mu(dz) - \theta(x,y)\mu(dx)\mu(dy)] .$$

For functions in $C_P(M_1(E))$, this becomes

$$\mathcal{G}F_{f,N}(\mu)$$

$$= F_{A^{(N)}f,N}(\mu) + \frac{\gamma}{2}\sum_{\substack{k,j=1\\k\neq j}}^{N}[F_{\Psi_{jk}f,N}(\mu) - F_{f,N}(\mu)] + \frac{\gamma}{2}N(N-1)F_{f,N}(\mu)$$

where $\Psi_{jk}:C(E^N) \longrightarrow C(E^N)$ is defined by

$$(\Psi_{jk}f)(x_1,\dots,x_N) = \theta(x_j,x_k)(\Theta_{jk}f)(x_1,\dots,x_N) - \theta(x_j,x_k)f(x_1,\dots,x_N)$$

and Θ_{jk} is defined as in Section 2.5.

The dual process, $\{Y(t)\}$ then has jumps $f \longrightarrow \Psi_{jk}f$ at rate $\gamma\cdot\#(f)(\#(f)-1)/2$ and satisfies $\#(Y(t)) = \#(Y(0))\ \forall t\geq 0$. Between jumps it evolves deterministically according to the semigroup $f \longrightarrow S_t^{(\#(f))}f$.

The uniqueness to the martingale problem follows from the FK-dual representation with $\beta(Y(t)) := \frac{\gamma}{2}\#(Y(t))(\#(Y(t)-1)$. The case $\theta(x,y) \equiv 1$ reduces to the usual Fleming-Viot model.

5.7.9. Multilevel Sampling.

In example 6.2.4 we discuss multilevel branching systems. The analogous multilevel sampling systems were introduced in Dawson (1986b).

5.8. Some Other Classes of Measure-valued Processes with Function-Valued Duals.

In order to further clarify both the usefulness and limitations of the duality method in this section we very briefly review some other examples of measure-valued processes. We will give additional examples in which duality can be used to establish uniqueness for martingale problems. For contrast, some closely related examples in which the method of duality is not applicable are also considered.

5.8.1. Exchangeable Diffusions in a Brownian Medium.

In order to motivate this model we first consider a system of particles in a Brownian medium in $\mathbb{R}$. The simplest such system has the form

$$dx_i(t) = \int R(y-x_i(t))W(dy,dt)$$

where W is a cylindrical Brownian motion (cf. Ex. 7.1.2) and R is a symmetric smooth function on $\mathbb{R}$.

$$\text{Then } \langle x_i,x_j\rangle_t = \int_0^t \rho(x_i(s)-x_j(s))ds \text{ where } \rho(z) := \int_{\mathbb{R}} R(z-y)R(y)dy.$$

Vaillancourt (1988) introduced a model of interacting diffusions which combines features of the standard mean-field model of interacting diffusions (cf. Gärtner (1988)) and this model of particles in a Brownian medium. It leads naturally to the notion of a family of exchangeable diffusions which are characterized in the following result.

Theorem 5.8.1.1. Consider a differential operator on $C^\infty((\mathbb{R}^d)^N)$ of the form

$$L := \tfrac{1}{2}(\mathbf{a}\nabla)^T\nabla + \mathbf{b}^T\nabla = \tfrac{1}{2}\sum_{i,j=1}^N (a_{ij}\nabla_i)^T\nabla_j + \sum_{i=1}^N b_i^T\nabla_i$$

where ∇ denotes the gradient on $\mathbb{R}^d$, and T denotes transpose.

Let **a** and **b** be continuous and assume that the martingale problem for L is well-posed. Now consider a system of N particles in $\mathbb{R}^d$ diffusing according to L. Then the solutions $\{P_y : y \in (\mathbb{R}^d)^N\}$ form an exchangeable family, (i.e. a system of N-exchangeable d-dimensional processes) if and only if

$b_i(\pi y) = b_{\pi i}(y)\ \forall i=1,\dots,m,\ y \in E^N,\ \pi \in \mathcal{P}er(N)$, and $a_{ij}(\pi y) = a_{\pi i,\pi j}(y)$.

Proof. See Vaillancourt (1988).

Remark: If the conclusion of Theorem 5.8.1.1 is satisfied then we can rewrite the above system in the form

(5.8.1.1)

$$b(y_i,\mu_N) := b_i(y), \quad \sigma(y_i,\mu_N) := a_{ii}(y) \text{ and } \rho(y_i,y_j,\mu_N) := a_{ij}(y), \quad 1 \le i \ne j \le N$$

where $\mu_N := N^{-1} \sum_{i=1}^{N} \delta_{y_i} \in M_1(E)$.

The empirical measure process is defined by

$$X^N(t) := \Xi_N(y_1(t),\dots,y_N(t))$$

where Ξ_N is defined as in Section 2.2. The generator of X^N is denoted by $\mathfrak{G}_N$.

If $F_{f,k}(.) := \int_{\mathbb{R}^d} \dots \int_{\mathbb{R}^d} f(x_1,\dots,x_k)\mu(dx_1)\dots\mu(dx_N)$

with $f \in C_c^\infty((\mathbb{R}^d)^k)$, then

$$\mathfrak{G}_N F_{f,k}(\mu) = F_{A_k(\mu)f,k}(\mu) + \frac{1}{2N} F_{B_k(\mu)f,k}(\mu)$$

where

$$A_k(\mu)f(y) := \sum_{\alpha=1}^{k} b^T(y_\alpha,\mu)\nabla_\alpha f(y) + \frac{1}{2}\sum_{\alpha=1}^{k} (\sigma(y_\alpha,\mu)\nabla_\alpha)^T \nabla_\alpha f(y)$$

$$+ \frac{1}{2} \sum_{\substack{\alpha,\beta=1 \\ \alpha\ne\beta}}^{k} (\rho(y_\alpha,y_\beta,\mu)\nabla_\beta)^T \nabla_\alpha f(y),$$

$$B_k(\mu)f(y) = \sum_{\substack{\alpha,\beta=1 \\ \alpha\ne\beta}}^{k} ((\sigma(y_\alpha,\mu)-\rho(y_\alpha,y_\alpha,\mu)\nabla_\beta)^T \nabla_\alpha f(y_{\alpha\beta})$$

and $y_{\alpha\beta} = (y_1,\dots,y_{\beta-1},y_\alpha,y_{\beta+1},\dots,y_k)$.

The limiting generator (as $N \longrightarrow \infty$) is given by

$$\mathfrak{G}_\infty F_f(\mu) = F_{A_k(\mu)f}(\mu).$$

Note that it is a second order differential operator of the form

$$\mathfrak{G}_\infty F(\mu) = \int_{\mathbb{R}^d} \left[b^T(z,\mu)\nabla + \tfrac{1}{2}(\sigma(z,\mu)\nabla)^T\nabla \right] \frac{\delta F(\mu)}{\delta\mu(z)}\,\mu(dz)$$
$$+ \frac{1}{2}\int_{\mathbb{R}^d}\int_{\mathbb{R}^d}\left[\nabla^T_{z_1}\nabla_{z_2}\frac{\delta^2 F(\mu)}{\delta\mu(z_1)\delta\mu(z_2)}\right]\rho(z_1,z_2,\mu)\mu(dz_1)\mu(dz_2).$$

Theorem 5.8.1.2. Assume that b, σ and ρ are continuous and satisfy

(i) $|\mathbf{b}(y,\mu)|^2+|\sigma(y,\mu)|^2 \le K(1+|y|^2)$

where $|.|$ denotes the Euclidean norm,

(ii) for each $\mathbf{y} \in (\mathbb{R}^d)^N$ the matrix $\mathbf{a}(\mathbf{y})$ defined in (5.8.1.1) is positive definite and symmetric,

(iii) at time zero $(y_1(0),\dots,y_N(0))$ are i.i.d. (μ) and $\int |y|^4\mu(dy) < \infty$.

Then the laws of the empirical measure-valued processes $\{X^N\}$ are tight and the limit points are solutions to the martingale problem for $\mathfrak{G}_\infty$.

Proof. See Vaillancourt (1988).

Remarks.

1. When the matrix a is diagonal, that is, $\rho \equiv 0$, the solution to the martingale problem for G_∞ is a deterministic measure-valued function which satisfies the so-called *McKean-Vlasov equation*. Under some mild conditions it is known that this martingale problem is well-posed (for example see Gärtner (1988), Léonard (1986)). However when $\rho \neq 0$, the resulting limiting process is a measure-valued stochastic process which formally solves a *random McKean-Vlasov stochastic partial differential equation*.

2. This leaves open the question of the uniqueness of the martingale problem when $\rho \neq 0$. If the $A_k(\mu)$ do not depend on μ, then the solution to the martingale problem of $\mathfrak{G}_\infty$ can be proved to be well-posed using duality (see Dawson and Kurtz (1982) and Vaillancourt (1988)). However for the more general case the duality method is not applicable and other methods are required.

The case in which duality does work, namely $A_k(\mu) \equiv A_k$, is of considerable interest and is in fact a generalization of a model of Chow which we discuss in the next example.

.8.2. A model of molecular diffusion with turbulent transport

Chow (1976) introduced a model of molecular diffusion with turbulent transport n $\mathbb{R}^3$. This process also arises in the study of partial differential equations with andom coefficients (see H. Watanabe (1989)).

However in order to illustrate the power of the duality method we will also add leming-Viot sampling to this mechanism (cf. Dawson and Kurtz (1982)). The resulting enerator for the process involving molecular diffusion, turbulent transport and leming-Viot sampling is given (in dual form) by

$$\mathfrak{G}F_{f,N}(\mu) = F_{G^{(N)}f,N}(\mu) + \frac{\gamma}{2}\sum_{\substack{i,j=1 \\ j\neq i}}^{N} (F_{\tilde{\Theta}_{ij}f}(\mu)-F_f(\mu)).$$

or $\alpha,\beta = 1,2,3$ let $\rho_{\alpha\beta}(x)$ be smooth functions on $\mathbb{R}^3$ with

$$\sum_{j=1}^{k}\sum_{i=1}^{k} \rho_{\alpha\beta}(x_j-x_i)\xi_{j\alpha}\xi_{k\beta} \geq 0 \quad \forall \text{ real } \{\xi_{\alpha\beta}\}.$$

f $\mathbb{R}^3 \ni x_j = (x_{j,1},x_{j,2},x_{j,3})$, then

$$G^{(N)}f(x_1,\dots,x_N) = \sum_{j=1}^{N}\sum_{i=1}^{N}\sum_{\alpha,\beta=1}^{3} \rho_{\alpha\beta}(x_j-x_i)\frac{\partial^2}{\partial x_{j,\alpha}\partial x_{i,\beta}} f(x_1,\dots,x_N)$$

$$+ \kappa\sum_{j=1}^{N}\sum_{\alpha=1}^{3}\frac{\partial^2}{\partial x_{j,\alpha}^2}f(x_1,\dots,x_N).$$

For sufficiently large κ the operators $G^{(N)}$ are uniformly elliptic and generate strongly continuous semigroups on $C_0((\mathbb{R}^3)^N)$. Then a $C_{sim}(E^{\mathbb{N}})$-valued dual process is constructed exactly as for the Fleming-Viot process except that the semigroup $S_t^{(N)}$ on $C((\mathbb{R}^3)^N)$ now has generator $G^{(N)}$ and then the proof that the martingale problem is well-posed follows from Lemma 5.5.1.

In the case $\gamma=0$ (the original model of Chow), the martingale problem is that associated with the bilinear stochastic partial differential equation

$$dX(t) = LX(t)dt + (\nabla X(t))^T\cdot dW(t)$$

where L is an elliptic operator and W is a Brownian motion with values in $(\mathcal{S}'(\mathbb{R}^3))^3$ and covariance structure given by ρ (see Chow (1978) for the details). A systematic development of stochastic partial differential equations of this form can be found in Pardoux (1975), Krylov and Rozovskii (1981) and Rozovskii (1990).

5.8.3. A Bilinear Stochastic Evolution Equation

Let $\{W(t):t\geq 0\}$ denote a Brownian motion such that for each t, W(t) is a generalized random field on $\mathbb{R}^d$. More precisely W is a Gaussian process with values in $\mathcal{S}'(\mathbb{R}^d)$ and covariance structure

$$E(\langle W(t),\phi\rangle\langle W(s),\psi\rangle) = \min(s,t)\int_{\mathbb{R}^d}\int_{\mathbb{R}^d}\phi(x)\psi(y)Q(x-y)dxdy$$

where Q is assumed to be continuous and non-negative definite. Now consider the *bilinear stochastic partial differential equation*

$$dX(t,x) = \frac{1}{2}\Delta X(t,x)dt + \sigma X(t,x)W(dt,x),$$

$$X(0,x) \in pbC(\mathbb{R}^d),$$

or in evolution form

$$X(t,x) = \int p(t,x,y)X(0,y)dy + \sigma\int\int_0^t p(t-s,x,y)X(s,y)W(ds,dy)$$

where $p(t,.,.)$ is the transition function of standard Brownian motion in $\mathbb{R}^d$. The moment equations for this model are closed and hence there is a function-valued dual. However in general the moments of solutions of bilinear stochastic partial differential equations may not satisfy the Carleman condition and so other methods are required to characterize the process. The existence and uniqueness of a strong evolution solution to this equation can be established (cf. Dawson and Salehi (1980), Noble (1992)).

5.8.4. Remark. Refer to Sections 10.2-10.4 for further examples of the application of the method of duality as well as other examples of stochastic evolution equations.

6. A MARTINGALE PROBLEM FORMULATION OF MEASURE-VALUED BRANCHING

6.1. The (A,Φ)-Superprocess Martingale Problem.

We have constructed a general class of measure-valued branching processes in Ch. 4 exploiting the method of Laplace functionals. In this section we verify that the resulting processes are in fact solutions of well-posed martingale problems. This fact will make possible the development of a stochastic calculus for superprocesses in the next chapter.

For technical reasons we will carry this out for a restricted class of measure-valued branching processes. Using the notation of sections 4.3, 4.4, we assume that

(i) $\kappa(dt) = dt$, and

(ii) E is a locally compact metric space and the motion process is a Markov process with Feller semigroup $\{S_t:t\geq 0\}$ on $C_0(E)$ with generator $(D(A),A)$,

(iii) the critical branching mechanism Φ is time homogeneous,

$$\Phi(x,\lambda) = -\tfrac{1}{2}c(x)\lambda^2 + \int_0^\infty (1 - e^{-\lambda u} - \lambda u)n(x,du)$$

where the measure $n(x,dy)$ satisfies $\sup_x \left[\int_0^1 u^2 n(x,du) + \int_1^\infty u\, n(x,du)\right] < \infty$, and $0 \leq c(x) \leq c_0 < \infty\ \forall x$.

(iv) $\{X(t):t\geq 0\}$ is the canonical process defined on $(\mathbf{D},\mathcal{D},(\mathcal{D}_t)_{t\geq 0})$ where $\mathbf{D} := D([0,\infty),M_F(E))$, and there exists a family of Markov probability laws $\{P_\mu:\mu\in M_F(E)\}$ on $\mathbf{D}$ with transition function given as in (4.4.1). This means that for $t>s$, $\mu\in M_F(E)$,

$$P_\mu\left(e^{-\langle X_t,\phi\rangle} \,|\, \mathcal{D}_s\right) = e^{-\langle X_s, V_{t-s}\phi\rangle}, \quad \phi\in pb\mathcal{E}. \tag{6.1.1}$$

where $\{V_t:t\geq 0\}$ is the (A,Φ)-nonlinear semigroup. In this case $v_t = V_t\phi$ solves

$$v_t = S_t\phi + \int_0^t S_{t-s}\Phi(v_s)ds, \tag{6.1.2}$$

where $\Phi(\phi)(x) := \Phi(x,\phi(x))$. We will refer to this as the *(A,Φ)-superprocess.*

An immediate consequence of (6.1.1) and the semigroup property is the fact that

$$E\left(\exp(-\langle X_t,V_{T-t}\phi\rangle)\,|\,\mathcal{D}_s\right) = \exp(-\langle X_s,V_{t-s}V_{T-t}\phi\rangle) = \exp(-\langle X_s,V_{T-s}\phi\rangle),$$

that is,

(6.1.3) for $\phi\in bp\mathcal{E}$, $\exp(-\langle X_t,V_{T-t}\phi\rangle)$ is a martingale on $[0,T]$.

We next give an expression for an MP-generator for the (A,Φ)-superprocess. The domain $\mathcal{D}(\mathfrak{G})$ will consist of functions of the form $F(\mu) = f(\langle\mu,\phi\rangle)$, $\phi\in D(A)$, $f\in C_b^\infty(\mathbb{R})$, and

(6.1.4)

$$\mathfrak{G}F(\mu) = \iint\mu(dx)c(x)\delta_x(dy)F''(\mu;x,y)$$
$$+ \int\mu(dx)\int_0^\infty n(x,du)[F(\mu+u\delta_x)-F(\mu) - uF'(\mu;x)]$$
$$+ \int\mu(dx)AF'(\mu;x)$$

where $F'(\mu;x) := \delta F(\mu)/\delta\mu(x)$, $F''(\mu;x,y) := \delta^2F(\mu)/\delta\mu(x)\delta\mu(y)$,

or equivalently,

$$\mathfrak{G}F\mu) = f'(\langle\mu,\phi\rangle)\langle\mu,A\phi\rangle + \tfrac{1}{2}f''(\langle\mu,\phi\rangle)\langle\mu,c\phi^2\rangle$$
$$+ \int\mu(dx)\int_0^\infty n(x,du)[f(\langle\mu,\phi\rangle+u\phi(x))-f(\langle\mu,\phi\rangle)-f'(\langle\mu,\phi\rangle)u\phi(x)]$$

We next briefly review some basic facts from stochastic calculus (standard references include Dellacherie and Meyer (1975, 1980), [LS], and Métivier (1982)).

An adapted càdlàg process, X, defined on a filtered probability space $(\Omega,\mathcal{F},(\mathcal{F}_t)_{t\ge0},P)$, is a *semimartingale* if it has a representation

(6.1.5) $$X_t = X_0 + M_t + A_t$$

where M is a local martingale with $M_0 = 0$, and A is a process of locally bounded variation. M can be taken to be a locally square integrable martingale. It is called a *special semimartingale* if A has locally integrable variation.

Lemma 6.1.1. A special semimartingale has a unique representation of the form (6.1.5) under the additional requirement that A be predictable.

Proof. See [LS, p. 85].

Let N be an adapted random point measure on $\mathbb{R}_+\times S$ where (S,d) is a Polish space, i.e. $N(t,B)$ denote the number of points in $[0,t]\times B$, $B\in\mathcal{B}(S)$. Assume that N satisfies the following:

(a) $E(N(t,B)) < \infty$ $\forall$ $t > 0$, $B\in \mathcal{B}(S)_b$ (i.e. bounded Borel subsets of S),

(b) there exists $\{\hat{N}(t,B):t\ge0,\ B\in\mathcal{B}(S)\}$ such that

(i) for $B\in\mathcal{B}(S)_b$, $t \longrightarrow \hat{N}(t,B)$ is a continuous $(\mathcal{F}_t)$-adapted increasing process,

(ii) for each t and a.e. $\omega\in\Omega$, $B \longrightarrow \hat{N}(t,B)$ is a locally finite measure on $(S,\mathcal{B}(S))$,

(iii) for $B\in \mathcal{B}(S)_b$, $t\longrightarrow \tilde{N}(t,B) := N(t,B) - \hat{N}(t,B)$ is an $(\mathcal{F}_t)$-local martingale,

that is, $\tilde{N}$ is a *martingale measure* (refer to section 7.1 for the basic definitions of martingale measures and stochastic integrals with respect to martingale measures). In fact $\tilde{N}$ is an L^2 orthogonal martingale measure (cf. Ikeda and Watanabe (1981, Ch. 2, Theorem 3.1)).

The predictable increasing process, $\hat{N}$, is called the *compensator* (cf. [LS, p. 172])).

Consider the semimartingale

$$Z(t) = Z(0) + A(t) + M_c(t) + \int_0^{t+}\int_S g_1(s,z)\tilde{N}(ds,dz) + \int_0^{t+}\int_S g_2(s,z)N(ds,dz)$$

where A is continuous, adapted, locally of bounded variation and $A_c(0)=0$, M_c is a continuous local martingale with increasing process $\langle M_c\rangle_t$ and $M_c(0)=0$, g_1 is predictable and satisfies the local version of $E[\int_0^t |g_1(s,z)|^2\hat{N}(ds,dx)] < \infty$, g_2 is predictable and $\int_0^{t+}\int_S |g_2(s,z)|N(ds,dz) < \infty$ a.s. for each t and $g_1g_2 = 0$.

Lemma 6.1.2. (Itô's Lemma) For $f\in C^{1,2}(\mathbb{R}\times\mathbb{R})$, $f(t,Z(t))$ is also a semimartingale and

(6.1.6)

$$\begin{aligned} &f(t,Z(t))-f(0,Z(0)) \\ &= \int_0^t f_1(s,Z(s))ds \\ &+ \int_0^t f_2(s,Z(s))dA(s) + \tfrac{1}{2}\int_0^t f_{22}(s,Z(s))d\langle M_c\rangle_s \\ &+ \int_0^{t+}\int_S [f(s,Z(s-)+g_2(s,z))-f(s,Z(s-))]N(ds,dz) \\ &+ \int_0^{t+}\int_S [f(s,Z(s)+g_1(s,z))-f(Z(s))-g_1(s,z)f'(s,Z(s))]\hat{N}(ds,dz) \\ &+ \text{a local martingale,} \end{aligned}$$

where $f_1(s,.) := \partial f(s,.)/\partial s$, $f_2(.,x) := \partial f(.,x)/\partial x$, $f_{22}(.,x) := \partial^2 f(.,x)/\partial x^2$.

Proof. See Ikeda and Watanabe (1981, p. 66), or [LS, p. 118].

The following result due to El Karoui and Roelly (1991) shows that the martingale problem for $\mathfrak{G}$ is well-posed.

Theorem 6.1.3 Assume (i), (ii), (iii), (iv) above. Let $D_0(A) := \{\phi: \phi=\phi_0+\varepsilon, \phi_0\in pD(A),\ \varepsilon>0\}$. Then

(a) for $F(\mu) = f(\langle\mu,\phi\rangle)$, $f\in C_b^\infty(\mathbb{R})$, $\phi\in D_0(A)$,

$M_F(t) := F(X(t)) - \int_0^t \mathfrak{G}F(X(s))ds$ is a martingale where $\mathfrak{G}$ is given by (6.1.4).

(b) Any solution of the $(\mathfrak{G},\mathfrak{D}(\mathfrak{G}))$-martingale problem satisfies (6.1.3) and therefore the martingale problem is well-posed.

Proof. (a) Let $Z_t(\phi) := e^{-\langle X_t,\phi\rangle}$. We first show that $Z_t(\phi)$ is a special semimartingale, that is, a locally square integrable martingale plus a process of locally integrable variation. The main idea is then to use the uniqueness of the representation of a special semimartingale to identify the terms in two semimartingale representations of $Z_t(\phi)$.

Step 1. We first prove that for $\phi\in D_0(A)$, $\phi \geq \kappa > 0$,

(6.1.7) $$H_t(\phi) = \exp\left(-\langle X_t,\phi\rangle\right) + \int_0^t \langle X_s, A\phi+\Phi(\phi)\rangle ds \text{ is a } P_{X(0)}\text{-local martingale.}$$

To prove this we will show that for $B\in\mathcal{D}_s$, $s\leq t$,

$$\frac{d}{dt} E(1_B\exp(-\langle X_t,\phi\rangle)) = -E(1_B\langle X_t,(A\phi+\Phi(\phi))\rangle\exp(-\langle X_t,\phi\rangle)).$$

But from (6.1.1)

$$\varepsilon^{-1}E\left(1_B\exp(-\langle X_{t+\varepsilon},\phi\rangle) - \exp(-\langle X_t,\phi\rangle)\right)$$

$$= \varepsilon^{-1}E\left(1_B(\exp(-\langle X_t,V_\varepsilon\phi\rangle) - \exp(-\langle X_t,\phi\rangle))\right).$$

To take the limit as $\varepsilon\downarrow 0$ inside the expectation we use dominated convergence and equation (6.1.2). To justify this it suffices to note that

$$D^{\varepsilon}(\mu) := \varepsilon^{-1}|\exp(-\langle\mu,V_\varepsilon\phi\rangle)-\exp(-\langle\mu,\phi\rangle)|$$
$$\leq \exp(-\kappa\langle\mu,1\rangle)\ \varepsilon^{-1}(\exp(K_T(\phi)\varepsilon\langle\mu,1\rangle) - 1)$$

and this is bounded function of $\langle\mu,1\rangle$ uniformly in $0< \varepsilon \leq 1$. To obtain the last inequality note that $v_t = S_t\phi + \int_0^t S_{t-s}\Phi(v_s)ds$ and $S_t\phi-\phi = \int_0^t S_s A\phi ds$ which yields

$$\|V_t\phi-\phi\|/t \leq (\|A\phi\| + \sup_{r<T}\|\Phi(V_r\phi)\|) \leq K_T(\phi).$$

(A similar argument works for $\varepsilon\uparrow 0$.) Taking the limit $\varepsilon\downarrow 0$ we obtain the derivative

$$\lim_{\varepsilon\downarrow 0} \varepsilon^{-1}E\left(1_B(\exp(-\langle X_{t+\varepsilon},\phi\rangle) - \exp(-\langle X_t,\phi\rangle))\right)$$

$$= -E\left(1_B\langle X_t,A\phi+\Phi(\phi)\rangle\exp(-\langle X_t,\phi\rangle)\right).$$

This shows that $\exp(-\langle X_t,\phi\rangle) - \int_0^t \langle X_r,A\phi+\Phi(\phi)\rangle\exp(-\langle X_r,\phi\rangle)dr$ is a local martingale. It then follows that $H_t(\phi)$ is a local martingale (cf. [EK, Ch. 2, Cor 3.3]) and this completes the proof that $H_t(\phi)$ is a martingale.

Step 2. The semimartingale representations.

Let $Y_t(\phi) = \exp\left(-\int_0^t \langle X_s, A\phi+\Phi(\phi)\rangle ds\right)$ and $Z_t(\phi) = \exp(-\langle X_t,\phi\rangle)$.

Note that $Y_t(\phi)$ is a continuous semimartingale with finite variation and $H_t(\phi)$ is a martingale. Therefore $Z_t(\phi) = H_t(\phi)Y_t(\phi)$ is a semimartingale and the (Itô) integration by parts formula yields

(6.1.8) (*Representation 1)*

$$\begin{aligned} dZ_t(\phi) &= Y_t(\phi)dH_t(\phi) + H_{t-}(\phi)dY_t(\phi) \\ &= Y_t(\phi)dH_t(\phi) - \langle X_t, A\phi+\Phi(\phi)\rangle Z_{t-}(\phi)dt. \end{aligned}$$

Note that the variation of $H_s(\phi)dY_s$ in $[0,t]$ is dominated by

$$\int_0^t \exp\left(\int_0^s |\langle X_u, A\phi+\Phi(\phi)\rangle| du\right) |\langle X_s, A\phi+\Phi(\phi)\rangle| ds$$

which is locally integrable under assumption (iii). Hence $Z_t(\phi)$ is a *special semimartingale.*

On the other hand, for $\phi\in D_0(A)$, using Itô's formula we conclude that $\langle X_t,\phi\rangle = -\log Z_t(\phi)$ is also a semimartingale. Let $N(ds,d\mu)$ be the adapted random point measure on $\mathbb{R}_+\times M_F(E)$ given by $\sum \delta_{(s,\Delta X_s)}$ (where $\Delta X_s := X_s - X_{s-}$), $\hat{N}(ds,d\mu)$ denote its compensator and $\tilde{N}(ds,d\mu)$ the corresponding martingale measure (in order that N satisfy the assumptions stated before Lemma 6.1.2 we equip $M_F(E)$ with a metric, d, under which all bounded subsets, B, have the property that $\{\langle\mu,1\rangle:\mu\in B\}$ is a compact subset of $(0,\infty)$).

Then for $\phi\in D_0(A)$, $\langle X_t,\phi\rangle$ has the canonical representation (cf. [LS. p. 191]) of the form

(6.1.9) $$\langle X_t,\phi\rangle = \langle X_0,\phi\rangle + U'_t(\phi) + M^c_t(\phi) + \tilde{N}_t(\phi) + N_t(\phi)$$

where $U'_t(\phi)$ is continuous and locally of bounded variation, $M^c_t(\phi)$ is a continuous local martingale with increasing process $C_t(\phi)$ and where

$$\tilde{N}_t(\phi) = \int_0^t\int_{M_F^{\pm}(E)} 1(|\langle\mu,\phi\rangle|\le\|\phi\|)\langle\mu,\phi\rangle\, \tilde{N}(ds,d\mu),$$

$$N_t(\phi) = \int_0^t\int_{M_F^{\pm}(E)} 1(|\langle\mu,\phi\rangle|>\|\phi\|)\langle\mu,\phi\rangle\, N(ds,d\mu),$$

where $M_F^{\pm}(E)$ denotes the space of signed measures on E of finite variation.

Applying Itô's lemma (Lemma 6.1.2) to $\exp(-\langle X_t,\phi\rangle)$ with $\langle X_t,\phi\rangle$ given by (6.1.9) we get

$$dZ_t(\phi) = Z_{t-}(\phi)\Big\{-dU'_t(\phi) + \tfrac{1}{2}dC_t(\phi)$$

$$+ \int_{M_F^{\pm}(E)} \Big((e^{-\langle\mu,\phi\rangle} - 1 + \langle\mu,\phi\rangle)1(|\langle\mu,\phi\rangle| \le \|\phi\|)\Big)\hat{N}(dt,d\mu)$$

$$+ \int_{M_F^{\pm}(E)} \Big((e^{-\langle\mu,\phi\rangle} - 1)1(|\langle\mu,\phi\rangle| > \|\phi\|)\Big)N(dt,d\mu)\Big\} + d(\text{loc. mart.})$$

It is easy to verify that under assumption (iii)

$$Z_{t-}(\phi)\Big(\exp(-\langle\Delta X_t,\phi\rangle) - 1 + \langle\Delta X_t,\phi\rangle\Big) \ge 0$$

is integrable and we obtain

(6.1.10) *(Representation 2)*

$$dZ_t(\phi) = Z_{t-}(\phi)\cdot\Big(-dU_t(\phi) + \tfrac{1}{2}dC_t(\phi) + \int_{M_F^{\pm}(E)} (e^{-\langle\mu,\phi\rangle} - 1 + \langle\mu,\phi\rangle)\hat{N}(dt,d\mu)\Big)$$

$$+ d(\text{loc. mart.})$$

Step 3. Identification of the Two Representations.

Since $Z_t(\phi)$ is a special semimartingale we can now identify the predictable components of locally integrable variation in the two decompositions (6.1.8) and (6.1.10) (e.g. [LS] 85) to obtain that

$$-Z_{t-}(\phi)\langle X_t, A\phi+\Phi(\phi)\rangle dt$$

$$= Z_{t-}(\phi)\Big(-dU_t(\phi) + \tfrac{1}{2}dC_t(\phi) + \int_{M_F^{\pm}(E)} (e^{-\langle\mu,\phi\rangle} - 1 + \langle\mu,\phi\rangle)\hat{N}(dt,d\mu)\Big)$$

where $U_t(\phi) := U'_t(\phi) + \int_{M_F^{\pm}(E)} \langle\mu,\phi\rangle 1(|\langle\mu,\phi\rangle| > 1)\hat{N}(dt,d\mu)$

is locally of bounded variation and $U_t(\theta\phi) = \theta U_t(\phi)$ for $\theta\in\mathbb{R}_+$; $C_t(\phi)$ is a continuous increasing process and $C_t(\theta\phi) = \theta^2 C_t(\phi)$. Then the process $\langle X_s, A\phi+\Phi(\phi)\rangle$ is integrable on $[0,t]$ and

$$\int_0^t \langle X_s, A\phi+\Phi(\phi)\rangle ds = U_t(\phi) - \tfrac{1}{2}C_t(\phi) - \int_0^t\int_{M_F^{\pm}(E)} (e^{-\langle\mu,\phi\rangle} - 1 + \langle\mu,\phi\rangle)\hat{N}(ds,d\mu).$$

Replacing ϕ by $\theta\phi$, $(\theta>0)$ in $\Phi(\phi)$, we obtain

$$\int_0^t <X_s, A\theta\phi+\Phi(\theta\phi)>ds$$

$$= \theta\int_0^t <X_s, A\phi>ds - \frac{\theta^2}{2}\int_0^t <X_s, c\phi^2>ds$$

$$- \int_0^t ds\int X_s(dx)\int_0^\infty n(x,du)\left(e^{-u\theta\phi(x)}-1+u\theta\phi(x)\right)$$

$$= \theta U_t(\phi) - \frac{\theta^2}{2}C_t(\phi) - \int_0^t\int_{M_F^{\pm}(E)} (e^{-<\mu,\theta\phi>}-1+<\mu,\theta\phi>)\hat{N}(ds,d\mu).$$

This allows us to conclude:

(i) in the semimartingale representation of $<X_t,\theta\phi>$ (6.1.10)

$$C_t(\phi) = \theta^2\int_0^t <X_s, c\phi^2>ds, \quad U_t(\phi) = \theta\int_0^t <X_s, A\phi>ds,$$

$$\int_0^t\int_{M_F^{\pm}(E)} (e^{-<\mu,\theta\phi>}-1+<\mu,\theta\phi>)\hat{N}(ds,d\mu)$$

$$= \int_0^t ds\int X_s(dx)\int_0^\infty n(x,du)\left(\exp(-u\theta\phi(x))-1+u\theta\phi(x)\right)$$

$$= \int_0^t ds\int X_s(dx)\int_0^\infty n(x,du)\left(\exp(-u<\delta_x,\theta\phi>) - 1 + u<\delta_x,\theta\phi>\right) \quad \forall\theta > 0,\ \phi\in D_0(A),$$

that is, the jump measure of the process X has compensator

$$\hat{N}(ds,d\mu) = dsX_s(dx)n(x,du)\cdot\delta_{u\delta_x}(d\mu), \quad \mu\in M_F(\mathbb{R}^d).$$

In particular this implies that the jumps of X are almost surely in $M_F(E)$, i.e. positive measures.

Step 4. Proof that the semimartingale representation of $\{X_t\}$ implies that it is a solution of the $\mathfrak{G}$-martingale problem.

<u>Proof.</u> Let $f \in C^\infty(\mathbb{R})$. Then by Ito's formula (Lemma 6.1.2)

$$f(<X_t,\phi>) = f(<X_0,\phi>)+\int_0^t f'(<X_{s-},\phi>)dU_s(\phi) + \frac{1}{2}\int_0^t f''(<X_s,\phi>)dC_s(\phi)$$

$$+ \int_0^t\int_{M_F^{\pm}(E)} \left(f(<X_s+\mu,\phi>)-f(<X_s,\phi>)-f'(<X_s,\phi>)<\mu,\phi>\right)\hat{N}(ds,d\mu)$$

$$+ \mathcal{M}_F(t) \text{ (local martingale)}.$$

Hence

$$M_F(t) = f(\langle X_t,\phi\rangle) - \int_0^t \left(f'(\langle X_s,\phi\rangle)\langle X_s,A\phi\rangle + \tfrac{1}{2}f''(\langle X_s,\phi\rangle)\langle X_s,c\phi^2\rangle\right) ds$$
$$- \int_0^t \langle X_s, \int_0^\infty n(.,d\lambda)\left[f(\langle X_s,\phi\rangle+\lambda\phi(.))-f(\langle X_s,\phi\rangle)-f'(\langle X_s,\phi\rangle)\lambda\phi(.)\right]\rangle ds$$

is a local martingale. If f, f' and f'' are bounded this is a martingale. This completes the proof that the process X_t is a solution to the martingale problem for $\mathfrak{G}$.

(b) Assume that $\{X(t)\}$ is a solution of the $(\mathfrak{G},\mathfrak{D}(\mathfrak{G}))$-martingale problem. It remains to prove that this implies (6.1.3). For this, assume that $\psi(.,x) \in C_b^1(\mathbb{R}_+)$, $\psi(t,.) \in D_0(A)$, $t \to A\psi(t,x)$ is continuous, $\psi>0$, and $\frac{\partial\psi}{\partial t}(t,.) \in C_b(E)$. If $f \in C_b^\infty(\mathbb{R})$, then by Lemma 6.1.2 we obtain that

(6.1.11)

$$f(\langle X_t,\psi(t)\rangle) - \int_0^t \left(f'(\langle X_s,\psi(s)\rangle)\langle X_s,A\psi(s)+ \frac{\partial\psi}{\partial s}\rangle + \tfrac{1}{2}f''(\langle X_s,\psi(s)\rangle)\langle X_s,c\psi^2(s)\rangle\right) ds$$
$$- \int_0^t \langle X_s, \int_0^\infty n(.,d\lambda)$$
$$\cdot\left(f(\langle X_s,\psi(s)\rangle+\lambda\psi(s,.))-f(\langle X_s,\psi(s)\rangle)-f'(\langle X_s,\psi(s)\rangle)\lambda\psi(s,.)\right)\rangle ds$$

is a local martingale.

If $\phi \in D_0(A)$, and and we assume that $\psi(t,x) := V_{T-t}\phi(x)$ satisfies the above conditions, then V_t satisfies the differential form of (6.1.2)

$$(6.1.12)\quad \left(\frac{\partial}{\partial t} + A\right)(V_{T-t}\phi) + \Phi(V_{T-t}\phi) = 0$$

(for example, this is satisfied for the (α,d,β)-superprocess).

If we now apply (6.1.11) to $\exp(-\langle X_t,V_{T-t}\phi\rangle)$ we can verify that it is a martingale. The result follows for general ϕ by taking limits. Therefore the marginal distribution of X_t is determined and by Theorem 5.1.2 we conclude that the martingale problem is well-posed. The argument in the case in which V_t satisfies (6.1.2) but not necessarily (6.1.12) is a little more involved. One first shows (using 6.1.2, the semigroup property of V_t, and $E(\langle X_t,\phi\rangle|\mathcal{D}_s) = \langle X_s,S_{t-s}\phi\rangle$, $t\ge s$) that $\langle X_t,V_0\phi\rangle - \langle X_0,V_t\phi\rangle + \int_0^t \langle X_u,\Phi(V_{t-u}\phi)\rangle du$ is a martingale and then applies Itô's lemma (see Fitzsimmons (1992, Corollary 2.23) for the details). □

<u>Corollary 6.1.4.</u> Under the assumptions of Theorem 6.1.3, the (A,Φ)-superprocess is $\{X(t):t\ge 0\}$ is a strong Markov process.

<u>Proof.</u> The arguments above remain valid if $D_0(A)$ is replaced by an appropriate countable subset. The strong Markov property then follows by Theorem 5.1.3. □

Remark: The analogue of Theorem 6.1.3 was obtained for (W,Φ)-superprocesses when W is a time homogeneous right processes by Fitzsimmons (1991).

6.2. Examples of Measure-valued Branching Processes

In order to give an indication of the richness of the class of branching systems we give some examples and variations of the general class dealt with in this chapter.

6.2.1. The B(A,c)-Superprocess.

Consider the special case in which E is a compact metric space, A is the generator of a Feller process on E and $\Phi(\lambda) \equiv -\frac{1}{2}c\lambda^2$. This process will be called the B(A,c)-superprocess. In Theorem 4.7.2 we proved that this process has almost surely continuous sample paths. The corresponding martingale problem will be studied in more detail in Chapters 7 and 8.

In the special case $E = \{e\}$, $A = 0$, the B(0,c)-superprocess reduces to the *Feller continuous critical branching process*, that is, a non-negative martingale, Z_t, with increasing process $c\int_0^t Z_s ds$.

6.2.2 The (α,d,β)-superprocess, $\beta \le 1$

Let $D_0(\Delta_\alpha,p) \subset C_p(\dot{\mathbb{R}}^d)$ denote a core of the semigroup $\{S_t^\alpha\}$ on $C_p(\dot{\mathbb{R}}^d)$ (refer to Section 4.5 for notation).

Let $\mathfrak{D}_0(\mathfrak{G})$ consist of functions of the form

$$F(\mu) := e^{-\langle\mu,\phi\rangle}, \quad \mu \in M_p(\dot{\mathbb{R}}^d), \quad \phi\in D_0(\Delta_\alpha,p)$$

$$:= 0 \text{ if } \langle\mu,\phi_p\rangle = \infty.$$

Then for $F \in \mathfrak{D}_0(\mathfrak{G})$,

if $0< \beta < 1$

$$\mathfrak{G}F(\mu) = \frac{\beta(1+\beta)\gamma}{\Gamma(1-\beta)}\int\mu(dx)\int_0^\infty \frac{1}{u^{\beta+2}}[F(\mu+u\delta_x)-F(\mu) - uF'(\mu;x)]d\mu$$

$$+ \int\mu(dx)\Delta_\alpha F'(\mu;x),$$

and if $\beta=1$,

$$\mathfrak{G}F(\mu) = \frac{1}{2}c\int\mu(dx)\delta_x(dy)F''(\mu;x,y)+ \int\mu(dx)\Delta_\alpha F'(\mu;x).$$

We will show in Prop. 7.3.1 that the $(\alpha,d,1)$ superprocess has continuous sample paths. On the other hand the arguments in step 3 of the proof of Theorem 6.1.3 show that for $\beta<1$, the (α,d,β)-superprocess has jumps corresponding to the creation of mass atoms (note that all jumps have the form $\mu \longrightarrow \mu + u\xi_x$ with $u > 0$, $x\in E$). The set of jump times is dense in $[0,\tau)$ where $\tau := \inf\{t:X_t(\mathbb{R}^d)=0\}$ (because $n((0,1)) = \infty$).

6.2.3. Spatially Inhomogeneous Super-Brownian Motion

Consider a system of critical binary branching Brownian motions in $\mathbb{R}$ but in which particles only branch at the origin. In particular we assume that particles branch only when they are located in the interval $(-\varepsilon/2,\varepsilon/2)$, $\varepsilon>0$, and then at rate ε^{-1}. In the measure-valued limit we obtain a superprocess of the above type in which the additive functional $\kappa(ds)$ is given by the local time at 0. It turns out that the log-Laplace function can also be obtained as the unique solution of the following singular nonlinear partial differential equation:

$$\frac{\partial v}{\partial t} = 2\Delta v - \delta_0 v^2 \tag{6.2.3.1}$$

where δ_0 denotes the delta function at the origin. For details refer to Dawson and Fleischmann (1992).

6.2.4. Superprocesses with Immigration and Superprocesses conditioned on Non-Extinction.

Consider an (A,Φ)-superprocess. In addition to fixed initial conditions one can incorporate the immigration of new mass. For example if $\{\nu(t):t\geq 0\}$ is a continuous $M_F(E)$-valued function, then the superprocess with immigration rate ν is given by the Laplace functional

$$P_{X_0}\left(e^{-\langle X_t,\phi\rangle}\right) = e^{-\langle X_0,V_t\rangle}\exp\left(-\int_0^t \langle \nu(s),V_{t-s}\rangle ds\right).$$

The Laplace functional can also be obtained when the immigration involves an infinitely divisible random measure (cf. Kawazu and Watanabe (1971)).

Let P^* be defined as the law of the process conditioned on non-extinction in the remote future, that is, for $B\in \mathcal{D}_t$,

$$P^*_{X_0}(B) := \lim_{s\to\infty} P_{X_0}(B\,|\,\langle X_{t+s},1\rangle > 0).$$

Roelly-Coppoletta and Rouault (1989) showed that

$$P^*_{X_0}|_{\mathcal{D}_t} = \frac{\langle X_t,1\rangle}{\langle X_0,1\rangle} e^{-\Phi'(0)t} P_{X_0}|_{\mathcal{D}_t}.$$

Furthermore they have shown that the conditioned process can be represented as a process with state dependent immigration.

6.2.5. Multitype Superprocesses.

At the particle level we consider a system consisting of particles which can be one of a finite number, K, of types. The process can be viewed as a $M(\cup_{j=1}^{K} E_j)$-valued process where $E_1,\ldots,E_K$ are copies of E. We consider a branching particle system in which the motion process consists of an A_j-motion on each E_j together with jumps between the different E_j. Taking the high density limit as above leads to the multitype superprocess. See Gorostiza and Lopez-Mimbela (1990), Gorostiza and Roelly (1990), Gorostiza, Roelly and Wakolbinger (1992), Gorostiza and Wakolbinger (1992), and Li (1992).

6.2.6. Multilevel Branching

The state space for the class of two-level branching process is $M_F(M_F(E_1))$ where E_1 is a Polish space. We can obtain the two level branching process as the limit of a system of branching superparticles which move in $M_F(E_1)$ according to a one level measure-valued branching process.

For example, the $(A_1;\beta_1,\beta_2)$-two-level superprocess is obtained from the construction of Ch. 4 with

$$E_2 := M_F(E_1), \quad \Phi_2(\mu,\lambda) = -\gamma_2\lambda^{1+\beta_2},$$

and the motion process on $M_F(E_1)$ is the (A_1,Φ_1)-superprocess (with $\Phi_1(x,\lambda) = -\gamma_1\lambda^{1+\beta_1}$) with linear semigroup $\{S_t^2 : t\geq 0\}$ and generator A_2 on $bC(M_F(E_1))$ given by:

$$A_2F(\mu) := c(\beta_1)\int\mu(dx)\int_0^\infty \frac{1}{u^{\beta_2+2}}[F(\mu+u\delta_x)-F(\mu) - uF'(\mu;x)]du + \int\mu(dx)A_1F'(\mu;x), \text{ if } 0<\beta_2<1$$

or

$$A_2F(\mu) := \gamma_2\int_{E_2}\int_{E_2}\mu(dx)\delta_x(dy)F''(\mu;x,y)+\int_{E_2}\mu(dx)A_1F'(\mu;x) \quad \text{if } \beta_2 = 1.$$

In this case the log-Laplace equation is given by

$$V_t^2\phi(\mu) = S_t^2\phi(\mu) + \gamma_2\int_0^t S_{t-s}^2[V_s^2\phi]^{1+\beta_2}ds, \quad \phi\in bp\mathcal{B}(M_F(E_1)).$$

The higher level processes can then be constructed by iterating this procedure. See Dawson, Hochberg and Wu (1990), Dawson and Hochberg (1991) and Wu (1991, 1992) for further information.

7. STOCHASTIC CALCULUS OF MEASURE-VALUED PROCESSES

7.1. Martingale Measures

Let $(\Omega,\mathcal{F},(\mathcal{F}_t)_{t\geq 0},P)$ be a probability space with a right continuous filtration. The σ-sub-algebra of $\mathcal{B}(\mathbb{R}_+)\otimes\mathcal{F}$ generated by the real-valued continuous adapted processes is called the $\mathcal{F}_t$*-predictable σ-algebra* and denoted by $\mathcal{P}$. $\mathcal{P}$ is generated by predictable rectangles of the form $(s,t]\times F$, $F\in\mathcal{F}_s$ and $\{0\}\times F$, $F\in\mathcal{F}_0$.

Let (E,d) be a Polish space and $\mathcal{E} = \mathcal{B}(E)$. Consider a random set function $U(A,\omega)$ defined on $\mathcal{A}\times\Omega$ where $\mathcal{A}$ is a subring of $\mathcal{E}$ which satisfies

$$\|U(A)\|_2^2 = E[U(A)]^2 < \infty \;\forall\; A\in\mathcal{A}$$

$$A\cap B = \emptyset \Longrightarrow U(A)+U(B) = U(A\cup B) \text{ a.s. } \forall A \text{ and } B \text{ in } \mathcal{A}.$$

The set function U is said to be *σ-finite* if there exists an increasing sequence (E_n) of E such that

(a) $\bigcup_n E_n = E$;

(b) $\forall n\; \mathcal{E}_n = \mathcal{E}|_{E_n} \subset \mathcal{A}$;

(c) $\forall\; n$, $\sup\,\{\|U(A)\|_2,\; A\in\mathcal{E}_n\} < \infty$.

Remark. In our applications $\mathcal{A}$ is usually assumed to be the ring of d-bounded Borel subsets of E. In this case the random set function is called *locally finite*.

The set function U is said to be *countably additive* if for each sequence (A_j) decreasing to $\emptyset$, $\|U(A_j)\|_2$ tends to zero. Then it is easy to extend U by $U(A) := \lim_n U(A\cap E_n)$ on every set A of $\mathcal{E}$ such that the limit exists in $L^2(\Omega,\mathcal{F},P)$. A set function which satisfies all these properties is called a *σ-finite L^2-valued measure.*

$\{M_t(A),t\geq 0,A\in\mathcal{A}\}$ is a L^2*-martingale measure* with respect to $\mathcal{F}_t$ if and only if:

(a) $M_0(A) = 0$, $\forall A\in\mathcal{A}$

(b) $\{M_t(A),t\geq 0\}$ is a $\mathcal{F}_t$-martingale $\forall A\in\mathcal{A}$

(c) $\forall t>0$, $M_t(.)$ is a L^2-valued σ-finite measure

The *covariance functional* $Q(t;.,.)$ of M is defined by

$$Q_M(t;A,B) := \langle M(A),M(B)\rangle_t$$

(the compensator of the process $M_t(A)M_t(B)$. It defines a covariance measure Q_M on $E\times E\times\mathbb{R}_+$.

The martingale measure M is *worthy* if there exists a random σ-finite measure $K(.,.,.;\omega)$, $\Lambda\in\mathcal{E}\times\mathcal{E}\times\mathcal{B}(\mathbb{R}_+)$, $\omega\in\Omega$, such that

(a) K is symmetric and positive definite, that is, for any $f\in b\mathcal{E}\times\mathcal{B}(\mathbb{R}_+)$,

$\iiint f(x,s)f(y,s)K(dx,dy,ds) \geq 0.$

(b) for fixed A,B, $\{K(A\times B\times(0,t]),\ t\geq 0\}$ is $\mathcal{F}_t$-predictable,

(iii) $\exists\ E_n\uparrow E$ such that $E(K(E_n\times E_n\times[0,T])) < \infty\ \forall\ n$,

(iv) $|Q_M(t;A,A)| \leq K(A\times A\times[0,t])$.

K is called the *dominating measure* for M..

M is an *orthogonal martingale measure* if it is a martingale measure and if $A\cap B = \emptyset \Longrightarrow M_t(A)$ and $M_t(B)$ are orthogonal martingales, that is, $M_t(A)M_t(B)$ is a martingale, $\forall A,B \in \mathcal{A}$.

If M is a martingale measure and if for all $A\in\mathcal{A}$, the map $t\longrightarrow M_t(A)$ is continuous we say that M is *continuous*.

Theorem 7.1.1 (a) A worthy martingale measure is orthogonal if and only if $Q_M(t)$ is supported by $\{(x,y):x=y\in E\}$ for all t.

(b) An orthogonal martingale measure, M, is worthy, and there exists a random σ-finite measure $\nu(ds,dx)$ on $\mathbb{R}_+\times E$, such that for each $A \in \mathcal{A}$ the process $(\nu(0,t]\times A)_t$ is predictable and satisfies:

$\forall A\in \mathcal{A},\ \forall t>0,\quad \nu((0,t]\times A) = \langle M(A)\rangle_t$, P.a.s.

The measure ν is called the *intensity* of the orthogonal martingale measure M.

(c) If the orthogonal measure M is continuous, then $\nu(\{t\}\times E_n)=0\ \forall t>0,\ \forall n\in\mathbb{N}$.

Proof. Walsh (1986, Chapt. 2).

Example 7.1.2 Gaussian White Noise (Cylindrical Brownian Motion)

Let $(E\times\mathbb{R}_+,\mathcal{E}\otimes\mathcal{B}(\mathbb{R}_+),\nu(dx)dt)$ be a σ-finite measure space. A *Gaussian white noise* based on $\nu(dx)dt$ is a random set function, W, on $\mathcal{E}\otimes\mathcal{B}(\mathbb{R}_+)$ such that

(a) $W(B\times[a,b])$ is a Gaussian random variable with mean zero and variance $\nu(B)\cdot|b-a|$,

(b) if $(A\times[a_1,a_2])\cap(B\times[b_1,b_2]) = \emptyset$, then $W(A\times[a_1,a_2]\cup(B\times[b_1,b_2])) = W(A\times[a_1,a_2]) + W((B\times[b_1,b_2])$ and $W(A\times[a_1,a_2])$ and $W(B\times[b_1,b_2])$ are independent.

Then for $B\in\mathcal{E}$, $W_t(B) := W(B\times[0,t])$ is an orthogonal martingale measure intensity ν. Walsh (1986) proved that a continuous orthogonal martingale measure is a white noise if and only if its covariance measure is deterministic.

Let $E = \mathbb{R}^d$, ν be Lebesgue measure and $\mathcal{S}(\mathbb{R}^d)$ denote the Schwartz space of C^∞-functions on $\mathbb{R}^d$ which are rapidly decreasing at infinity. Consider the martingale problem: $\forall\phi\in\mathcal{S}(\mathbb{R}^d)$,

$W_t(\phi)$ is a continuous square integrable martingale

with increasing process $\langle W(\phi)\rangle_t = t\int\phi^2(x)dx$.

This martingale problem has a unique solution on $C([0,\infty),\mathcal{S}'(\mathbb{R}^d))$ which is a Gaussian process with $\mathrm{Cov}(W_t(\phi),W_s(\psi)) = (t\wedge s)\int\phi(x)\psi(x)dx$. Then $W_t(\phi)$ can be extended by L^2-continuity to $L^2(\mathbb{R}^d)$ and then $\{W_t(B):t\geq 0,\ B$ bounded measurable$\}$ is a white noise on $\mathbb{R}_+\times\mathbb{R}^d$ with intensity measure dtdx. The $\mathcal{S}'(\mathbb{R}^d)$-valued process $\{W_t:t\geq 0\}$ is called *cylindrical Brownian motion.*

Example 7.1.3 The B(A,c)-Superprocess.

Consider the B(A,c)-superprocess as described in section 6.2.1. Let $\phi\in pD(A)$, and $\theta\in\mathbb{R}$. Then by (6.1.6),

$$\exp\left(\theta\left[\langle X_t,\phi\rangle-\int_0^t\langle X_s,A\phi\rangle ds\right]-\tfrac{1}{2}\theta^2c\cdot\left[\int_0^t\langle X_s,\phi^2\rangle ds\right]\right)$$

is a continuous local martingale. Therefore for $\phi\in D(A)$,

$$M_t^B(\phi) := \langle X_t,\phi\rangle-\langle X_0,\phi\rangle-\int_0^t\langle X_s,A\phi\rangle ds$$

is a $(\mathcal{D}_t)$-martingale with increasing process

$$\langle M^B(\phi)\rangle_t = c\int_0^t\langle X_s,\phi^2\rangle ds$$

and

$$\langle M^B(\phi),M^B(\psi)\rangle_t = c\int_0^t\langle X_s,\phi\psi\rangle ds,\quad \phi,\psi\in D(A).$$

Since D(A) is dense in C(E) and therefore bp dense in $b\mathcal{E}$, by taking limits we can extend this to all of $b\mathcal{E}$. This yields a continuous orthogonal martingale measure $M^B(ds,dx)$ with intensity $\nu((0,t]\times A) = c\int_0^t X_s(A)ds$.

Example 7.1.4 The (A,Φ)-Superprocess.

Consider the (A,Φ)-superprocess $\{X_t\}$ as formulated in section 6.1. Recall that for any $\phi\in D(A)$,

$$M_t(\phi) := \langle X_t,\phi\rangle - \langle X_0,\phi\rangle - \int_0^t\langle X_s,A\phi\rangle ds$$

is a martingale with decomposition

$$M_t(\phi) = M_t^c(\phi) + M_t^d(\phi)$$

where $M_t^c(\phi)$ is a continuous martingale with increasing process $\int_0^t\langle X_s,c\phi^2\rangle ds$ and $M_t^d(\phi)$ is a purely discontinuous martingale with predictable compensator $\int_0^t ds\int X_s(dx)\int_0^\infty n(x,du)\left(\exp(-u\phi(x))-1+u\phi(x)\right)$. Note that in general this does not yield an L^2-martingale measure since $M_t^d(\phi)$ need not belong to $L^2(P_\mu)$.

Example 7.1.5. The $S(A,\gamma)$ Fleming-Viot Process.

Consider the $S(A,\gamma)$-Fleming-Viot process $\{Y_t:t\geq 0\}$ introduced in Section 5.7.1 with $\gamma\in C_+$. Then for $\phi\in D(A)$,

$$M_t^{S,\gamma}(\phi) := \langle Y_t,\phi\rangle-\langle\mu,\phi\rangle-\int_0^t \langle Y_s,A\phi\rangle ds$$

is a $\{\mathcal{G}_t\}$-martingale (cf. Section 8.1 for notation) with increasing process

$$\langle M^{S,\gamma}(\phi)\rangle_t = \int_0^t [\langle Y_s,\phi^2\rangle-\langle Y_s,\phi\rangle^2]\gamma(s)ds.$$

Again this can be extended to all $\phi\in b\mathcal{E}$ and yields a continuous martingale measure. Note that for $A,B\in\mathcal{E}$,

$$Q(t;A,B) := \langle M^{S,\gamma}(A),M^{S,\gamma}(B)\rangle_t = \int_0^t [Y_s(A\cap B)-Y_s(A)Y_s(B)]\gamma(s)ds$$

and therefore this martingale measure is not orthogonal. However

$$0 \leq Q(t;A,A) = \int_0^t [Y_s(A)-Y_s(A)^2]\gamma(s)ds \leq \int_0^t Y_s(A)\gamma(s)ds.$$

Therefore, $K(dx\times dy\times[0,t]) := \int_0^t \delta_x(dy)Y_s(dx)ds$ is a dominating measure and the martingale measure, $M^{S,\gamma}(ds,dx)$, is worthy.

We next define stochastic integrals with respect to martingale measures. Let M be a worthy measure with dominating measure K. Define

$$(f,g)_K := \iiint_{E\times E\times\mathbb{R}_+} f(x,s)f(y,s)K(dx,dy,ds),\quad f,g\in b\mathcal{E}\times\mathcal{B}(\mathbb{R}_+),$$

and for predictable f, $\|f\|_M := E(|f|,|f|)_K$. Let

$$\mathcal{P}_M := \{f(\omega,s,x) \text{ is } \mathcal{P}\otimes\mathcal{E}\text{-measurable} : \|f\|_M<\infty\}.$$

If M is an orthogonal martingale measure with intensity ν, then

$$\mathcal{P}_M = L^2_\nu \text{ where}$$

$$L^2_\nu := \left\{f(\omega,s,x)\ \mathcal{P}\otimes\mathcal{E} \text{ measurable},\ E\left(\int_{\mathbb{R}_+\times E} f^2(\omega,s,x)\nu(\omega,ds,dx)\right) < \infty\right\}.$$

The collection of *simple predictable functions* is given by

$\mathcal{S} := \{h(\omega,s,x) = \sum_{i=1}^n h_i(\omega)1_{(u_i,v_i]}(s)1_{B_i}(x),\ B_i\in\mathcal{A},\ h_i\in b\mathcal{F}_{u_i}\}$. Then $\mathcal{S}$ is dense in $\mathcal{P}_M$ (Walsh (1986, Prop. 2.3)).

If h is a function of $\mathcal{S}$, we can define a new martingale measure

(7.1.1) $h\cdot M_t(A) = \left(\int_0^t h dM\right)(A) := \sum_{i=1}^{n} h_i(M_{v_i\wedge t}(A\cap B_i) - M_{u_i\wedge t}(A\cap B_i)) \quad \forall A\in \mathcal{A}.$

and we can show that $E((h\cdot M_t(A))^2) \le \|h\|_M^2$.
Since $\mathcal{S}$ is dense in $\mathcal{P}_M$ the linear mapping $h \longrightarrow \{h.M_t(A), t\ge 0, A\in\mathcal{A}\}$ can be extended to $h\in\mathcal{P}_M$ by taking a sequence $\{h_m\}$ in $\mathcal{S}$, with $\|h_m - h\| \longrightarrow 0$ and defining $h\cdot M$ as the limit in L^2 of $h_m\cdot M$. (It is easy to verify that the limit is independent of the sequence $\{h_m\}$.) If $h\in \mathcal{P}_M$, $h\cdot M$ is called the *stochastic integral martingale measure* of h with respect to M.

The *usual stochastic integral of the predictable process* h *with respect to the martingale measure* M is defined to be $h\cdot M_t(E)$ and is denoted by

$$\int_0^t\int_E h(s,x)M(ds,dx).$$

Theorem 7.1.6. Let M be a worthy martingale measure and $f\in \mathcal{P}_M$. Then
(a) $f\cdot M$ is a worthy martingale measure. If M is continuous, then $f.M$ is continuous.
(b) If f and g belong to $\mathcal{P}_M$, A and B $\in \mathcal{A}$, then

(7.1.2) $\langle f\cdot M(A), g\cdot M(B)\rangle_t = \int_{(0,t]}\int_{A\times B} f(s,x)g(s,y)Q_M(ds;dx,dy).$

if M is orthogonal,

(7.1.3) $\langle f\cdot M(A), g\cdot M(B)\rangle_t = \int_{(0,t]}\int_{A\cap B} f(s,x)g(s,x)\nu(ds,dx).$

Proof. See Walsh (1986, Chapt.2).

Theorem 7.1.7
(a) Let A_t be a continuous increasing adapted process. Then the following two statements are equivalent:

(7.1.3) M_t is a continuous local martingale with increasing process A_t,

(7.1.4) $Z_t(\theta) := \exp\left[\theta M_t - \frac{1}{2}\theta^2 A_t\right]$ is a continuous local martingale for each $\theta\in\mathbb{R}$.

Furthermore, $Z_t(\theta)$ is a supermartingale, and it is a martingale if and only if $E[Z_t(\theta)] = 1$ for every $t\ge 0$.
(b) If M_t is a local martingale and if either $\exp(\frac{1}{2}\theta M_t)$ is a submartingale or $E[\exp(\frac{1}{2}\theta^2 A_t)] < \infty$ for every t, then $Z_t(\theta)$ is a martingale.
(c) Let M be a continuous worthy martingale measure on E, with covariance functional $Q(t;.,.)$ and dominating measure K and $f\in \mathcal{P}_M$. Then

(7.1.5)

$$Z_t(\theta) := E\left(\exp\left\{\int_0^t\int_E \theta f(s,x)M(ds,dx) - 1/2\int_0^t\int_E\int_E \theta^2 f(s,x)f(s,y)Q(ds;dx,dy)\right\}\right)$$

is a continuous local martingale (and also a supermartingale) $\forall\theta\in\mathbb{R}$ and $\forall f\in \mathcal{P}_M$. If $E[Z_t(\theta)] = 1$ $\forall$ $t\geq 0$, then it is a martingale.

Proof. (a) cf. Priouret (1974, Ch. 1, Prop. 17 and Ikeda and Watanabe (1981, p. 142)).

(b) cf. Revuz and Yor (1990, p. 309).

(c) follows from (a), (b) and Theorem 7.1.6. □

7.2 Cameron-Martin-Girsanov Formula for Measure-valued Diffusions

In this section we will consider the Cameron-Martin-Girsanov formula in the setting of continuous measure-valued martingale problems. We begin by recalling the equivalent formulations of the martingale problem for a continuous $M_F(E)$-valued process with E a Polish space.

Assume that D(A) is a dense subspace of bC(E) and $A:D(A)\longrightarrow bC(E)$, $Q:M_F(E)\longrightarrow \mathcal{Q}_F(E\times E)$ where $\mathcal{Q}_F(E)$ denotes the collection of signed measures, ν, on E×E satisfying $0\leq \nu(A\times A) < \infty$ $\forall$ $A\in\mathcal{E}$.

Let $\mathfrak{D}(\mathfrak{G}) := \{F:F(\mu):= f(\langle\mu,\phi\rangle),\ \phi\in D(A),\ f\in C_b^\infty(\mathbb{R})\}$ and for $F\in \mathfrak{D}(\mathfrak{G})$,

$$\mathfrak{G}F(\mu) := f'(\langle\mu,\phi\rangle)\langle\mu,A\phi\rangle + \frac{1}{2}\iint f''(\langle\mu,\phi\rangle)\phi(x)\phi(y)Q(\mu;dx,dy).$$

Lemma 7.2.1 Assume that $Q(\mu;A,A) \leq K\mu(A)$ $\forall$ $A\in\mathcal{E}$ where K is a constant. Let P be a probability measure on $(D_E,\mathcal{D},(\mathcal{D}_t)_{t\geq 0})$ with $D_E = D([0,\infty),E)$. Let $\{X_t\}$ be a continuous $M_1(E)$-valued process. Then the following are equivalent:

(i) $\forall F\in \mathfrak{D}(\mathfrak{G})$, $\bar{M}_t(F) := F(X_t)-F(X_0) - \int_0^t \mathfrak{G}F(X_s)ds$ is a P-local-martingale,

(ii) $\forall$ $\phi\in D(A)$, $M_t(\phi) := \langle X_t,\phi\rangle-\langle X_0,\phi\rangle-\int_0^t \langle X_s,A\phi\rangle ds$ is a P-local martingale with increasing process

$$\int_0^t\iint\phi(x)\phi(y)Q(X_s;dx,dy)ds,$$

(iii) $\forall$ $\phi\in D(A)$,

$$Z_t(\phi) := \exp\left\{\langle X_t,\phi\rangle-\langle X_0,\phi\rangle-\int_0^t\langle X_s,A\phi\rangle ds - \frac{1}{2}\int_0^t\iint\phi(x)\phi(y)Q(X_s;dx,dy)ds\right\}$$

is a P-local martingale.

Proof. The equivalence of (i) and (ii) follows from Itô's lemma (Lemma 6.1.2).

The equivalence of (ii) and (iii) follows from Theorem 7.1.7. □

Note that under the conditions of Theorem 7.2.1, $M_t(.)$ extends to a worthy martingale measure.

We next introduce the Cameron-Martin-Girsanov formula for measure-valued processes (cf. Dawson (1978a)). In order to keep the exposition as simple as possible we will state it here for probability measure-valued processes on compact metric spaces. The extension to $M_F(E)$-valued processes is discussed in Section 10.1.

Let $R:M_1(E) \longrightarrow M_1(E)$ be a measurable mapping and let

$$\mathfrak{G}_R F(\mu) := f'(\langle\mu,\phi\rangle)\langle\mu,A\phi\rangle + f'(\langle\mu,\phi\rangle)\langle R(\mu),\phi\rangle$$
$$+ \frac{1}{2}\iint f''(\langle\mu,\phi\rangle)\phi(x)\phi(y)Q(\mu;dx,dy).$$

Note that the statements (i)-(iv) of Theorem 7.2.1 are also valid with $\mathfrak{G}_R$ in place of $\mathfrak{G}$.

<u>Theorem</u> <u>7.2.2</u> (<u>Cameron-Martin-Girsanov</u>)

Assume that $R(\mu)$ can be written in the integral form

$$R(\mu,dx) = \int r(\mu,y)Q(\mu;dx,dy)$$

where $r:M_1(E) \longrightarrow C(E)$ is bounded and measurable. Then the initial value martingale problem associated with $(\mathfrak{D}(\mathfrak{G}),\mathfrak{G}_R)$ is well-posed if and only if the initial value martingale problem for $(\mathfrak{D}(\mathfrak{G}),\mathfrak{G})$ is well-posed. If $\{\Pi_{[0,T]}P_\mu:\mu\in M_1(E)\}$ and $\{\Pi_{[0,T]}P^R_\mu:\mu\in M_1(E)\}$ denote the respective solutions on $(D([0,\infty),M_1(E)),(\mathcal{D}_t)_{t\ge 0},\mathcal{D})$ restricted to $\mathcal{D}_T$, then for each μ, $\Pi_{[0,T]}P_\mu$ and $\Pi_{[0,T]}P^R_\mu$ are equivalent measures and the Radon-Nikodym derivative

$$R_r(T) := \frac{d\,\Pi_{[0,T]}P^R_\mu}{d\,\Pi_{[0,T]}P_\mu} > 0, \quad \Pi_{[0,T]}P_\mu\text{-a.s.}$$

is given by $R_r(T) = Z_r(T)$

where for $g:M_1(E) \longrightarrow bC(E)$ bounded and continuous,

$$Z_g(t) := \exp\left\{\int_0^t\int_E g(X_s,y)M(ds,dy) - 1/2\int_0^t\int_E\int_E g(X_s;x)g(X_s;y)Q(X_s;dx,dy)ds\right\}$$

where $M(ds,dy)$ is the martingale measure obtained from $M_t(\phi)$.

<u>Proof.</u> First assume that the initial value martingale problem for $(\mathfrak{D}(\mathfrak{G}),\mathfrak{G})$ is well posed with solution $\{P_\mu\}$.

Let $g:M_1(E)\longrightarrow C(E)$ be bounded and measurable. Since $\{X_s\}$ is continuous and g is bounded and measurable, $g(X_s,.) \in \mathcal{P}_M$ and by Theorem 7.1.6 the stochastic integral $\int_0^t\int g(X_s,y)M(ds,dy)$ is a continuous martingale with increasing process

$\int_0^t\int_E\int_E g(X_s;x)g(X_s;y)Q(X_s;dx,dy)ds.$

Therefore by Theorem 7.1.7, $Z_g(t)$ is a continuous P_μ-local martingale. Since under our assumptions $\int_0^t\int_E\int_E g(X_s;x)g(X_s;y)Q(X_s;dx,dy)ds$ is a.s. bounded, we conclude that $Z_g(t)$ is a P_μ-martingale from Theorem 7.1.7(b).

Define $P^R_\mu(B) := Z_r(T)P_\mu(B)$ for $B\in \mathcal{D}_T$.

Then

$$\frac{d\,\Pi_{[0,T]}P^R_\mu}{d\,\Pi_{[0,T]}P_\mu} = Z_r(T)$$

and P^R_μ is a probability measure on $\mathcal{D}_T$ since $Z_r(t)$ is a martingale.

If we define $g(\mu) := \phi + r(\mu)$ with $\phi\in D(A)$, then

$$Z_g(t) = Z_r(t).\exp\left\{\int_0^t\int\phi(x)M(ds,dx) - \int_0^t\iint\phi(x)r(X_s;y)Q(X_s;dx,dy)ds - \frac{1}{2}\int_0^t\iint\phi(x)\phi(y)Q(X_s;dx,dy)ds\right\}$$

$$= Z_r(t)\cdot\exp\left\{\int_0^t\phi(x)M(ds,dx) - \int_0^t\iint\phi(x)R(X_s;dx)ds - \frac{1}{2}\int_0^t\iint\phi(x)\phi(y)Q(X_s;dxdy)ds\right\}$$

is a P_μ-martingale.

Hence for each $\phi \in D(A)$,

$$\exp\left\{\int_0^t\phi(x)M(ds,dx) - \int_0^t\iint\phi(x)R(X_s;dx)ds - \frac{1}{2}\int_0^t\iint\phi(x)\phi(y)Q(X_s;dx,dy)ds\right\}$$

is a P^R_μ-martingale. To verify this one should note that

$$dP^R_\mu|_{\mathcal{D}_t}|_{\mathcal{D}_s} = dP^R_\mu|_{\mathcal{D}_s},\quad s < t.$$

But since

$$dP^R_\mu|_{\mathcal{D}_t} = Z_r(t)dP_\mu|_{\mathcal{D}_t},$$

it follows that

$$dP^R_\mu|_{\mathcal{D}_t}|_{\mathcal{D}_s} = P_\mu[Z_r(t)|_{\mathcal{D}_s}]dP_\mu|_{\mathcal{D}_s} = dP^R_\mu|_{\mathcal{D}_s}$$

since $Z_r(s)$ is a P_μ-martingale.

Hence by Theorem 7.2.2 P^R_μ is a solution to the initial value martingale problem for $(\mathfrak{D}(\mathfrak{G}),\mathfrak{G}_R)$. It remains to verify the uniqueness. Let P^R_μ denote a solution to the initial value martingale problem for $(\mathfrak{D}(\mathfrak{G}),\mathfrak{G})$. Let

$$Z^R_{-r}(t) := \exp\left\{\int_0^t\int_E -r(X_s,y)M_R(ds,dy) - 1/2\int_0^t\int_E\int_E r(X_s;x)r(X_s;y)Q(X_s;dx,dy)ds\right\},$$

here $M_R(.)$ denotes the martingale measure coming from the martingale

$$M_{R,t}(\phi) := \langle X_t,\phi\rangle - \langle X_0,\phi\rangle - \int_0^t (\langle X_s,A\phi\rangle + \langle R(X_s),\phi\rangle)ds$$

corresponding to the martingale problem for $(\mathfrak{D}(\mathfrak{G}),\mathfrak{G}_R)$.

Then the same argument as above yields

$$\Pi_{[0,T]}P_\mu = Z^R_{-r}(T)\Pi_{[0,T]}P^R_\mu \quad \text{and} \quad Z^R_{-r}(T) > 0,\ P^R_\mu\text{-a.s.}$$

which means that

$$\frac{dP_\mu}{dP^R_\mu}\Big|_{\mathcal{D}_T} = Z^R_{-r}(T).$$

Therefore if two distinct solutions P^R_μ and $\tilde{P}^R_\mu$ exist, then this would give two distinct solutions P_μ and $\tilde{P}_\mu$ of the initial value martingale problem for $(\mathfrak{D}(\mathfrak{G}),\mathfrak{G})$ leading to a contradiction. Hence the uniqueness is proved. □

7.3. Application to Sample Path Continuity

Using the martingale structure we can now improve the sample path continuity result which was obtained in Theorem 4.7.2.

Proposition 7.3.1

(a) Let $\{X(t):t\geq 0\}$ be a $B(A,c)$-superprocess where $(A,D(A))$, and d are as in Theorem 4.7.2. Then with probability one, $\{X(t):t\geq 0\}$ is β-Hölder continuous in the metric d for any $\beta < 1/2$.

(b) The $(\alpha,d,1)$-superprocess has trajectories in $C([0,\infty),M_p(\dot{\mathbb{R}}^d))$ (with p as in Section 4.6) so that $\phi_p \in D(A)$ P_μ-a.s. for all $\mu \in M_p(\mathbb{R}^d)$.

(c) Assume that $D(A)$ is convergence determining in $M_1(E)$. Then the $S(A,\gamma)$-Fleming-Viot process (cf. ex. 7.1.5) has trajectories in $C([0,\infty),M_1(E))$ a.s.

Proof. (a) Let $\phi\in D(A)$. Then it follows from Theorems 6.1.3 and 7.2.1 that

$$M^B_t(\phi) := \langle X_t,\phi\rangle - \langle X_0,\phi\rangle - \int_0^t \langle X_s,A\phi\rangle ds$$

is a martingale and Theorem 4.7.2 implies that it is continuous. Then M^B extends to a continuous martingale measure $M^B(ds,dx)$ and applying Itô's formula (Lemma 6.1.2) to $\psi(s,x) = S_{t-s}\phi(x)$, $0\leq s\leq t$, we obtain

$$\langle X_t,\phi\rangle - \langle X_0,S_t\phi\rangle = \int_0^t\int S_{t-s}\phi(x)M^B(ds,dx) = \hat{X}_t(\phi) \quad \text{for } \phi\in D(A).$$

By bounded pointwise convergence this can be extended to $\phi\in b\mathcal{E}$. Further if

$\psi:\mathbb{R}_+\times E\to\mathbb{R}$ is bounded and measurable, then $\int_0^t\int\psi(s,x)M^B(ds,dx)$ is a continuous martingale with increasing process $\int_0^t\int\psi^2(s,x)X(s,dx)ds$, and by the Burkholder-Davis-Gundy inequality (cf. Revuz and Yor (1991, p. 151)),

$$E\left|\int_0^t\int\psi(s,x)M^B(ds,dx)\right|^q \le C_q\, E\left(\left(\int_0^t\int\psi^2(s,x)X(s,dx)ds\right)^{q/2}\right),\quad q>0. \tag{7.3.1}$$

Therefore for $q\ge 2$, $0\le t<u\le T$, $\phi\in D(A)$,

$$E(|\hat{X}_u(\phi)-\hat{X}_t(\phi)|^q)^{1/q}$$

$$\le\left[E\left(\left|\int_t^u\int_{\mathbb{R}^d}S_{u-s}\phi(x)M^B(ds,dx)\right|^q\right)^{1/q}+E\left(\left|\int_0^t\int_{\mathbb{R}^d}(S_{u-s}\phi(x)-S_{t-s}\phi(x))M^B(ds,dx)\right|^q\right)^{1/q}\right]$$

$$\le C_q\left[E\left(\left(\int_t^u\langle X_s,(S_{u-s}\phi)^2\rangle ds\right)^{q/2}\right)^{1/q}+E\left(\left(\int_0^t\langle X_s,S_{t-s}[S_{u-t}\phi-\phi]^2\rangle ds\right)^{q/2}\right)^{1/q}\right]$$

(by (7.3.1))

$$\le C_q\left[\left(\sup_{s\le u}E(\langle X_s,1\rangle^{q/2})\|\phi\|^q(u-t)^{q/2}\right)^{1/q}+\left((u-t)^{q/2}t^{q/2}\sup_{s\le u}E(\langle X_s,1\rangle^{q/2})\;\|A\phi\|^q\right)^{1/q}\right]$$

$$\le \text{const }(\|\phi\|+\|A\phi\|)(u-t)^{1/2}.$$

The remaining argument is similar to that of Theorem 4.7.2.

(b) (Sketch of proof). From (a) we know that there exists a continuous $M_F(\mathbb{R}^d)$-valued process. To extend the process to $M_p(\dot{\mathbb{R}}^d)$ we first define $X_n(t)\in M_F(\mathbb{R}^d)$ using the initial condition $X_n(0):=X(0)1(0\le|x|<n)$. It is then easy to check that if $\phi\in D(A)\cap C_p$ and $A\phi\in C_p$, then using Doob's maximal inequality

$$\sup_{t\le t_0}|\langle X(t),\phi\rangle-\langle X_n(t),\phi\rangle|\longrightarrow 0\quad\text{a.s. as } n\to\infty.$$

Recalling that $\phi_p\in D(A)$ we conclude that $\langle X(t),\phi\rangle$ is continuous a.s. for ϕ belonging to a $M_p(\dot{\mathbb{R}}^d)$-convergence determining class. Thus $X(.)\in C([0,\infty),M_p(\dot{\mathbb{R}}^d))$, $P_{X(0)}$-a.s.

(c) The proof of (c) is similar to the proof of (a). □

<u>Remarks 7.3.2</u> (a) Reimers (1989) and Perkins (1991) proved that under additional conditions on the semigroup S_t (e.g. analyticity) $\langle X(t),\phi\rangle$ is a.s. continuous on $(0,\infty)$ for any bounded measurable function ϕ and hence $X(t)$ is continuous in the τ-topology (i.e. the $\sigma(M_F(E),b\mathcal{E})$-topology). The idea is that under analyticity, the

above inequality in (a) may be replaced by one involving only $\|\phi\|$.

(b) Fitzsimmons (1988) established path regularity (right process, Hunt process) for the (W,Φ)-superprocesses when W is a right (Hunt) process by using the Ray compactification and Rost's theorem on Skorohod embedding. He also obtained a.s. continuity when $\eta\equiv 0$ in his general setting using the results of Bakry and Emery.

(c) It is reasonable to ask whether $\langle X(t),\phi\rangle$ is a semimartigale for $\phi\in\mathcal{E}$. The fact that this is not necessarily so was established by Tribe in the following theorem.

Theorem 7.3.3 Let $\mu\in M_F(\mathbb{R}^d)$, $d\geq 1$, and X_t be the $(\alpha,d,1)$ superprocess. Let H be the indicator of a halfspace. Define

$\Gamma(\alpha) := 2\alpha/(\alpha+1)$ if $\alpha>1$.

Then for any $T > 0$ we have the following decomposition:

$X_t(H) = \mu(H) + V_t + M_t$ for $0\leq t\leq T$

where M_t is a continuous L^2 martingale satisfying $\langle M\rangle_t = \int_0^t X_s(H)ds$ and V_t is continuous on $(0,T]$.

If $0 < \alpha < 1$ and μ has a bounded density then V_t has integrable variation on $[0,T]$.

If $1 < \alpha \leq 2$ and μ has a bounded density, u, then V_t has integrable $\Gamma(\alpha)$ variation on $[0,T]$. If in addition the density is uniformly Hölder continuous and satisfies $u(0,x) > 0$ for some x on the boundary of the halfspace then with probability one V_t has strictly positive $\Gamma(\alpha)$ variation on $[0,T]$ and hence X_t fails to be a semimartingale.

Proof. Tribe (1989, Theorem 3.2).

7.4. The Weighted Occupation Time Process

Let $\{X_t\}$ denote the (A,Φ)-superprocess on E (cf. Sect. 6.1). Following Iscoe (1986) we define the associated *weighted occupation time process*, $\mathcal{O}_{s,t}$ as follows:

$$(7.4.1)\qquad \mathcal{O}_{s,t}(A) := \int_s^t X_u(A)du,\quad A\in\mathcal{E}.$$

Theorem 7.4.1. Let $\psi,\phi\in C_0(E)$, $\mu\in M_F(E)$. Then

$$(7.4.2)\qquad E_\mu\left(\exp\{-[\langle X_t,\phi\rangle+\langle\mathcal{O}_{0,t},\psi\rangle]\}\right) = \exp\left(-\int u(t,x)\mu(dx)\right)$$

where u is the unique solution of the initial value problem

$$(7.4.3)\qquad \frac{\partial u}{\partial t} = (Au + \Phi(u)) + \psi,\quad u(0,x) = \phi(x)$$

where Φ is as in Section 6.1.

Proof. Assume that $\psi(.,x)\in C_b^1(\mathbb{R}_+)$, $\psi(t,.)\in D_0(A)$, $t\to A\psi(t,x)$ is differentiable, $\psi>0$, and $\frac{\partial\psi}{\partial t}(t,.)\in D_0(A)$. If $f\in C_b^\infty(\mathbb{R})$, then by the time inhomogeneous form of

Itô's lemma we obtain

(7.4.4)

$$f(\langle X_t,\psi(t)\rangle) - \int_0^t \left(f'(\langle X_s,\psi(s)\rangle)\langle X_s, A\psi(s)+ \frac{\partial\psi}{\partial s}\rangle + \frac{1}{2}f''(\langle X_s,\psi(s)\rangle)\langle X_s, c\psi^2(s)\rangle\right) ds$$

$$- \int_0^t \langle X_s, \int_0^\infty n(.,d\lambda)\left(f(\langle X_s,\psi(s)\rangle+\lambda\psi(s,.))-f(\langle X_s,\psi(s)\rangle)-f'(\langle X_s,\psi(s)\rangle)\lambda\psi(s,.)\right)\rangle ds$$

is a martingale.

For $\phi\in D(A)$, let $\xi(t,x)$ be the unique evolution solution of

$$(7.4.5) \qquad \frac{\partial}{\partial t}\xi(t,x) = A\xi(t,x) + \Phi(\xi(t,x)) + \psi(x), \qquad \xi(0,x) = \phi(x).$$

Now we apply (7.4.4) to the function $f(u) = e^{-u}$ and $\psi(s,x) := \xi(t-s,x)$ where ξ is given by (7.4.5) to verify that

$\exp\{-[\langle X_t,\psi(t)\rangle+\langle \mathcal{O}_{0,t},\psi\rangle\}$ is a P_μ-local martingale.

But then it follows that it is a martingale by bounded convergence. Therefore

$$P_\mu\left(\exp\{-[\langle X_t,\phi\rangle+\langle \mathcal{O}_{0,t},\psi\rangle]\}\right) = \exp\{-\langle\mu,\xi(t)\rangle\} \quad \forall\ \phi\in D(A),\psi\in C(E).$$

Then we obtain it for general ϕ by taking limits. □

<u>Remark 7.4.2.</u> It is also possible to obtain the weighted occupation time process by an appropriate rescaling of the analogous process for the branching particle systems (cf. Gorostiza and López-Mimbela (1992)). Since particles in a branching particle system with branching rate n have lifetimes of order 1/n, it turns out that $\mathcal{O}_{st}(A)$ can also be interpreted as the (renormalized) limiting mass of particles which die in the time interval [s,t) and whose location at time of death is in the set A.

<u>Remark 7.4.3.</u> The weighted occupation time of a set is a special case of the class of *J-functionals* introduced by Dynkin (1990). To explain this notion consider a (W,Φ)-superprocess $\{X_t\}$ and let A be an *additive functional* of W (i.e. a random measure A on $\mathcal{B}(\mathbb{R}_+)$ such that $A[r,t) \in \sigma\{W_s: r\le s<t\}$). Let $\mathcal{A}^o_W$ denote the set of all finite additive functionals with compact support. Convergence of additive functionals is defined by $A_n \Longrightarrow A$ if $P_{r,x}(A_n[r,\infty)\longrightarrow A[r,\infty)) = 1\ \forall\ (r,x)$ and there exists b such that $A_n[b,\infty) = 0\ \forall n$ and $\sup_{\omega,n} A_n[0,b] < \infty$.

For $u\in\mathbb{R}_+$, and $f\in \mathcal{E}^u$, $A_t = 1_{u<t}f(W_u)$ defines an element of $\mathcal{A}^o_W$. Denote by $\mathcal{A}^\ell_W$ the minimal subcone of $\mathcal{A}^o_W$ which contains all these functionals. Let $A\in\mathcal{A}^1_W$ if there exists $A_n\in\mathcal{A}^\ell_W$ such that $A_n(B)\uparrow A(B)\ \forall\ B\in\mathcal{B}(\mathbb{R}_+)$.

If $A \in \mathcal{A}^1_W$ has the form $\sum_{t\in\Lambda} f^t(W_t)\delta_t$ where Λ is finite then we define $I(A) =$

$\sum_{t\in\Lambda} \langle f^t, X_t \rangle$. For general $A \in \mathcal{A}^1_W$, with $A_n \Rightarrow A$, (A_n as above), set $I(A) := \text{med}\lim_{n\to\infty} I(A_n)$ (for the definition of the medial limit, cf. Dellacherie and Meyer). $I(A)$ is called the *J-functional* of the superprocess $\{X_t\}$ associated with the additive functional A.

For example, if W is continuous and B is an open set in E, let

$$A_n(\{\tfrac{k}{n}\}) := \tfrac{1}{n}1_B(W_{k/n})1(k\le nt),\ k,n \in \mathbb{N}.$$

Then $A_n \Rightarrow A_{s\wedge t}$ where $A_s = \int_0^s 1_B(W_u)du$ (occupation time in B up to $s \le t$).

Then $I(A_n) = \sum_{k\le nt} X_{k/n}(B)$ converges in probability to $\mathcal{O}_{0,t}(B)$ and hence the weighted occupation time of the set B at time t, $\mathcal{O}_{0,t}(B)$, is the J-functional associated with the additive functional $s \longrightarrow \int_0^{s\wedge t} 1_B(W_u)du$.

8. STRUCTURAL PROPERTIES OF BRANCHING AND SAMPLING MARTINGALE PROBLEMS

8.1. Relations between Branching and Sampling Measure-valued Processes

Let E be a compact metric space and $(A,D(A))$ be the generator of a Feller process on E. In this chapter we explore the relations between the continuous A-superprocess and the Fleming-Viot process.

Let us first restrict our attention to the class, $B(A,c)$, $c > 0$, of continuous critical branching superprocesses, that is, (W,Φ)-superprocesses in which W is a Feller process with generator $(D(A),A)$, $\Phi(\lambda) \equiv -\frac{1}{2}c\lambda^2$. By Theorem 4.7.2 the resulting $M(E)$-valued process can be realized as a continuous process

$$\left\{C([0,\infty),M(E)),\{\mathcal{F}_t\}_{t\in\mathbb{R}_+},\{X_t\}_{t\in\mathbb{R}_+},\{P_\mu^{B(A,c)}:\mu\in M(E)\}\right\}$$

where $\mathcal{F}_t = \bigcap_{s>t}\mathcal{F}_s^0$, $\mathcal{F}_s^0 = \sigma\{X_u: u\le s\}$. $\{P_\mu^{B(A,c)}:\mu\in M(E)\}$ is characterized as the unique solution of the martingale problem: for $\phi\in D(A)$,

$$M_t^B(\phi) = \langle X_t,\phi\rangle - \langle\mu,\phi\rangle - \int_0^t \langle X_s,A\phi\rangle ds$$

is a $P_\mu^{B(A,c)}$, $\{\mathcal{F}_t\}$-martingale with increasing process $\langle M^B(\phi)\rangle_t = c\int_0^t \langle X_s,\phi^2\rangle ds$.

The second class of processes we will consider are the time inhomogeneous Fleming-Viot sampling processes, $S(\sigma A,\gamma)$, $\gamma^{-1}\in C_+$, $\sigma\in pC(\mathbb{R}_+)$, constructed in Theorem 5.7.1. They are denoted by

$$\left\{C([0,\infty),M_1(E)),\{\mathcal{G}_t\}_{t\in\mathbb{R}_+},\{Y_t\}_{t\in\mathbb{R}_+},\{P_\mu^{S(\sigma A,\gamma)}:\mu\in M_1(E)\}\right\}$$

where $\mathcal{G}_t = \bigcap_{s>t}\mathcal{G}_s^0$, $\mathcal{G}_s^0 = \sigma\{Y_u: u\le s\}$, and $\mathcal{G} = \bigvee_t \mathcal{G}_t$.

$\{P_\mu^{S(\sigma A,\gamma)}:\mu\in M_1(E)\}$ is characterized as the unique solution of the martingale problem: for $\phi\in D(A)$,

$$M_t^{S,\gamma}(\phi) = \langle Y_t,\phi\rangle - \langle\mu,\phi\rangle - \int_0^t \sigma(s)\langle Y_s,A\phi\rangle ds,\quad t<\tau_\gamma$$

$$= M_{\tau_\gamma-}^{S,\gamma}(\phi)\quad \text{for } t\ge\tau_\gamma,\ \text{where } \tau_\gamma := \inf\{t:\gamma(t)=0\}$$

is a $\{P_\mu^{S(\sigma A,\gamma)}\}$, $\{\mathcal{G}_t\}$-martingale with increasing process

$$\langle M^{S,\gamma}(\phi)\rangle_t = \int_0^t [\langle Y_s,\phi^2\rangle - \langle Y_s,\phi\rangle^2]\gamma(s)ds,\quad t<\tau_\gamma$$

and $Y_t = Y_{\tau_\gamma}$, $t\ge\tau_\gamma$.

The main difference between the two processes is that the total mass of the Fleming-Viot process, $\langle Y_t,1\rangle$, is constant in time but the total mass process for the critical branching process, $\langle X_t,1\rangle$, is a martingale with increasing process $t\to$

$c\int_0^t \langle X_s,1\rangle ds$. In particular the B(A,c) process suffers extinction, that is, $\tau_\infty < \infty$ a.s. where $\tau_\infty := \inf\{t:\langle X_t,1\rangle = 0\}$ (extinction of critical branching). Konno and Shiga (1988) and Shiga (1990a) discovered a relationship between the normalized process associated with a *time changed version*, $\tilde{X}(t)$, of the B(A,c)-superprocess and the $S(\sigma(.)A,\gamma)$-process (with $\sigma(t) = \langle\tilde{X}_t,1\rangle$). Etheridge and March (1990) discovered that the S(A,c) process arises by conditioning the normalized B(A,c)-superprocess to have constant total mass. Perkins (1991) established that the conditional law of the B(A,c)-superprocess given the total mass process $\langle X_t,1\rangle$ is given by a $S(A,\gamma(.))$-process with $\gamma(t) = \langle X_t,1\rangle^{-1}$.

To explain these results let us begin with some formal calculations based on the results of Shiga (1990). Consider the B(A,c)-process X_t and define

$$C_t := \int_0^t \langle X_s,1\rangle^{-1}\, ds.$$

Let D_t denote the inverse of C_t on $[0,C_{\tau_\infty-})$. Then $\tilde{Z}_t := \langle X_{D_t},1\rangle$ is a martingale with increasing process $c\int_0^t (\tilde{Z}_s)^2 ds$. Hence $\tilde{Z}_t = \tilde{Z}_0\exp\left(c^{1/2}B_t - \frac{1}{2}ct\right)$ for some Brownian motion B_t and ($\tilde{Z}_t$ is the only martingale whose increasing process is $c\cdot\int_0^t \tilde{Z}_s^2 ds$).

Then $\langle X_t,1\rangle = \tilde{Z}_0\exp\left(c^{1/2}B_{C_t} - \frac{1}{2}cC_t\right)$. Therefore $C_t\uparrow\infty$ a.s. since otherwise $\lim_{t\uparrow\tau_\infty} \langle X_t,1\rangle > 0$, which yields a contradiction.

Hence C_t is a homeomorphism between $[0,\tau_\infty)$ and $[0,\infty)$. Let $D_t:[0,\infty) \longrightarrow [0,\tau_\infty)$ be the continuous strictly increasing inverse to C_t.

Now define $\tilde{X}_t := X_{D_t}$, $\tilde{Y}_t = \tilde{X}_{nor}(t) := \tilde{X}_t/\langle\tilde{X}_t,1\rangle$ and $\mathcal{G}_t := \mathcal{F}_{D_t}$. Then $\{\tilde{Y}_t\}$ is a probability measure-valued process. We now derive the martingale problem for $\tilde{Y}_t$. For $\phi\in\mathcal{D}(A)$,

$$\begin{aligned}\langle\tilde{X}_t,\phi\rangle &= \langle\mu,\phi\rangle + \int_0^{D_t} \langle X_s,A\phi\rangle ds + M^B_{D_t}(\phi)\\ &= \langle\mu,\phi\rangle + \int_0^t \langle\tilde{X}_s,A\phi\rangle\langle\tilde{X}_s,1\rangle ds + \tilde{N}_t(\phi)\end{aligned}$$

where, since D_t is a continuous time change, $\tilde{N}_t(\phi)$ is a continuous $\mathcal{G}_t$-local martingale with increasing process

$$\begin{aligned}\langle\tilde{N}(\phi)\rangle_t &= c\int_0^{D_t} \langle X_s,\phi^2\rangle ds\\ &= c\int_0^t \langle\tilde{X}_s,\phi^2\rangle\langle\tilde{X}_s,1\rangle ds.\end{aligned}$$

Then

$$<\tilde{X}_t,1> = <\mu,1> + \tilde{N}_t(1)$$

$$<\tilde{N}(1)>_t = \int_0^t c<\tilde{X}_s,1>^2 ds$$

$$<\tilde{N}(\phi),\tilde{N}(1)>_t = \int_0^t c<\tilde{X}_s,\phi><\tilde{X}_s,1>ds.$$

Applying Ito's formula and noting that $<\tilde{X}_t,1> > 0$ for all $t > 0$ we have

$$<\tilde{Y}_t,\phi> = <\mu_{nor},\phi> + \int_0^t <\tilde{X}_s,1><\tilde{Y}_s,A\phi>ds + N_t^{S,1}(\phi)$$

where $\mu_{nor} := \mu/\mu(E)$, and $N_t^{S,1}(\phi)$ is a continuous $\mathcal{G}_t$-local martingale satisfying

$$<N(\phi)>_t = c\int_0^t [<\tilde{Y}_s,\phi^2>-<\tilde{Y}_s,\phi>^2]ds.$$

Hence $\{\tilde{Y}_t\}$ is a $S(\sigma A,\gamma)$-process with "random evolution operator", that is, with $\gamma=c$, $\sigma(s) = <\tilde{X}_s,1>$.

We can also reverse the process and obtain a "skew product" representation for $\tilde{X}_t$ with initial measure μ. To do this we begin with $\tilde{Z}_t = \tilde{Z}_0\exp(c^{1/2}B_t - \frac{1}{2}ct)$, with $\tilde{Z}_0 = <\mu,1> > 0$. Note that $\tilde{Z}_s$ is a martingale with increasing process

$$<\tilde{Z}>_t = \int_0^t c(\tilde{Z}_s)^2 ds.$$

Conditioned on $\tilde{Z}_t$, let $\tilde{Y}_t$ be a $S(\sigma A,\gamma)$ process with initial measure μ_{nor}, $\gamma=c$, $\sigma(s) = \tilde{Z}_s$. We can then verify that $\tilde{Z}_t\tilde{Y}_t \overset{\mathcal{D}}{=} \tilde{X}(t)$.

We can then do the reverse time change as follows. Let $D_t := \int_0^t \tilde{Z}_s ds$ and note that $D_t \uparrow \tau_\infty < \infty$. Now let C_t denote the inverse of D_t on $[0,\tau_\infty-)$. Then $\tilde{Z}_{C_t}\tilde{Y}_{C_t} \overset{\mathcal{D}}{=} X_t$, thus yielding a skew product representation of the $B(A,c)$-superprocess. Moreover note that $Z_t := \tilde{Z}_{C_t}$ is a Feller critical continuous branching process (cf. Sect. 6.2.1) and $Y_t = \tilde{Y}_{C_t}$ is a $S(A,\gamma)$-process with $\gamma(s) = c/Z_s$. These formal calculations suggest the following result which will now be proved following Perkins (1991).

<u>Theorem 8.1.1.</u> Let $Q_y := P_\mu^{B(A,c)}\circ(<X_\cdot,1>)^{-1}$ with $y:=<\mu,1>$.
If $\mu\in M_F(E)-\{0\}$, and $\mu_{nor}(.) := \mu(.)/\mu(E)$, then $\forall$ $B\in \mathcal{G}$

$$P_\mu^{B(A,c)}(X_{nor} \in B|<X_\cdot,1>=\gamma^{-1}) = P_{\mu_{nor}}^{S(A,\gamma)}(B), \quad \text{for } Q_{<\mu,1>}\text{-a.a. } \gamma^{-1}.$$

Hence $P_{\mu_{nor}}^{S(A,\gamma)}$ is a regular conditional distribution for $P_\mu^{B(A,c)}\circ(X_{nor})^{-1}$ given $\gamma(.) = <X_\cdot,1>^{-1}$.

Proof. Step 1. Representation of the Normalized Process.

Let $\tau_n := \inf\{t:\langle X_t,1\rangle \le 1/n\}$ and $\tau_\infty := \inf\{t:\langle X_t,1\rangle=0\}$. Let $\phi\in D(A)$ be fixed. Then applying Itô's lemma to $\langle X_t,\phi\rangle/\langle X_t,1\rangle$ we obtain

(8.1.1) $$\langle X_{nor}(t\wedge\tau_n),\phi\rangle = \langle\mu_{nor},\phi\rangle + \int_0^t 1(s\le\tau_n)\langle X_{nor}(s),A\phi\rangle ds + \bar{M}_t^n(\phi).$$

where

(8.1.2)

$$\bar{M}_t^n(\phi) := \int_0^t 1(s\le\tau_n)\langle X_s,1\rangle^{-1}dM_s^B(\phi) - \int_0^t 1(s\le\tau_n)\langle X_s,\phi\rangle\langle X_s,1\rangle^{-2}dM_s^B(1).$$

and

$$\langle\bar{M}_t^n\rangle = \int_0^t 1(s<\tau_n)[\langle X_{nor}(s),\phi^2\rangle - \langle X_{nor}(s),\phi\rangle^2]\langle X_s,1\rangle^{-1}ds.$$

From (8.1.1)

(8.1.3) $$\sup_{t\le T,n\in\mathbb{N}} |\bar{M}_t^n(\phi)| \le 2\|\phi\|_\infty + T\|A\phi\|_\infty.$$

Using Doob's maximal inequality we obtain

(8.1.4) $$E\Big[\sup_{t\le T}(\bar{M}_t^n(\phi)-\bar{M}_t^m(\phi))^2\Big] \le 4E(\bar{M}_T^n(\phi)-\bar{M}_T^m(\phi))^2.$$

For each ϕ and T, $\{\bar{M}_T^n(\phi):n\in\mathbb{N}\}$ is a bounded martingale (in n) (cf. (8.1.2) and (8.1.3)). Therefore by the martingale convergence theorem $\bar{M}_T^n(\phi)$ converges a.s. and in L^2 as $n\to\infty$. Using (8.1.4) and the usual Borel-Cantelli argument we can choose a subsequence $\{n_k\}$ such that almost surely $\{\bar{M}_t^{n_k}(\phi)\}$ is a Cauchy sequence in $C([0,T])$ with continuous limit $\bar{M}_t(\phi)$. In addition $\{\bar{M}_t(\phi):t\ge0\}$ is a continuous martingale that satisfies

(8.1.5) $$\bar{M}_t^n(\phi) = \bar{M}_{t\wedge\tau_n}(\phi) \text{ and } \bar{M}_t(\phi) = \bar{M}_{t\wedge\tau_\infty}(\phi), \quad \forall t\ge0, \text{ a.s.}$$

(8.1.6) $$\sup_{t\le T} |\bar{M}_t(\phi)| \le 2\|\phi\|_\infty + T\|A\phi\|_\infty, \text{ a.s.}$$

Letting $n\to\infty$ in (8.1.2) we obtain

(8.1.7)

$$\langle X_{nor}(t),\phi\rangle = \langle\mu_{nor},\phi\rangle + \int_0^t 1(s<\tau_\infty)\langle X_{nor}(s),A\phi\rangle ds + \bar{M}_t(\phi)\ , \quad t\ge0, \text{ a.s.}$$

Step 2. Martingale Problem in Enlarged Filtration.

Let $\mathcal{H}_t = \mathcal{F}_t\vee\sigma\{\langle X_s,1\rangle:s\ge0\}$. We will now prove that $\bar{M}_t(\phi)$ is an $\mathcal{H}_t$-martingale.

Let s<t and let F be a bounded $\sigma(\langle X_s,1\rangle: s\geq 0)$-measurable random variable. Recall that $\langle X_t,1\rangle$ is the unique solution to the martingale problem:

$\langle X_t,1\rangle$ is a continuous martingale with increasing process $c\int_0^t \langle X_s,1\rangle ds$.

Then the predictable representation theorem (cf. Theorem 5.1.1(ii))) applied to the martingale $E[F|\sigma(\langle X_s,1\rangle: 0\leq s\leq t)]$ provides the representation

$$(8.1.8)\qquad F = P_\mu(F) + \int_0^\infty f(s)d\langle X_s,1\rangle$$

for some $\sigma(\langle X_s,1\rangle: s\leq t)$-predictable f. Therefore

$$P_\mu^{B(A,c)}((\bar{M}_{t\wedge\tau_n}(\phi)-\bar{M}_{s\wedge\tau_n}(\phi))F|\mathcal{F}_s)$$

$$= P_\mu^{B(A,c)}((\bar{M}_t^n(\phi)-\bar{M}_s^n(\phi))\int_0^\infty f(u)dM_u^B(1)|\mathcal{F}_s) \text{ by (8.1.5) and (8.1.8)}$$

$$= P_\mu^{B(A,c)}\Bigg(\Bigg[\int_s^t 1(u\leq\tau_n)\langle X_u,1\rangle^{-1}dM_u^B(\phi)$$

$$- \int_s^t 1(u\leq\tau_n)\langle X_u,\phi\rangle\langle X_u,1\rangle^{-2}dM_u^B(1)\Bigg]\int_s^t f(u)dM_u^B(1)|\mathcal{F}_s\Bigg)$$

(by definition of $\bar{M}_t^n(\phi)$)

$$= P_\mu^{B(A,c)}\Bigg(\int_s^t 1(u\leq\tau_n)(\langle X_u,\phi\rangle\langle X_u,1\rangle^{-1}-\langle X_u,\phi\rangle\langle X_u,1\rangle^{-1})f(u)du|\mathcal{F}_s\Bigg)$$

(by definition of $M_u^B(\phi)$, $M_u^B(1)$)

$$= 0.$$

Let $n\to\infty$ in the above formula, then an application of (8.1.5) and (8.1.6) yields$P_\mu((\bar{M}_t(\phi)-\bar{M}_s(\phi))F|\mathcal{F}_s) = 0$. This completes the proof that $\bar{M}_t(\phi)$ is an $\mathcal{H}_t$-martingale. Moreover it follows from (8.1.2) and (8.1.5) that

(8.1.9)

$$\langle\bar{M}(\phi)\rangle_t = \int_0^t 1(s<\tau_\infty)(\langle X_{nor}(s),\phi^2\rangle-\langle X_{nor}(s),\phi\rangle^2)\langle X_s,1\rangle^{-1}ds,\ P_\mu^{B(A,c)}\text{-a.s.}$$

<u>Step 3.</u> The Conditional Martingale Problem. Let $\{P(B|\gamma): B\in\mathcal{G}, \gamma^{-1}\in C_+\}$ be a regular conditional probability for X_{nor} given $\{\langle X.,1\rangle = \gamma^{-1}\}$ (under $P_\mu^{B(A,c)}$). We will now verify that $P(B|\gamma)$ satisfies the $S(A,\gamma)$-martingale problem for $Q_{\langle\mu,1\rangle}$-a.e. γ^{-1}.

If $\gamma^{-1}\in C_+$ and $\phi\in D(A)$, define

$$M_t^{S,\gamma}(\phi) := \langle Y_t,\phi\rangle - \langle\mu,\phi\rangle - \int_0^t \langle Y_s,A\phi\rangle ds,\quad t<\tau_\gamma,$$

$$M_t^{S,\gamma}(\phi) := M_{\tau_\gamma -}^{S,\gamma}(\phi)\quad \text{for}\quad t\geq\tau_\gamma,$$

on $\left\{C([0,\infty),M_1(E)),\{\mathcal{G}_t\}_{t\in\mathbb{R}_+},\{Y_t\}_{t\in\mathbb{R}_+},\{P_\mu^{S(A,\gamma)}:\mu\in M_1(E)\}\right\}$.

By (8.1.5) and (8.1.7) we conclude that

(8.1.10) $$\bar{M}_t(\phi) = M_t^{S,<X.,1>^{-1}}(\phi)(X_{nor}) \ \forall\ t\geq 0,\ P_\mu^{B(A,c)}\text{-a.s.}$$

If $G\in b\mathcal{G}_s^o$, and $s<t$, then the $\mathcal{H}_t$-martingale property of $\bar{M}_t(\phi)$ shows that

$$P_\mu^{B(A,c)}((\bar{M}_t(\phi)-\bar{M}_s(\phi))G(X_{nor})\,|\,<X.,1>) = 0, \qquad P_\mu^{B(A,c)}\text{-a.s.}$$

and hence by (8.1.10),

(8.1.11) $$P((M_t^{S,\gamma}(\phi)-M_s^{S,\gamma}(\phi))G\,|\,\gamma) = 0, \quad \text{for } Q_{<\mu,1>}\text{-a.e. } \gamma^{-1}.$$

We can choose Λ, with $Q_{<\mu,1>}(\Lambda) = 0$ such that (8.1.11) holds on Λ^c for all rational $s<t$ and all G in a countable set in $b\mathcal{G}_s^o$ with pointwise bounded closure $b\mathcal{G}_s^o$. Taking limits in s and G we obtain that

(8.1.12) $\{M_t^{S,\gamma}(u):t\geq 0\}$ is an $(\mathcal{G}_t)$-martingale under $P(.\,|\,\gamma)$ for $\gamma^{-1}\in\Lambda^c$.

If $\tau_n^\gamma = \inf\{u:\gamma(u)\geq n\}$, then (8.1.9) implies that

(8.1.13)

$$M_{t\wedge\tau_n^\gamma}^{S,\gamma}(\phi)^2-\int_0^t 1(s<\tau_n^\gamma)[<Y_s,\phi^2>-<Y_s,\phi>^2]\gamma(s)ds \quad \text{is an } \mathcal{G}_t\text{-martingale}$$

under $P(.\,|\,\gamma)$ $\forall n\in\mathbb{N}$, $Q_{<\mu,1>}$-a.e. γ^{-1}. Now consider a countable core, D_0, for A and fix γ^{-1} outside a $Q_{<\mu,1>}$-null set so that (8.1.12) and (8.1.13) hold for all $\phi\in D_0$. Take uniform limits in $(\phi,A\phi)$ to see that (8.1.12) and (8.1.13) hold for all ϕ in D(A) and for $Q_{<\mu,1>}$-a.e. γ^{-1}. Therefore $P(.\,|\,\gamma)$ solves the $S(A,\gamma)$-martingale problem for $Q_{<\mu,1>}$-a.e. γ^{-1} and hence $P(.\,|\,\gamma) = P_{\mu_{nor}}^{S(A,\gamma)}$ for $Q_{<\mu,1>}$-a.e. γ^{-1} by Theorem 5.7.1. □

<u>Theorem</u> <u>8.1.2</u> (Etheridge and March (1990))

For $1>\varepsilon>0$ let $\tau(\varepsilon) := \inf\{t>0:|<X_t,1>-1|\geq\varepsilon\}$ and fix $0<T<T'<\infty$. Let $\varepsilon_n\to 0$, and $\mu_n\longrightarrow\mu$ in $M_F(E)$ with $\mu\in M_1(E)$. Let $B_n := \{\tau(\varepsilon_n) > T'\}$ and note that $P_{\mu_n}^{B(A,c)}(B_n) > 0$.
Then the conditional law $(P_{\mu_n}^{B(A,c)}(X_{nor}\in\cdot\,|\,B_n))|_{\mathcal{F}_T}$ is well-defined and

$$(P_{\mu_n}^{B(A,c)}(X_{nor}\in\cdot\,|\,B_n))|_{\mathcal{F}_T} \Rightarrow P_\mu^{S(A,\gamma)}|_{\mathcal{G}_T} \quad \text{as } n\to\infty, \text{ with } \gamma(.)\equiv c$$

where $P_\mu^{S(A,\gamma)}|_{\mathcal{G}_T}$ denotes the restriction of $P_\mu^{S(A,\gamma)}(B)$ to $\mathcal{G}_T$.

<u>Proof.</u> Let $\phi:C([0,T],M_1(E))\longrightarrow\mathbb{R}$ be bounded and continuous. Then by Theorem

8.1.1,

$$|P^{B(A,c)}_{\mu_n}(\phi(X_{nor})|B_n) - P^{S(A,c)}_{\mu}(\phi)| = |\int_{B_n}\left(P^{S(A,\gamma')}_{\mu_{n,nor}}(\phi) - P^{S(A,c)}_{\mu}(\phi)\right)Q^{(n)}(d\gamma')|$$

where $Q^{(n)}(d\gamma') := Q_{<\mu_n,1>}(d\gamma')/Q_{<\mu_n,1>}(B_n)$.

Note that $B_n = \{<X_t,1> \in \tilde{B}_n\}$ with $\tilde{B}_n := \{f\in C_+, \sup_{0\le t\le T'} |f(t)-1| < \varepsilon_n\}$.

Hence

$$|\int_{B_n}\left(P^{S(A,\gamma')}_{\mu_{n,nor}}(\phi) - P^{S(A,c)}_{\mu}(\phi)\right)Q^{(n)}(d\gamma')| \le \sup_{f\in\tilde{B}_n} |P^{S(A,\gamma')}_{\mu_{n,nor}}(\phi) - P^{S(A,c)}_{\mu}(\phi)|.$$

But $\sup_{f\in\tilde{B}_n} |P^{S(A,\gamma')}_{\mu_{n,nor}}(\phi) - P^{S(A,c)}_{\mu}(\phi)| \longrightarrow 0$ as $n \to \infty$ by Theorem 5.7.1(b).This completes the proof. □

Remark. We can use the above conditional process together with a more general total mass process which is a one dimensional diffusion to build a larger class of measure-valued diffusions.

8.2. Pure Atomic Measure-valued Processes

In the remainder of this chapter we will use the tools of stochastic calculus developed in the last chapter to identify classes of measure-valued processes which belong to one of two extreme cases, namely, pure atomic processes and absolutely continuous processes.

In this section we consider the S(A,γ)-Fleming-Viot process on a compact metric space E with $\gamma(.) \equiv \gamma > 0$ and show that the process is pure atomic if the mutation generator is bounded.

Refer to Section 3.4.1 for the definitions of $M_a(E)$, ζ_a, $\mathfrak{A}_\infty$, $\bar{\mathfrak{A}}_\infty$, ϑ and the partition family $\{E^n_j, x^n_j; n,m,j\in\mathbb{N}\}$.

Theorem 8.2.1 Assume that $\gamma(.) \equiv \gamma > 0$.

(a) If the mutation operator A is a bounded linear operator on $\mathcal{E}$ of the form

$$Af(x) = \theta(x)\int(f(y)-f(x))P(x,dy)$$

where $\theta\in p b\mathcal{E}$ and $P(x,.)$ is a transition function on E, then

$$P^{S(A,\gamma)}_{\mu}\{Y_t\in M_a(E) \text{ for all } t > 0\} = 1.$$

(b) Assume in addition that θ is constant and for each $x \in E$, $P(x,.)$ has no atoms. Let $\vartheta(t) := \vartheta(Y_t)$ for $t\ge0$. Then $\vartheta(.)$ is a solution of the $C([0,\infty);\bar{\mathfrak{A}}_\infty)$-martingale problem for the infinite dimensional differential operator

(8.2.1) $$G_\infty = \frac{\gamma}{2}\sum_{i=1}^{\infty}\sum_{j=1}^{\infty} a_i(\delta_{ij}-a_j)\frac{\partial^2}{\partial a_i \partial a_j} - \theta\sum_{i=1}^{\infty} a_i\frac{\partial}{\partial a_i}$$

with $\mathcal{D}(G_\infty)$ taken to be the algebra generated by $\{1,f^2,f^3,\ldots\}$ where for $\eta > 0$,

(8.2.2) $f^\eta(a) := \sum a_i^\eta$.

Then Y is the size ordered atom process for the infinitely-many-neutral-alleles diffusion (cf. Example 5.7.3).

Proof. (cf. Ethier and Kurtz (1981, and 1987 p. 441))

We first reduce (a) to (b). Let $\tilde{E} = E\times[0,1]$, $\bar{\theta} = \sup \theta(x)$,

$$\tilde{P}(x,u;dydv) := \frac{\theta(x)}{\bar{\theta}} P(x,dy)dv + \frac{\bar{\theta}-\theta(x)}{\bar{\theta}}\delta_x(dy)dv, \text{ and set}$$

$$\tilde{A}\phi(x,u) := \bar{\theta}\Big[\int_0^1\int(\phi(y,v)-\phi(x,u))\tilde{P}(x,u;dy)dv\Big].$$

Then the $\tilde{A}$ martingale problem is of type of (b). Let $\tilde{Y}_t$ be the solution to the $\tilde{A}$-martingale problem. Then $Y(t,A) = \tilde{Y}(t,A\times[0,1])$ and the atomicity result for $Y(.)$ clearly follows from the corresponding result for $\tilde{Y}(.)$. Hence it suffices to prove (b).

(b) Let $\{Y_t\}$ be a solution of the martingale problem for $\mathfrak{G}$ (defined by (5.4.1)). Since A is a bounded operator we have that for each $\phi\in b\mathcal{E}$ $\langle X_t,\phi\rangle$ is a semimartingale and

(8.2.3) $$\tilde{M}_t(\phi) := \langle Y_t,\phi\rangle - \int_0^t [\langle Y_s,\theta(.)\int(\phi(y)-\phi(.))P(.,dy)\rangle]ds$$

is a continuous martingale with increasing process

$$\langle\tilde{M}(\phi)\rangle_t = \gamma\int[\langle Y_s,\phi^2\rangle-\langle Y_s,\phi\rangle^2]ds.$$

Let $\{E_j^n : n,j\in\mathbb{N}\}$, $n\in\mathbb{N}$, be the sets of a partition family as defined in Section 3.4.1. For $\mu\in M_1(E)$ let

$$\xi_n(\mu) := \sum_j \mu(E_j^n)\delta_{x_j^n}.$$

If $\vartheta^n(t) := \vartheta(\xi_n(Y(t)))$, that is, the descending order statistics of $\{Y(t,E_1^n),Y(t,E_2^n),\ldots\}$, then $\vartheta(t) := \vartheta(Y_t) = \lim_{n\to\infty}\vartheta^n(t)$.

For $\eta>1$, let

$$f^\eta(a) := \sum_j a_i^\eta \quad \text{for } a\in\bar{\mathfrak{A}}_\infty, \text{ and}$$

$$F_n^\eta(\mu) := \sum_j \mu(E_j^n)^\eta = f^\eta(\xi_n(\mu)).$$

For $\eta \geq 2$, by Itô's formula

$$(8.2.4) \qquad M_{\eta,n}(t) = F_n^{\eta}(Y_t) - \int_0^t \mathfrak{G}F_n^{\eta}(Y_s)ds$$

is a bounded (uniformly in n) continuous square integrable martingale where

$$\mathfrak{G}F_n^{\eta}(\mu) = \sum_{i=1}^{\infty}\Big[\eta\theta(\langle P(.,E_i^n),\mu\rangle - \mu(E_i^n))\mu(E_i^n)^{\eta-1} + \tfrac{\gamma}{2}\eta(\eta-1)(\mu(E_i^n)-\mu(E_i^n)^2)\mu(E_i^n)^{\eta-2}\Big].$$

Since $P(x,.)$ has no atoms, then

$$(8.2.5) \qquad \mathfrak{G}F_n^{\eta}(\mu) = G_{\infty}f^{\eta}(\xi_n(\mu)) + R_n(\mu) \quad \text{where} \quad |R_n(\mu)| \xrightarrow{bp} 0 \text{ as } n\to\infty.$$

Therefore as $n \to \infty$, $M_{\eta,n}(t)$ converges to

$$(8.2.6) \quad M_{\eta}(t) = f^{\eta}(\vartheta(t)) - \int_0^t G_{\infty}f^{\eta}(\vartheta(s))ds$$

which is therefore a $\{\mathcal{F}_t^{\vartheta}\}$-martingale where $\mathcal{F}_t^{\vartheta} := \sigma\{\vartheta(Y_s):0\leq s\leq t\}$. Similar calculations show that the analogues of (8.2.5), (8.2.6) are valid for all elements of $\mathfrak{D}(G_{\infty})$ and therefore $\vartheta(.)$ is a solution of the martingale problem for G_{∞}. The result (b) will then follow if we verify that

$$(8.2.7) \qquad P_{\mu}^{S(A,\gamma)}(\vartheta(t) \in \mathfrak{A}_{\infty} \text{ for all } t > 0) = 1.$$

Using Itô's lemma we can verify that for $\eta \geq 2$,

$$(8.2.8) \; Z_{\eta}(t) := f^{\eta}(\vartheta(t))-f^{\eta}(\vartheta(0)) - \int_0^t \Big\{\tfrac{1}{2}\gamma\eta(\eta-1)(f^{\eta-1}(\vartheta(s)) - f^{\eta}(\vartheta(s))) + \tfrac{1}{2}\eta\theta(f^{\eta-1}(\vartheta(s))\Big\}ds$$

is a square integrable continuous martingale with increasing process

$$(8.2.9) \qquad I_{\eta}(t) = \eta^2\gamma^2\int_0^t (f^{2\eta-1}(\vartheta(s))-f^{\eta}(\vartheta(s))^2)ds.$$

In fact (8.2.8) and (8.2.9) remain valid for $\eta \in (1,2)$. To verify this consider the C^2-function

$$f_{\varepsilon}^{\eta}(x) := \sum h_{\varepsilon}(x_i) \quad \text{with} \quad h_{\varepsilon}(u) := (u+\varepsilon)^{\eta}-\varepsilon^{\eta} - \eta\varepsilon^{\eta-1}u,$$

apply Itô's lemma to $f_{\varepsilon}^{\eta}(\vartheta(s))$ for $\varepsilon>0$ and then take the limit $\varepsilon \to 0$.

(The details can be found in Ethier and Kurtz (1981).)

Now let $f^{1+}(\mu) := \text{bp-}\lim_{\eta\downarrow 1} f^{\eta}(\mu) = \sum \mu(\{x\})$, and note that $f^1(\mu) := 1$.

Then

$$(8.2.10)\quad 0 = \lim_{\eta\to 2+} E(Z_\eta(t)-Z_2(t)) = E\left[\int_0^t \gamma(1-f^{1+}(\vartheta(s)))ds\right] \text{ for } t\geq 0.$$

Then (8.2.10) yields $P_\mu^{S(A,c)}(\vartheta(t) \in \mathfrak{A}_\infty \text{ a.e. } t>0) = 1$. To show that $\vartheta(t) \in \mathfrak{A}_\infty$ for all t, a.s. note that

$$(8.2.11)\qquad \lim_{\eta\downarrow 1} E[Z_\eta(t_0)]^2 = \lim_{\eta\downarrow 1} E[I_\eta(t_0)]$$

$$= \gamma^2 E\left[\int_0^{t_0} f^{1+}(\vartheta(s))[1-f^{1+}(\vartheta(s))]ds\right] = 0.$$

Therefore by Doob's maximal inequality

$$(8.2.12)\quad \sup_{0\leq t\leq t_0} |Z_\eta(t)| \longrightarrow 0 \quad \text{in probability as} \quad \eta \longrightarrow 1+,$$

and hence there exists a sequence $\eta_n \to 1+$ such that

$$(8.2.13)\quad \sup_{0\leq t\leq t_0} |Z_{\eta_n}(t)| \longrightarrow 0 \quad \text{a.s.}$$

Letting $\rho(t) := \limsup_{n\to\infty} \frac{1}{2}\gamma\eta_n(1-\eta_n)\int_0^t f^{\eta_n-1}(\vartheta(s))ds$

we obtain from (8.2.8)

$$(8.2.14)\qquad f^{1+}(\vartheta(t))-f^{1+}(\vartheta(0)) - \rho(t) - \tfrac{1}{2}\theta t = 0 \quad \text{for} \quad 0\leq t\leq t_0.$$

Since $\rho(t)$ is nondecreasing in t, (8.2.13) implies that $P(\vartheta(t) \in \mathfrak{A}_\infty \ \forall\, t>0) = 1$. □

Remarks. (1) In the previous result we can replace the constant function γ by any function γ such that $\gamma^{-1} \in C_+$. This together with Theorem 8.1.1 allows us to conclude that the B(A,c)-superprocess (c > 0), X_t, t>0, is a.s. pure atomic if A is a bounded operator satisfying the hypotheses of Theorem 8.2.1.

(2) Ethier and Kurtz (1992) introduced a stronger topology on $M_1(E)$, called the *weak atomic topology* such that convergence in this topology implies weak convergence as well as the convergence of the sizes and locations of the atoms. They also showed under the conditions of Theorem 8.2.1 that the Fleming-Viot process has continuous sample paths in this topology. Under the same conditions Shiga (1990) proved continuity in the total variation norm (except at t=0).

(3) Consider the special case of Theorem 8.2.1 in which E = [0,1] and

$$Af(x) = \theta\int_{[0,1]} (f(y)-f(x))dy.$$

In this case there exits a unique ergodic reversible stationary distribution (cf. Ethier and Kurtz (1992)). It is then natural to pose the question as to *how many* non-zero atoms there are. Schmuland (1991) proved using Dirichlet form techniques that the stationary process $\vartheta(t)$ hits the the (k-1) dimensional simplex $\{a: \sum_{i=1}^{k} a_i = 1\}$ if and only if $\gamma > 4\theta$.

8.3 Absolutely Continuous Superprocesses and Stochastic Evolution Equations

In this section we consider the canonical (α,d,β)-superprocess $(D,\mathcal{D},(\mathcal{D}_t)_{t\ge0},(X_t)_{t\ge0},(P_\mu)_{\mu\in M_F(\mathbb{R}^d)})$. The objective is to determine when the random measures X_t on $\mathbb{R}^d$ are absolutely continuous and in this case to discuss the formulation in terms of stochastic evolution equations.

Theorem 8.3.1 For $t > 0$, $X(t,dx)$ is absolutely continuous with continuous density $X(t,x)$, P_μ-a.s. $\forall\ \mu \in M_F(\mathbb{R}^d)$ if and only if $d < \alpha/\beta$.

Proof. Some special cases of this result were proved in Dawson and Hochberg (1979), and Roelly-Coppoletta (1986). We will now sketch the main steps of the proof following Fleischmann (1988).

Step 1. Log-Laplace Equation with Measure-valued Initial Condition.

Let $\{S_t\}$ denote the α-symmetric stable semigroup and consider the log-Laplace equation

$$(8.3.1) \qquad v_t = S_t\varphi - \int_0^t S_{t-s}v_s^{1+\beta}ds.$$

Given $\varphi\in pC_c(\mathbb{R}^d)$ the mapping $\varphi \longrightarrow v$ is denoted by $V[\varphi]$.

For $\rho\ge1$, let $(L^{\rho,T},\|.\|_{\rho,T})$ denote the Banach space of measurable real-valued functions, f, on $[0,T]\times\mathbb{R}^d$ with

$$\|f\|_{\rho,T} := \int_0^T\int_{\mathbb{R}^d} |f(s,x)|^\rho dsdx < \infty.$$

If $\varphi\in pC_c(\mathbb{R}^d)$, then it is easy to check that $V[\varphi]|_{[0,T]} \in L^{1,T}\cap L^{1+\beta,T}$. The mapping $V:pC_c(\mathbb{R}^d) \longrightarrow L^{1,T}\cap L^{1+\beta,T}$ extended by continuity from $pC_c(\mathbb{R}^d)$ to $M_F(\mathbb{R}^d)$ (equipped with the weak topology) such that

(8.3.2) $\forall\varphi\in M_F(\mathbb{R}^d)$, $V[\varphi]$ is a solution to (8.3.1),

$\varphi \longrightarrow V[\varphi]$ is continuous, and

$\lim_{t\downarrow0} V[\varphi]_t = \varphi$ (in $M_F(\mathbb{R}^d)$).

Moreover the function $\theta \longrightarrow V[\theta\delta_0]_t$ for $\theta\ge0$, $t>0$ is differentiable in θ at $\theta=0$ and

$$(8.3.3) \qquad -\frac{\partial V[\theta\delta_0]_t}{\partial\theta}\Big|_{\theta=0} = S_t\delta_0.$$

For the proofs of (8.3.2) and (8.3.3) refer to Fleischmann (1988) or [D, Lemma

7.1.2].

Step 2. Proof of Absolute Continuity

Using Lemma 3.4.2.2 the proof of absolute continuity in the case $d < \alpha/\beta$ reduces to verifying certain analytical properties for the log-Laplace equation (8.3.1). In particular condition (i) of Lemma 3.4.2.2 follows from (8.3.2) which implies that

$$V[\theta\varphi_n] \longrightarrow L_z(t,\theta,\cdot) := V[\theta\delta_z],$$

provided that $\varphi_n \longrightarrow \delta_z$ as $n \to \infty$,

and condition (ii) of Lemma 3.4.2.2 follows from (8.3.3) which implies that

$$-\partial L_z(t,\theta,x)/\partial\theta|_{\theta=0} = p_\alpha(t,z-x).$$

Step 3. Proof of Singularity.

The a.s. singularity of X_t for $t > 0$, $d > \alpha/\beta$ will be proved in the case $\alpha = 2$ in Cor. 9.3.3.5. (The proof for the case $\alpha<2$ can be found in Dawson (1992, Theorem 7.3.4)). For the case $d = \alpha/\beta$ see Dawson and Hochberg (1979) and Fleischmann (1988). □

We next consider in more detail the $B(\Delta_\alpha,c)$-superprocess on $\mathbb{R}^d$, $\{X_t:t\geq 0\}$. By Theorem 6.1.3 this process is characterized as the solution of the martingale problem: for $\psi\in C^{1,2}([0,\infty),\mathbb{R}^d)$,

$$\langle X_t,\psi_t\rangle-\langle X_0,\psi_0\rangle-\int_0^t \left\langle X_s, \left(\frac{\partial\psi}{\partial s} + \Delta_\alpha\psi\right)\right\rangle ds = M_t(\psi)$$

where $M_t(\psi)$ is a continuous martingale with increasing process

$$\langle M(\psi)\rangle_t = \int_0^t \psi^2(s,x)X_s(dx).$$

We denote by $M_t(dx)$ the corresponding orthogonal martingale measure and $\int_0^t\int_{\mathbb{R}^d} f(s,x)M(ds,dx)$ the corresponding stochastic integral constructed in Section 7.1. The covariance functional of M_t is given by $Q(t;dx,dy) = \int_0^t \delta_x(dy)X_s(dx)ds$.

The following theorem was proved independently by Konno and Shiga (1986), and Reimers (1986). It establishes the fact that in one dimension the $(\alpha,1,1)$-superprocess has a density which in an appropriate sense satisfies the stochastic partial differential (or pseudo differential if $\alpha < 2$) equation:

$$(8.3.4)\quad d\tilde{X}_t(x) = \Delta_\alpha\tilde{X}_t(x)dt + \sqrt{\tilde{X}_t(x)}\; W(dt,dx)$$

where $W(.,.)$ is space-time white noise.

Theorem 8.3.2. Let $d = 1$, $\beta = 1$, $A = \Delta$ or Δ_α, $0<\alpha\le 2$ and $\{S_t\}$ denote the corresponding semigroup. Let $\mu\in M_F(\mathbb{R}^d)$ and P_μ denote the law of the $B(\Delta_\alpha,1)$-superprocess. Assume that $\mu = X_0(dx) = \tilde{X}_0(x)dx$ where $\tilde{X}_0(x)$ is bounded and continuous. Then there exist a jointly continuous density process $\tilde{X}_t(x)$ and a Gaussian white noise, $W(ds,dx)$ (or cylindrical Brownian motion (cf. Example 7.1.2)) defined on an extension of the canonical probability space $(D,\mathcal{D},(\mathcal{D}_+)_{t\ge 0},P_\mu)$ such that $\forall\ \phi \in C_c^\infty(\mathbb{R}^1)$ and $t > 0$,

(8.3.5a)

$$\int_{\mathbb{R}} X_t(x)\phi(x)dx = \int_{\mathbb{R}} X_0(x)\phi(x)dx + \int_0^t\int_{\mathbb{R}} X_s(x)(A\phi)(x)dxds + \int_0^t\int_{\mathbb{R}} \sqrt{\tilde{X}_s(x)}\ \phi(x)W(ds,dx),$$

P_μ-a.s., that is,. $\tilde{X}_t(x)$ is a *weak* $D'(\mathbb{R})$-valued *solution* of (8.3.4).
In addition, for fixed $t\ge 0$, and $x \in \mathbb{R}$,

$$\text{(8.3.5b)}\quad \tilde{X}_t(x) = \int p_t(y-x)X_0(y)dy + \int_0^t\int_{\mathbb{R}} p_{t-s}(y-x)\sqrt{\tilde{X}_s(y)}\ W(ds,dy),\ P_\mu\text{-a.s.}$$

where $p_t(.)$ is the appropriate α-stable (or Brownian) transition density, that is, $\tilde{X}_t(x)$ is also a *weak evolution (mild) solution* of (8.3.4).

In addition, P_μ-a.s. $\tilde{X}_t(x)$ is Hölder continuous in t with any modulus less than 1/4 and Hölder continuous in x with any modulus less than 1/2.

Sketch of Proof. We will not give a complete proof but will sketch the main ideas. We begin with the the canonical probability space $(D,\mathcal{D},(\mathcal{D}_t)_{t\ge 0},P_\mu)$.

Step 1. Construction of the Gaussian White Noise.

Lemma 8.3.3 There exist a Gaussian white noise, W, and an $(\mathcal{F}_t)$-predictable function $\tilde{X}_t(x,\omega):[0,\infty)\times\mathbb{R}\times\Omega \longrightarrow [0,\infty)$ defined on an extension of the canonical probability space, denoted by $(\Omega,\mathcal{F},\mathcal{F}_t,\bar{P}_\mu)$, such that

$$\text{(8.3.6)}\quad \int_{\mathbb{R}} \tilde{X}_t(x)\phi(x)dx - \int_{\mathbb{R}} \tilde{X}_0(x)\phi(x)dx$$

$$= \int_0^t\int_{\mathbb{R}} \sqrt{\tilde{X}_s(x)}\ \phi(x)W(ds,dx) + \int_0^t\int_{\mathbb{R}} \tilde{X}_s(x)(A\phi)(x)dxds$$

holds for every $\phi\in C_c^\infty(\mathbb{R})$, P_μ-a.s.

Proof. By Theorem 8.3.1 $X_t(dx)$ is almost surely absolutely continuous. Therefore $X_t(dx) = \tilde{X}_t(x)dx$, P_μ-a.s., where $\tilde{X}_t(z) := \lim_{\varepsilon\downarrow 0} \dfrac{X_t(B(z,\varepsilon_n))}{\varepsilon_n}$. Since X_t is a continuous process $\tilde{X}_t(x)$ is $\mathcal{F}_t$-predictable (cf. Section 7.1).

We first construct a cylindrical Brownian motion $\bar{W}_t$ and associated white noise $\bar{W}(ds,dx)$ on a probability space $(\bar{\Omega},\bar{\mathcal{F}},\bar{\mathcal{F}}_t,\bar{P}^W)$ independent of $(D,\mathcal{D},(\mathcal{D}_t)_{t\geq 0},P_\mu)$ and define the extended probability space $(\Omega,\mathcal{F},\mathcal{F}_t,\bar{P}_\mu)$ to be the product space $(\Omega,\mathcal{F},\mathcal{F}_t,P^W)\otimes(D,\mathcal{D},(\mathcal{D}_t)_{t\geq 0},P_\mu)$. Then for $\phi\in C_c^\infty(\mathbb{R})$ define

$$(8.3.7)\qquad W_t(\phi) := \int_0^t\int_{\mathbb{R}} \frac{1}{\sqrt{\tilde{X}_s(x)}}\, 1(\tilde{X}_s(x)\neq 0)\phi(x)M(ds,dx) + \int_0^t\int_{\mathbb{R}} 1(\tilde{X}_s(x)=0)\phi(x)\bar{W}(ds,dx)$$

where $M(ds,dx)$ is the martingale measure associated with the Fleming-Viot process. Then W_t is a cylindrical Brownian motion which extends to a Gaussian white noise $W(ds,dx)$ defined on $(\Omega,\mathcal{F},\mathcal{F}_t,P_\mu)$, and

$$(8.3.8)\qquad M_t(\phi) = \int_0^t\int_{\mathbb{R}} \sqrt{\tilde{X}_s(x)}\ \phi(x)W(ds,dx)\quad \forall\ \phi\in C_c^\infty(\mathbb{R}).$$

Thus

$$(8.3.9)\qquad \int_{\mathbb{R}}\tilde{X}_t(x)\phi(x)dx - \int_{\mathbb{R}}\tilde{X}_0(x)\phi(x)dx = \int_0^t\int_{\mathbb{R}}\sqrt{\tilde{X}_s(x)}\ \phi(x)W(ds,dx) + \int_0^t\int_{\mathbb{R}}\tilde{X}_s(x)(A\phi)(x)dxds$$

P_μ-a.s. $\forall\ t\geq 0$. □

<u>Step 2.</u> $\tilde{X}$ is an evolution solution and a.s. jointly continuous.

<u>Proof.</u> For $\phi,\psi\in C_c^\infty(\mathbb{R})$, we have applying (8.3.6) that

$$P_\mu\left(\left(\int_{\mathbb{R}}\tilde{X}_t(x)\phi(x)dx\right)\cdot\int_0^t\int_{\mathbb{R}}\sqrt{\tilde{X}_s(x)}\ \psi(x)W(ds,dx)\right)$$

$$= P_\mu\left(\int_{\mathbb{R}}\tilde{X}_t(x)\phi(x)dx\ \left(\int_{\mathbb{R}}\tilde{X}_t(x)\psi(x)dx-\int_{\mathbb{R}}\tilde{X}_0(x)\psi(x)dx - \int_0^t\int_{\mathbb{R}}\tilde{X}_s(x)(A\psi)(x)dxds\right)\right).$$

But then by Lemma 4.7.1,

$$(8.3.10)\qquad P_\mu(\langle X_t,\phi\rangle) = \langle\mu,S_t\phi\rangle,$$
$$P_\mu(\langle X_t,\phi\rangle\langle X_t,\psi\rangle) = \int_0^t ds\langle\mu S_{t-s},(S_s\phi)(S_s\psi)\rangle + \langle\mu,S_t\phi\rangle\langle\mu,S_t\psi\rangle$$

and therefore

$$(8.3.11)\qquad P_\mu\left(\left(\int_{\mathbb{R}}\tilde{X}_t(x)\phi(x)dx\right)\cdot\int_0^t\int\sqrt{\tilde{X}_s(x)}\ \psi(x)W(ds,dx)\right) = \int_0^t ds\langle\mu S_s,\psi S_t\phi\rangle.$$

Using (8.3.11) and a piecewise constant approximation (cf. Konno and Shiga (1986) for the details) we can obtain

(8.3.12) $P_\mu\left(\left(\int_{\mathbb{R}}\tilde{X}_t(x)\phi(x)dx\right)\cdot\int_0^t\int\sqrt{\tilde{X}_s(x)}\; S_{t-s}\phi(x)W(ds,dx)\right) = \int_0^t ds\langle\mu S_s,(S_{t-s}\phi)^2\rangle.$

Using (8.3.10), (8.3.11) and (8.3.12) we can then verify by direct calculation that

$$P_\mu\left(\int_{\mathbb{R}}\tilde{X}_t(x)\phi(x)dx - \int_{\mathbb{R}}\tilde{X}_0(x)S_t\phi(x)dx - \int_0^t\int_{\mathbb{R}}\int_{\mathbb{R}}\sqrt{\tilde{X}_s(x)}\; p_{t-s}(x-y)\phi(y)dyW(ds,dx)\right)^2 = 0.$$

This proves that for $\phi\in C_c^\infty(\mathbb{R})$,

$$\int_{\mathbb{R}}\tilde{X}_t(x)\phi(x)dx = \int_{\mathbb{R}}\tilde{X}_0(x)S_t\phi(x)dx + \int_0^t\int_{\mathbb{R}}\sqrt{\tilde{X}_s(x)}\;(S_{t-s}\phi)(x)W(ds,dx),\quad P_\mu\text{-a.s.}$$

This completes the construction of the white noise W and the verification of the fact that $\tilde{X}$ is both a weak $D'(\mathbb{R})$ solution and an evolution solution ofequation (8.3.4).

To prove the a.s. joint continuity of the mapping $(t,x)\to\tilde{X}_t(x)$ it suffices to show the a.s. joint continuity of

$$Z_t(x) := \int_{t_0}^t\int_{\mathbb{R}}p_{t-s}(y-x)\sqrt{\tilde{X}_s(y)}\; W(ds,dy).$$

This is proved by verifying that (cf. argument in proof of Proposition 7.3.1) for some $\delta,\delta'>0$,

$$P_\mu\left([Z_t(x)-Z_s(y)]^{2n}\right) \le \text{const}\cdot(|t-s|^\delta+|x-y|^{\delta'})^{n-1}$$

and then using Lemma 3.7.2(b). The Hölder continuity is obtained by a refinement of this argument (see Reimers (1986)). □

<u>Corollary 8.3.4.</u> X_t is absolutely continuous for all $t>0$, with probability one.
<u>Proof.</u> This follows immediately from the a.s. continuity as a function of s of both the measure-valued process $X_s(dx)$ and the function-valued density process $\tilde{X}_s(.)$. □

8.4. Some Remarks on the Stochastic Evolution Equation

<u>Remarks.</u> 1. Konno and Shiga (1986) also established that a similar stochastic evolution equation can be obtained for the $S(A,\gamma)$-Fleming-Viot process $\{Y_t:t\ge0\}$ in $\mathbb{R}^1$. In this case $Y_t(dx) = \tilde{Y}_t(x)dx$ and $\tilde{Y}_t(x)$ satisfies the stochastic integral equation (evolution solution)

$$\tilde{Y}_t(x) = \int_{\mathbb{R}}p_t(x-y)\tilde{Y}_0(y)dy + \gamma\int_0^t\int_{\mathbb{R}}p_{t-s}(y-x)\sqrt{\tilde{Y}_s(y)}\; W(ds,dy)$$
$$-\gamma\int_0^t\int_{\mathbb{R}}p_{t-s}(x-y)\tilde{Y}_s(y)dy\int_{\mathbb{R}}\sqrt{\tilde{Y}_s(z)}\; W(ds,dz).$$

In particular this implies that the Fleming-Viot process is also a.s. absolutely continuous in $\mathbb{R}$.

2. An important method for establishing that certain finite dimensional martingale problems are well-posed involves the notion of pathwise uniqueness. We will explain this in the present context. Starting with any solution to the martingale problem we have obtained a weak solution, X, of a stochastic evolution equation driven by Gaussian white noise, W(ds,dx).

Given any two such weak solutions (X^1,W) and (X^2,W) of the stochastic evolution equation defined on the same probability space $(\Omega,\mathcal{F},P)$ with the *same* Wiener process W, assume that we can show that

$$P(\{\omega: X_t^1(\omega)=X_t^2(\omega),\ 0\le t\le T\}) = 1, \text{ for each } T<\infty.$$

Then we say that the stochastic evolution equation has the *pathwise uniqueness property*.

Theorem 8.4.1 The pathwise uniqueness property implies that the initial value martingale problem is well-posed.

Proof. See Yamada and Watanabe (1971).

Pathwise uniqueness has been established for some classes of stochastic partial differential equations. Of particular importance is the case in which the coefficients are Lipschitz (see e.g. Walsh (1986)). Another important class of stochastic partial differential equations for which pathwise uniqueness can be established is that satisfying the coercivity and monotonicity conditions formulated in Pardoux (1975) and Krylov and Rozovskii (1981).

The question of pathwise uniqueness for the stochastic evolution equation (8.3.4) associated with $(\alpha,1,1)$-superprocess is open. However in the case in which E is finite (and A is a matrix) Watanabe (1969) proved pathwise uniqueness for the resulting system of stochastic differential equations.

Remark. Theorem 8.3.2 suggests (in the one dimensional case, d=1) the study of non-negative function-valued (that is, absolutely continuous measure-valued) solutions of the broader class of stochastic partial differential equations:

$$(8.4.1)\quad X_t(x) = \int_{\mathbb{R}} p_t(x-y)X_0(y)dy + \int_0^t\int_{\mathbb{R}} p_{t-s}(y-x)q(X(s,y))W(ds,dy)$$

where $q(0)=0$, q is monotone increasing and continuous and W(ds,dy) is the white noise.

In the cases in which q is Lipschitz continuous or local Lipschitz existence and pathwise uniqueness of solutions to equation (8.4.1) was studied by many authors (for example, Dawson (1972), Walsh (1986), Iwata (1987), Kotelenez (1989), Gyöngy and Pardoux (1991)).

In the special case $q(u) = u^{\gamma}$ with $\gamma>0$ equation (8.4.1) was studied in detail by Mueller (1991a,b) using comparison methods, scaling ideas, etc. Another special case of interest is that in which q is Lipschitz and has compact support - see for example Theorem 10.4.1.

8.5 Absolute Continuity of the Occupation Time Measure.

Let Y be the occupation time process associated to the $B(\Delta,c)$-superprocess in $\mathbb{R}^d$ constructed in Section 7.4.

Theorem 8.5.1 Let $d \leq 3$. Then the occupation time $\mathcal{O}_{0,t}(dx)$ has a density $\mathcal{O}(t,x)$ (the *local time*) which is jointly continuous in (t,x) a.s.

Proof. Iscoe (1986, Theorem 4) and Fleischmann (1988) proved the existence of a density and Sugitani (1987) proved that it is a.s. jointly continuous using Kolmogorov's criterion. □

Adler and Lewin (1990) also obtained a Tanaka-type formula for the local time of both super-Brownian motion and super-stable processes.

9. SAMPLE PATH PROPERTIES OF SUPER-BROWNIAN MOTION

9.1 Some Introductory Remarks

In this chapter we will survey results on the sample path behavior of super-Brownian motion. Throughout the chapter we consider the $(2,d,\beta)$-superprocess $\{X_t:t\geq 0\}$ with probability law $(\{P^{\beta}_{\mu}:\mu\in M_F(\mathbb{R}^d)\})$, that is, the measure-valued branching process $(\mathbf{D},\mathcal{D},(\mathcal{D}_t)_{t\geq 0},(X_t)_{t\geq 0},\{P_{\mu}:\mu\in M_F(E)\})$, associated with the log-Laplace equation

$$(9.1.1)\quad \frac{\partial v}{\partial t} = \frac{1}{2}\Delta v - \frac{1}{1+\beta}v^{1+\beta},\quad \Delta = \text{d-dimensional Laplacian},$$

where $0<\beta\leq 1$. The case $\beta=1$ will be referred to as *continuous super-Brownian motion.*

Outlines of proofs are included in some cases as illustrations of the ideas and techniques involved but many results are described without proof. However we hope that this incomplete survey will convince the reader that super-Brownian motion is a rich subject for research and will provide an introduction to the growing literature.

9.2. Probability Estimates from the log-Laplace Equation

The study of Brownian motion is intimately linked with the heat equation and Laplace equation. It is therefore not surprising that the study of $(2,d,\beta)$-superprocesses is intimately related to the study of the nonlinear parabolic equation (9.1.1) and the associated non-linear elliptic equation

$$(9.2.1)\quad \Delta v = \frac{2}{1+\beta}\, v^{1+\beta}.$$

For example, equation 9.1.1 was used to derive the exact form of the log-Laplace function of the total mass (4.5.2) and the scaling relation (4.5.3).

Another important role of the log-Laplace functional is the determination of the probability of the events $\{X_t(A) = 0\}$ and $\{X_s(A) = 0\ \forall\ 0\leq s<t\}$. This method was developed by Iscoe (1988) (in the case $\beta=1$) and has as its starting point the following.

Lemma 9.2.1. Let $B := B(x_0,r) := \{y: d(y,x_0) < r\}$ and for $\varepsilon < r$ let $\phi_\varepsilon \in C(E)$ be defined by

$$\begin{aligned}\phi_\varepsilon(x) &= 1 \text{ if } d(x_0,x) \leq r-\varepsilon\\ &= (r-d(x_0,x))/\varepsilon \text{ if } r-\varepsilon < d(x_0,x) < r\\ &= 0 \text{ if } d(x_0,x) > r.\end{aligned}$$

Then

(a) For fixed $t > 0$,

$$(9.2.2)\qquad P^{\beta}_{\mu}(X_t(B)=0) = \lim_{\theta\to\infty}\lim_{\varepsilon\to 0} P^{\beta}_{\mu}\left(e^{-\langle\mu,V_t(\theta\phi_\varepsilon)\rangle}\right),$$

where $V_t\phi$ is the unique solution of the initial value problem

$$(9.2.3)\qquad \frac{\partial v}{\partial t} = \frac{1}{2}\Delta v - \frac{1}{1+\beta}v^{1+\beta},\quad v(0,x) \equiv \phi.$$

(b) For $t > 0$,

$$(9.2.4)\quad P^{\beta}_{\mu}(X_s(B)=0\ \forall\ 0\le s<t) = \lim_{\theta\to\infty}\lim_{\varepsilon\to 0} P^{\beta}_{\mu}\left(e^{-\langle\mu,U_t(\theta\phi_\varepsilon)\rangle}\right).$$

where $U_t\phi$ is the unique solution of the inhomogeneous equation

$$\frac{\partial v}{\partial t} = \frac{1}{2}\Delta v - \frac{1}{1+\beta}v^{1+\beta} + \phi,\quad v(0,x) \equiv 0.$$

Proof. (a) We obtain from the properties of the Laplace functional and the monotone convergence theorem that

$$P^{\beta}_{\mu}(X_t(B)=0) = \lim_{\theta\to\infty} P^{\beta}_{\mu}\left(e^{-\theta X_t(B)}\right) = \lim_{\theta\to\infty}\lim_{\varepsilon\to 0} P^{\beta}_{\mu}\left(e^{-\theta\langle X_t,\phi_\varepsilon\rangle}\right)$$

$$= \lim_{\theta\to\infty}\lim_{\varepsilon\to 0} P^{\beta}_{\mu}\left(e^{-\langle\mu,V_t(\theta\phi_\varepsilon)\rangle}\right).$$

(b) Since $\{X_s:s\ge 0\}$ is right continuous in $M_F(E)$ and B is open in E, then $\liminf_{r\downarrow s} X_r(B) \ge X_s(B)$ and hence $X_s(B) > 0$ for some $0\le s<t$ implies that $\int_0^t X_s(B)ds > 0$. Therefore

$$P^{\beta}_{\mu}(X_s(B)=0\ \forall\ 0\le s<t) = P^{\beta}_{\mu}\left(\int_0^t X_s(B)ds = 0\right)$$

$$= \lim_{\theta\to\infty} P^{\beta}_{\mu}\left(\exp\left(-\theta\int_0^t X_s(B)ds\right)\right)$$

$$= \lim_{\theta\to\infty}\lim_{\varepsilon\to 0} P^{\beta}_{\mu}\left(\exp\left(-\theta\int_0^t \langle X_s,\phi_\varepsilon\rangle ds\right)\right). \quad \square$$

A similar result can be obtained for the complement of a closed ball. As an illustration we have the following result.

Theorem 9.2.2. Consider the $(2,d,\beta)$-superprocess $\{X_t:t\ge 0\}$ and let $\mu(\mathbb{R}^d)<\infty$ with supp

$\mu \subset B(0,R_0)$ and let $R_0 < R$. Then

$$P^{\beta}_{\mu}(X_{\cdot} \text{ ever charges } \overline{B(0,R)^c}) = 1 - \exp(-R^{-2/\beta}\langle u(R^{-1}.),\mu\rangle)$$

where u is the unique positive solution of the singular elliptic boundary value problem

$$\Delta u(x) = \frac{2}{1+\beta} u^{1+\beta}(x), \quad x \in B(0,1)$$

$$u(x) \longrightarrow \infty \quad \text{as} \quad x \to \partial B(0,1).$$

In particular if $\mu = \delta_0$, then

$$P^{\beta}_{\delta_0}(X_{\cdot} \text{ ever charges } \overline{B(0,R)^c}) = 1 - \exp(-u(0)R^{-2/\beta}).$$

Proof. Iscoe (1988, Theorem 1) for $\beta=1$. The proof for $0<\beta<1$ is essentially the same.□

Using the same technique we now derive an estimate of the probability that a propagating initial point mass charges the exterior of a small ball in a short interval. But first we state a preliminary result on the solution of the related non-linear elliptic problem.

Lemma 9.2.3 Let $u(.,R)$ denote the solution of the equation

$$(9.2.1) \qquad \Delta u = \frac{2}{1+\beta} u^{1+\beta}, \quad u(x) \uparrow \infty \text{ as } |x| \uparrow R. \quad \text{Then}$$

$$(9.2.2) \quad u(x,R) = R^{-2/\beta} u(x/R,1),$$

and

$$(9.2.3) \qquad \lim_{|x|\uparrow 1} \frac{u(x,1)}{1/(1-|x|^{2/\beta})} = \left(\frac{(2+\beta)(1+\beta)}{\beta^2}\right)^{1/\beta}.$$

Proof. This is obtained by an o.d.e. argument - [Dawson, Iscoe and Perkins, (1989, Lemma 3.6), Dawson and Vinogradov (1992a, Appendix 1, Lemma 4.4)]. □

Theorem 9.2.4. Let $a>0$, and $R > (2/\beta)t^{1/2}$. Then

$$(9.2.4) \quad P^{\beta}_{a\delta_0}(\, X_s(\overline{B(0;R)^c}) > 0 \text{ for some } s \le t)$$

$$\le c(\beta,d)\, aR^{-2/\beta}\left[\frac{R}{t^{1/2}}\right]^{d+(4/\beta)-2} \exp\left\{-\frac{R^2}{2t}\right\}.$$

Outline of Proof. (See Dawson, Iscoe and Perkins (1989, Theorem 3.3(b)) or Dawson and Vinogradov (1992a, Proposition 1.1) for details.) Set

$$P := P_{a\delta_0}(X_s(\overline{B(0,R)}^c) > 0 \text{ for some } s\le t)$$

$$= \lim_{\theta\to\infty} 1 - e^{-u^\theta(t,0)a} \le \lim_{\theta\to\infty} a\, u^\theta(t,0,R)$$

where u^θ satisfies

$$\frac{\partial u^\theta}{\partial t} = \frac{1}{2}\Delta u^\theta - \frac{1}{1+\beta}(u^\theta)^{1+\beta} + \theta\psi$$

$$u^\theta(0)=0, \quad \psi \approx 1\{|x|>R\}.$$

In order to apply a Feynman-Kac type representation of the solution of this equation let $\{B_t:t\ge 0\}$ denote a Brownian motion in $\mathbb{R}^d$ and set $\tau := \inf\{t:B_t\notin B(0;R_1)\}$, $R_1<R$. Then using the strong Markov property of Brownian motion,

$$u^\theta(t,0) = E_0\Big\{1(\tau<t)\exp\Big[-\int_0^\tau (1+\beta)^{-1}(u^\theta(t-v,B_v))^\beta dv\Big]$$

$$\cdot E_{B(\tau)}\Big\{\int_0^{t-\tau} \theta\psi(B_s)\exp\Big[-\int_0^s (1+\beta)^{-1}(u^\theta(t-\tau-v,B_v))^\beta dv\Big] ds\Big\}\Big\}$$

$$\le E_0\Big[1(\tau<t)\, u^\theta(t-\tau,R_1)\Big], \quad \text{since } |B(\tau)|=R_1.$$

But $u^\theta(t,x,R) \le u(x,R)$ where $u(x,R)$ is the solution of (9.2.1).

By Lemma 9.2.1 $u(x,R) = R^{-2/\beta}u(x/R,1)$ and

(9.2.5) $$\lim_{|x|\uparrow R} \frac{R^{-2/\beta}u(x/R,1)}{1/(1-|x/R|^{2/\beta})} = R^{-2/\beta}\left(\frac{(2+\beta)(1+\beta)}{\beta^2}\right)^{1/\beta}.$$

Hence $P \le a\, P(\tau_{R_1}\le t)\cdot u(R_1,R).$

Then optimizing on R_1 we obtain

$$P \le c(\beta,d)\, aR^{-2/\beta}\left[\frac{R}{t^{1/2}}\right]^{d+(4/\beta)-2} e^{-R^2/2t}. \quad \square$$

9.3. The support process of super-Brownian motion

9.3.1. The Support Map, Support Dimension and Carrying Dimension.

Let (E,d) be a separable metric space. Without loss of generality we may assume that d is bounded (e.g. $d(x,y)\le 1$ $\forall$ x,y). Let $\mathfrak{F}(E)$ (resp. $\mathfrak{K}(E)$) denote the space of closed (resp. compact) subsets of E.

The induced *Hausdorff metric* ρ on $\mathfrak{F}(E)$ is defined by

$\rho(A,B) := \max(\rho_1(A,B),\rho_1(B,A))$, if A and B are non-empty, where

$$\rho_1(A,B) := \sup_{x\in A} d(x,B), \qquad d(x,B) := \inf_{y\in B} d(x,y).$$

$$\rho(A,\emptyset) := 1 \text{ if } A\neq\emptyset.$$

Note that $\rho_1(A,B) \le r$ implies $A \subset B^r$ where $B^r := \{x:d(x,B)\le r\}$.

N.B. ρ_1 is not symmetric and hence not a metric. However it satisfies the triangle inequality

$$\rho_1(A_1,A_3) \le \rho_1(A_1,A_2) + \rho_1(A_2,A_3).$$

If $\mu\in M_F(E)$, $\mu\neq 0$, the *support of* μ, denoted by $S(\mu)$, is defined to be the smallest closed set $C\subset E$ satisfying $\mu(E\setminus C)=0$. Equivalently,

$S(\mu) := \{x\in E:\mu(B(x)) > 0$ for every open ball $B(x)$ centered at $x\}$, if $\mu\neq 0$,

$S(\mu) := \emptyset$ if $\mu = 0$.

Now consider $\mathfrak{F}(E)$, (resp. $\mathfrak{K}(E)$) where $E = \mathbb{R}^d$ (or any locally compact separable metric space). The *compact topology* on $\mathfrak{F}(E)$ is defined by the subbasis consisting of the sets $\mathfrak{F}_G$, $\mathfrak{F}^K$, where G runs over all the open sets of E and K runs over all the compact sets of E. Here

$\mathfrak{F}_A := \{F\in\mathfrak{F}:F\cap A\neq\emptyset\}$, A is a subset of E,

$\mathfrak{F}^A := \{F\in\mathfrak{F}:F\cap A=\emptyset\}$, A is a subset of E.

Since $\mathfrak{F}^{K_1}\cap\mathfrak{F}^{K_2} = \mathfrak{F}^{K_1\cup K_2}$ a basis element in the compact topology is of the form:

$\mathfrak{F}^K_{G_1\ldots G_n} = \{F\in\mathfrak{F}:F\cap K=\emptyset,\ F\cap G_i\neq\emptyset,\ i=1,\ldots,n\}$ where

$n\ge 0$, K is compact and G_i is open.

Theorem 9.3.1.1 (a) $\mathfrak{F}(E)$ is second-countable, compact and Hausdorff in the compact topology.

(b) The compact topology on $\mathfrak{F}(E)$ is coarser than the Hausdorff topology.

Proof. Cutler (1984, p. 74, 80), Matheron (1975).

Theorem 9.3.1.2. (a) Suppose E is a separable metric space. Then the mapping $S:M_F(E)\longrightarrow (\mathfrak{F}(E),\rho)$ is measurable.

(b) If E is also locally compact, then $S:M_F(E)\longrightarrow (\mathfrak{F}(E),\text{compact})$ is also *lower semi-continuous* (in the sense of Matheron), i.e. $\psi^{-1}(\mathfrak{F}_G)$ is open for every open set G.

(c) Let $\mu,\mu_n\in M_F(E)$ and suppose $\mu_n \xrightarrow{w} \mu$. Then

$$S(\mu) \le \underline{\lim}_{n\to\infty} S(\mu_n).$$

Proof. (a) Cutler (1984, Theorem 4.4.1, Cor. 4.4.1.2, Theorem 4.4.2.)

Let φ be a continuous, strictly increasing function on $[0,\infty)$ with $\varphi(0)=0$. Let $A \subset E$.

The *Hausdorff φ-measure* of A is defined as

$$\varphi\text{-m}(A) := \lim_{\delta\downarrow 0} \varphi_\delta\text{-m}(A)$$

where $\varphi_\delta\text{-m}(A) := \inf\left\{\sum_k \varphi(r_k) : \bigcup_k B(x_k,r_k) \supset A,\ r_k < \delta/2\right\}$,

(where $\{B(x_k,r_k)\}$ are open balls with centres x_k and radii r_k.

In particular set $\varphi_\alpha(r) := r^\alpha$, for $0 < \alpha < \infty$, and

$$\mathcal{H}^\alpha(A) = \varphi_\alpha\text{-m}(A) := \lim_{\delta\downarrow 0} \mathcal{H}^\alpha_\delta(A)$$

where $\mathcal{H}^\alpha_\delta = \inf\left\{\sum_k (r_k)^\alpha : \bigcup_k B(x_k,r_k) \supset A,\ r_k < \delta/2\right\}$.

The *Hausdorff-Besicovitch* dimension of a set A, dim(A), is defined by means of

$$\mathcal{H}^\alpha(A) := +\infty \quad \text{if} \quad \alpha < \dim(A)$$

$$:= 0 \quad \text{if} \quad \alpha > \dim(A),$$

i.e. $\dim(A) := \inf\{\alpha\ge 0: \mathcal{H}^\alpha(A)=0\}$.

Note that if $\mathcal{H}^\alpha(A) < \infty$, then $\dim(A) \le \alpha$.

Definition. A measure $\mu \in M_F(E)$ is said to have *carrying dimension* $\text{cardim}(\mu) = d$, if there exists a d-dimensional Borel set A such that $\mu(A^c) = 0$ and $\dim(A) = d$, and this fails for $d' < d$. It is said to have *support dimension* $\text{supdim}(\mu) = d$ if $\dim(S(\mu)) = d$.

Remarks. 1. Clearly $\text{supdim}(\mu) \ge \text{cardim}(\mu)$.

2. If $E = \mathbb{R}^d$ equipped with the Euclidean metric, then $\dim(A) \le d\ \forall\ A\subset E$.

Theorem 9.3.1.3. (a) The dimension map $\dim: \mathfrak{F}(E) \to \mathbb{R}_+$ is measurable (with respect to the Borel σ-algebra generated by the compact (and also the Hausdorff) topology.

(b) The map $\mu \to \dim(S(\mu))$ from $M_F(E) \to \mathbb{R}_+$ is measurable.

Proof. (a) Cutler, (1984, Theorem 4.5.1).

(b) follows from (a) and Theorem 9.3.1.2. □

Remark 9.3.1.4. Consider a measure $\mu \in M_F(\mathbb{R}^d)$. If $\text{cardim}(\mu) < d$, then μ is singular with respect to Lebesgue measure.

9.3.2. A Modulus of Continuity for the Support Process

Let $\{X_t:t\geq 0\}$ and $S(X_t)$ be a $(2,d,\beta)$-superprocess and the closed support of X_t, respectively.

Recall Lévy's classical result on the modulus of continuity for standard Brownian motion:

$$(9.3.2.1)\qquad \lim_{\delta\downarrow 0} \frac{\sup_{0\leq s\leq 1-\delta}\ \sup_{0<u\leq\delta} |W(s+u)-W(s)|}{\sqrt{2\delta\ \log 1/\delta}} = 1 \quad \text{a.s.}$$

In this section we will derive a partial analogue to this result for the support process.

Lemma 9.3.2.1. Let $X_0=\mu\in M_F(\mathbb{R}^d)$, and $R>(2/\beta)s^{1/2}$. Then

$$P^\beta_{X_0}\left(\{\rho_1(S(X_{t+u}),S(X_t))>R\}\cap\{<X_{t+u},1>\neq 0\} \text{ for some } u\leq s\,|\,\mathcal{D}_t\right)$$

$$= P^\beta_{X_t}\left(\{\rho_1(S(X_u),S(\mu))>R\}\cap\{<X_{t+u},1>\neq 0\} \text{ for some } u\leq s\right)$$

$$\leq c(\beta,d)\ R^{-2/\beta}\left[\frac{R}{s^{1/2}}\right]^{d+(4/\beta)-2} \exp\left\{-\frac{R^2}{2s}\right\}\cdot X_t(\mathbb{R}^d),\ P^\beta_{X(0)}\text{-a.s.}$$

Proof. We first approximate $X_t(dx)$ by atomic measures of the form

$$\mu_n=\sum_{i=1}^{N(n)} a_i^n\cdot\delta_{x_i^n},\quad \text{with } x_i^n\in S(X_t) \text{ and } \sum a_i^n = X_t(\mathbb{R}^d),$$

$$\mu_n \Rightarrow X_t(dx) \text{ as } n\to\infty.$$

Then $P^\beta_{\mu_n}\Rightarrow P^\beta_{X_t}$ (by Feller property of $(2,d,\beta)$-superprocesses).

Since $S(\mu)$ is closed, then for $\eta>0$,

$$\{\{\rho_1(S(X_u),S(\mu))>R\}\cap\{<X_u,1>\neq 0\} \text{ for some } 0\leq u<s+\eta\}$$

$$=\{\nu(.)\in D([0,\infty),M_F(\mathbb{R}^d)):\ \nu(u,S(\mu)^c))>0 \text{ for some } 0\leq u<s+\eta\}$$

is an open set in the Skorohod topology.

Therefore by the Markov property

$$P^\beta_{X_0}\left(\rho_1(S(X_{t+u}),S(X_t))>R \text{ for some } 0\leq u\leq s\,|\,\mathcal{D}_t\right)$$

$$= P^{\beta}_{\mu}\left(\rho_1(S(X_u),S(\mu)) > R \text{ for some } 0\le u\le s\right) \quad \text{with} \quad \mu = X_t$$

$$\le \liminf_{n\to\infty} P^{\beta}_{\mu_n}\left(\rho_1(S(X_u),S(\mu)) > R \text{ for some } 0\le u<s+\eta\right).$$

Since $P^{\beta}_{\mu_n} = \mathop{*}\limits_{i=1}^{N(n)} P^{\beta}_{a^n_i\delta_{x^n_i}}$, then $X_t \overset{\mathcal{D}}{=} X^1_t+\ldots X^{N(n)}_t$ where $\{X^i_u: i=1,\ldots,N(n)\}$ are independent and X^i_t is a version of the $(2,d,\beta)$-superprocess with initial measure $a^n_i\cdot\delta_{x^n_i}$. (Here $*$ denotes the operation of convolution.) Moreover

$$\{\rho_1(S(X(u),S(\mu)) > R \text{ for some } 0\le u<s+\eta\}$$

$$\subset \left\{\bigcup_{i=1}^{N(n)} \rho_1(S(X^i_u),\{x_i\})) > R \text{ for some } 0\le u<s+\eta\right\}.$$

Therefore, by Theorem 9.2.4,

$$\liminf_{n\to\infty} P^{\beta}_{\mu_n}\left(\rho_1(S(X_u),S(\mu)) > R \text{ for some } 0\le u<s+\eta\right)$$

$$\le \liminf_{n\to\infty} \sum_{i=1}^{N(n)} P^{\beta}_{a^n_i\cdot\delta_{x^n_i}}\left(\rho_1(S(X^i_u),\{x_i\}) > R \text{ for some } 0\le u<s+\eta\right)$$

$$\le c(\beta,d)\, R^{-2/\beta}\left[\frac{R}{(s+\eta)^{1/2}}\right]^{d+(4/\beta)-2} \exp\left\{-\frac{R^2}{2(s+\eta)}\right\}\cdot X_t(\mathbb{R}^d).$$

The result follows since $\eta>0$ is arbitrary. □

Let

(9.3.2.2) $$h_\kappa(t) := \left[\frac{2}{\beta}(1+\beta)\min\left(1,\ t\left(\log\frac{1}{t} + \kappa\log\log\frac{1}{t}\right)\right)\right]^{1/2}.$$

The following result is an analogue of Lévy's global modulus of continuity for the path behavior of the $(2,d,\beta)$-superprocess on the interval $[0,T]$.

<u>Theorem</u> <u>9.3.2.2</u> Let $\mu\in M_F(\mathbb{R}^d)$, $\kappa > \frac{\beta d+2}{2(1+\beta)}$ and $B_K := \{X_T\neq 0\}\cap\{\sup_{0\le t\le T} X_t(\mathbb{R}^d) \le K\}$. Then for P^{β}_{μ}-a.s. $\omega\in B_K$, there is a $\delta = \delta(\omega,\kappa,K,T)>0$ such that if $0 \le s,t \le T$ satisfy $0 < t-s < \delta$, then

(9.3.2.3) $$S(X_t) \subset S(X_s)^{h_\kappa(t-s)}, \text{ i.e. } \rho_1(S(X_s),S(X_t)) < h_\kappa(t-s).$$

<u>Remarks:</u> Note the difference in the constant $\sqrt{2(1+\beta)/\beta}$ in $h_\kappa(t)$ and $\sqrt{2}$ in the

classical Lévy modulus of continuity result for Brownian motion (in the latter case this is sharp). Theorem 9.3.2.2 can be stated in a more precise form in the setting of the historical process (which will be introduced in Ch. 12). In the case of $\beta=1$, the constant $\sqrt{4}$ (=2) is known to be sharp (cf. [DP, Theorem 8.7]).

Proof of Theorem 9.3.2.2. For simplicity we take T=1. Since by Theorem 4.6.2(c) $\{X_t:t\geq 0\} \in D([0,\infty),M_F(\mathbb{R}^d))$, P^{β}_{μ}-a.s., then $\sup_{0\leq t\leq 1} X_t(\mathbb{R}^d) < \infty$. It therefore suffices to show that for any $K > 0$,

$$P^{\beta}_{\mu}\left(\left\{S(X_t) \subset S(X_s)^{h_\kappa(t-s)}\right\}^c \cap B_K\right) = 0.$$

Given $\kappa > \frac{\beta d+2}{2(1+\beta)}$ we choose any $0 < \varepsilon(\beta,\kappa) < (\kappa\cdot\frac{1+\beta}{\beta} - d/2-1/\beta)/2$.
We first derive an almost-sure result for the grid:

(9.3.2.4)

$$P^{\beta}_{\mu}\left\{B_K \cap \max_{\substack{0<k=j-i\leq(\log N)^{\varepsilon}\\ 0\leq i<j\leq N}} \frac{\rho_1(S(X_{j/N}),S(X_{i/N}))}{h_\kappa(k/N)} \leq 1 \text{ for all } N=2^n \text{ large enough}\right\}$$
$$= P^{\beta}_{\mu}(B_K).$$

To get (9.3.2.4), we first obtain upper estimates for the probabilities

$$(9.3.2.5)\quad p_N := P^{\beta}_{\mu}\left\{B_K \cap \max_{\substack{0<k=j-i\leq(\log N)^{\varepsilon}\\ 0\leq i<j\leq N}} \frac{\rho_1(S(X_{j/N}),S(X_{i/N}))}{h_\kappa(k/N)} > 1\right\}.$$

It can be shown by use of Lemma 9.3.2.1 that

$$p_N \leq C_1(\beta,d,K)\cdot \sum_{0<k\leq(\log N)^{\varepsilon}} N\cdot\left(\frac{k}{N}\cdot\log\frac{N}{k}\right)^{-1/\beta}\cdot\left(\log\frac{N}{k}\right)^{d/2-1+2/\beta}$$
$$\cdot\exp\left\{-\frac{1+\beta}{\beta}\left(\log\frac{N}{k}+\kappa\log\log\frac{N}{k}\right)\right\}$$
$$\leq C_2(\beta,d,K)\cdot\sum_{0<k\leq(\log N)^{\varepsilon}} k\cdot(\log N)^{-1-(\kappa\cdot\frac{1+\beta}{\beta}-\frac{d}{2}-\frac{1}{\beta})}$$
$$\leq C_3(\beta,d,K)\cdot(\log N)^{-1+2\varepsilon-(\kappa\cdot\frac{1+\beta}{\beta}-\frac{d}{2}-\frac{1}{\beta})}.$$

Recall that $0 < \varepsilon < (\kappa\cdot\frac{1+\beta}{\beta} - d/2 - 1/\beta)/2$, i.e. the power of the logarithm in the latter expression is less than -1. This implies that (taking $N = 2^n$)

$$p_{2^n} \le C_4(\beta,d,\kappa,\delta,K)\cdot n^{-1+2\varepsilon-(\kappa\cdot\frac{1+\beta}{\beta} - \frac{d}{2} - \frac{1}{\beta})}$$

i.e. it is the general term of a convergent series. Then the Borel-Cantelli Lemma implies that for the subsequence $N = 2^n$ at most a finite number of events

$$\left\{B_K \cap \max_{\substack{0<k=j-i\le(\log N)^\varepsilon \\ 0\le i<j\le N}} \frac{\rho_1(S(X_{j/N}),S(X_{i/N}))}{h_\kappa(k/N)} > 1\right\}$$

occurs. This yields (9.3.2.4).

The remainder of the proof is carried out by standard argument (cf. McKean (1969) p.16) and is based on estimate (9.3.2.4) for maximum over the grid, and the triangle inequality $\rho_1(A_1,A_3) \le \rho_1(A_1,A_2) + \rho_1(A_2,A_3)$.

Given $0 < \varepsilon(\beta,\kappa) < (\kappa\cdot\frac{1+\beta}{\beta} - d/2 - 1/\beta)/2$ choose $n_0(\varepsilon)$ so that

(9.3.2.6) $$\max_{\substack{0<k=j-i\le(\log N)^\varepsilon \\ 0\le i<j\le N}} \frac{\rho_1(S(X_{j/N}),S(X_{i/N}))}{h_\kappa(k/N)} \le 1$$

$\forall\ N = 2^n$ with $n \ge n_0$. Now let

$$\delta(\beta,\kappa) \le (\log 2)^\varepsilon\cdot n_0^\varepsilon \,/\, 2^{n_0}.$$

(We will pick $\delta(\beta,\kappa)$ later.)

Now consider a pair (s,t): $0 \le s < t \le 1$ such that $u := t-s < \delta(\beta,\kappa)$.

Pick $n = n(u) \ge n_0$ so that $(\log 2)^\varepsilon(n+1)^\varepsilon/2^{n+1} \le u < (\log 2)^\varepsilon n^\varepsilon/2^n$ and note that $n(u) \longrightarrow \infty$ as $u \to 0$.

We can then choose sequences $n < p_1 < p_2 < \ldots$, and $n < q_1 < q_2 < \ldots$ such that

(9.3.2.7) $$s_k := i\cdot 2^{-n} - 2^{-p_1} - 2^{-p_2} - \ldots - 2^{-p_k},$$

(9.3.2.7') $$t_k := j\cdot 2^{-n} + 2^{-q_1} + 2^{-q_2} + \ldots + 2^{-q_k},$$

$$|s_k - s| \le 2^{-p_k}, \quad |t_k - t| \le 2^{-q_k},$$

$$s \le s_k \le i\cdot 2^{-n} < j\cdot 2^{-n} \le t_k \le t, \text{ and } 0 < k = j - i \le u\cdot 2^n < (\log 2)^\varepsilon\cdot n^\varepsilon.$$

The triangle inequality then yields

(9.3.2.8)

$$\rho_1(S(X_t),S(X_s)) \le \rho_1(S(X_{i/2^n}),S(X_s)) + \rho_1(S(X_{j/2^n}),S(X_{i/2^n}))$$

$$+ \rho_1(S(X_t),S(X_{j/2^n})).$$

To estimate the middle term on the right-hand side of (9.3.2.8) we apply monotonicity of function $h_\kappa(\cdot)$ and (9.3.2.4):

(9.3.2.9) $\rho_1(S(X_{j/2^n}),S(X_{i/2^n})) \le h_\kappa(k/2^n) \le h_\kappa(t-s)$ for $n \ge n_0$.

Also note that

(9.3.2.10') $$\rho_1(S(X_{i/2^n}),S(X_s)) \le \sum_{k=1}^{\infty} \rho_1(S(X_{s_k}),S(X_{s_{k+1}}))$$

(9.3.2.10') $$\rho_1(S(X_t),S(X_{j/2^n})) \le \sum_{k=1}^{\infty} \rho_1(S(X_{t_{k+1}}),S(X_{t_k})).$$

By (9.3.2.4) and the monotonicity of $h_\kappa(.)$ it is easily seen that the each of these two expressions does not exceed

(9.3.2.11) $$\sum_{\ell=n+1}^{\infty} h_\kappa(1/2^\ell).$$

Therefore

$$\rho_1(S(X_{i/2^n}),S(X_s)) \le \sum_{\ell=n+1}^{\infty} h_\kappa(1/2^\ell)$$

(9.3.2.12)

$$\le C_1(\beta,\kappa) \sum_{\ell=n+1}^{\infty} \frac{\sqrt{\ell}}{2^{\ell/2}}$$

$$\le C_1(\beta,\kappa) \int_n^{\infty} x^{1/2} \cdot 2^{-x/2} \cdot dx.$$

The latter integral is not difficult to estimate by use of Laplace's method since it is equivalent as $n \to \infty$ (up to some positive constant) to $n^{1/2} \cdot 2^{-n/2}$. The latter expression in its turn is equivalent as $n \to \infty$ (up to some positive constant) to $h_\kappa(1/2^n)$. These arguments yield the upper estimate for the first term on the right-hand side of (9.3.2.8):

(9.3.2.13) $\rho_1(S(X_{i/2^n}),S(X_s)) \le C_2(\beta,\kappa)\cdot h_\kappa(1/2^n)$

and a similar estimate can be obtained for the third term on the right hand side of (9.3.2.8).

Note that due to our choice of n, $1/2^n \le \dfrac{2}{(\log 2)^\varepsilon (n+1)^\varepsilon}\cdot(t-s)$. Then by the monotonicity of $h_\kappa(\cdot)$ we obtain $h_\kappa(1/2^n) \le h_\kappa(2\cdot(t-s)/((\log 2)^\varepsilon(n+1)^\varepsilon))$. Combining (9.3.2.8, 9.3.2.9, 9.3.2.13) we obtain

(9.3.2.14) $\rho_1(S(X_t),S(X_s)) \le h_\kappa(t-s) + C_2(\beta,\kappa)\cdot h_\kappa(C(\varepsilon)(t-s)/(n+1)^\varepsilon)$

for $n \ge n_0$, where $C(\varepsilon) := 2/(\log 2)^\varepsilon$ (< 3).

Given any $\eta > 0$, the expression $C_2(\beta,\kappa)\cdot h_\kappa(C(\varepsilon)(t-s)/(n+1)^\varepsilon)$ can be made less than $\eta\cdot h_\kappa(t-s)$, by choosing $\delta(\beta,\kappa) > 0$ sufficiently small. Then (9.3.2.14) yields

$$P^\beta_\mu\left\{ \exists\, \delta > 0: \rho_1(S(X_t),S(X_s)) \le h_\kappa(t-s)\ \forall\ 0<t-s<\delta \right\} = 1. \ \square$$

<u>Remark</u> Note that this result yields the *compact support property*, that is, if X_0 has compact support, so does X_t for $t> 0$. It also implies that $P^\beta_{\delta_0}(\lim_{t\downarrow 0} \mathrm{diam}(S(X_t)) = 0) = 1$. This can be sharpened as follows:

<u>Theorem</u> <u>9.3.2.3</u> Let $r(t) := \inf\{R: S_t \subset B(0,R)\}$. Then

$$P^\beta_{\delta_0}\left\{ \lim_{t\downarrow 0} \frac{\sup_{0\le u\le t} r(u)}{\sqrt{\frac{2}{\beta}\, t \log \frac{1}{t}}} = 1 \right\} = 1.$$

<u>Proof.</u> Tribe (1989, Theorem 2.1) for the case $\beta=1$, Dawson and Vinogradov (1992a, Formula (1.16")).

<u>Remarks.</u> 1. In fact an exact almost-sure rate of convergence of $\dfrac{\sup_{0\le u\le t} r(u)}{\sqrt{\frac{2}{\beta}\, t \log \frac{1}{t}}}$ to 1 as $t\to 0$ can be obtained (see Dawson and Vinogradov (1992a) Theorem 1.4); this result is analogous to the Chung, Erdös and Sirao (1959) result for standard Brownian motion (cf. Itô and McKean (1965), Section 1.9).

2. A local version of Theorem 9.3.2.2 is also proved in Dawson and Vinogradov

(1992a, Theorem 1.3) with constant $\sqrt{2/\beta}$ instead of $\sqrt{2(1+\beta)/\beta}$. The fact that this result is sharp follows from Theorem 9.3.2.3.

Theorem 9.3.2.4. If $\mu \in M_F(\mathbb{R}^d)$ then for P_μ^β-a.e. ω

(a) $\{S(X_t):t\geq 0\}$ is a $\mathfrak{K}(\mathbb{R}^d)$-valued process having right continuous paths with left limits.

If in addition $\beta=1$, then

(b) $\lim_{s\uparrow t} S(X_s) \supset S(X_t)$ $\forall t>0$

(c) If $\beta = 1$, then $S_{t-}-S(X_t)$ is empty or a singleton for all $t>0$ where $S_{t-} := \lim_{s\uparrow t} S(X_s)$.

Proof. Theorem 9.3.2.1 shows that if $t\downarrow s$, then $\rho_1(S_t,S_s) \longrightarrow 0$. On the other hand the compactness of S_s and right continuity of X_t imply that

$$\rho_1(S_s,S_t) \longrightarrow 0 \text{ if } t \to s.$$

This proves the right continuity.

The existence of left limits is a consequence of Theorem 9.3.2.1 and the following criterion of Perkins (1990, Lemma 4.1):

if $f:(0,\infty) \longrightarrow \mathfrak{K}(\mathbb{R}^d)$ is such that

$$\forall \varepsilon>0 \ \exists \ \delta>0 \text{ such that } 0\leq u-t<\delta \text{ implies } f(u)\subset f(t)^\varepsilon,$$

then f possesses left and right limits at all t in $(0,\infty)$.

(b) If $\beta=1$, then the process X_t is continuous and therefore if $X_t(B(x,\varepsilon)) > 0$, then $X_u(B(x,\varepsilon)) > 0$ for u near t by the continuity of X. Then (b) follows immediately.

(c) is proved in Perkins (1990, Prop. 4.6).□

Remarks: (1) In the case $\beta=1$, the countable set of discontinuities of $\{S_t:t>0\}$ occurs when an isolated colony becomes extinct (cf. Perkins (1990, Theorem 4.7). Note that Perkins (1990, Theorem 4.8) also proved that the space-time set of these extinction points form a dense subset of the graph of S if $d\geq 3$.

Tribe (1989) obtained a detailed description of the behavior of continuous super-Brownian motion (and super-stable-processes) near extinction using the Shiga's result on the relation between Fleming-Viot process and the superprocess (cf. Section 8.1).

(2) In the case of (α,d,β)-superprocess with $\alpha<2$ (i.e. super-stable processes) the support behavior is completely different. In fact Perkins has proved the following *instantaneous propagation of support* property.

Theorem 9.3.2.5. Consider the $(\alpha,d,1)$ process with $\alpha < 2$. Then $S_t \neq \emptyset$ implies that $S_t = \mathbb{R}^d$, Q_μ-a.s. for all $t>0$ and $\mu \in M_F(\mathbb{R}^d)$.

Proof. Perkins (1990, Cor. 1.6).

9.3.3. Application to the carrying and support dimensions.

Theorem 9.3.3.1. Let $\{X(t):t\geq 0\}$ be an (α,d,β)-superprocess. Then for each $t>0$,

$$P^\beta_\mu(\text{cardim}(X_t) = \alpha/\beta \mid X_t(\mathbb{R}^d)>0) = 1.$$

Proof. See [D, Section 7.3].

For continuous super-Brownian motion the following sharp result was obtained by Perkins (as well as a weaker result in d=2).

Theorem 9.3.3.2. If $d\geq 3$, then there exist constants $0 < c(d)\leq C(d)<\infty$ such that for any $\mu\in M_F(\mathbb{R}^d)$, and P^1_μ-a.e. ω

$$c(d)\varphi_{2,2}\text{-}m(A\cap S(X_t))\leq X_t(A)\leq C(d)\varphi_{2,2}\text{-}m(A\cap S(X_t)) \quad \forall\ A\in\mathcal{B}(\mathbb{R}^d) \text{ and all } t>0$$

where $\varphi_{2,2}(r) = r^2 \log\log 1/r$.

Proof. Perkins (1989, Theorem 1).

For fixed times this can be even further refined as follows.

Theorem 9.3.3.3 If X_t is the $(2,d,1)$-superprocess, with $d\geq 3$, $X_0=\mu\in M_F(\mathbb{R}^d)$, and $t > 0$, then P^1_μ-a.s. there is a $c(d)\in(0,\infty)$ such that

$$X_t(A) = c(d)\ \varphi_{2,2}\text{-}m(A\cap S(X_t)) \quad \forall\ A\in \mathcal{B}(\mathbb{R}^d).$$

Proof. The proof (see [DP, Theorem 5.2]) is based on a zero-one law for the of the historical process (cf. Ch. 12 for an introduction to the historical process).□

Remark 9.3.3.4. By Theorem 8.1.1, that is, the identification of the $(\frac{1}{2}\Delta,f)$-Fleming-Viot process as the conditional law of the branching process conditioned to have $\langle X_t,1\rangle = f(t)$, and Theorem 9.3.3.1 we deduce: for $d\geq 3$, for $P_\mu\circ\langle .,1\rangle^{-1}$-a.a. f,

$$Y_t(A) = f(t)^{-1}c_d\phi\text{-}m(A\cap S(Y_t)) \quad \forall\ A\in \mathcal{B}(\mathbb{R}^d),\quad P^{FV}_{\mu,f}\text{-a.s.}$$

Perkins (1991b Formula (15)) conjectures that this is also true for $f(t)\equiv 1$ but with another unknown constant c'_d. However we will remark at the end of Ch. 12 that one cannot obtain an immediate proof of this from the corresponding result for the continuous superprocess.

The proofs of Theorems 9.3.3.2-9.3.3.3 are too long to be included here. An introduction to the main tools and in particular the lower bound method is given in [D, Section 7.2]. In this section we will derive an uniform in time upper bound on the support dimension for the $(2,d,\beta)$-superprocess using the upper estimate of Theorem 9.2.4.

Theorem 9.3.3.5. Let $\{X(t):t\geq 0\}$ be a $(2,d,\beta)$-superprocess and $\{S_t\}$ the corresponding support process.

(a) If $\varphi_\beta(x) := x^{2/\beta}\left(\log(1/x)\right)^{-1/\beta}$, then

$$P_\mu^\beta(\varphi_\beta\text{-}m(S_t)<\infty\ \forall t>0) = 1.$$

(b) $P_\mu^\beta(\dim(S_t) \leq 2/\beta\ \forall t>0) = 1.$

(c) If $d > 2/\beta$ and $t > 0$, then

$$P_\mu^\beta(X_t \text{ is singular} | X_t(\mathbb{R}^d) > 0) = 1$$

(where singularity is understood to be with respect to d-dimensional Lebesgue measure).

Proof. (b) follows immediately from (a) and the definition of dimension.

(c) follows immediately from (b) and Remark 9.3.1.4.

The proof of (a) is based on the representation of $X_{t+\varepsilon}$ as a Cox cluster random measure with intensity $\frac{X(t)}{\varepsilon^{1/\beta}}$ (cf. Corollary 11.5.3 and Theorem 12.4.3). Although the formal development of this approach is deferred to Sections 11 and 12, we will outline the intuitive ideas here and then proceed to the relevant computations.

Let N(s,t) denote the number of clusters starting at time s whose lifetimes are greater than t. Intuitively each cluster represents a subpopulation alive at time (s+t) and having a single common ancestor located at a point x at time s. We will denote by $P^{\varepsilon *}_{s,x}$ the law of this cluster. We will obtain a covering of S_t in the time interval $(i2^{-n},(i+1)2^{-n})$ by open balls by decomposing X_t into a finite number, $N_n(i)$, of clusters starting at time $(i-1)2^{-n}$ at points $\{x_\ell\}_{\ell=1,\ldots,N_n(i)}$ of radius r_n. We can then use Theorem 9.2.4 to obtain an estimate of the probability that the ball $B(x_\ell,r_n)$ covers the evolving cluster during the time interval $(i2^{-n},(i+1)2^{-n})$. To carry this out we first formulate and prove two lemmas.

Lemma 9.3.3.6. Let $t > 0$ and $\tilde{X}_s$ denote a cluster with lifetime greater than ε with cluster law $P^{\varepsilon *}_{s,x}$ and $R > (2/\beta)t^{1/2}$. Then for $0 < \varepsilon < t$, $P^{\varepsilon *}_{s,x}(\tilde{X}_u(\overline{B(x;R)}^c) > 0$ for some $s\leq u\leq s+t)$

$$\leq c(\beta,d)(\varepsilon/t)^{1/\beta}\left[\frac{R}{t^{1/2}}\right]^{d+(2/\beta)-2} \exp\left\{-\frac{R^2}{2t}\right\}.$$

Proof. We obtain this by letting $a \downarrow 0$ in (9.2.4) and conditioning on non-extinction by time $s+\varepsilon$. □

We denote the number of clusters at instant $i/2^n$ of age 2^{-n} by $N_n(i) := N((i-1)2^{-n}, 2^{-n})$. Note that conditioned on the total mass $\langle X((i-1)2^{-n}),1\rangle \le M$ at time $t=(i-1)2^{-n}$, $N_n(i)$ is Poisson with mean

$$\lambda = \frac{\langle X((i-1)2^{-n}),1\rangle}{c_\beta 2^{-n/\beta}} \le 2^{n/\beta}M/c_\beta$$

where $c_\beta = (\beta/(\beta+1))^{1/\beta}$.

Lemma 9.3.3.7. Let $\tau_M := \min(\inf\{t:\langle X_t,1\rangle \ge M\},1)$.
(i) If $u>2M$, then we have the following estimate for the number of clusters of age 2^{-n} at instants $j/2^n < \tau_M$:

$$(9.3.3.1) \qquad P_\mu^\beta(N_n(j) > u2^{n/\beta},\ j/2^n < \tau_M,\ j\in\mathbb{N}) \le 2^n \exp\left\{-2^{n/\beta}\left(u \log\frac{c_\beta u}{M} - u + \frac{M}{c_\beta}\right)\right\}.$$

(ii) There exists $n_0(w)$ such that for $n> n_0(\omega)$,

$$(9.3.3.2) \qquad \max_{i\le 2^n\tau_M} N_n(i) \le M\cdot\left(1+\frac{1}{\beta}\right)^{1/\beta} 2^{n/\beta}, \quad P_\mu^\beta\text{-a.s.}$$

Proof. (i) Obviously

$$P_\mu^\beta(N_n(j) > u2^{n/\beta},\ j/2^n < \tau_M,\ j\in\mathbb{N}) \le \sum_{j=1}^{2^n[\tau_M]} P\{N_n(j) > u2^{n/\beta}\}.$$

To estimate the tail probabilities of the Poisson distributions we note that the cumulant generating function for the Poisson distribution with parameter λ and its Legendre transform are equal to $\lambda(e^s-1)$ and $H_\lambda(v) = v\log\frac{v}{\lambda} - v + \lambda$ respectively. Applying the exponential Chebyshev inequality we obtain

$$P\{N_n(j) > u2^{n/\beta}\} \le \exp\left\{-H_{\lambda_{n,j}}(u2^{n/\beta})\right\}$$

where $\lambda_{n,j} = E(N_n(j)) \le 2^{n/\beta}M/c_\beta$. Differentiating with respect to λ shows that $-H_\lambda(v)$ is increasing in λ if $\lambda\le v$. Hence

$$P\{N_n(j) > u2^{n/\beta}\} \le \exp\left\{-2^{n/\beta}\left(u\log\frac{c_\beta u}{M} - u + \frac{M}{c_\beta}\right)\right\}.$$

(ii) Due to (9.3.3.1) the probability of the event

$\left\{\max_{i\le 2^n\tau_M} N_n(i) > M\cdot\left(1+\frac{1}{\beta}\right)^{1/\beta} 2^{n/\beta}\right\}$ is the general term of a convergent series. Applying the Borel-Cantelli lemma we obtain (9.3.3.2). □

Completion of the Proof Theorem 9.3.3.5(a)

We will cover each of the $N_n(i-1)$ cluster birth points at time $(i-1)2^{-n}$ that have descendents at time $i2^{-n}$ by balls of radius $r_n = C_1 \cdot 2^{-n/2} n^{1/2}$.

Recall that the number of such cluster birth points does not exceed $C_2(M)\, 2^{n/\beta}$, at each $j/2^n < M$. Set

$$\Xi_n(i) := \bigcup_{j=1}^{N_n(i)} B(x_{i,j}, r_n) .$$

Applying Lemma 9.3.3.6 to each of these clusters we obtain

$$P\left(X_u(\overline{\Xi_n(i)}^c > 0 \text{ for some } 0 \le u \le t \ \forall\ t \in (i2^{-n},(i+1)2^{-n}) \forall\ i/2^n \le \tau_M \right\}$$

$$\le C_3 \cdot 2^{n(1+1/\beta)} \cdot n^{((d-2)/2+1/\beta)} \cdot \exp\left(-C_1 2^n \cdot 2^{-n} n \right)$$

$$\le C_3 2^{n(1+1/\beta)} \cdot n^{[(d-2)/2+1/\beta-2\kappa/\beta]} . \exp\left(-C_1 n \right)$$

Hence it is the general term of a convergent series for sufficiently large C_1.
Therefore

$$P^{\beta}_{\mu}\left\{ S(X_t) \subset \Xi_n(i)\ \forall t \in (i2^{-n},(i+1)2^{-n})\ \forall\ i/2^n \le \tau_M,\ i \in \mathbb{N}\ \forall \text{ suff. large } n \right\} = 1.$$

But then for all sufficiently large n,

$$\varphi_\beta\text{-m}(S(X_t)) \le \lim_{n\to\infty} \varphi_{\beta,2r_n}\text{-m}(S_t) \le \lim_{n\to\infty} \varphi_{\beta,2r_n}\text{-m}(\Xi_n(i))$$

$$\le \lim_{n\to\infty} c_4(M) \cdot 2^{n/\beta} \varphi_\beta(2r_n)$$

$$\le C_4(M) \cdot \lim_{n\to\infty} \frac{2^{n/\beta} 2^{-n/\beta} n^{1/\beta}}{(\log 2^n/n)^{1/\beta}} < \infty. \quad □$$

9.4 Charging and Hitting Sets.

9.4.1. Some Basic Definitions and Probability Estimates.

Consider the $(2,d,\beta)$-superprocess with log-Laplace equation (9.1.1). The canonical process X is said to *charge* the Borel subset B of $\mathbb{R}^d$ if $X_t(B) > 0$ for some $t>0$.

It turns out that the notion of charging a set is not the natural analogue of the notion of hitting a set for Brownian motion. To introduce the appropriate definition let $\bar{\mathcal{R}}_+(0,t) := \bigcup_{t>s>0} \bar{\mathcal{R}}(s,t)$ for $0<t\le\infty$,

where

$$\bar{\mathcal{R}}(s,t) = \text{closure of } \mathcal{R}(s,t) := \Big[\bigcup_{s\le u\le t} S(X_u)\Big].$$

Equivalently, $\bar{\mathcal{R}}(s,t) := S(\mathcal{O}_{s,t})$ where $\mathcal{O}_{s,t}$ denotes the occupation time measure as defined in Section 7.4. The *range* of super-Brownian motion is defined by $\bar{\mathcal{R}} := \bar{\mathcal{R}}_+(0,\infty)$. Theorem 9.2.2 implies that the range is bounded P^{β}_{μ}-a.s. provided that μ has compact support.

The canonical process X is said to *hit* the set $A \subset \mathbb{R}^d$ if $A\cap\bar{\mathcal{R}} \ne \phi$. If A is an analytic subset of $\mathbb{R}^d$, then the event $\{\omega:\bar{\mathcal{R}}(\omega)\cap B\ne\emptyset\}$ belongs to the universal completion $\mathcal{D}^{uc}$ of the σ-algebra $\mathcal{D}$.

<u>Theorem</u> <u>9.4.1.1.</u> Let $\varepsilon > 0$ and $x_0\in\mathbb{R}^d$.

(a) Let $d>4$. Then there exists a constant $c_2(d)$ such that

(9.4.1.1) $$P^1_{\delta_x}[X_t(B(x_0;\varepsilon)) > 0 \text{ for some } t\ge0\,] \le c_2(d)\varepsilon^{d-4}|x-x_0|^{2-d}$$

provided that $x\ne x_0$.

(b) $$P^1_{\delta_0}(X \text{ charges } B(x;\varepsilon)) \sim \begin{cases} 6/x^2 & d=1 \\ 4/x^2 & d=2 \\ 2/x^2 & d=3 \\ 2/[x^2\log|x|] & d=4. \end{cases}$$

<u>Proof.</u> (a) We will outline the main ideas of the proof and refer the reader to Dawson, Iscoe and Perkins (1989, Section 3) for the details.

First note that

$$P^1_{\delta_x}\exp\Big\{-\int_0^t\!\!\int \theta 1_{B(0;\varepsilon)}(x)X(s,dx)ds\Big\} = \exp\{-u(t,x,\theta,\varepsilon)\}$$

where us is the solution of the initial value problem

$$\partial u/\partial t = 1/2(\Delta u-u^2) + \theta 1_{B(0;\varepsilon)}(.),\quad u(0) = 0.$$

Note that

$$P^1_{\delta_x}\Big\{\int_0^\infty X(s,B(0;\varepsilon))ds = 0\Big\} = \lim_{\theta\uparrow\infty}\lim_{T\uparrow\infty}\exp\{-u(T,x,\theta,\varepsilon)\} = \exp\{-u_\infty(x,\varepsilon)\}$$

where $\Delta u_\infty(x,\varepsilon) = u_\infty(x,\varepsilon)^2$ for $x\notin B(0;\varepsilon)$,

$$u_\infty(x,\varepsilon) \longrightarrow \infty \text{ as } x \to \partial B(0;\varepsilon).$$

By a scaling argument $u_\infty(x,\varepsilon) = \varepsilon^{-2}u_\infty(x/\varepsilon,1)$.

Therefore

$$P^1_{\delta_x}(X \text{ charges } B(0;\varepsilon)) = 1 - \exp\{-u_\infty(x,\varepsilon)\} = 1 - \exp\{-\varepsilon^{-2}u_\infty(x/\varepsilon,1)\}.$$

But in d>4 an analytic argument yields

$$u_\infty(x,1) \sim \frac{1}{|x|^{d-2}} \qquad \text{as } |x| \longrightarrow \infty.$$

Therefore

$$P^1_{\delta_x}(X \text{ charges } B(0;\varepsilon)) \sim 1 - \exp\left\{-\frac{\varepsilon^{-2}\varepsilon^{d-2}}{|x|^{d-2}}\right\}.$$

(b) See Iscoe (1988, Theorem 2). □

If $d \le 3$, and $\beta=1$, then by Theorem 9.4.1.1(b),

$$P^1_{\delta_x}(X \text{ charges } B(0;\varepsilon)) \sim \frac{\text{const}}{|x|^2} \quad (\text{independent of } \varepsilon > 0).$$

This, as well as the the existence of a jointly continuous local time (cf. Theorem 8.5.1), implies that X hits points. This is made precise as follows.

<u>Theorem</u> <u>9.4.1.2</u>. Let $d \le 3$, $\beta=1$. Then

$$P^\beta_\mu(X \text{ hits } \{x\}) = 1 - \exp\left\{-2(4-d)\int |x-y|^{-2}\mu(dy)\right\} > \text{ if } \mu\neq 0.$$

<u>Proof.</u> The proof of this result is given in [Dawson, Iscoe and Perkins (1989) Theorem 1.3].

On the other hand if $d > 4$ and A is sufficently small then it will not be hit. For example the following result is an easy consequence of (9.4.1.1).

<u>Proposition</u> <u>9.4.1.3.</u> Let d>4 and $A \subset \mathbb{R}^d$. Then if $x^{(d-4)}$-m(A) = 0, then $P^1_\mu(X \text{ hits } A) = 0$.

<u>Proof.</u> Let $\{B_j\}$ be a covering of A. Then

$$\begin{aligned} P^1_\mu(X \text{ hits } A) &\le P^1_\mu(X \text{ hits } \textstyle\bigcup_j B_j) \\ &\le \sum_j P^1_\mu(X \text{ hits } B_j) \\ &\le c_2(d) \sum_j (\text{diam}(B_j))^{d-4} \quad (\text{by Theorem 9.4.1.1(a)}). \end{aligned}$$

But since the x^{d-4}-m(A) = 0, this can be made arbitrarily small. □

We should emphasize the difference between the hitting a set and *charging* a set, B. By Remark 7.3.2(a) $X_t(B) = 0 \ \forall \ t>0$, a.s. if $|B|=0$ and hence X does not charge any Lebesgue null sets.

9.4.2. $\mathcal{R}$-Polar Sets

For d-dimensional Brownian motion there is a classical necessary and sufficient condition for a set to be not hit in terms of capacity. In this section we will formulate an analogous necessary and sufficient condition for super-Brownian motion which is due to Dynkin. The main step is to reduce the hitting problem to an analytical problem for the nonlinear elliptic equation (9.2.1).

Definition. The analytic set A is said to be R_β-*polar* if $A^c \neq \emptyset$ and $P^\beta_\mu(A \cap \bar{\mathcal{R}} = \emptyset) = 1$, $\forall \mu \in M_F(\mathbb{R}^d)$. (It is actually sufficient to check that $P^\beta_{\delta_x}(A \cap \bar{\mathcal{R}} = \emptyset) = 1$ $x \notin A$, (cf. Dynkin (1991c, Lemma 2.4)). An arbitrary set is called R_β-polar if it is a subset of a Borel R_β-polar set B.

Let $p_t(.,.)$ denote the standard Brownian transition density in $\mathbb{R}^d$ and

$$G_d(x,y) = \int_0^\infty e^{-t/2} p_t(x,y)dt$$
$$\sim |x-y|^{2-d} \text{ for small } |x-y| \text{ if } d>2.$$
$$\sim \log^+ |x-y|^{-1} \text{ if } d=2.$$

For $\beta \geq 0$ the $(1+\beta)$-*capacity* of a subset of $\mathbb{R}^d$ is defined as follows:

$$\mathrm{Cap}_{d-2,1+\beta}(A) := \sup \ \nu(A)$$

where the supremum is taken over $\nu \in M(\mathbb{R}^d)$ with $\nu(A^c)=0$ and such that

$$\int_{\mathbb{R}^d} \left[\int_{\mathbb{R}^d} \nu(dx) G_d(x,y) \right]^{1+\beta} dy \leq 1$$

If $\beta=0$, $\mathrm{Cap}_{d-2,1}$ coincides with the *classical Newtonian capacity*.

Theorem 9.4.2.1 (Kakutani) Let $d \geq 2$ and $\{B_t : t \geq 0\}$ denote the standard d-dimensional Brownian motion. Then the following assertions are equivalent

(i) $P_x(\tau_A = +\infty) = 1$ $\forall x \notin A$, where $\tau_A := \inf\{t>0 : B_t \in A\}$,

(ii) $\mathrm{Cap}_{d-2,1}(A) = 0$.

Proof. See Itô and McKean (1965, Section 7.8).

Theorem 9.4.2.2 (Dynkin(1991c)). A set $A \subset \mathbb{R}^d$ is R_β-polar if and only if $C_{d-2,1+\beta}(A) = 0$.

Corollary 9.4.2.3. A single point $\{x\}$ is R-polar if and only if $d \geq \frac{2(1+\beta)}{\beta}$.

Remarks on the Proof. This is proved in Dynkin (1991c). The proof is based on probabilistic arguments which reduce the problem to an analytical problem for semi-linear elliptic partial differential equations and then by an application of analytical results of Baras and Pierre (1984), and Brezis and Veron (1980).

We will give an indication of the probabilistic step which involves an interesting aspect of superprocesses which we have not yet considered. To explain this in heuristic terms let us go back to the construction of the superprocess in terms of the branching particle approximations.

Let B be a bounded regular domain in $\mathbb{R}^d$ and for $w \in C([0,\infty),\mathbb{R}^d)$, let $\tau_B := \inf\{t: w_t \notin B\}$. Let us introduce the additive functional $\kappa_B(dt) = 1_{[0,\tau_B]}(t)$ and let W_B denote Brownian motion in $\mathbb{R}^d$ stopped at time τ_B. Now consider the $(W_B,\kappa_B,\Phi_\beta)$-superprocess $\{X_t^B: t\geq 0\}$ with $\Phi_\beta(\lambda) = -\lambda^{1+\beta}$.

We define $X_{\tau_B} := \lim_{t\to\infty} X_t^B$. Then X_{τ_B} is a random measure and for $\phi \in bpC(\partial B)$

$$v(x) := -\log\left(P_{\delta_x} \exp(-\langle X_{\tau_B},\phi\rangle)\right)$$

is the unique solution of

$$\Delta v = v^{1+\beta} \quad \text{in } B \tag{9.4.2.1}$$

$$v(x) \longrightarrow \phi(a) \text{ as } x\to a \in \partial B,\ x\in B.$$

With a little more work one can construct an enriched version giving the joint law of X_{τ_B} and the full superprocess $\{X_t: t\geq 0\}$. In fact in Dynkin's proof of the result below, an increasing sequence of open sets $B_n \uparrow B$ and a sequence of random measures $\{X_{\tau_{B_n}}: n=1,2,\dots\}$ defined on a common probability space (cf. Section 12.3.5) is involved.

The main step in the reduction of the proof of Theorem 9.4.2.2 to an analytical problem is summarized in the following.

Theorem 9.4.2.4.

(a) Let B be a bounded regular domain in $\mathbb{R}^d$. Then

$$v(x) = -\log P^\beta_{\delta_x}\{X_{\tau_B}=0\}$$

is the minimal positive solution to the boundary problem

$$\Delta v(x) = v(x)^{1+\beta} \quad \text{in } B$$

$$v(x) \longrightarrow \infty \text{ as } x\to a\in\partial B,\ x\in B.$$

(b) For an arbitrary open set B

$$v(x) = -\log P^{\beta}_{\delta_x} \{\bar{\mathcal{R}} \subset B\}$$

determines the maximal positive solution of

(9.4.2.2) $\Delta v(x) = v(x)^{1+\beta}$ in B.

(c) A necessary and sufficient condition for a closed set F to be R_β-polar is that one of the following equivalent statements holds

(i) if $v \geq 0$ satisfies (9.4.2.1) in $B = F^c$, then $v=0$,

(ii) the maximal solution of (9.4.2.1) in $B = F^c$ is bounded.

Proof. See Dynkin (1991c, Theorems 1.2, 1.3, 1.4). □

Remark: For a probabilistic approach to these questions in the case of continuous branching, see Perkins (1990). Also Le Gall (1991b) has recently developed an alternative approach to the study of these questions for continuous branching using his path-valued Markov process. For a survey of the relations between superprocesses and partial differential equations, see Dynkin (1992b).

9.5 Multiple Points and Intersections

In this section we consider *multiple points*, that is, points visited at two or more distinct times by a given (2,d,1)-superprocess and *intersections* (or *collisions*) of independent (2,d,1)-superprocesses, that is, points visited by two or more superprocesses at the same time. One of the main ideas involved is to regard the range of a superprocess over a given time interval as a *target* and to use the criteria of section 9.4.2 to determine if this is a R-polar set. For the sake of comparison we begin with the corresponding results for ordinary Brownian motion.

Theorem 9.5.1 Let $\{B_t : t \geq 0\}$ be a d-dimensional Brownian motion andset

$$\phi_0(x) := x^2 \log \frac{1}{x} \log\log\log \frac{1}{x} \quad \text{and} \quad \phi(x) := x^2 \log\log \frac{1}{x}.$$

Then

(a) In the case $d \geq 3$, ϕ-$m(B[0,t]) = c_d t \ \forall t$ a.s.

(b) In the case $d=2$, ϕ_0-$m(B[0,t]) = c_2 t \ \forall t>0$ a.s.

Proof. (a). (Ciesielski and Taylor (1962)

(b) Taylor (1964). □

Theorem 9.5.2 (a) Two independent d-dimensional Brownian motions intersect if and only if $d \leq 3$.

(b) Let Γ_k := set of k-multiple points of a d-dimensional Brownian motion. Then

(i) $\dim \Gamma_k(B) = d-k(d-2)$ for $d>2$.

(ii) $\dim \Gamma_k(B) = 2$ for all k if $d=2$.

(iii) dim $\Gamma_2(B) = 1$ if $d=3$

(iv) dim $\Gamma_3(B) = \emptyset$ if $d=3$.

Proof. (a) follows from Kakutani's criterion (Theorem 9.4.2.1).

(b) See Taylor (1966), and Le Gall (1987). □

The analogues of these results for super-Brownian motion are contained in the following three theorems.

Theorem 9.5.3. Let $\beta=1$ and $\bar{\mathcal{R}}_+(0,s)$ denote the range of the continuous super-Brownian motion in $\mathbb{R}^d$ with $d > 4$. Let $\phi(x) := x^4\log^+\log^+(1/x)$.
Then there are positive constants $c_1(d)$ and $c_2(d)$ such that

$$c_1\phi\text{-m}(\bar{R}_+(0,s)\cap A) \le Y_{0,s}(A) \le c_2\phi\text{-m}(\bar{R}_+(0,s)\cap A)$$

$\forall\ A\in \mathcal{B}(\mathbb{R}^d)$ w.p.1.

Proof. (cf. [Dawson, Iscoe and Perkins (1989), Theorem 1.4]). □

Definition. The set of *k-multiple points* of X is denoted by

$$\bar{R}_k := \bigcup \left\{ \bigcap_{j=1}^{k} \bar{R}(I_j)\colon I_1,\dots,I_k \text{ disjoint, compact} \subset (0,\infty)\right\}$$

Theorem 9.5.4. Let $d\ge4$. Then

(a) If $t>0$, then $\dim(\bar{R}_+(0,t)) = 4$ w.p.1.

(b) $\bar{R}_k = \phi$ if $k \ge \dfrac{d}{d-4}$, and

$\dim \bar{R}_k \le d - k(d-4)$ a.s.

Proof. (b) [Dawson, Iscoe and Perkins (1989), Theorem 1.5]

Remark: This means that there are no double points if $d\ge8$, triple points if $d\ge6$ or quintuple points if $d\ge5$!

The *closed graph* of the (2,d,1)-superprocess is defined to be the closure of the random space-time set

$$\bar{G}(X) := \{(t,x)\colon t>0,\ x\in S(X_t)\} \in \mathcal{B}((0,\infty)\times\mathbb{R}^d).$$

An analytic subset of $\mathbb{R}_+\times\mathbb{R}^d$ is called *G-polar* if $P^1_\mu(\bar{G}(X)\cap A=\emptyset)=1\ \forall\ \mu\in M_F(\mathbb{R}^d)$. Dynkin (1992a) established a criterion for G-polarity following a program similar to that described in section 9.4 for R-polarity. An important application of the notion of G-polarity is the following result of Barlow, Evans and Perkins.

Theorem 9.5.5 Let $X^{(1)}$ and $X^{(2)}$ denote two independent (2,d,1)-superprocesses with $X^{(1)}(0) = \mu_1$ and $X^{(2)}(0) = \mu_2$. If $d\ge6$, then $\bar{G}(X^{(1)})\cap\bar{G}(X^{(2)}) = \emptyset$ for all $t > 0$, $P^{(1)}_{\mu_1}\times P^{(2)}_{\mu_2}$-a.s.

Proof. Barlow, Evans and Perkins (1991, Theorem 3.6).

Remark. An immediate consequence of Theorem 9.5.5 is the *non-intersection* property $S(X_t^{(1)}) \cap S(X_t^{(2)}) = \emptyset \ \forall \ t>0$. Barlow, Evans and Perkins (1991) also establish the existence of a *collision local time* in dimensions $d \le 5$.

9.6 Some Further Comments

The objective of this section has been to give an introduction to the sample path properties of $(2,d,\beta)$-superprocesses. As we have mentioned at several places the known results for (2,d,1)-superprocesses, that is, continuous super-Brownian motion are considerably more complete. References for some recent developments in this area are Adler and Lewin (1992), Barlow, Evans and Perkins (1991), Dawson, Iscoe and Perkins (1989), Dawson and Perkins (1991), Dynkin (1991a,b,c,d, 19921,b), Evans and Perkins (1991), Iscoe (1988), Le Gall (1991a,b), Perkins (1988, 1989, 1990, 1991a,b), and Tribe (1989, 1991, 1992).

There still remain some open questions concerning the structure of $S(X_t)$ for the (2,d,1)-superprocess even at fixed times. For example is $S(X_t)$ a totally disconnected set? In this direction, Tribe (1989) showed that at fixed times $t > 0$ for X_t-a.e. x, the connected component containing x is simply $\{x\}$.

Theorem 9.3.3.3 implies that in dimensions $d \ge 3$, the finite dimensional distributions of $\{X_t : t>0\}$ can be obtained from those of the support process $\{S(X_t) : t>0\}$. This implies that $\{S(X_t) : t>0\}$ is a $\mathfrak{K}(\mathbb{R}^d)$-valued Markov process. However the question as to whether $\{S(X_t)\}$ is strong Markov is open and is in fact closely linked to the question as to whether or not the statement of Theorem 9.3.3.3 is valid for all $t>0$, P_μ^1-a.s. which is also open. To explain the possibility that $\{S(X_t)\}$ might not be strong Markov consider the same question for the measure-valued process in $\mathbb{R}$, $Z_t = \delta_{B_1(t)} + 2\delta_{B_2(t)} + 4\delta_{B_3(t)}$ where $B_1(.)$, $B_2(.)$ and $B_3(t)$ are independent Brownian motions. In this case $\{Z_t\}$ is strong Markov but the continuous $\mathfrak{K}(\mathbb{R})$-valued process, $\{S(Z_t)\}$, is Markov but not strong Markov.

10. BUILDING MEASURE-VALUED PROCESSES WITH INTERACTIONS. VARIOUS EXAMPLES.

To this point we have explored measure-valued branching processes and Fleming-Viot processes and have discovered that they have a rich mathematical structure. However from the point of view of population modelling, they are rather oversimplified. In order to have models exhibiting some of the more interesting features of real populations it is necessary to introduce interactions and additional spatial structure. For example in population models phenomena such as competition for resources, predation, genetic selection and ecological niches are of considerable interest. In this chapter we consider the basic measure-valued branching process and Fleming-Viot process as building blocks from which we can build more complex models. In particular we will consider certain natural types of interaction that can be studied via the Cameron-Martin-Girsanov formula and the FK-dual representation.

10.1. Application of the Cameron-Martin-Girsanov Theorem.

10.1.1 Fleming-Viot with Selection

The class of Fleming-Viot processes arise from the *neutral theory of evolution* (e.g. Kimura (1983)). However *selection (competitive interaction)* plays the central role in the Darwinian theory of evolution and in this theory variability arises from mutation and recombination. From this viewpoint finite population sampling provides a random perturbation which can lead to *tunnelling* between locally optimal points in the selective landscape (cf. Wright (1949)). In order to study such questions in a mathematical setting it is necessary to consider a modification of the Fleming-Viot process to include these effects.

We begin by considering the Fleming-Viot model with *selection*.

Let $\mathcal{D}(\mathfrak{G}) := \{F:F(\mu) = f(\langle\mu,\phi\rangle),\ \phi\in D(A),\ f\in C_b^\infty(\mathbb{R})\}$ and for $F\in\mathcal{D}(\mathfrak{G})$,

$$\mathfrak{G}_R F(\mu) := f'(\langle\mu,\phi\rangle)\langle\mu,A\phi\rangle + f'(\langle\mu,\phi\rangle)\langle R(\mu),\phi\rangle + \frac{1}{2}\int\int f''(\langle\mu,\phi\rangle)\phi(x)\phi(y)Q(\mu;dx,dy)$$

with

$$Q(\mu;dx,dy) = \gamma[\delta_x(dy)\mu(dx)-\mu(dx)\mu(dy)],$$

and

$$R(\mu,dx) = \mu(dx)\left[\int_E V(x,y)\mu(dy)\right] - \mu(dx)\left[\int_E\int_E V(y,z)\mu(dy)\mu(dz)\right]$$

$$= \gamma^{-1}\int_E\left[\int_E V(y,z)\mu(dz)\right]Q(\mu;dx,dy).$$

The function V is called the *fitness function* and it is assumed to belong to $b\mathcal{E}\times\mathcal{E}$.

By applying Theorem 7.2.2 (the Cameron-Martin-Girsanov theorem) it follows that the corresponding martingale problem for $(\mathcal{D}(\mathfrak{G}),\mathfrak{G}_R)$ is well-posed. The resulting Markov process is called the Fleming-Viot process with mutation operator A, sampling intensity γ, and fitness function V.

10.1.2. Branching with Interaction

Let us now return to the B(A,c)-superprocess, X_t, described in section 4.7. Recall that X_t is $M_F(E)$-valued and

$$M_t^B(\phi) = \langle X_t,\phi\rangle - \langle X_0,\phi\rangle - \int_0^t \langle X_s, A\phi\rangle ds, \quad \phi\in D(A),$$

is a square integrable martingale with increasing process

$$\langle M^B(\phi)\rangle_t = c\int_0^t \langle X_s,\phi^2\rangle ds$$

that is, $Q(\mu;dx,dy) = c\delta_x(dy)\mu(dx)$ with $c > 0$.

Let

$$R(\mu,dx) = c\int_E r(\mu,y)\delta_x(dy)\mu(dx)$$

$$= \int_E r(\mu,y)Q(\mu;dx,dy) = cr(\mu,x)\mu(dx)$$

and assume that $r \in \mathcal{B}(M_F(E)\times E)$ satisfies

(10.1.2.1) $c\cdot\sup_{\mu,x} r(\mu,x) := K < \infty$, and

$\sup\{|r(\mu,x)|: x\in E, \langle\mu,1\rangle \le n\} < \infty$ for each $n \in \mathbb{N}$.

Let $\mathcal{D}(\mathfrak{G}) = \{F(\mu) := f(\langle\mu,\phi\rangle)$ with $\phi\in D(A)$, $f\in C_b^\infty(\mathbb{R})\}$ and for $F\in \mathcal{D}(\mathfrak{G})$, let

$$\mathfrak{G}_R F(\mu) = f'(\langle\mu,\phi\rangle)\langle\mu,A\phi\rangle + f'(\langle\mu,\phi\rangle)\langle R(\mu),\phi\rangle + \frac{1}{2}\int\int f''(\langle\mu,\phi\rangle)\phi(x)\phi(y)Q(\mu;dx,dy).$$

We would like to apply the same argument as in Section 10.1.1 to establish that the martingale problem is well-posed. To do this we must extend the result of Theorem 7.2.2 to cover processes with values in $M_F(E)$ in which case $\langle X(t),1\rangle$ is not necessarily bounded. This means that

$$Z_r(t) := \exp\left\{\int_0^t\int_E r(X(s),y)M(ds,dy) - \frac{c}{2}\int_0^t\int_E [r(X(s),x)]^2 X(s,dx)ds\right\}$$

is a non-negative P_μ-local martingale (and thus a supermartingale) but an additional argument is necessary to prove that it is actually a martingale.

Recall that $Z_r(.)$ is a martingale if and only if $P_\mu[Z_r(t)] = 1\ \forall t\ge 0$ (cf. Theorem 7.1.7) and this will now be established.

Lemma 10.1.2.1 Assume that $R(\mu,dx) = cr(\mu,x)\mu(dx)$ satisfies (10.1.2.1). Then $Z_r(t)$ is a P_μ-martingale and the martingale problem for $(\mathfrak{D}(\mathfrak{G}),\mathfrak{G})$ is well-posed.

Proof. Let $\tau_n := \inf\{u:\langle X_u,1\rangle+\int_0^u \langle X_s,1\rangle ds \ge n\}$. By the assumption (10.1.2.1), $\tau_n \uparrow \infty$, P_μ-a.s. Fix t and let $B_n := \{\tau_n \le t\}$. Then $Z_r(u\wedge\tau_n)$ is a continuous local P_μ-martingale and

$$P_\mu\left[\exp\left(\frac{1}{2}\int_0^{t\wedge\tau_n}\int_E\int_E r(X(s);x)r(X(s);y)Q(X(s);dx,dy)ds\right)\right] < \infty.$$

Using the criterion of Theorem 7.1.7(b) we conclude that $Z_r(u\wedge\tau_n)$ is a P_μ-martingale and therefore

$$P_\mu[Z_r(t\wedge\tau_n)] = P_\mu[1_{B_n} Z_r(\tau_n)] + P_\mu[1_{B_n^c} Z_r(t)] = 1.$$

Note that $P_\mu[1_{B_n^c} Z_r(t)] \uparrow P_\mu[Z_r(t)]$ as $n\to\infty$ since $\tau_n \uparrow \infty$, P_μ-a.s.

Consider the solution, $P_\mu^r = Z_r(t)P_\mu$, to the local martingale problem for $\mathfrak{G}_R$. If

$$M_u(\phi) := \langle X_u,\phi\rangle - \langle X_0,\phi\rangle - \int_0^u \langle X_s,A\phi\rangle ds - \int_0^u \langle R(X_s),\phi\rangle ds, \quad \phi\in D(A),$$

then $M_{u\wedge\tau_n}(\phi)$ is a P_μ^r-martingale with $\langle M(\phi)\rangle_{u\wedge\tau_n} = c\int_0^{u\wedge\tau_n}\langle X_s,\phi^2\rangle ds$.

If $\phi = 1$ we get

$$P_\mu^r(\langle X_{u\wedge\tau_n},1\rangle) = \langle\mu,1\rangle + P_\mu^r\left(\int_0^{u\wedge\tau_n}\langle R(X_{s\wedge\tau_n}),1\rangle ds\right)$$

$$\le \langle\mu,1\rangle + K\cdot\left(\int_0^u P_\mu^r(\langle X_{s\wedge\tau_n},1\rangle)ds\right).$$

Hence $P_\mu^r(\langle X_{t\wedge\tau_n},1\rangle) \le \langle\mu,1\rangle e^{Kt}$ by Gronwall's inequality. Therefore we also have

$$P_\mu^r\left(\int_0^t \langle X_{u\wedge\tau_n},1\rangle du\right) \le K^{-1}\langle\mu,1\rangle(e^{Kt}-1).$$

Next, note that

$$P_\mu^r(B_n) := P_\mu[1_{B_n} Z_r(\tau_n)] \le P_\mu^r\left\{\int_0^t \langle X_{s\wedge\tau_n},1\rangle ds \ge n/2\right\} + P_\mu^r\left\{\sup_{u\le t\wedge\tau_n}\langle X_u,1\rangle \ge n/2\right\},$$

$$P_\mu^r\left\{\int_0^t \langle X_{s\wedge\tau_n},1\rangle ds \ge n/2\right\} \le 2n^{-1}P_\mu^r\left\{\int_0^t \langle X_{s\wedge\tau_n},1\rangle ds\right\} \le 2n^{-1}K^{-1}\langle\mu,1\rangle(e^{Kt}-1).$$

Since $P^r_\mu\left\{\left(\sup_{u\le t\wedge\tau_n} M_u(1)\right)^2\right\} \le$ *const*$\cdot K^{-1}\langle\mu,1\rangle(e^{Kt}-1)$ by the Burkholder-Davis-Gundy inequality, therefore $P^r_\mu\left\{\sup_{u\le t\wedge\tau_n}\langle X_u,1\rangle \ge n/2\right\} \longrightarrow 0$ as $n\to\infty$. Hence $P_\mu[1_{B_n} Z_r(\tau_n)] \longrightarrow 0$ as $n\to\infty$. Therefore $P_\mu[Z_r(t)] = 1\ \forall\ t$ and therefore $Z_r(t)$ is a martingale. □

Example 10.1.2.2. Noncritical Branching

It was convenient in the earlier sections to restrict our attention to *critical branching*, i.e. $P_\mu[\langle X_t,1\rangle] = \langle\mu,1\rangle\ \forall t\ge0$. However the above result applied to $R(\mu) \equiv c\mu$, $c\in\mathbb{R}$, shows that the corresponding martingale problems for subcritical and supercritical branching are also well-posed. It is then an easy exercise to show that the corresponding log-Laplace equation is also satisfied. Moreover the resulting law $\Pi_{[0,T]}P^c_\mu \ll \Pi_{[0,T]}P_\mu$ where the latter is the law of the critical B(A,c)-superprocess. This shows that all a.s. sample path properties in [0,T] derived for the critical B(A,c)-superprocess are also true for the subcritical and supercritical cases.

Example 10.1.2.3. Branching with Nonlinear Death Rates

Let $h\in pbC(E\times E)$, $\mu\in M_F(E)$ and let $P_{0,\mu}$ denote the law of the B(A,c)-superprocess. Consider the following B(A,c;h)-martingale problem:

(10.1.2.2a)

$$M_t(\phi) = \langle X_t,\phi\rangle - \langle X_0,\phi\rangle - \int_0^t \langle X_s,A\phi\rangle ds + \int_0^t\int_E\left(\int_E h(x,y)X_s(dy)\right)\phi(x)X_s(dx)$$

is a continuous martingale with increasing process

(10.1.2.2b)

$$\langle M(\phi)\rangle_t = c\int_0^t \langle X_s,\phi^2\rangle ds.$$

We can then obtain from Theorem 7.2.2 and Lemma 10.1.2.1 the following result.

Theorem 10.1.2.4. The martingale problem (10.1.2.2) is well-posed and the unique solution $\{P_h\}$ is characterized by

$$\frac{\Pi_{[0,T]}\, dP_{h,\mu}}{\Pi_{[0,T]}\, dP_{0,\mu}} = \exp\left(M^h_T - \tfrac{1}{2}\langle M^h\rangle_T\right),$$

where $$M^h_t := -\int_0^t\int_E\left(\int_E h(x,y)X_s(dy)\right)M(ds,dx)$$

and $M(ds,dx)$ is the orthogonal martingale measure extending M_t.

Example 10.1.2.5. Multitype Branching with Nonlinear Death Rates

Consider the following martingale problem M_{h_1,h_2}:

(10.1.2.3a)

$$M_t^i(\phi) = \langle X_t^i,\phi\rangle - \langle X_0^i,\phi\rangle - \int_0^t \langle X_s^i, A\phi\rangle ds + A_t^i(\phi), \quad i=1,2,$$

where

(10.1.2.3b) $$\langle M^i(\phi),M^j(\psi)\rangle_t = \delta_{ij} c\int_0^t \langle X_s^i,\phi\psi\rangle ds$$

and $$A_t^i(\phi) = \int_0^t\int_E\int_E \phi(x)\int h_i(x-y)X_s^1(dy)X_s^2(dx)ds, \quad h_i \in pbC(E).$$

Theorem 10.1.2.6. The martingale problem (10.1.2.3) is well-posed and the unique solution is characterized by

$$\frac{\Pi_{[0,T]}\, dP_{h_1h_2}}{\Pi_{[0,T]}\, dP_{0,0}} = \exp((M_1^h+M_2^h)_T - \tfrac{1}{2}(\langle M_1^h\rangle_T+\langle M_2^h\rangle_T),$$

where $$M_1^h(t) := -\int_0^t\iint h_1(x,y)X_s^2(dy)M^1(ds,dx)$$

$$M_2^h(t) := -\int_0^t\iint h_2(x,y)X_s^1(dy)M^2(ds,dx)$$

Proof. See Evans and Perkins (1992). □

Now consider the case $h_i(x-y) = p_\varepsilon(x-y)$, $\varepsilon > 0$, and note that $\langle M_1^h,M_1^h\rangle_t = \int_0^t\int\left[\int p_\varepsilon(x-y)X_s^2(dy)\right]^2 X_s^1(dx)ds \longrightarrow \infty$ if $\varepsilon\to 0$.

This suggests that the limit, if it exists, will be singular if $d\geq 2$. In fact,in dimensions $d = 2,3$, Evans and Perkins (1992) have established the existence of a non-trivial limit (at least in the case $h_2 \equiv 0$). This involves the notion of *collision local time* (cf. Barlow, Evans and Perkins (1991)). On the other hand in dimension $d\geq 6$ Theorem 9.5.5 implies that two independent superprocesses never intersect (collide) and consequently the limit is degenerate (i.e. the interaction disappears and we are left with two independent super-Brownian motions). In fact asimilar phenomenon occurs in dimensions $d = 4$ and 5 but for more subtle reasons (related to the lack of a collision local time between an ordinary Brownian particle and a super-Brownian motion). Thus the study of *local interactions* of this type involves a strong interplay with the sample path properties of the underlying super-processes.

10.2 An FK-Dual Representation for Fleming-Viot with Recombination and Selection.

In this section we build new processes from the basic Fleming-Viot process studied in previous chapters by introducing interactions corresponding to the mechanisms of selection and recombination in population genetics. The existence of the corresponding process can be obtained by a weak convergence argument similar to that in Chap. 2. However in this case the moment equations are *not closed* and uniqueness cannot be obtained by the arguments used there. The uniqueness is established by the use of a dual process which we will now introduce along with the description of the generator for the measure-valued process. (*See Section 5.6 for the appropriate notation - in particular* Θ_{ij} is as defined in Ch. 2 and $\{S_t\}$ is as in Section 5.6.)

Let $\rho \geq 0$ and let $\eta(x_1,x_2,\Gamma)$ be a transition function from $E\times E$ to E. For $i=1,\ldots,m$ define $R_{im}:\mathcal{B}(E^m)\to \mathcal{B}(E^{m+1})$ by

$$R_{im}f(x_1,\ldots,x_{m+1}) = \int f(x_1,\ldots,x_{i-1},z,x_{i+1},\ldots,x_m)\eta(x_i,x_{m+1},dz)$$

and assume that $R_{im}:C_b(E^m) \to C_b(E^{m+1})$. The R_{im} are called the *recombination operators* for the process and ρ is called the *recombination rate*.

For $V \in \mathcal{B}_{sym}(E\times E)$, set $\bar{V} = \sup_{x,y,z} |V(x,y)-V(y,z)|$, and for $i=1,\ldots,m$ define the *selection operators* $V_{im}:\mathcal{B}(E^m)\to\mathcal{B}(E^{m+2})$ by

$$V_{im}f(x_1,\ldots,x_{m+2}) = \frac{V(x_i,x_{m+1})-V(x_{m+1},x_{m+2})}{\bar{V}} f(x_1,\ldots,x_m).$$

For $f\in \mathcal{D}(A^{(n)})\cap\mathcal{B}(E^n)$, define $F(f,\mu) := \langle\mu^n,f\rangle$ and

$$\mathfrak{G}F(f,\mu) := F(A^{(n)}f,\mu) + \gamma \sum_{1\leq i<j\leq n} \left(F(\Theta_{ij}f,\mu)-F(f,\mu)\right)$$
$$+ \rho\sum_{i=1}^{n}\left(F(R_{in}f,\mu)-F(f,\mu)\right) + \bar{V}\sum_{i=1}^{n} F(V_{in}f,\mu).$$

For $f\in C_{slm}(E^{\mathbb{N}})$, with $\#(f)=n$, and $f\in \mathcal{D}(A^{(n)})\cap\mathcal{B}(E^n)$ let

$$Kf := \sum_{i=1}^{n} A_i f + \gamma \sum_{j=1}^{n}\sum_{k\neq j} [\Theta_{jk}f-f] + \rho\sum_{i=1}^{n}[R_{in}f-f] + \bar{V}\sum_{i=1}^{n}[V_{in}f-f]$$

If $\beta(f) := \bar{V}\#(f)$, then

$$\mathfrak{G}F(f,\mu) = F(Kf,\mu) + \bar{V}(\#(f))F(f,\mu)$$

and $\sup_{\mu\in M_1(E)} |F(Kf,\mu)| \le const\cdot\#(f)$.

Remark. There is considerable current interest in *genetic algorithms* (cf. Goldberg (1989)). These are a class of random optimization algorithms. They are discrete time analogues of Fleming-Viot processes with a selection operator describing *pure search* and the latter is defined by

$$V_{im}f(x_1,\ldots,x_{m+2}) := \frac{V(x_i)-V(x_{m+1})}{\bar{V}} f(x_1,\ldots,x_m).$$

In addition recombination also plays a central role in the study of this class of algorithms.

Theorem 10.2.1. Let $\mathfrak{G}$ satisfy the above conditions and suppose that the mutation process with generator A has a version with sample paths in $D_E[0,\infty)$. Then for each initial distribution in $M_1(M_1(E))$ there exists a unique solution of the martingale problem for $\mathfrak{G}$.

Proof. The proof (due to Ethier and Kurtz (1987)) is based on duality. We will first construct the appropriate function-valued process.

Let N be a jump Markov process taking non-negative integer values with transition intensities $q_{m,m-1} = \gamma m(m-1)$, $q_{m,m+2} = \bar{V}m$, $q_{m,m+1} = \rho m$, and $q_{i,j} = 0$ otherwise. Let $\{\tau_k\}$ be the jump times of N ($\tau_0=0$) and let $\{\Gamma_k\}$ be a sequence of random operators which are conditionally independent given M and satisfy

$$P(\Gamma_k = \Theta_{ij}|N) = [2/(N(\tau_k-)N(\tau_k))]1_{\{N(\tau_k-)-N(\tau_k)=1\}},\quad 1\le i<j\le N(\tau_k-),$$

$$P(\Gamma_k=R_{im}|N) = m^{-1}1_{\{N(\tau_k-)=m,\ N(\tau_k)=m+1\}}$$

$$P(\Gamma_k=V_{im}|N) = m^{-1}1_{\{N(\tau_k-)=m,\ N(\tau_k)=m+2\}}$$

for $1\le i\le m$.

For $f\in C_{sym}(E^{\mathbb{N}})$, define the $C_{sym}(E^{\mathbb{N}})$-valued process Y with $Y(0)=f$ by

$$Y(t) = S_{t-\tau_k}\Gamma_k S_{\tau_k-\tau_{k-1}}\Gamma_{k-1}\cdots\Gamma_1 S_{\tau_1}f,\quad \tau_k\le t<\tau_{k+1}.$$

Then for any solution P_μ of the martingale problem for $\mathfrak{G}$ and $f\in C_{sim}(E^{\mathbb{N}})$ we have the FK-dual representation

$$P_\mu[F(f,X(t))] = Q_f\left[F(Y(t),\mu)\exp\left(\bar{V}\int_0^t \#(Y(u))du\right)\right]$$

which establishes that the martingale problem for $\mathfrak{G}$ is well-posed. Since the function $\beta(f) = \tilde{V}\#(f)$ is not bounded we must verify the condition (ii) of Corollary 5.5.3. This follows from the following lemma due to Ethier and Kurtz (1990). □

Lemma 10.2.2. Let $N(t) = \#(Y(t))$ be as above, $\tau_K := \inf\{t:N(t)\geq K\}$ and $\theta>0$. Then there exists a function $F(n) \geq const\cdot n^2$ and a constant $L > 0$ such that

$$E\left[F(N(t\wedge\tau_K))\exp\left\{\theta\int_0^{t\wedge\tau_K} N(s)ds\right\}|N(0)=n\right] \leq F(n)e^{Lt} \quad \forall\ K \geq 1,$$

and given $N(0)=n$, $\left\{N(t\wedge\tau_K)\exp\left\{\overline{V}\int_0^{t\wedge\tau_K} N(s)ds\right\}:\ K\geq 1\right\}$ are uniformly integrable.

Proof. Take $F(m) := (m!)^\beta$, with $\beta < 1/2$. Then

$$QF(m) + \theta mF(m) = \gamma m(m-1)(F(m-1)-F(m)) + \rho m(F(m+1)-F(m))$$

$$+ \theta m(F(m+2)-F(m)) + \theta mF(m)$$

$$= \gamma m(m-1)(((m-1)!)^\beta - (m!)^\beta) + \rho m(((m+1)!)^\beta-(m!)^\beta) + \theta m((m+2)!)^\beta + \theta m$$

$$\leq m(m!)^\beta[\rho((m+1)^\beta-1) + \theta\sigma(m+1)^\beta(m+2)^\beta + \theta(m!)^{-\beta} - \gamma(m-1)(1-m^{-\beta})]$$

$$\leq ((m-1)!)^\beta\{m^\beta[\rho(m+1)^\beta-1] + \theta m[m(m+1)(m+2)]^\beta + \gamma m(m-1)[1-m^\beta]\}.$$

Since the last negative term dominates for large m if $\gamma>0$, we can choose $L > 0$ such that

$$QF(m) + \theta mF(m) \leq L.$$

The optional sampling theorem (cf. [EK, p 61]) implies that for $\tau_K = \inf\{t:N(t)\geq K\}$ and $N(0)=m$

$$E\left[\exp\left\{\theta\int_0^{t\wedge\tau_K} N(s)ds\right\}\middle|N(0)=m\right]$$

$$\leq E\left[F(N(t\wedge\tau_K))\exp\left\{\theta\int_0^{t\wedge\tau_K} N(s)ds\right\}\middle|N(0)=m\right]$$

$$= F(m) + E\left[\int_0^{t\wedge\tau_K} \exp\left\{\theta\int_0^u N(s)ds\right\}(QF(N(u))+\theta N(u)F(N(u)))du\middle|N(0)=m\right]$$

$$\leq F(m) + L\ E\left[\int_0^{t\wedge\tau_K} \exp\left\{\theta\int_0^u N(s)ds\right\}du\middle|N(0)=m\right]$$

and the lemma follows by Gronwall's inequality. □

10.3 Stepping Stone Models in Population Genetics

In this section we give a brief introduction to the class of stepping stone models of spatially distributed populations and indicate the role of duality in their study.

Stepping stone models describe multitype populations which are divided into geographically separated colonies in which the total population size of each colony does not change in time. The set of colonies will be indexed by a countable set S. Each colony consists of a population described by an element of $M_1(E)$ (the distribution of types within the colony). These models can also be considered as mathematical idealizations of a spatially distributed population with locally finite carrying capacity (such as would arise if we considered a supercritical branching model in which the individual death rate is proportional to the population within the colony). The idealization that the population in each colony is fixed simplifies the analysis but it is reasonable to expect that it will have the same long-time qualitative behavior as the more complex branching-interaction model.

10.3.1. The Stepping Stone Model with Two Types.

We first consider the case in which $E = (E_1,E_2)$. In this case it suffices to prescribe $\{x_k(.):k\in S\}$ where

$x_k(t)$:= proportion of type 1 individuals in colony.

The $\{x_k(.)\}$ are assumed to satisfy the following system of Itô stochastic differential equations:

$$(10.3.1)\quad dx_k(t) = \sqrt{\gamma x_k(t)(1-x_k(t))}\, dw_k(t) + b_k(x(t))dt, \quad k\in S,$$

where the $\{w_k(.):k\in S\}$ are independent standard Brownian motions and

$$b_k(x) = a_2(1-x_k) - a_1x_k + Vx_k(1-x_k) + \rho \sum_{j\in S} q_{jk}(x_j-x_k).$$

Here $a_i\geq 0$ denotes the mutation rate of type i, V denotes the selective advantage of type 1 and $\rho q_{jk}\geq 0$ denotes the migration rate from colony j to colony k. The existence of a solution to this system of equations was obtained in Notohara and Shiga (1980).

The corresponding generator is given by

$$\mathfrak{G} = \frac{\gamma}{2} \sum_{i\in S} x_i(1-x_i)\frac{\partial^2}{\partial x_i^2} + \sum_{i\in S} b_i(x)\frac{\partial}{\partial x_i}$$

where $b_i(x) = \sum_{j\neq i} \rho q_{ji}(x_j-x_i) + a_2 - (a_1+a_2)x_i + Vx_i(1-x_i).$

The dual process. The state space for the dual process is

$$\mathbb{Z}_+^S := \{\mathbf{n}= (n_i)_{i\in S}: n_i=0,1,2,\ldots, |\mathbf{n}|=\sum n_i < \infty\}.$$

Let $(\mathbf{e}_i)_j := 1$ if $i=j$

$:= 0$ if $i \neq j$.

The dual process is a finite particle system on S with births, deaths and particle motion having the following rates:

$\mathbf{n} \to \mathbf{n} + \mathbf{e}_i$ rate $(-V)n_i$ (assume $-V \geq 0$) *(birth at site i)*

$\mathbf{n} \to \mathbf{n} - \mathbf{e}_i$ rate $a_2 n_i + \frac{1}{2}\gamma n_i(n_i-1)$ *(death at site i)*

$\mathbf{n} \to \mathbf{n} - \mathbf{e}_i + \mathbf{e}_j$ rate $\rho q_{ij} n_i$ *(particle moves from i to j)*

(N.B. If $V>0$ we reverse the roles of x_i and $(1-x_i)$.)

Let $\mathbb{L}$ denote the generator of this finite particle system.

Define $F \in \mathcal{B}(\mathbb{N}^S \times [0,1]^S)$ by

$$F(\mathbf{n},\mathbf{x}) := \prod_{i \in S} (x_i)^{n_i} \text{ and } F(0,\mathbf{x}) \equiv 1.$$

Then

$$\begin{aligned} \mathfrak{G}F(\mathbf{n},\mathbf{x}) &= \sum_i (\tfrac{\gamma}{2} n_i(1-n_i)+a_2)[F(\mathbf{n}-\mathbf{e}_i,\mathbf{x})-F(\mathbf{n},\mathbf{x})] \\ &\quad + \sum_i n_i \Big\{ \sum_j \rho q_{ji}[F(\mathbf{n}+\mathbf{e}_j-\mathbf{e}_i,\mathbf{x})-F(\mathbf{n},\mathbf{x})] + (-V)[F(\mathbf{n}+\mathbf{e}_i,\mathbf{x})-F(\mathbf{n},\mathbf{x})] - a_1 F(\mathbf{n},\mathbf{x}) \Big\} \\ &= \mathbb{L}F(\mathbf{n},\mathbf{x}) - a_1 F(\mathbf{n},\mathbf{x}). \end{aligned}$$

The following result of Shiga (1982) establishes the uniqueness for the martingale problem for $\mathbf{x}(t)$ as well as a useful tool in studying the ergodic behavior of measure-valued processes.

<u>Theorem</u> <u>10.3.1</u> Let $\{\mathbf{x}(t)\}$ be a solution of the martingale problem for this system and $\{\mathbf{n}(t)\}$ the $\mathbb{N}^S$-valued process constructed above. Then for each $\mathbf{n} \in \mathbb{N}^S$

$$E_{\mathbf{x}}(F(\mathbf{n},\mathbf{x}(t))) = E_{\mathbf{n}}\left[F(\mathbf{n}(t),\mathbf{x}) \exp\left(\int_0^t -a_1 |\mathbf{n}(s)| ds\right)\right].$$

<u>Proof.</u> This again follows from Theorem 5.1.2. □

<u>Remark:</u> If V=0 (neutral case) the dual process can be considered to be a system of (slowly) coalescing random walks. The voter model (cf. Holley and Liggett (1975)) can be viewed as a limiting case of this model - the dual process has infinite *coalescing rate* γ, that is, it is a coalescing random walk in which particles coalesce as soon as they meet.

<u>10.3.2. The general stepping stone model.</u>

In this section we extend the stepping stone model to the case in which the space of types is a general compact metric space E. This was done for finite systems of colonies by Vaillancourt (1987, 1990) and for countable systems of colonies by Handa (1990). In this section we give an outline of this development.

The state space is now $M_1(E)^S$ where again S is a countable set. Let Π^m denote the set of all mappings from $\{1,2,\ldots,m\}$ to S and $\mathcal{A} := \bigcup_{m\geq 1}(D(A^{(m)})\cap C(E^m)\times\Pi^m)$, and for each $(f,\pi)\in\mathcal{A}$,

$$(10.3.2.1)\quad F((f,\pi),\underline{\mu}) := \int_{E^m} f(x_1,\ldots,x_m)\prod_{i=1}^{m}\mu_{\pi(i)}(dx_i),\qquad \underline{\mu}\in(M_1(E))^S.$$

Let $\mathcal{D}(\mathfrak{G})$ denote the class of functions on $(M_1(E))^S$ of the form (10.3.2.1).

For $F\in\mathcal{D}(\mathfrak{G})$, we define the operator

$$\mathfrak{G}F(\underline{\mu}) := \sum_{j\in S}\left\{\int_E A\left(\frac{\delta F(\underline{\mu})}{\delta\mu_j(x)}\right)\mu_j(dx) + \gamma\int_E\int_E\left(\frac{\delta^2F(\underline{\mu})}{\delta\mu_j(x)\delta\mu_j(y)}\right)Q_S(\mu_j;dx,dy)\right.$$
$$\left. + \rho\sum_{k=1}^{m} q_{jk}\int_E\left(\frac{\delta F(\underline{\mu})}{\delta\mu_j(x)}\right)(\mu_k(dx)-\mu_j(dx))\right\}$$

where $Q_S(\mu;dx,dy) = \mu(dx)\delta_x(dy) - \mu(dx)\mu(dy)$. Here $\{q_{jk}\}$ denotes the rate of migration from the jth to the kth colony. Here A is the mutation operator, γ is the sampling rate and ρ is the migration rate.

The existence of an $(M_1(E))^S$-valued solution, $\{\Xi(t)\}$ to the initial value martingale problem for $\mathfrak{G}$ can again be obtained using approximating particle systems (cf. Vaillancourt (1987, 1990) when $|S|$ is finite). The uniqueness is again proved by duality. We next describe an $\mathcal{A}$-valued process, denoted by $(f,\pi)_t = (f_t,\pi_t)$, which serves as the dual to the stepping stone process. Intuitively, jumps in π_t correspond to migration. If the cardinality of the range of π_{t_0} is k, then up to the next jump of π_t, f_t is the dual process associated with k independent $M_1(E)$-valued Fleming-Viot processes.

We first recursively build the process π_t, started at $\pi_0\in\Pi^m$, consisting of independent coordinates $\pi_t(j)$, j=1,...,m. Each of these is a continuous time Markov chain on S with transition matrix Q and successive $\mathcal{E}xp(1/\rho)$ holding times $\tau_{j,k+1}$, for k=0,1,2,... and with $\tau_{j,0}=0$.

We now turn to the construction of the process f_t, obtained via the construction of an operator-valued process as follows.

Each $\pi\in\Pi^m$ generates a partition $\{\pi^{-1}(k):k\in S\}$ of $\{1,2,\ldots,m\}$. Considering every $p\in S$ such that $\pi^{-1}(p)$ has cardinality not less than 2, and for every pair (i,j) of distinct elements of $\pi^{-1}(p)$, we construct a family $\{\sigma^p_{(i,j)}:p\in S;i,j\in$

$\pi^{-1}(p), i \neq j\}$ of independent $\mathcal{E}xp(1/\gamma)$-random variables, and order it as follows.

Define a finite set of stopping times $\mathcal{U} := \{v_\ell^p : p=1,2,..,r; \ell=1,..,u(p)-1\}$ where $u(p) = \mathrm{card}(\pi^{-1}(p)) \vee 1$ denotes the number of elements of $\pi^{-1}(p)$, with $u(p) = 1$ in the case $\pi^{-1}(p) = \emptyset$, by

$$v_1^p := \min\{\sigma^p_{(i,j)} : i,j \in \pi^{-1}(p),\ i \neq j\} := \sigma_{(i_1^p, j_1^p)}$$

$$v_2^p := \min\{\sigma^p_{(i,j)} : i,j \in \pi^{-1}(p) \setminus \{j_1^p\},\ i \neq j\} := \sigma_{(i_2^p, j_2^p)}$$

$$\dots$$

$$v_{\ell+1}^p := \min\{\sigma^p_{(i,j)} : i,j \in \pi^{-1}(p) \setminus \{j_1^p, \dots, j_\ell^p\},\ i \neq j\} := \sigma_{(i_{\ell+1}^p, j_{\ell+1}^p)}$$

for $\ell = 0,1,\dots,u(p)-2$, where $i_1^p, i_2^p, \dots$ and $j_1^p, j_2^p, \dots$ are well-defined random indices.

Let $0 < \zeta_1 < \dots < \zeta_u < \zeta_{u+1} = \infty$ (with $u = \sum_{p \in S} (u(p)-1)$) be a reordering of the set $\mathcal{U}$, and define $\Phi_\ell f := \Theta_{ij} f$, where $(i,j) = (i(\ell), j(\ell))$ is the pair of random indices associated with jump ζ_ℓ, $\ell=1,2,\dots,u$.

The operator-valued process R_t^π is then defined recursively on $C(E^m)$ by

$$R_t^\pi := \begin{cases} H_t^\pi & \text{for } t \in [0, \zeta_1) \\ H^\pi_{t-\zeta_t} \circ \Phi_\ell \circ R^\pi_{\zeta_\ell -} & \text{for } t \in [\zeta_\ell, \zeta_{\ell+1}), \ell=1,2,\dots,u, \end{cases}$$

where $H_t^\pi : C(E^m) \longrightarrow C^\infty(E^m)$ is a semigroup defined by

$$H_t^\pi g(x_1,\dots,x_m) := (S_t^{(m)} g)(x_{\pi 1},\dots,x_{\pi m}).$$

Then R^π is a well-defined stochastic bounded linear operator since $\|R_t^\pi g\| \leq \|g\|$ for all $t \geq 0$ and H_t^π is strongly continuous in t.

Let $0=\tau_0<\tau_1<\tau_2<\dots$ denote the successive jump times of process π. Given a countable collection $\{{}^kR^\pi : k=0,1,2,..; \pi \in \Pi^m\}$ of independent copies of each process R^π, define recursively f_t, started at $f_0 \in C(E^m)$ by

$$f_t := {}^kR^{\pi_{\tau_k}}_{t-\tau_k}(f_{\tau_k -}) \quad \text{for } t \in [\tau_k, \tau_{k+1}),\ k=0,1,2,\dots$$

with $f_{0-} = f_0$.

Theorem 10.3.2.1. The $(\mathfrak{D}(\mathfrak{G}),\mathfrak{G},(M_1(E))^S)$-martingale problem is well-posed.

Proof. If K is the generator of the $\mathcal{A}$-valued process $(f,\pi)_t$ defined above, then it can be verified that

$$\mathfrak{G}F((f,\pi),\underline{\mu}) = KF((f,\pi),\underline{\mu}).$$

Then applying Theorem 5.5.2 we obtain the duality relation

$$Q_{(f,\pi)_0}F((f,\pi)_t,\Xi(0),) = P_{\{\mu_j:j\in S\}}F((f,\pi)_0,\Xi(t)).$$

(For details consult Vaillancourt (1987, 1990)). □

Remark: It is also possible to incorporate selection into the general stepping stone model combining the ideas of sections 10.2 and 10.3. The appropriate uniqueness argument in this case is proved in Handa (1990, Theorem 3.5).

10.4 A Two-type Model in a Continuous Environment and a Stochastic Evolution Equation

Finally we will consider the situation in which there are two types and in which the geographical space is S (e.g. $\mathbb{R}^d$ or a domain in $\mathbb{R}^d$). In this case $X(t,x)$ denotes the proportion of the population of type 1 at $x\in S$. The proposed state space for the process is $C(S;[0,1])$. A study of this model with $S = \mathbb{R}^d$ will provide some additional insight into one of the sources of dimension-dependent qualitative behavior. To describe it we will introduce a family of nonlinear stochastic evolution equations.

Let $W(t,dx)$ be a cylindrical Brownian motion, that is, W has independent increments and $W(t+s)-W(s)$ is Gaussian with covariance structure given by

$$E[\langle W(t),\phi\rangle\langle W(t),\psi\rangle] = t\int_{\mathbb{R}^d}\phi(x)\psi(x)dx \quad \text{for} \quad \phi,\psi\in L^2(\mathbb{R}^d).$$

We denote the resulting Gaussian white noise martingale measure by $W(ds,dx)$.

Suppose $a(y):[0,1]\to\mathbb{R}$ is continuous and $a(0)=a(1)=0$, and $b(y):[0,1]\to\mathbb{R}$ is continuous and $b(0)\ge 0$, $b(1)\le 0$. We assume that $(A,\mathcal{D}(A))$ is the generator of a strongly continuous Markov semigroup on $C(\mathbb{R}^d)$ and that a subset of $C_c(\mathbb{R}^d)$ is a core. We also assume that this process has a transition density $p(t,x;y)$ which is jointly continuous in $(0,\infty)\times\mathbb{R}^d\times\mathbb{R}^d$.

Consider the formal stochastic integral equation

$$(10.4.1)\quad X(t,x) = X(0,x) + \int_0^t a(X(s,x))W(ds,dx) + \int_0^t b(X(s,x))ds + \int_0^t AX(s,x)ds.$$

Since we are looking for solutions with general values in $C(\mathbb{R}^d;[0,1])$ and not necessarily belonging to $\mathcal{D}(A)$ the last term may be undefined. For this reason we formulate the following *evolution (mild) form* of this equation:

$$(10.4.2)\quad X(t,x) = \int p(t,x,y)X(0,y)dy + \int_0^t\int p(t-s,x,y)a(X(s,y))W(ds,dy)$$
$$+ \int_0^t\int p(t-s,x,y)b(X(s,y))dsdy.$$

We first consider the case in which the coefficients are Lipschitz and the process with generator A satisfies a certain condition that guarantees the existence of a local time at the diagonal for the two particle motion of independent A-processes.

Theorem 10.4.1. *(Stochastic evolution equation)*

Suppose that a(.) and b(.) are Lipschitz and A satisfies one of

(10.4.3) $S = \mathbb{R}$ and $A = -(-\Delta)^{\alpha/2}$, $0 < \alpha \le 2$,

or

(10.4.4)

(i) for some $0<\alpha<1$, $\sup_{0\le y\le T} \sup_x p_t p_t^*(x,x)/t^\alpha < \infty$.

(ii) for every compact K, every $h_0 > 0$ and every $t > 0$ there exist constants $C > 0$ and $\beta > 0$ such that

$$\int_0^t\int (p(s+h,x;y)-p(s,x';y))^2 dsdy \le C(h^\beta + |x-x'|^\beta)$$

for every $0 < h < h_0$, $x\in K$, $x'\in K$. The the stochastic evolution equation (10.4.2) has a unique solution. Furthermore, the associated martingale problem is well-posed.

Proof. The existence and pathwise uniqueness under condition (10.4.3) (as well as under weaker conditions) has been intensively studied (see e.g. Walsh (1986), Gyöngy and Pardoux (1991). The fact that the associated martingale problem is well-posed follows by an extension of an argument due to Yamada and Watanabe (1971) (cf. Theorem 8.4.1). The proof under condition (10.4.4) is given in Shiga (1988) and under a closely related condition in Kotelenez (1989). □

Following Shiga (1988) we next consider an important special case in which the method of duality applies.

Theorem 10.4.2. Consider Equation (10.4.2) under the assumption (10.4.3) or (10.4.4) and with the following (stepping stone with mutation and selection) coefficients:

(10.4.5) $a(y) = \sqrt{y(1-y)}$ and $b(y) = c_0(1-y)-c_1y+c_2y(1-y)$,

$c_0, c_1, c_2 \ge 0$. Then the associated martingale problem has a unique solution.

Proof. The key step is again a duality argument which we only sketch. The state space for the dual process is the collection of (unordered) finite subsets

$\mathbf{x}=(x_1,\ldots,x_n)$ of $\mathbb{R}^d$. The function F is defined by

$$F(\mathbf{x},X) := \prod_{i=1}^{n} (1-X(x_i)), \quad F(\emptyset,X) = 1.$$

The formal generator of the dual process is

$$Lf(\mathbf{x}) = \sum_{i=1}^{n} A_i f(\mathbf{x}) + c_1 \sum_{i=1}^{n} [f(\Phi_i \mathbf{x})-f(\mathbf{x})] + c_2 \sum_{i=1}^{n} [f(\Psi_i \mathbf{x})-f(\mathbf{x})]$$
$$+ \frac{1}{2} \sum_{i\neq j}\sum \delta(x_i,x_j)[f(\Phi_i \mathbf{x})-f(\mathbf{x})]$$

where Φ_i denotes the deletion of x_i, Ψ_i denotes the addition of a second copy of x_i, and $\delta(x,y)$ is a delta function at the diagonal set $\{(x,y)\in\mathbb{R}^d\times\mathbb{R}^d, x=y\}$. The duality relation

$$E_{X(0)}(F(\mathbf{x},X(t,.)) = E_{\mathbf{x}}\left(F(\mathbf{x}(t),X(0))\cdot \exp\left(-c_0\int_0^t |\mathbf{x}(s)|ds\right)\right)$$

is then obtained using Corollary 5.5.3. □

Remarks: The condition (10.4.3) guarantees the existence of the *local time at the diagonal for the two particle motion* of independent A-processes and it is this fact that leads to a nontrivial behavior. For example if $A = \Delta_d$, the d-dimensional Laplacian, then côndition (10.4.4) is satisfied if and only if d=1. It then makes sense to ask what happens in higher dimensions if we start with the corresponding particle systems. In fact it turns out that the resulting particle systems are tight and converge but that the limiting process is deterministic and given by the solution of the partial differential equation obtained by deleting the cylindrical Brownian motion term in (10.4.2) (cf. Reimers (1986)). This can be interpreted as a law of large numbers phenomenon which in part results from the independence of particles at different locations.

There are at least two ways in which non-trivial stochastic behavior can occurs in higher dimensions. The first is the removal of the assumption that densities are bounded and this then leads to the singular measure-valued processes such as measure-valued branching and Fleming-Viot processes. The second is the removal of the assumtion that particles at different locations are stochastically independent. If this assumption is removed stochastic evolution equations of the above type can arise in higher dimensions - however the cylindrical Brownian motion is then replaced by one with (for example) nuclear covariance. The existence of a unique strong solution to a class of equations of this type was established by Viot (1976).

10.5. Some Other Interaction Models

It is clear that there are many further possible types of interaction which are not covered by the above examples. This is an active area of research having relations with the theory of interacting particle systems.
In particular there have been a number of recent developments concerning superprocesses with interaction. A partial list of these is the following:

1) Shiga (1990) introduced a strong equation for an atomic measure-valued process incorporating branching and interaction in a Poisson setting.
2) Perkins (1992) introduced a strong equation with respect to the historical process for a superprocess in which the particle motions are influenced by the mass distribution. This is based on the development of a new stochastic calculus for historical processes.
3) Adler (1991) introduced and simulated the so-called "goat" superprocess in which the particles move towards or against the gradient of the occupation time measure.
4) Dawson and March (1992) studied the Fleming-Viot process with non-constant coefficients.

11. DE FINETTI AND POISSON CLUSTER REPRESENTATIONS, PALM DISTRIBUTIONS

11.1 Campbell Measures and Palm Distributions.

Let E be a Polish space and $M_{LF}(E)$ denote the set of locally finite measures on E. Now let X be a random measure with law $P \in M_1(M_{LF}(E))$ and locally finite intensity (mean) measure $I(B) := \int \mu(B)\, P(d\mu)$, $B \in \mathcal{E}_b$.

The associated *Campbell measure*, $\bar{P}$ is a measure on $\mathcal{B}(M_{LF}(E)) \times \mathcal{E}_b$ defined by

(11.1.1) $$\bar{P}(B \times A) := \int_B \mu(A)\, P(d\mu), \quad B \in \mathcal{B}(M_{LF}(E)), \ A \in \mathcal{E}_b.$$

The associated *Palm distributions* $\{(P)_x : x \in E\}$ are a version of the Radon-Nikodym derivatives

$$(P)_x = \frac{\bar{P}(d\mu \times dx)}{I(dx)}, \quad I\text{-a.s.}$$

and can be assumed to form a family of regular conditional probabilities (for the random measure X given $x \in E$) for the Campbell measure $\bar{P}$, that is, $(P)_x \in M_1(M_{LF}(E))$ for every x. Then for any bounded $\mathcal{E} \times \mathcal{B}(M_{LF}(E))$-measurable function, g, satisfying $\{x : \sup_\mu g(x,\mu) \neq 0\} \in \mathcal{E}_b$,

(11.1.2) $$\int_E I(dx)\left(\int_{M_{LF}(E)} g(x,\mu)(P)_x(d\mu)\right) = \int_{M_{LF}(E)} \int_E g(x,\mu)\mu(dx)P(d\mu)$$

and this system of identities uniquely characterizes the $(P)_x$, I-a.s.

In order to get some intuitive feeling for Palm measures we first consider the case of a random probability measure. Consider a two stage experiment in which at the first stage a $X \in M_1(M_1(E))$ is chosen with law P and in the second stage a point Z in E is chosen according to the law X. The Palm distribution $(P)_z$ is then simply the conditional (Bayesian posterior) distribution of X given $Z=z$. More generally, if X is a finite random measure, then there is an additional weighting on the posterior measures proportional to their total mass. This is indicated by the following consequence of (11.1.2)

$$\int (P)_x(\langle \mu,1\rangle \geq a) I(dx) = \int_a^\infty r \cdot P(\langle \mu,1\rangle \in dr).$$

11.2 De Finetti's Representation of a Random Probability Measure.

Let E be a Polish space and $M_1(E)$ the space of probability measures on E. A sequence of E-valued random variables $\{Z_n : n \in \mathbb{N}\}$ defined on a probability space $(\Omega, \mathcal{F}, P_0)$ is said to be *exchangeable* if $(Z_1,\ldots,Z_n,Z_{n+1},Z_{n+2},\ldots)$ has the same joint distribution as $(Z_{\pi(1)},\ldots,Z_{\pi(n)},Z_{n+1},Z_{n+2},\ldots)$ for every $\pi \in \mathcal{P}er(n)$ and $n \in \mathbb{N}$. Given $\{Z_n\}$, let $\mathcal{G}_n = \sigma\{f_n(Z_1,\ldots,Z_n) : f_n \in C_{sym}(E^n)\}$, $\mathcal{H}_n := \sigma\{\mathcal{G}_n, Z_{n+1}, Z_{n+2},\ldots\}$ and $\mathcal{H}_\infty :=$

$\bigcap_n \mathcal{H}_n$. Note that $\mathcal{H}_\infty$ is the σ-algebra of *exchangeable events*.

Theorem 11.2.1 (de Finetti's Theorem)

(a) Let $P \in M_1(M_1(E))$. Then there exists a sequence $\{Z_n\}$ of E-valued exchangeable random variables defined on a probability space $(E^{\mathbb{N}}, \mathcal{E}^{\mathbb{N}}, P_{dF})$ such that $(Z_1,\ldots,Z_n)$ has joint distribution

$$P^{(n)}(dx_1,\ldots,dx_n) := \int_{M_1(E)} \mu(dx_1)\ldots\mu(dx_n)P(d\mu), \quad n \in \mathbb{N},$$

(b) Consider the sequence $\{Z_n\}$ of E-valued exchangeable random variables. Let $X_n(\omega) := \Xi_n(Z_1,\ldots,Z_n)$. Then

$$X(\omega) = \lim_{n\to\infty} X_n(\omega) \in M_1(E) \quad \text{exists for a.e. } \omega,$$

where the limit is taken in the weak topology on $M_1(E)$ and X has probability law P.

(c) X is $\mathcal{H}_\infty$-measurable and conditioned on $\mathcal{H}_\infty$, $\{Z_n\}$ is a sequence of i.i.d. random variables with marginal distribution X.

Proof. (See e.g. Aldous (1985), Dawson and Hochberg (1979))

(a) Given $P \in M_1(M_1(E))$, we can construct a consistent family $\{P^{(n)} \in M_1(E^n)\}$ of probability measures $P^{(n)}(dx_1,\ldots,dx_n) := \int_{M_1(E)} X(dx_1)\ldots X(dx_n)P(d\mu)$, $n \in \mathbb{N}$.

By Kolmogorov's extension theorem we can then construct a canonical process $(E^{\mathbb{N}}, \mathcal{E}^{\mathbb{N}}, \{Z_n\}_{n\in\mathbb{N}}, P_{dF})$ such that $\{Z_n\}$ form a sequence of exchangeable E-valued random variables and the joint distribution of $(Z_1,\ldots,Z_n)$ is given by $P^{(n)}$.

(b) By exchangeability we have $(Z_i,Y) \overset{\mathcal{D}}{=} (Z_1,Y)$, $1\le i\le n$, when Y is of the form $(f_n(Z_1,\ldots,Z_n),Z_{n+1},Z_{n+2},\ldots)$ and f_n is symmetric. Therefore conditioned on $\mathcal{H}_n$ the $Z_1,\ldots,Z_n$ are a.s. identically distributed . Hence for $\phi \in b\mathcal{E}$

$$\text{(11.2.1)} \qquad P_{dF}(\phi(Z_1)|\mathcal{H}_n) = P_{dF}\left(n^{-1}\sum_{i=1}^{n}\phi(Z_i)|\mathcal{H}_n\right)$$

$$= \langle X_n,\phi\rangle \quad \text{since this is } \mathcal{H}_n\text{-measurable.}$$

Since $\mathcal{H}_n \supset \mathcal{H}_{n+1}$, and $P_{dF}(\phi(Z_1)|\mathcal{H}_n)$ is a reverse martingale, we can apply the reverse martingale convergence theorem to obtain

$$\text{(11.2.2)} \quad \langle X_n,\phi\rangle = n^{-1}\sum_{i=1}^{n}\phi(Z_i) \longrightarrow P_{dF}(\phi(Z_1)|\mathcal{H}_\infty) \text{ a.s.}$$

Thus in particular for each compact set, $K\subset E$, $X_n(\omega,K) \longrightarrow X(\omega,K)$ for a.e. ω and $P_{dF}(X(K)) = P_{dF}(Z_1\in K)$. Then $P_{dF}(X(K^c) > 2^{-n}) \le 2^n P_{dF}(Z_1\in K^c)$ and we can find a sequence of compact sets K_n such that $X(K_n^c) < 2^{-n}$ for all sufficiently large n, a.s., that is, given $\varepsilon > 0$, and ω there exists let $n(\omega)$ such that $X(\omega,K_n^c) < 2^{-n}$ for all

$n \geq n(\omega)$. Given $\varepsilon > 0$ we choose $n_1 \geq n(\omega)$ so that $2^{-n_1} < \varepsilon/2$, and then we choose $n_2 \geq n_1$ such that $X_n(\omega, K_{n_1}^c) \leq \varepsilon$ for all $n \geq n_2$. This gives the almost sure tightness of the $\{X_n\}$. Then applying (11.2.2) to a countable subset of $b\mathcal{E}$ which is convergence determining for $M_1(E)$ establishes the a.s. convergence of X_n in the weak topology on $M_1(E)$. Denoting the limit by X, we conclude that the random probability measure, $X(\cdot)$, defined in this way is a version of the conditional distribution $P_{dF}(Z_1 \in \cdot \,|\, \mathcal{H}_\infty)$.

(c) Fix $k \in \mathbb{N}$ and let $\phi \in bC(E^k)$. Then an argument similar to the one at the beginning of the proof of (b) shows that for $n>k$,

$$P_{dF}(\phi(Z_1,\ldots,Z_k)\,|\,\mathcal{H}_n)$$
$$= |\Lambda_{n,k}|^{-1} \cdot \sum \cdots \sum_{(j_1,\ldots,j_k)\in\Lambda_{n,k}} \phi(Z_{j_1},\ldots,Z_{j_k})$$

where $\Lambda_{n,k} := \{(j_1,\ldots,j_k): 1 \leq j_r \leq n;\ j_m \neq j_r \text{ if } m \neq r\}$, $|\Lambda_{n,k}| = [n(n-1)\ldots(n-k+1)]$. Again using the reverse martingale convergence theorem and the fact that $|\Lambda_{n,k}|/n^k \to 1$ as $n \to \infty$, we obtain

$$(11.2.3) \qquad n^{-k} \sum_{j_1=1}^{n} \cdots \sum_{j_k=1}^{n} \phi(Z_{j_1},\ldots,Z_{j_k}) \longrightarrow P_{df}(\phi(Z_1,\ldots,Z_k)\,|\,\mathcal{H}_\infty) \text{ a.s.}$$

Taking $\phi(x_1,\ldots,x_k) := \prod_{j=1}^{k} \phi_j(x_j)$, and using (11.2.2) and (11.2.3) we conclude

$$P_{dF}\Big(\prod_{j=1}^{k} \phi_j(Z_j)\,|\,\mathcal{H}_\infty\Big) = \prod_{j=1}^{k} P_{dF}(\phi_j(Z_1)\,|\,\mathcal{H}_\infty) \quad \text{a.s.}$$

This implies that the $\{Z_i\}$ conditioned on $\mathcal{H}_\infty$ are i.i.d. □

Corollary 11.2.2 (a) Let $\{Z_n(.): n \in \mathbb{N}\}$ be an exchangeable sequence of $D([0,\infty),E)$-valued random variables and $X_n := \Xi_n(Z_1(.),Z_2(.),\ldots,Z_n(.))$. Then

(i) $P_{dF}\left\{m^{-1}\sum_{n=1}^{m} \delta_{Z_n(.)} \Rightarrow X\right\} = 1$ where X is a random probability measure on $D([0,\infty),E)$, and

(ii) for $k \in \mathbb{N}$, $0 \leq t_1 < \ldots < t_k$,

$$P_{dF}\left\{m^{-1}\sum_{n=1}^{m} \delta_{Z_n(t_1),\ldots,Z_n(t_k)} \Rightarrow X_{\{t_1,\ldots,t_k\}}\right\} = 1$$

where $X_{\{t_1,\ldots,t_k\}}$ is a random probability measure on E^k.

(b) If for each $t \geq 0$, $\{Z_n(t): n \in \mathbb{N}\}$ is exchangeable, then

$$\left\{m^{-1}\sum_{n=1}^{m}\delta_{Z_n(t_1)},\ldots, m^{-1}\sum_{n=1}^{m}\delta_{Z_n(t_k)}\right\} \Rightarrow \{(X_{t_1},\ldots,X_{t_k})\},$$

where the laws of $\{(X_{t_1},\ldots,X_{t_k})\}$ form a set of finite dimensional distributions of a $M_1(E)$-valued stochastic process.

Proof. (a) and (b) both follow from the almost sure convergence in Theorem 11.2.1. □

Corollary 11.2.3. Let $P\in M_1(M_1(E))$. Then the Palm distributions $\{(P)_z : z\in E\}$ are given by

$$(P)_z(B) = P_{dF}(X\in B\,|\,Z_1=z), \quad B\in \mathcal{E}.$$

Proof. First note that the mean measure $I(B) = P_{dF}(Z_1\in B)$. Then in view of the characterization (11.1.2) it suffices to prove that for $\phi\in b\mathcal{E}$, $f\in b\mathcal{E}^k$, $k\in \mathbb{N}$,

$$\int_E \phi(x_1)\int_{M_1(E)}\left(\int_E\cdots\int_E f(x_2,\ldots,x_{k+1})\mu(dx_2)\ldots\mu(dx_{k+1})\right)P_{dF}(X\in d\mu\,|\,Z_1=x_1)P_{dF}(Z_1\in dx_1)$$

$$= \int_{M_1(E)}\left(\int_E\cdots\int_E \phi(x_1)f(x_2,\ldots,x_{k+1})\mu(dx_1)\ldots.\mu(dx_{k+1})\right)P_{dF}(d\mu).$$

But this is true since both sides are equal to $P_{dF}[\phi(Z_1)f(Z_2,\ldots,Z_{k+1})]$. □

11.3 An Infinite Particle Representation of Fleming-Viot Processes

In view of the results of the last section it is reasonable to look for an exchangeable infinite particle system representation for the Fleming-Viot process. This idea was used in Dawson and Hochberg (1982) to study the support dimension of the Fleming-Viot process at fixed times. Donnelly and Kurtz (1992) introduced an infinite particle system which is not exchangeable but which does yield a representation of the full Fleming-Viot process. We will now outline the main ideas of their construction.

For simplicity we return to the setting of Ch. 2. Let (E,d) be a compact metric space, $P(t,x,\Gamma)$ be the transition function, and $(A,D(A))$ the generator of a Feller process with sample paths in $D_E = D([0,\infty),E)$. We begin with a description of an n-particle process, $\{Y^n(t): t\geq 0\}$ called by Donnelly and Kurtz (1992) the n-particle *look-down process*. This is a Markov jump process with state space E^n and generator:

(11.3.1)

$$K_n f(x_1,\ldots,x_n) = \sum_{i=1}^{n} A_i f(x_1,\ldots,x_n) + \gamma\sum_{i<j}[\Theta_{ij}f(x_1,\ldots,x_n)-f(x_1,\ldots,x_n)].$$

<u>Theorem</u> <u>11.3.1.</u> Let $\{Z^n(t):t\geq 0\}$ be the n-particle Moran process with generator L_n defined by (2.5.2) and $\{Y^n(t):t\geq 0\}$ the n-particle look-down process with generator K_n defined by (11.3.1). Then the $M_1(E)$-valued Markov processes $\{\Xi_n(Y^n(t)):t\geq 0\}$ and $\{\Xi_n(Z^n(t)):t\geq 0\}$ have the same distribution provided that $Y^n(0) \overset{\mathcal{D}}{=} Z^n(0)$ is exchangeable.

<u>Proof.</u> (cf. Donnelly and Kurtz (1992) for different proof.) Let $f \in C_{sym}(E^n)\cap D(A^n)$. Then $f(x_1,\dots,x_n) = (n!)^{-1} \sum_{\pi\in\mathcal{P}er(n)} f(x_{\pi 1},\dots,x_{\pi n})$. Substituting this expression into the generator L_n we obtain

$$K_nf(x_1,\dots,x_n) = (n!)^{-1} \sum_{\pi} K_nf(x_{\pi 1},\dots,x_{\pi n})$$

$$= \sum_{i=1}^{n} A_if(x_1,\dots,x_n) + \tfrac{1}{2}\gamma \sum_{i\neq j} \Theta_{ij}[f(x_1,\dots,x_n)-f(x_1,\dots,x_n)].$$

$$= L_nf(x_1,\dots,x_n) \in C_{sym}(E^n).$$

Hence the restricted initial value martingale problem for $(K_n,C_{sym}(E^n),M_{1,ex}(E^n))$ martingale problem is identical to the restricted initial value martingale problem for $(L_n,C_{sym}(E^n),M_{1,ex}(E^n))$ where $M_{1,ex}(E^n)$ denotes the set of exchangeable probability laws on E^n. The uniqueness of the marginal distributions follows by solving the moment equations as in Theorem 2.5.1 and Section 2.6 (or using a function-valued dual). The identity in law of the processes then follows from Theorem 5.1.2. □

<u>Remark.</u> Let $\tilde{\mu} \in M_{1,ex}(E^{\mathbb{N}})$. The n-particle look-down process with initial law given by the appropriate marginal of $\tilde{\mu}$ can be realized as $(D([0,\infty),E^n),\mathcal{D},P_{n,\tilde{\mu}})$.

<u>Corollary</u> <u>11.3.2.</u> Let $\tilde{\mu} \in M_{1,ex}(E^{\mathbb{N}})$. Let $\mathbf{D}_\infty := D([0,\infty),E^{\mathbb{N}})$, $\mathcal{D}_\infty$ denote the Borel σ-algebra on $\mathbf{D}_\infty$ and for $n\in\mathbb{N}$ let $Y_n(.)$ denote the nth coordinate of the canonical process $\mathbf{Y}^\infty$. Then there exists a probability measure $P^{DK}_{\tilde{\mu}}$ on $(\mathbf{D}_\infty,\mathcal{D}_\infty)$ such that the distribution of $(Y_1(.),\dots,Y_n(.))$ coincides with that of the process $\mathbf{Y}^n(.)$ and $\mathbf{Y}^\infty(0)$ has law $\tilde{\mu}$. (This infinite particle system is called the *infinite look-down process* of Donnelly-Kurtz.)

<u>Proof.</u> Observe that if we consider functions f depending only on $x_1,\dots,x_{n-1}$, then $K_nf(x_1,\dots,x_{n-1}) = K_{n-1}f(x_1,\dots,x_{n-1})$. This implies that the laws of $\{Y_1(.),\dots,Y_n(.)\}$, $n\in\mathbb{N}$, form a consistent family of probability measures. Then the infinite look-down process with initial law $\tilde{\mu}$, can be realized by taking the projective limit $(D([0,\infty),E^n),P_{n,\tilde{\mu}})$ which will be denoted by $(\mathbf{D}_\infty,\mathcal{D}_\infty,P^{DK}_{\tilde{\mu}})$. □

<u>Remarks.</u> (1) Let $\mathbf{Y}(t) := (Y_1(t),Y_2(t),\dots:t\geq 0)$ denote the infinite particle look-

down process of Donnelly and Kurtz and assume that it has exchangeable initial condition $\mathbf{Y}(0)$. It is then easy to verify that $\mathbf{Y}(t)$ is an exchangeable system for each $t\ge 0$.

(3) Donnelly and Kurtz (1992) observed that if the mutation process has a stationary distribution then there is a stationary version of the infinite particle system $\{Y_n(t):-\infty<t<\infty,\ n\in\mathbb{N}\}$ and if the mutation process has stationary independent increments, then there is a stationary version of $\{0,Y_2-Y_1,Y_3-Y_1,\dots\}$.

Given $\{Y_n(.)\}$, consider the empirical process

$$X_m(t) := \Xi_m(Y_1(t),\dots,Y_m(t)) = \frac{1}{m}\sum_{i=1}^{m}\delta_{Y_i(t)},$$

$$\mathcal{H}_{m,t} := \sigma(X_m(s),(Y_{m+1}(s),Y_{m+2}(s),\dots),s\le t)$$

and $\mathcal{H}_{\infty,t} := \bigcap_{m=1}^{\infty}\mathcal{H}_{m,t}$.

<u>Theorem</u> <u>11.3.3</u> (Donnelly and Kurtz (1992))

Let $\{Y_k(0):k\in\mathbb{N}\}$ have law $\tilde{\mu}\in M_{1,ex}(E^{\mathbb{N}})$. Then $X_m(t)$ converges, as $m\to\infty$, uniformly on bounded intervals to a càdlàg $M_1(E)$-valued process $\{X_t:t\ge 0\}$ with probability one. The limit process $\{X_t\}$ is a Fleming-Viot process with generator $\mathfrak{G}$ (given by (5.4.1)).

Moreover, for $B\in\mathcal{E}$, $t\ge 0$, and $k\in\mathbb{N}$,

$$\text{(11.3.2))}\quad P^{DK}_{\tilde{\mu}}\{Y_1(t)\in B_1,\dots,Y_k(t)\in B_k\,|\,\mathcal{H}_{\infty,t}\} = \prod_{j=1}^{k}X(t,B_j).$$

<u>Proof.</u> By Theorem 11.2.1 , for fixed t, $X_m(t)$ converges almost surely in the weak topology on $M_1(E)$ and for fixed $k\ge 1$

$$\text{(11.3.3)}\quad P^{DK}_{\tilde{\mu}}\{Y_k(t)\in B\,|\,\mathcal{H}^m_t\} = X_m(t,B)\quad\text{for}\quad m\ge k.$$

Then letting $m\to\infty$ and again using the reverse martingale convergence theorem on the left hand side we obtain

$$P^{DK}_{\tilde{\mu}}\{Y_1(t)\in B\,|\,\mathcal{H}_{\infty,t}\} = X(t,B).$$

(11.3.2) then follows from Theorem 11.2.1(c).

The fact that the resulting $M_1(E)$-valued process is Fleming-Viot process follows from Theorems 11.3.1 and 2.7.1.

It remains to prove the uniform convergence on bounded intervals. Let $f\in bC(E)$, $t>0$ and $\varepsilon>0$. By Theorem 11.2.1 $\{Y_n(t):n\in\mathbb{N}\}$ conditioned on $\mathcal{H}_{\infty,t}$ are i.i.d. Therefore by Cramér's Theorem (cf. Varadhan (1984, Section 3)) applied to the resulting independent random variables there exist C and $D>0$ such that

$$P^{DK}_{\tilde{\mu}}\left\{ \left|\int f(x)X_m(t,dx) - \int f(x)X(t,dx)\right| > \varepsilon \,|\, \mathcal{H}_{\infty,t}\right\} \le e^{C-Dm}, \quad P^{DK}_{\tilde{\mu}}\text{-a.s.}$$

where C,D are $\mathcal{H}_{\infty,t}$-measurable. However since the random variables $\{f(Z_n):n\in\mathbb{N}\}$ are uniformly bounded by $\|f\|$, in fact there exist constants c,δ (depending only on $\|f\|$) such that $C \le c$, $D \ge d$, and such that

$$(11.3.4)\qquad P^{DK}_{\tilde{\mu}}\left\{ \left|\int f(x)X_m(t,dx) - \int f(x)X(t,dx)\right| > \varepsilon \,|\, \mathcal{H}_{\infty,t}\right\} \le e^{c-\delta m}, \quad P^{DK}_{\tilde{\mu}}\text{-a.s.}$$

Since Y^m is the solution of the martingale problem for K_n,

$$M_f(t) := m^{-1}\sum_{i=1}^{m}\left(\left[f(Y_i(s))-f(Y_i(t)-\int_s^t Af(Y_i(u))du\right]1(\tau_i>t)\right)$$

where τ_i denotes the time of the first look-down of Y_i after time s, is a martingale. Then Doob's maximal inequality applied to the submartingales $\exp[\pm\lambda M_f(t)]$, with $\lambda>0$, the exponential Chebyshev inequality (cf. [LS, 13.2,13.27]) and an appropriate choice of λ yield an inequality of the form

$$(11.3.5)\qquad P^{DK}_{\tilde{\mu}}\left\{ \sup_{t\le T}\ \sup_{t\le s\le t+h_m} \left|\int f(x)X_m(s,dx)-\int f(x)X_m(t,dx)\right| > \varepsilon \right\}$$

$$\le e^{c-\delta m} + P^{DK}_{\tilde{\mu}}\left(m^{-1}\sum_{j=1}^{m} 1(\tau_j\le h_m) > \frac{\varepsilon}{2(\|f\|+\|Af\|)}\right).$$

But then by the exponential Chebyshev inequality we obtain

$$P^{DK}_{\tilde{\mu}}\left(\sum_{j=1}^{m} 1(\tau_j\le h) > m\cdot a\right) \le e^{-ma}\cdot P^{DK}_{\tilde{\mu}}\left(\exp\left(a\cdot\sum_{j=1}^{m} 1(\tau_j\le h)\right)\right)$$

$$= e^{-ma}\prod_{i=1}^{m}\left[e^{a}(1-e^{-(j-1)\gamma h})+e^{-(j-1)\gamma h}\right].$$

$$\text{Hence}\quad \log\left(P^{DK}_{\tilde{\mu}}\left(\sum_{j=1}^{m} 1(\tau_j\le h) > m\cdot a\right)\right) \le \sum_{j=1}^{m}\log\left(1-(1-e^{-a})e^{-(j-1)\gamma h}\right)$$

$$\le -\sum_{i=1}^{m}(1-e^{-a})e^{-(j-1)\gamma h}$$

$$= -\left((1-e^{-a})\,\frac{(1-e^{-mh})}{(1-e^{-h})}\right).$$

Therefore

$$P^{DK}_{\tilde{\mu}}\left(\sum_{j=1}^{m} 1(\tau_j \le h) > m\cdot a\right) \le \exp\left(-(1-e^{-a})\,\frac{(1-e^{-mh})}{(1-e^{-h})}\right).$$

If $a > 0$, we can choose c and δ so that

$$(11.3.6)\qquad P^{DK}_{\tilde{\mu}}\left(\sum_{j=1}^{m} 1(\tau_j \le h_m) > m\cdot a\right) \le e^{c-m\delta}$$

for all m provided that $mh_m \to 0$.

The almost sure uniform convergence result is obtained by choosing a grid $\{kh_m,\ k<T/h_m\}$, with $m\cdot h_m \longrightarrow 0$ and $\sum e^{-\delta m}/h_m < \infty$. Using the continuity of $\{X(t):t\ge 0\}$, and (11.3.4)-(11.3.6) it then follows that for $T < \infty$ and $\varepsilon>0$,

$$P^{DK}_{\tilde{\mu}}\left(\limsup_{m\to\infty}\ \sup_{t\le T}\ \left|\int f(x)X_m(t,dx)-\int f(x)X(t,dx)\right| > \varepsilon\right) = 0$$

(cf. Donnelly and Kurtz (1992) for the details). □

11.4 Campbell Measures and Palm Distributions for Infinitely Divisible Random Measures.

In this section we review the notions of Campbell and Palm measures associated with an infinitely divisible random measure. Standard sources for this material are Kallenberg (1983), Matthes, Kerstan and Mecke (1978), and Daley and Vere-Jones (1988).

Let E be a Polish space and $M_{LF}(E)$ the space of locally finite random measures introduced in section 3.1. We recall from Theorem 3.3.1 that an $M_{LF}(E)$-valued infinitely divisible random measure has the following canonical representation of its log-Laplace functional:

$$(11.4.1)\quad \log(L(\phi)) = \langle M,\phi\rangle + \int_{M_{LF}(E)} (1-e^{-\langle\nu,\phi\rangle})R(d\nu),\quad \phi\in pb\mathcal{E}_b,$$

where $(M,R) \in M_{LF}(E)\times M_{2,LF}(E)$ and $M_{2,LF}(E)$ denotes the set of measures ν on $M_{LF}(E)$ such that

(i) $\nu(\{0\}) = 0$

(ii) $\displaystyle\int_{M_{LF}(E)} (1-e^{-\mu(A)})\nu(d\mu) < \infty\ \forall\ A\in \mathcal{E}_b.$

The notions of Campbell measure and Palm distribution defined in Section 11.1 for probability measures on $M_{LF}(E)$ can be extended to $M_{2,LF}(E)$ and in particular will be defined for the canonical measure R of an infinitely divisible random measure with locally finite intensity.

Let $R \in M_{2,LF}(E)$ in (11.4.1) and $I(A) := \int \mu(A)R(d\mu)$ be the corresponding intensity measure. For $A \in \mathcal{E}_b$, $B \in \mathcal{B}(M_{LF}(E))$ the *Campbell measure* is defined as follows

$$\bar{R}(A\times B) := \int \mu(A)1_B(\mu)R(d\mu) \le I(A) = \int \mu(A)R(d\mu) < \infty.$$

Lemma 11.4.1. Assume that $I \in M_{LF}(E)$. Then

(a) $\bar{R}$ extends to a measure on $\mathcal{E}_b \otimes \mathcal{B}(M_{LF}(E))$.

(b) There exists a mapping $(e,B) \longrightarrow R_e(B)$ from $E\times\mathcal{B}(M_{LF}(E))$ to $[0,1]$ such that for each e, $R_e(.)$ is a probability measure on $\mathcal{B}(M_{LF}(E))$ and for each $B \in \mathcal{B}(M_{LF}(E))$ $R_\cdot(B)$ is $\mathcal{E}$-measurable such that the following disintegration applies

$$\bar{R}(A\times B) = \int_A R_e(B)I(de), \quad A\in\mathcal{E}_b.$$

(c) Further for any $\mathcal{E}_b \otimes \mathcal{B}(M_{LF}(E))$-measurable function g,

$$\int_E \int_{M_{LF}(E)} g(e,\mu)\bar{R}(de\times d\mu) = \int_E \left[\int_{M_{LF}(E)} g(e,\mu)R_e(d\mu)\right] I(de).$$

Proof. (a) Since by definition $\bar{R}$ is a bimeasure, this follows the lines of [EK, Appendix, Theorem 8.1].

(b) The existence of a version of the Radon-Nikodym derivatives

$$R_e(B) = \frac{\bar{R}(de\times B)}{I(de)}, \quad I\text{-a.e. } e,$$

which yields "regular conditional probabilities" follows by a standard argument (see [DP] for details).

(c) follows by approximating by simple functions and passing to the limit. □

The family of probability measures $\{R_e: e\in E\}$ whose existence is established in Lemma 11.4.1 are known as the *Palm distributions* associated with R.

Lemma 11.4.2. If X is an infinitely divisible random measure with law $P \in M_1(M_{LF}(E))$, with canonical representation (11.4.1) with $M\equiv 0$ and $R \in M_{2,LF}(E)$, then the relation between the Palm distributions $(P)_x$ of the random measure X, and those associated with the canonical measure, $(R)_x$, is given by

$$(P)_x(A) = \int_{M_{LF}(E)} \int_{M_{LF}(E)} 1_A(\mu_1+\mu_2)\ P(d\mu_1)(R)_x(d\mu_2),$$

$\forall A \in \mathcal{B}(M_{LF}(E))$ for I-a.e. x.

Proof. See Kallenberg [1983, Lemma 10.6].

11.5. Cluster Representation for (A,Φ)-Superprocesses.

Let $\{X_t: t\ge 0\}$ denote an (A,Φ)-superprocess with $X_0 = \mu \in M_F(E)$. Then for each t, X_t is an infinitely divisible random measure on E. In this section we will see that for a certain class of Φ, X_t can be represented as a Poisson cluster random measure and we will identify the cluster law. This will be carried out starting from the canonical representation of X_t. To show that we have a Poisson cluster representation we will show that the canonical measure R, which in general is σ-finite, is a finite measure. It will turn out (cf. Chapter 12) that each clusters obtained in this way can be interpreted as a submass having a common ancestor thus involving the family structure of the population.

Recall that

(11.5.1) $$P_{s,\mu}(e^{-\langle X_t,\phi\rangle}) = e^{-\langle \mu, V_{s,t}\phi\rangle}$$

and by the canonical representation theorem (cf. Theorem 3.3.1)

(11.5.2) $$\langle \mu, V_{s,t}\phi\rangle = \langle m(s,t,\mu),\phi\rangle + \int_{M_{LF}(E)} (1-e^{-\langle \nu,\phi\rangle})R_{s,t}(\mu,d\nu)$$

where

$$m(s,t,\mu) \in M_{LF}(E), \quad R_{s,t}(\mu,.) \in M_{2,LF}(E).$$

Remark By an application of (11.5.1) and (11.5.2), it is easy to verify that

(11.5.3) $$I_t(A) := E_{s,\mu}(X_t(A)) = \int (T_{s,t}1_A)(x)\mu(dx)$$
$$= \int \nu(A)R_{s,t}(\mu,d\nu) + m(s,t,\mu)(A) < \infty, \qquad A \in \mathcal{E}, \text{ and}$$
$$V_{s,t}\phi(x) = \langle m(s,t,x),\phi\rangle + \int (1-e^{-\langle \nu,\phi\rangle})R_{s,t}(x,d\nu)$$

where $m(s,t,x) = m(s,t,\delta_x)$, $R_{s,t}(x,.) = R_{s,t}(\delta_x,.)$.

The measurability follows from the measurability of $(s,t,x) \longrightarrow V_{s,t}\phi(x)$.

Let $\Phi(x,\lambda)$ be defined as in (4.3.2); recall that $\Phi(x,\lambda) \le 0$ and is monotone decreasing in λ. We next obtain a condition on Φ which guarantees that the canonical measure is finite.

Let $\Psi(\lambda) := \inf_x |\Phi(x,\lambda)|$ and assume that Ψ satisfies (4.3.6). Consider the log-Laplace equation

$$\partial u(t,y)/\partial t = Au(t,y) + \Phi(y,u(t,y)), \quad u(0,y) = \theta,$$

and the ordinary differential equation

$$dv(\theta,t)/dt = -\Psi(v(\theta,t)), \quad v(\theta,0) = \theta.$$

Under rather general conditions on A and Φ the *comparison theorem*: $u(t,y) \le v(\theta,t)$ $\forall$ y can be proved (see, for example Pardoux and Gyöngy (1991)). We will not state explicit hypotheses for the comparison theorem but will assume that it holds.

Lemma 11.5.1. Let $\Psi(\lambda) := \inf_x |\Phi(x,\lambda)|$. Assume that

(11.5.4) $$\int_\theta^\infty \frac{1}{\Psi(\lambda)}\,d\lambda < \infty \quad \text{for } \theta>0.$$

(a) If $t > s$ and $\theta>0$, then $\limsup_{\theta\to\infty} \sup_y V_{s,t}(\theta 1)(y) < \infty$.

(b) $m(s,t,y) \equiv 0$ and $R_{s,t}(y,M_F(E)) = R_{s,t}(y,M_{LF}(E)) < \infty$.

(c) $P_{s,\mu}(\langle X_t,1\rangle = 0) > 0 \ \forall\ \mu\in M_F(E)$.

Proof. (a) Let $u(t,y) = V_{s,t}(\theta 1)(y)$, $\theta>0$, and recall that for $t>s$,

$$\partial u(t,y)/\partial t = Au(t,y) + \Phi(y,u(t,y)), \quad u(0,y) = \theta.$$

Consider the ordinary differential equation

$$dv(\theta,t)/dt = -\Psi(v(\theta,t)), \qquad v(\theta,0) = \theta.$$

Then $\tilde{v}(t) := \lim_{\theta\to\infty} v(\theta,t)$ is given by

$$\int_{\tilde{v}(t)}^\infty 1/\Psi(s)ds = t.$$

Using (11.5.4) and noting that $\int_0^\infty \frac{1}{\Psi(\lambda)}\,d\lambda = \infty$ (by (4.3.6)), we conclude that $\tilde{v}(t)$ is well defined $\forall\ t > 0$ and that $0 < \tilde{v}(t) < \infty$. (a) then follows from the comparison theorem.

(b) follows from (a), (11.5.3) and the assumption that $P_{s,\mu}(\langle X_t,1\rangle) < \infty$.

(c) $$P_{s,\mu}(\langle X_t,1\rangle = 0) = \lim_{\theta\to\infty} P_{s,\mu}(e^{-\theta\langle X_t,1\rangle}) = \lim_{\theta\to\infty} e^{-\langle \mu, V_{s,t}(\theta 1)\rangle} > 0. \ \square$$

Remark 11.5.2. A sufficient condition for (11.5.4) is that

$$\inf_x \liminf_{\lambda\to\infty} -\Phi(x,\lambda)/\lambda^{1+\delta} > 0 \text{ for some } \delta > 0$$

and this is satisfied if $c(x) \ge c_0 > 0$ or there exists $\beta_0 > 0$, $u_0 > 0$ such that $n(x,du) = n(x,u)du$ with $n(x,u) \ge cu^{-(\beta+2)}du\ \forall\ u\le u_0$ near zero for some $\beta > 0$.

Corollary 11.5.3. *Cox Cluster Representation*

Assume (11.5.4), $X_s = \mu \in M_F(E)$ and $t > s$. Then

(a) X_t can be represented as a Cox cluster random measure (cf. (3.3.4)) with $E_1 = E_2 = E$, intensity $X_s(dx)R_{s,t}(x,M_F(E))$ and cluster probability law

$$P^*_{s,t;y}(e^{-\langle\mu,\phi\rangle}) = \frac{R_{s,t}(x,d\nu)}{R_{s,t}(x,M_F(E))} = 1 - \frac{V_{s,t}\phi(y)}{R_{s,t}(y,M_F(E))}.$$

(b)

$$(11.5.5)\qquad X_t(A) \overset{\mathcal{D}}{=} \int_E\int_{M_F(E)\setminus\{0\}} \nu(A)\, N(dx,d\nu),\ A\in\mathcal{E}, \qquad \textit{(cluster representation)}$$

where N is a Poisson random measure on $E\times(M_F(E)\setminus\{0\})$ with intensity measure $n(dx,d\nu) = X_s(dx)R_{s,t}(x,d\nu)$. Thus X_t is represented as the superposition of a random number, N, of clusters where N is Poisson with mean $\int X_s(dx)R_{s,t}(x,M_F(E)) < \infty$.

11.6. Palm Distributions of (Φ,κ,W)-Superprocesses: Analytical Representation.

To introduce the intuitive content of this section let us choose a cluster, ν, with law $R_{s,t}(x,d\nu)/R_{s,t}(x,M_F(E))$ and then choose a point $y \in E$ with law $\nu/\nu(E)$. The corresponding Palm distribution is the conditional distribution of ν given y. The purpose of this section is to obtain two representations of this conditional distribution. Roughly speaking, this will be obtained by first setting down a path of the A-process conditioned to begin at x at time s and end at y at time t. Then along this path mass is produced which then evolves according to the measure-valued branching mechanism yielding sub-clusters of mass at time t. Intuitively we can think of a subcluster that breaks off at time $s\le u<t$ as the submass having a last common ancestor with the mass at the point y at time u. This intuitive picture will be made mathematically precise in Chapter 12.

We next give a simplified version of the calculations of this section. Consider the log-Laplace functional $V(\phi) = \int(1-e^{-\langle\nu,\phi\rangle})R(d\nu)$ of an infinitely divisible random measure with canonical measure R. Consider

$$\begin{aligned} U(\phi;\psi) &:= \frac{\partial V(\phi+\varepsilon\psi)}{\partial\varepsilon}\Big|_{\varepsilon=0} = \int\langle\nu,\psi\rangle e^{-\langle\nu,\phi\rangle}R(d\nu)\\ &= \iint\psi(x)e^{-\langle\nu,\phi\rangle}\bar{R}(dx,d\nu)\\ &= \iint\psi(x)e^{-\langle\nu,\phi\rangle}R_x(d\nu)I(dx)\end{aligned}$$

and hence formally the Laplace functional of the Palm distribution R_x is given by

$$\int e^{-\langle\nu,\phi\rangle}R_x(d\nu) = U(\phi,\delta_x) \quad \text{for } I\text{-a.e. } x.$$

The second observation is that if V_t satisfies an equation of the form (which for simplicity we assume to be time homogeneous)

$$\frac{\partial V_t(\phi)}{\partial t} = AV_t+\Phi(V_t(\phi)),\quad V_0(\phi) = \phi,$$

then U_t satisfies the linear equation

$$\frac{\partial U_t(\phi,\psi)}{\partial t} = AU_t(\phi,\psi)+\Phi'(V_t(\phi))U_t(\phi,\psi),\quad U_0(\phi,\psi) = \psi.$$

If A is the generator of a Markov process then this linear equation can be solved by the Feynman-Kac formula. We will now apply these ideas to determine the Laplace functional for the Palm distributions for the general (Φ,κ,W)-superprocess associated with log-Laplace equation 4.3.8 and canonical representation 11.5.2.

Theorem 11.6.1. Assume that the canonical measure of the (W,κ,Φ)-superprocess starting at δ_x at time s is given by $m \equiv 0$ and $R_{s,t,x} = R_{s,t}(x,.)$.
Then

$$\int e^{-\langle\mu,\phi\rangle}(R_{s,t,x})_y(d\mu)$$

$$= E_{s,x}\left(\exp\left\{ \int_s^t \Phi_2(r,W_r,V_{r,t}\phi(W_r))\kappa(dr)\right\} \Big| W_t=y\right), \quad P_{s,x}(W_t\in .)\text{-a.e. } y$$

where $\Phi_2(r,x,\lambda) = \frac{\partial}{\partial\lambda}\Phi(r,x,\lambda)$.

Proof. It suffices to show that for all $\psi\in bp\mathcal{E}$,

$$\int\psi(y)\int e^{-\langle\mu,\phi\rangle}(R_{s,t,x})_y(d\mu)P_{s,x}(W_t\in dy) = E_{s,x}\left[\exp\left(\int_s^t \Phi_2(r,W_r,V_{r,t}\phi(W_r))\kappa(dr)\right)\psi(W_t)\right].$$

Using the fact that $P_{s,x}(W_t\in dy) = \int\mu(dy)R_{s,t,x}(d\mu)$, Lemma 11.4.1(c) and (11.5.2), it follows that the left hand side is equal to

$$Z_{s,t}(\phi,\psi,z) := \int\langle\mu,\psi\rangle e^{-\langle\mu,\phi\rangle}R_{s,t,x}(d\mu) = \frac{\partial}{\partial\varepsilon}(V_{s,t}(\phi+\varepsilon\psi))|_{\varepsilon=0}.$$

Using equation (4.3.8) we can verify that Z is the (non-negative) solution of the linear equation,

$$Z_{s,t}(\phi,\psi,x) = E_{s,x}\left[\psi(W_t) + \int_s^t \kappa(dr)\Phi_2(r,W_r,V_{r,t}\phi(W_r))Z_{r,t}(\phi,\psi,W_r)\right].$$

We now verify that

$$Z_{s,t}(\phi,\psi,x) = E_{s,x}\left[\exp\left(\int_s^t \Phi_2(r,W_r,V_{r,t}\phi(W_r))\kappa(dr)\right)\psi(W_t)\right]$$

by substitution in the right hand side. Using the Markov property of W_t this yields

$$E_{s,x}\left[\psi(W_t) + \int_s^t \kappa(dr)\Phi_2(r,W_r,V_{r,t}\phi(W_r))Z_{r,t}(\phi,\psi,W_r)\right]$$

$$= E_{s,x}\left[\psi(W_t) + \int_s^t \kappa(dr)\Phi_2(r,W_r,V_{r,t}\phi(W_r))\exp\left(\int_r^t \Phi_2(u,W_u,V_{u,t}\phi(W_u))\kappa(du)\right)\psi(W_t)\right]$$

$$= E_{s,x}\left[\psi(W_t) - \int_s^t \kappa(dr)\ \frac{d}{d\kappa(r)}\exp\left(\int_r^t \Phi_2(u,W_u,V_{u,t}\phi(W_u))\kappa(du)\right)\psi(W_t)\right]$$

$$= E_{s,x}\left[\exp\left(\int_s^t \Phi_2(r,W_r,V_{r,t}\phi(W_r))\kappa(dr)\right)\psi(W_t)\right]. \quad \square$$

Remark 11.6.2. For the historical process, $\{W_u : s\le u\le t\}$ is measurable with respect o $\sigma(W_t)$ and hence we obtain

$$\int e^{-\langle\mu,\phi\rangle}(R_{s,t})_{y^t}(d\mu) = \exp\left(\int_s^t \Phi_2(r,y^r,V_{r,t}(y^r))\kappa(dr)\right)$$

or $P_{s,m}$ a.a. y (cf. [DP, Section 4]).

1.7 A Probabilistic Representation of the Palm Distribution

Note that $-\Phi_2(r,x,\lambda) = 2c(r,x)\lambda + \int_0^\infty (1-e^{-\lambda u})\, u\cdot n(r,x,du)$ is the log-Laplace unction of an infinitely divisible positive random variable for each pair (r,x). Given $\{W_r : s\le r\le t\}$ with $W_s=x$, $W_t=y$, we construct a random measure on $\mathbb{R}$, $T(W,dr)$ with independent increments and with Laplace functional

$$E\left(e^{-\int f(r)T(W,dr)}\right) = \exp\left(\int_s^t \Phi_2(r,W_r,f(r))\kappa(dr)\right).$$

Then given W and $T(W,dr)$ we define a Poisson measure, $N(W,.)$, on $\mathbb{R}\times(M_F(E)\setminus\{0\})$ with intensity measure

$$n(W,dr,d\nu) = R_{r,t}(W_r,d\nu)T(W,dr).$$

We assume that (W,T,N) are defined on a common probability space with law $P_{s,x}(dW)P(W,.)$.

Theorem 11.7.1. For $A \in \mathcal{B}(M_F(E))$

$$(R_{s,t,x})_y(A) = E_{s,x}\left\{ P\left(\int_s^t\int_{M_F(E)\setminus\{0\}} \nu\, N(W,dr,d\nu) \in A\right) \Big| W_t=y\right\}.$$

Proof. By the construction of (T,N) given W we obtain

$$E_{s,x}\left(E\left(\exp\left(-\int_s^t\int_{M_F(E)\setminus\{0\}} \langle\nu,\phi\rangle N(W,dr,d\nu)\right)\Big| W_u : s\le u\le t\right)\Big| W_t=y\right)$$

$$= E_{s,x}\left(E\left(\exp\left(-\int_s^t\int_{M_F(E)\setminus\{0\}} (1-e^{-\langle\nu,\phi\rangle})R_{r,t}(W_r,d\nu)T(W,dr)\right)\Big| W_u : s\le u\le t\right)\Big| W_t=y\right)$$

$$= E_{s,x}\left(E\left(\exp\left(-\int_s^t (V_{r,t}\phi)(W_r)T(W,dr)\right)\Big| W_u : s\le u\le t\right)\Big| W_t=y\right)$$

$$= E_{s,x}\left(\exp\left(\int_s^t \Phi_2(r,W_r,V_{r,t}\phi(W_r))\kappa(dr)\right)\Big| W_t=y\right)$$

$$= \int e^{-\langle\mu,\phi\rangle}(R_{s,t,x})_y(d\mu)$$

by Theorem 11.6.1 and the proof is complete. □

Remarks. The analogue of the representations for the (Φ,κ,W)-superprocess for the approximating branching particle systems are established in Gorostiza and Wakolbinger (1991). For applications of the representation of the Palm distribution to both the local and long-time behavior of the superprocess refer to [DP] and [D].

12. FAMILY STRUCTURE AND HISTORICAL PROCESSES

12.1 Some Introductory Remarks

The purpose of this chapter is to introduce a richer structure possessed by both the branching and sampling classes of processes. This structure involves not only the current distribution of the population but also the family relationships and past history of the population and plays an important role in studying both the sample path behavior and long time behavior of superprocesses.

Although the basic ideas of family structure in population models go back to the origins of the subject, in the context of measure-valued processes the subject is undergoing rapid development in several different directions at this time. This chapter is an attempt to give a unified introduction to these developments. We will outline the main ideas and make frequent references to the recent literature on this subject and parallel arguments in earlier chapters of these notes rather than attempt a complete discussion.

Since the precise formulation of the historical process involves a number of technicalities we begin with a brief informal explanation of the main ideas. In the setting of the N-particle Moran process (Chapt. 2), we can consider an enriched version of the empirical process, namely,

$$G_t^N := N^{-1} \sum_{i=1}^{N} \delta_{\hat{Z}_i(t)} \qquad \textit{(empirical measure on particle histories)}$$

where $\hat{Z}_i \in D_E$ is the path in E followed by the ith particle alive at time t and its ancestors *(its history)*. Note that $\hat{Z}_i(t)(s) := \hat{Z}_i(s \wedge t)$ (path stopped at t) and that when a particle jumps it assumes the entire past history of the particle onto which it jumps.

Then $G_t^N \in M_1(D_E)$ and describes the law of the path which would have been followed by one particle chosen at random at time t from the population. Now considertwo particles chosen at random (with replacement) from the probability distribution G_t^N. Even though these particles are chosen independently there is a positive probability that both will follow a common path in [0,s] and then follow separate paths in (s,t]. If the motion process with generator A is such that two independent particles never follow identical paths, then this would be exactly the situation which would occur if and only if they had a common ancestor up to time s but were descendents of two different offspring of that ancestor at time s. Taking larger and larger random samples from $G^N(t)$ reveals more and more information about the *family structure or genealogy* of the particles alive at time t.

If we consider the branching particle system instead of the Moran model, conditioned by non-extinction at time t and consider the corresponding normalized measure on D_E, then exactly the same situation would occur.

For both sampling and branching systems, if we consider the measure-valued limits ($N\to\infty$, resp. $\varepsilon\to 0$) as in Chapter 2, we obtain nonatomic measures, G_t, (resp.H_t) on D_E). *The key point is that even in these limiting cases,* G_t *restricted to* $[0,s]$ for $s<t$, *is almost surely concentrated on a finite number of paths.* This reflects the fact that, in the $N\to\infty$ limit, the number of individuals alive at time $s<t$ having descendents alive at time t converges in distribution to a finite random variable. This means that we can represent G_t in terms of a branching tree structure and that, as the family tree branches, the mass (representing the total population alive at time t) is divided into smaller and smaller atoms (representing more and more closely related groups of individuals). The main objective of this chapter is to outline the mathematical formulation of this representation.

Much of the recent development of historical process has its origins in the papers of Durrett (1978), Fleischmann and Prehn (1974, 1975) Fleischmann and Sigmund-Schultze (1977, 1978), Gorostiza (1981), Jagers and Nerman (1984), Kallenberg (1977), and Neveu (1986) on branching processes and Cannings (1974), Kingman (1982), Tavaré (1984) and Watterson (1984) on population genetics.

12.2 The Historical Branching Particle System.

In order to keep things as simple as possible in this section we consider a critical binary branching particle system with motion process W given by a Feller process on a compact metric space E and with representation $(\hat{\Omega},\hat{\mathcal{F}},\hat{Q}_A)$ as in section 2.2 (see remark 12.2.2.2). The family structure of a branching particle system is naturally represented by a tree and this structure was used in an essential way by Neveu (1986), Chauvin (1986a,b), Perkins (1988) and Le Gall (1989b, 1991a,b).

12.2.1 The Binary Tree System

Let us consider a critical binary branching particle system in which the particle motions are given by W. In order to keep track of the family structure of the particles, we introduce Neveu's formulation of the *binary tree system.*

Let $K := \bigcup_{n=0}^{\infty} \{1,2\}^n$ where by convention $\{1,2\}^0 = \{\partial\}$. $k\in K$ is represented as $k := k_1k_2\ldots k_n$ where each $k_i\in\{1,2\}$. Set $|k|=n$, $|\partial|=0$, and if $|k|\geq 1$, set $\bar{k} := k_1\ldots k_{n-1}$. For $k=k_1\ldots k_n$, $h=h_1\ldots h_m \in K$ set $hk = h_1\ldots h_m k_1\ldots k_n$ and $\partial k=k\partial=k$. Finally, $h < h'$ if there exists some $k\in K$ such that $h'=hk$.

A (finite binary) tree is a finite subset κ of K that satisfies

(i) $\partial \in \kappa$

(ii) $\bar{k} \in \kappa$ whenever $k \in \kappa$ and $|k| \geq 1$

(iii) if $k = k_1 \dots k_n \in K$, then either $k_1 \dots k_n 1 \in \kappa$, and $k_1 \dots k_n 2 \in \kappa$ or $k_1 \dots k_n 1 \notin \kappa$, and $k_1 \dots k_n 2 \notin \kappa$.

Denote by $\mathfrak{I}$ the set of all (finite binary) trees. If $\kappa \in \mathfrak{I}$ and $k \in \kappa$, we set $v_k(\kappa) = 1_{\{k1 \in \kappa\}} = 1_{\{k2 \in \kappa\}}$.

We next introduce the notion of a marked tree. A *marked tree* is a pair (κ, ϑ) where κ is a tree and ϑ is a map from κ into a *space of marks* E^*.

12.2.2. The A-Path Process.

Assume that the motion process $Y = (D_E, \mathcal{D}, \mathcal{D}_t, Y_t, \{P^x\}_{x \in E})$, filtration defined as in Sect. 3.6 and coordinate process Y_t) is an E-valued Feller process with generator $(D(A), A)$ and laws $\{P^x\}_{x \in E}$.

Notation. If $y, w \in D_E = D([0,\infty), E)$, and $s \geq 0$ let $Y_s(y) := y(s)$ and let $(y/s/w) \in D_E$ be defined by

$$(y/s/w) := \begin{cases} y(u) & \text{if } u < s \\ w(u-s) & \text{if } u \geq s. \end{cases}$$

Let $\pi_t : D_E \to E$ be defined by $\pi_t y := y(t)$. Also let $D_E^t := \{y : y = y^t\}$ where $y^t(\cdot) := y(\cdot \wedge t)$ (path stopped at t).

We now define the *path process associated with* Y to be the D_E-valued process defined on $(D_E, \mathcal{D}, \mathcal{D}_t, Y_t, P^x)$ as follows:

$$W_t(y) := \{Y_{t \wedge s}(y) : s \in \mathbb{R}_+\} \in D_E^t.$$

Then $y(t) = \pi_t W_t(y)$.

Of course with appropriate regularity conditions the path process is a Markov process even when the underlying Y process is not Markov. We do not develop this point of view here which goes back to Harris (1963, Sect. 24 and 27) but refer to Wentzell (1985, 1989, 1992) and Bulycheva and Wentzell (1989) for this as well as for a more systematic study of path processes.

Let $\hat{D}_{E,s} := \{(t,y) : t \in [s,\infty), y \in D_E^t\}$ and $\hat{D}_E := \hat{D}_{E,0}$. Let $M_F(D_E)^s := \{\mu \in M_F(D_E) : \mu((D_E^s)^c) = 0\}$.

The corresponding *canonical path process* is defined by

$(D_E, \mathcal{D}, \mathcal{D}_{s,t}, \{W_t\}_{t \geq 0}, (P_{s,y})_{(s,y) \in \hat{D}_E})$ with $\mathcal{D}_{s,t} := \bigcap_{\varepsilon > 0} \sigma\{W_u : s \leq u \leq t + \varepsilon\}$, and

$$P_{s,y}(B) := P^{y(s)}(y/s/W \in B). \tag{12.2.2.1}$$

Similarly,

$$P_{s,\mu}(B) := \int P^{y(s)}(y/s/W \in B) \mu(dy) \quad \text{if} \quad \mu \in M_F(D_E)^s.$$

Remark: Note that the path process, W, has the important property that
(12.2.2.2) for t fixed the σ-algebras of subsets of D_E generated by
$\sigma\{W_s : s\le t\}$ and $\sigma\{W_t\}$ are identical.

Although the path process W is not Feller, it does possess nice properties from the point of view of the general theory of Markov processes in the following sense (cf. Section 3.5 for the appropriate definitions).

Theorem 12.2.2.1. Assume that Y and W are as above.
Then $(D_E, \mathcal{D}, \mathcal{D}_{s,t}, \{W_t\}_{t\ge 0}, (P_{s,y})_{(s,y)\in \hat{D}_E})$ is a canonical inhomogeneous Borel strong Markov process with càdlàg paths in $D_E^t \subset D_E$ and inhomogeneous semigroup

(12.2.2.3) $S_{s,t}f(y) = P^{y(s)}(f(y/s/Y^{t-s})), \quad (s,y)\in\hat{D}_E, \ t\ge s, \quad f\in b\mathcal{D}.$

Proof. [DP Prop. 2.1.2]. (In fact this is true for a much wider class of underlying processes, Y.)

We now consider a martingale problem which characterizes the path process. Consider the algebra of functions, denoted by $D_0(\tilde{A})$, described as follows: $\phi \in D_0(\tilde{A})$ if for some $n\in\mathbb{N}$, $t_1<t_2<\ldots<t_n$, g_j and $(\frac{\partial}{\partial s}+A)g_j \in bC([0,\infty)\times E)$ for $j=1,\ldots,n$, and

$$\phi(s,y) = \prod_{j=1}^n g_j(s,y(s\wedge t_j))$$

Note that $\{\phi(t,.):\phi\in D_0(\tilde{A})\}$ is bp dense in $\mathcal{D}_t$.

For $\phi\in D_0(\tilde{A})$, set

(12.2.2.4) $$(\tilde{A}\phi)(s,y) := \prod_{\ell=1}^k g_\ell(s,y(t_\ell))(\frac{\partial}{\partial s}+A)\prod_{\ell=k+1}^n g_\ell(s,y(s)) \quad \text{if } t_k\le s<t_{k+1}$$
$$:= 0 \quad \text{if } s>t_n.$$

If $\phi\in D_0(\tilde{A})$, then $(\tilde{A}\phi)(s,y) \in bC([0,\infty)\times D_E)$ and

(12.2.2.5) $$M_t(\phi) := \phi(t,W_t)-\phi(s,W_s) - \int_s^t (\tilde{A}\phi)(r,W_r)dr \quad \text{is a } P_{s,W_s}\text{-martingale.}$$

$\tilde{A}$ is called the *compensating operator* by Wentzell (1985, 1992). If the path process can be characterized as the unique solution of the $(D_0(\tilde{A}),\tilde{A})$-martingale problem then we call $(D_0(\tilde{A}),\tilde{A})$ an *MP-generator*.

Remark 12.2.2.2. It can be verified that the $(D_0(\tilde{A}),\tilde{A})$-martingale problem is well posed if A is a Feller generator (apply [EK, Ch. 4, Th. 4.1] to verify that the finite dimensional distributions of the t-marginal distribution of W coincide with with those of Y stopped at t and then using Theorem 5.1.2). Also by extending the arguments used in the proof of Proposition 2.2.1, in the Feller case we can construct a D_E-valued measurable random function, ζ, on $\hat{D}_E$ (defined on a standard probability space $(\hat{\Omega},\hat{\mathcal{F}},\hat{Q}_A)$ such that $\zeta((s,y))$ has law $P_{s,y}$.

For certain applications it is desirable to identify a larger class of functionals for which an analogue of (12.2.2.5) is valid. Mueller and Perkins (1991) defined a natural extension of $(D_0(\tilde{A}),\tilde{A})$ as follows. Let

$$F_{s,\mu} := \{\phi:\hat{D}_{E,s}\to\mathbb{R},\ \phi \text{ is Borel measurable, } \phi(s,W_s) \text{ is right continuous, } P_{s,\mu}\text{-a.s. and } |\phi(s,W_s)|\le K,\ P_{s,\mu}\text{-a.s. for some } K\},$$

and define

$$D(\tilde{A}_{s,\mu}) := \Big\{\phi\in F_{s,\mu}:\exists\tilde{A}_{s,\mu}\phi\in F_{s,\mu} \text{ such that } M^{\phi}(t,W) \equiv \phi(t,W_t)-\phi(s,W_s)-\int_s^t(\tilde{A}_{s,\mu}\phi)(r,W_r)dr \text{ is a } P_{s,\mu}\text{-martingale on the } P_{s,\mu}\text{-completion of the filtration } (\mathcal{D}_t)_{t\ge s}\Big\}.$$

For $s\ge 0$, $\mu\in M_F(D_E)^s$, $\tilde{A}_{s,\mu}$ is called the $P_{s,\mu}$*-weak generator* of W by Mueller and Perkins (1991). A wide class of functionals which belong to the domain of the weak generator was identified by Wentzell (1992).

Definition: The path process $\{W_t\}$ is said to have *Property S* if

(12.2.2.6) $P_{s,y}\times P_{s,y}(W^1_u=W^2_u) = 0\ \forall\ u>s,\ y\in D^s.$

Note that we can always consider an enriched version of the original process for which the path process will satisfy this property. For example, we can take $E' = E\times\mathbb{R}^1$ and generator $A' := A + \Delta$ (i.e. the components are independent, the first component is an A-process and the second component is a standard Wiener process).

12.2.3. Binary branching A-motions.

We will now formulate a binary branching A-motion in terms of a probability measure on a set of marked trees.

As the appropriate space of marks we take $E^* = \mathbb{R}_+\times D_E$. The set of all E^*-marked binary trees is denoted by $\mathfrak{T}^*$, that is, $\mathfrak{T}^* = (E^*)^{\mathfrak{T}}$ and a typical element of $\mathfrak{T}^*$ will be written as

$$\mathfrak{t} = (\kappa,(\tau_k,W_k)_{k\in\kappa})$$

where $\kappa\in\mathfrak{T}$ and for every $k\in\kappa$, $\tau_k\in\mathbb{R}_+$, $W_k\in D_E$. For $\mathfrak{t} = (\kappa,(\tau_k,W_k)_{k\in\kappa})\in\mathfrak{T}^*$ let $s_h(\mathfrak{t}) := \sum_{k<h}\tau_k$, and $t_h(\mathfrak{t}) := s_h(\mathfrak{t}) + \tau_h$.

Finally we need to introduce translation operators T_h. For any $h\in K$, the mapping T_h is defined on the subset $\{h\in\kappa\}$ of $\mathcal{I}^*$, by

$$T_h((\kappa,(\tau_k,W_k)_{k\in\kappa}) := (\kappa_h,(\tau_{hk},W_{hk})_{k\in\kappa_h})$$

where $\kappa_h = \{k\in K:\ hk\in\kappa\}$.

Proposition 12.2.3.1 Let $\{N^\nu(s):s\geq 0\}$ be a time inhomogeneous Poisson process with intensity measure $\nu(ds) \in M_{LF}([0,T))$, $T\leq\infty$ and corresponding law Q_N and $0\leq p\leq 1$. Then there exists a unique measureable mapping $(t,z) \to \Lambda_{t,z}$ from $[0,\infty)\times D^t$ to $M_1(\mathcal{I}^*)$, such that the following properties hold:

(i) for every $z \in D^t$, the random variables v_∂,τ_∂ are independent and

$$\text{(12.2.3.1)}\quad \Lambda_{t,z}[v_\partial=1] = p,\ \Lambda_{t,z}[v_\partial=0] = 1-p,$$

(that is, the offspring generating function is $G(z) = 1-p+pz^2)$, $0\leq z\leq 1)$,

$$\text{(12.2.3.2)}\quad \Lambda_{t,z}[\tau_\partial>t+s] = P(N^\nu(t+s)-N^\nu(t) = 0),$$

(ii) the conditional law of W_∂, given $(v_\partial,\tau_\partial)$, is $P_{t,z}(W(t+\tau_\partial)\in .)$, where $P_{t,z}$ is defined by (12.2.2.1),

(iii) the conditional distribution under $\Lambda_{t,z}(dw)$ of the pair (T_1w,T_2w), given $v_\partial=1$ and $(\tau_\partial,W_\partial)$ is $\Lambda_{\tau_\partial,W_\partial(\tau_\partial)}\otimes\Lambda_{\tau_\partial,W_\partial(\tau_\partial)}$.

Proof. (cf. Neveu (1986), Chauvin (1986a), Le Gall (1991a)).

The process can be constructed in the natural way as in Ch. 2 on a probability space $(\Omega,\mathcal{F},P)$ containing three sequences of independent random variables, $\Big\{\{v_k\},\ \{N_k\},\ \{\zeta_k\}:\ k\in K\Big\}$, $P(v_k=0) = 1-p$, $P(v_k=1) = p$, $\{N_k\}$ have distribution Q_N, and $\{\zeta_k\}$ have distribution $\hat{Q}_A$. □

Remark. Note that for $t = (\kappa,(\tau_k,W_k)_{k\in\kappa}) \in \mathcal{I}^*$, $W_k \in D_E^{t_k}$ and $W_k(t,t) = W_h(t,t\wedge t_k)$ if $k<h$. The random variables $W_h(t)$ for all $h\in\kappa$ such that $s_h\leq t<t_h$ represent the paths followed by the particles alive at time t and before the instants of their births by their ancestors, up to time t. Let $\pi_t:D_E\to E$ be defined by $\pi_t(y) := y(t)$. Then $\pi_tW_h(t)$ denotes the locations of the particles at time t.

We then consider the following measure-valued processes

$$H_t^{*\nu}(t) := \sum_{\substack{h\in\kappa\\ s_h\leq t<t_h}} \delta_{W_h(t)} \quad \text{(counting measure on } D_E^t),$$

$$X_t^{*\nu}(t) := \sum_{\substack{h\in\kappa \\ s_h\le t<t_h}} \delta_{\pi_t W_h(t)}.$$

In the case in which $\nu(ds) := cds$ is a homogeneous Poisson process with $c>0$, and $p = 1/2$, the $M_F(D_E)$-valued process $\{H_t^{*\nu}: t\ge 0\}$ is called the *historical critical binary branching particle system.* It is an enriched version of the process $\{X_t^{*\nu}\}$ where the latter is the branching particle system introduced in Section 2.10. Note that this is a *trimmed tree* in which all the branches which have become extinct by time t (i.e. $v_h = 0$ for some $s_h<t$) are trimmed off.

Remark: If $\{W_t\}$ satisfies Property S, then P-a.s. we can reconstruct the marked subtree, $(\kappa,(\tau_k,W_k(t))_{k\in\kappa}: k<h,$ for some h with $s_h\le t<t_h)$ (up to relabelling of the two offspring as 1 and 2) from $H_t^{*\nu}(t)$. In particular, if $p = 1$ (i.e. no deaths) we can view this as follows. Let $N_t(u)$ denote the number of atoms in $r_u H_t^{*\nu}(t)$, $0\le u\le t$. Then $\{N_t(u): 0\le u\le t\}$ is a monotone increasing integer-valued function. Furthermore under Property S, there is a one-to-one correspondence between the jumps of $N_t(u)$ and the set $\{t_k: k<h$ for some h with $s_h\le t<t_h\}$. Then in order to determine which path split at t_h we look for the unique $W|_{[0,t_h)}$ which has two distinct extensions on $[0,t_h+\varepsilon)$ $\forall\ \varepsilon>0$.

12.2.4. The Generating Function Equation.

Let $N_F(D_E^t)$ denote the set of finite integer-valued measures on D_E^t. Then $\{H_t^{*\nu}\}$ is a càdlàg $N_F(D_E^t)$-valued process.

Let $\Omega^* := D(N_F(D_E))$ with its Borel σ-field $\mathcal{G}^*$, $H_t^{*\nu}(\omega) = \omega(t)$, and $\mathcal{G}^*_{[s,r]} := \sigma(H_t^*: s\le t\le r)$, $\mathcal{G}^*_{[s,\infty)} := \bigvee_{r\ge s} \mathcal{G}^*_{[s,r]}$. If $H_s^{*\nu} = m\in N_F(D_E^s)$, then the resulting law of $H_\cdot^{*\nu}$ on $\mathcal{G}^*_{[s,\infty)}$ is denoted by $Q_{s,m}^{*\nu}$.

If $H_s^{*\nu} = \delta_y$, $y\in D^s$, then the probability generating function of $H_t^{*\nu}$ is defined by

$$(12.2.4.1)\qquad G_{s,t}\xi(y) := Q_{s,\delta_y}^{*\nu}\left(e^{\langle H_t^{*\nu},\log\xi\rangle}\right),\quad \xi\in\mathcal{D},\ 0<\xi\le 1,\ t>s.$$

If $H_s^{*\nu} = \sum \delta_{y_i}$, then from each of these initial particles, δ_{y_i}, we construct independent copies of the above process and $H_t^{*\nu}$ is then given by the superposition of these. Therefore the probability generating function of $H_t^{*\nu}$ is given by

(12.2.4.2) $\prod_i G_{s,t}\xi(y_i)$.

Theorem 12.2.4.1 (a) $G_{s,t}\xi$, $0\le s\le t$, is the unique solution of the equation

(12.2.4.3) $$G^{\nu}_{s,t}\xi(y) = e^{-\nu((s,t])} S_{s,t}\xi(y) + \int_s^t S_{s,u} e^{-\nu((s,u])}\left[G(G^{\nu}_{u,t}\xi(y))\right]\nu(du).$$

If the intensity measure $\nu(ds) = \nu(s)ds$, then this is the evolution solution of the formal equation

(12.2.4.4) $$G^{\nu}_{t,t}\xi(y) = \xi(y),$$
$$-\frac{\partial G^{\nu}_{s,t}}{\partial s} = (A_s - \nu(s))G^{\nu}_{s,t} + \nu(s)G(G^{\nu}_{s,t}).$$

(b) Let $\mathcal{P}ois_m$ denote the law of a Poisson random measure on D^S_E with intensity m. If we assume that $H^{*\nu}_s$ has distribution $\mathcal{P}ois_m$, then $H^{*\nu}_t$ has Laplace functional

(12.2.4.5)
$$\int Q^{*\nu}_{s,\mu}\left[e^{-\langle H^{*\nu}_t,\phi\rangle}\right]\mathcal{P}ois_m(d\mu) = \exp\left\{-\int\left(1 - (G^{\nu}_{s,t}e^{-\phi})(y)\right)m(dy)\right\},$$

that is, $H^{*\nu}_t$ is a Poisson cluster random measure with intensity m and cluster Laplace functional $(G^{\nu}_{s,t}e^{-\phi})(y)$.

Proof. (a) Equation (12.2.4.3) is obtained by conditioning on the time and place of the first branching event. The uniqueness then follows as in Lemma 4.3.3.

(b) This follows from (a) and the Poisson cluster formula (3.3.2). □

Notation. In the case $\nu(ds) := (c/\varepsilon)ds$, with $\varepsilon>0$, we denote the process $H^{*\nu}$ by $H^{*\varepsilon}$.

12.3. The (A,c)-Historical Process

12.3.1 Construction and Characterization.

Consider the collection of historical critical binary branching particle systems, $\{H^{*\varepsilon}_t:\varepsilon>0\}$, and Poisson random field initial condition at time s with intensity μ/ε, $\mu\in M_F(D_E)^S$. Finally we consider the measure-valued process obtained by letting each particle to have mass ε:

$$H^{\varepsilon}_t := \varepsilon H^{*\varepsilon}_t.$$

Note that this is a special case of the general time inhomogeneous set-up of Chapter 4 in which the motion process is the A-path process W, $G(z) = \frac{1}{2}(1+z^2)$, $0\le z\le 1$, and $\kappa(ds) = \varepsilon^{-1}ds$. For the sake of brevity herafter we denote the (A,c)-historical process by H(A,c).

Theorem 12.3.1.1. There exists a transition function on $M_F(D_E)$ with Laplace transition functional

(12.3.1.1) $Q^{H(A,c)}_{s,\mu}\left(\exp(-\langle H_t,\phi\rangle)\right) = \exp\{-\langle\mu,V_{s,t}\phi\rangle\}$

$\forall\ \phi \in bp\mathcal{D}$, $\mu\in M_F(D)^s$, $s\le t$, where $V_{s,t}\phi(y) := v_{s,t}(y)$ is the unique solution of

(12.3.1.2) $$V_{s,t}\phi(y) = S_{s,t}\phi(y) - \frac{1}{2}c\int_s^t S_{s,r}((V_{r,t}\phi)^2)dr$$

which is Borel measurable in $(s,y,t) \in \{(s,y,t),\ s\in[0,\infty),\ y\in D^s,\ t\ge s\}$ and is bounded if $(t-s)$ is bounded.

<u>Proof.</u> Exactly as in the proof of Theorem 4.4.1 (cf. Lemma 4.4.3) we conclude that $\{H^\varepsilon_t\}$ converges, as $\varepsilon\to0$, to a time inhomogeneous $M_F(D_E)$-valued Markov process $\{H_t\}$ (in the sense of convergence of finite dimensional distributions) and that the transition function is given by (12.3.1.1, 12.3.1.2). □

The resulting process $\{H_t\}$ is called the (A,c)-*historical process* and is denoted by H(A,c). In fact it coincides with the (W,κ,Φ)-superprocess with W given by the A-path process, $\kappa(ds) = ds$, and $\Phi(\lambda) = -\frac{1}{2}c\lambda^2$.

Since D_E is not locally compact and the process is time inhomogeneous we cannot directly apply the tightness argument of Section 4.6 to obtain weak convergence on $D([0,\infty),M_F(D_E))$. The question of the tightness of the sequence of approximating historical branching particle systems was discussed in [DP, Section 7]. The main obstacle is to verify the first condition in Theorem 3.6.4, namely, given T>0 and $\eta>0$ to verify the existence of a compact subset, $K_{T,\eta}$, of $M_F(D_E)$ such that $P(H^\varepsilon_\cdot \in D([0,T],K_{T,\eta})) > 1-\eta\ \forall\ \varepsilon>0$. Since $\sup_\varepsilon P(\sup_{0\le t\le T} H^\varepsilon_t(M_F(D_E)) > k) \to 0$ as $k\to\infty$ (cf. Lemma 4.6.1), it suffices to find a sequence of compact subsets, $\{K_{T,n}\}$, of D_E such that

$$\sup_\varepsilon P(\forall n\in\mathbb{N},\ \sup_{0\le t\le T} H^\varepsilon_t(K^c_{T,n}) \le 1/n\ ) > 1-\eta.$$

Hence it suffices to find for each n a compact set $K_{T,n} \subset D_E$ such that $\sup_\varepsilon P(\sup_{0\le t\le T} H^\varepsilon_t(K^c_{T,n}) > 1/n) < \eta/2^n$. The existence of such a compact subset is established in [DP, Prop. 7.7]. Thus it can be established that the historical branching particle systems converge weakly to the historical process $\{H_t:t\ge0\}$ and the latter has càdlàg paths in $M_F(D_E)$. The existence of nice version of the historical process is summarized in the following theorem.

<u>Theorem</u> <u>12.3.1.2.</u> There is an inhomogeneous Borel strong Markov process with càdlàg

paths in $M_F(D_E)$, $\{H_t:t\geq s\}$ with laws $\{Q_{s,\mu}\}_{s\geq 0,\mu\in M_F(D_E)^s}$, whose finite dimensional distributions are determined by the transition function of Theorem 12.3.1.1.

Proof. This can be proved using the construction of Fitzsimmons (1988) and is given in [DP, Theorems 2.1.5, 2.2.3]. It can also be proved by the above weak convergence argument together with the uniqueness to the martingale problem (see Theorem 12.3.3.1). □

The *canonical* (A,c)-*historical process* is denoted by $H := (\Omega,\mathcal{G},\mathcal{G}_{s,t},(H_t)_{t\geq 0},(Q^{H(A,c)}_{s,\mu})_{s\geq 0,\mu\in M_F(D_E)^s})$ where $\Omega = D([0,\infty),M_F(D_E))$, $H_t(\omega) := \omega(t)$ for $\omega\in\Omega$, $\mathcal{G}_{s,t} := \bigcap_{u>t}\sigma(H_r:s\leq r\leq u)$ and $Q^{H(A,c)}_{s,\mu}$ is a probability measure on $\mathcal{G}$. Then $H_t \in M_F(D_E)^t$ $\forall t\geq s$, $Q^{H(A,c)}_{s,\mu}$-a.s. If $\phi:\hat{D}_E \rightarrow \mathbb{R}$ is measurable, then $\langle H_t,\phi(t)\rangle := \int_{D_E} \phi(t,y)H_t(dy)$.

In Theorem 12.3.4.2 below we will verify that $X_t(B) := H_t(\{y:y(t)\in B\})$, $B\in \mathcal{B}(E)$, is a version of the (A,c)-superprocess. We can thus regard H_t as an *enriched* version of X_t which contains information on the genealogy or family structure of the population.

12.3.2. Remarks on Alternative Formulations of the Historical Process

The above construction of historical processes as limits of branching particle systems was introduced in Dynkin (1991a) and their characterization as inhomogeneous Borel strong Markov processes was established in [DP]. There have also been alternative formulations of historical processes.

Perkins (1988) (also see Dawson, Iscoe and Perkins (1989)) developed a nonstandard model of superprocesses. His idea was to construct a binary branching Brownian motion with infinitely many particles having infinitesimal mass and infinite branching rate. The nonstandard approach has the advantage of allowing one to work with the intuitive particle picture and in particular to incorporate the historical information on each particle. Perkins used this formulation to obtain the precise results on Hausdorff measure function which were described in Section 9.3.3.

Another approach is that due to Le Gall (1991a). His approach to the construction of continuous superprocesses is based on relation between continuous branching and standard Brownian motion as reflected for example in the Ray-Knight Theorem. He has constructed a path-valued Markov process as a tool in studying the problem of polar sets for super-Brownian motion. In particular his construction is based on a detailed representation of the tree of excursions of a standard Brownian motion and is related to the following result of Neveu and Pitman (1989). Let w be a Brownian excursion from 0 which reaches level $h > 0$ and let $(N_x)_{x\geq 0}$ be the number of upcrossings from x to x+h of the process w. Then Neveu and Pitman (1989) showed that N_x is a continuous time Galton-Watson critical binary branching process (with particle lifetime exponentially distributed with mean $\frac{1}{2}h$). An h-minimum exists at time t if there exists $s<t<u$ such that $w_s=w_u=w_t+h$ and $w_v\geq w_t$ for $s\leq v\leq u$. The probability that there exists an h-minimum is $\frac{1}{2}$ and conditioned on the existence of an h-minimum the processes $\{w(\sigma_\alpha+s):0\leq s\leq\rho_\alpha-\sigma_\alpha\}$ and $\{w(\rho_\alpha+s):0\leq s\leq\tau_\alpha-\rho_\alpha\}$ are independent copies of w. Here ρ_α is the time when the lowest h-minimum is attained, $\sigma_\alpha=\sup\{u<\rho_\alpha:w(u)=w(\rho_\alpha)\}$ and $\tau_\alpha = \inf\{t>\rho_\alpha:w(t)=w(\rho_\alpha)\}$. Le Gall (1991a,b) used these objects to yield an explicit construction of the (2,d,1)-superprocess and also the (2,d,1)-historical process. In this setting the tree structure of the historical process at a fixed time is directly related to a set of excursions of the Brownian motion above a level $a-\varepsilon$ to hit a (upcrossings from $a-\varepsilon$ to a with a fixed and $\varepsilon>0$) (see Le Gall (1991a, Lemma 8.5)).

12.3.3. Martingale Problem

In the spirit of Chapter 6, it is natural to ask if the historical process can be characterized as the unique solution of a martingale problem. In this subsection we will simply give an outline of the main steps in establishing that this is in fact possible.

Let $H = (\Omega,\mathcal{G},\mathcal{G}_{s,t},\{H_t\},\{Q^{H(A,c)}_{s,\mu}\})$ denote the canonical historical process defined above and let $(D_0(\tilde{A}),\tilde{A})$ denote the MP-generator for the A-path process defined in Section 12.2.2. Using arguments similar to those developed in Chapter 6 we can verify the following: for each $\phi\in D_0(\tilde{A})$, and for $t\geq s$,

$$(12.3.3.1)\qquad Z_t(\phi) := \langle H_t,\phi(t)\rangle - \langle H_s,\phi(s)\rangle - \int_s^t \langle H_r,(\tilde{A}\phi)(r)\rangle dr$$

is a $Q^{H(A,c)}_{s,H_s}$,$\{\mathcal{G}_{s,t}:t\geq s\}$-martingale with increasing process

$$(12.3.3.2)\qquad \langle Z(\phi)\rangle_t = c\int_0^t\int_{D_E} \phi(r,y)^2 H_r(dy)dr.$$

For example, if $\psi\in b\mathcal{D}_u$, then $\{\langle H_t,\psi\rangle:t\geq u\}$ is a continuous martingale with increasing process

$$\langle H(\psi)\rangle_t = \int_u^t\int\psi^2(y)H_s(dy)ds.$$

In other words $\{Q_{s,\mu}\}$ is a solution to the martingale problem (12.3.3.1), (12.3.3.2). Using an extension of the argument of the proof of Theorem 6.1.3(b) it can be verified that any solution $Q_{s,\mu}$ to the martingale problem for $(D_0(\tilde{A}),\tilde{A})$ starting from μ at time s satisfies

$$Q_{s,\mu}^{H(A,c)}\left(\exp(-\langle H_t,\phi(t)\rangle)\right) = \exp\{-\langle\mu,V_{s,t}\phi\rangle\} \quad \text{for} \quad \phi \in D_0(\tilde{A})$$

and consequently this holds for all $\phi\in \mathcal{D}_t$.

Thus the one dimensional marginal distributions to this martingale problem are uniquely determined and applying Theorem 5.1.2 we conclude that the martingale problem is well-posed. Since in the above argument $D_0(\tilde{A})$ can clearly be replaced with a countable subset we can apply Theorem 5.1.2 to establish the strong Markov property. Thus we have the following.

Theorem 12.3.3.1. The law $Q_{s,\mu}^{H(A,c)}$ is the unique probability measure on $\mathcal{G}_{s,\infty}$ such that $\forall\ \phi\in D_0(\tilde{A})$,

$$Z_t(\phi) = \langle H_t,\phi(t)\rangle-\langle\mu,\phi(s)\rangle - \int_s^t \langle H_r,(\tilde{A}_{s,\mu}\phi)(r)\rangle dr, \quad t\geq s,$$

is a continuous $\{\mathcal{G}_{s,t}:t\geq s\}$-martingale such that $Z_0(\phi)=0$ and

$$\langle Z(\phi)\rangle_t = c\int_0^t\int_{D_E} \phi(r,y)^2 H_r(dy)dr.$$

Proof. The uniqueness follows as above. □

Mueller and Perkins (1991) established the analogue of (12.3.3.1) for the richer class of functions, $D(\tilde{A}_{s,\mu})$, which was defined in section 12.2.2 (and also proved that this is valid for a much wider class of underlying processes Y).

12.3.4. Transition Function - Finite Dimensional Distributions.

The purpose of this subsection is to give an analytical description of the transition function for the (A,c)-historical process hereafter denoted by $\{H_t\}$. Note that if $f_t(y) := g(y(t))$, $g\in bp\mathcal{E}$, then

(12.3.4.1) $\qquad S_{s,t}f_t(y) = P^{y(s)}g(y(t-s))$, and

(12.3.4.2) $\qquad V_{s,t}f_t(y) = U_{t-s}g(y(s)) \quad \forall(s,y)\in \hat{D}_E,\ s\leq t.$

Notation. $U_t^{(n)}:bp\mathcal{E}^n \longrightarrow bp\mathcal{E}^{n-1}$ is defined by

$$(U_t^{(n)}g)(x_1,\ldots,x_{n-1}) = U_t g(x_1,\ldots,x_{n-1},.)(x_{n-1}).$$

We define $\bar{U}_t^{(2)}:pb\mathcal{D}_{0,s}\times\mathcal{E} \longrightarrow pb\mathcal{D}_{0,s}$ by

$$\bar{U}_t^{(2)}g(y) = U_t g(y^s,.)(y(s)).$$

Theorem 12.3.4.1. (a) Let $f_{s,t}(y) = g(y,y(t))$, $g\in bp\mathcal{D}_{0,s}\times\mathcal{E}$. Then

(12.3.4.3) $$V_{s,t}f_{s,t}(y) = (\bar{U}^{(2)}_{t-s}\, g)(y) \in \mathcal{D}_{0,s} \quad \forall (s,y)\in \hat{D}_E,\ s\le t.$$

(b) If $g\in bp\mathcal{E}^n$, $t_1\le t_2\le \ldots \le t_n$, and $f_{t_1,\ldots,t_n}(y) = g(y(t_1),\ldots,y(t_n))$,

then for any $1\le k\le n$,

(12.3.4.4) $$V_{t_k,t_n} f_{t_1,\ldots,t_n}(y) = (U^{(k+1)}_{t_{k+1}-t_k} \ldots U^{(n)}_{t_n-t_{n-1}}\, g)(y(t_1),y(t_2),\ldots,y(t_k)).$$

Outline of Proof. (a) For $t\ge s$, $V_{s,t}f_{s,t}(y)$ satisfies (12.3.1.2).
We will verify that the right hand side of (12.3.4.3) also satisfies (12.3.1.1).
Let $\tilde{v}_{s,t}(y) := U^{(2)}_{t-s}g(y)$.
Since

$$(U^{(2)}_{t-s}g)(y) = P_{t-s}g(y,.)(y(s)) - \tfrac{1}{2}c\int_0^{t-s} P_u((U^{(2)}_{t-s-u}(g(y,.))^2)(y(s))du$$

it follows that

$$\tilde{v}_{s,t}(y) = P^{y(s)}(f_{s,t}(y/s/Y^{t-s})) - \tfrac{1}{2}c\int_0^{t-s} P^{y(s)}((\tilde{v}_{u+s,t}(y/s/Y^u))^2)du.$$

The result follows by uniqueness to 12.3.1.2.
(b) The proof follows by induction using repeatedly $V_{t_k,t_n} = V_{t_k,t_{n-1}}V_{t_{n-1},t_n}$ and (a). □

We next verify that $\{X_t\}$ can be embedded in $\{H_t\}$. Let $\Pi_t:M_F(D_E)\to M_F(E)$ be defined by $\Pi_t\mu(B) := \mu(\{y\in D_E:y(t)\in B\})$.

Theorem 12.3.4.2. $X_t = \Pi_t(H_t)$ is a version of the (A,c)-superprocess. Moreover if $\mu\in M_F(D_E)^s$, $\tau\ge s$ is a $\{\mathcal{G}_{s,t}:t\ge s\}$-stopping time and $\psi \in b\mathcal{B}(D(M_F(E)))$, then

(12.3.4.5) $$Q^{H(A,c)}_{s,\mu}(\psi(X(\tau+.)\,|\,\mathcal{G}_{s,\tau}) = P^{B(A,c)}_{X(\tau)}(\psi),\quad Q^{H(A,c)}_{s,\mu}\text{-a.s.}$$
on $\{\tau < \infty\}$ where $P^{B(A,c)}_{\mu}$ is defined as in section 8.1.

Remark on Proof. This follows by an argument similar to the proof of Theorem 12.3.4.1. See [DP Theorem 2.2.4] for the details. □

12.3.5. A Class of Functionals of the Historical Process

Since the historical process involves random measures on D_E it is natural to consider $\int F(y)H_t(dy)$ where F is a measurable functional of interest. In this section we will consider a class of functionals which were introduced by Dynkin (1991b).

Let B be a domain in E and define $\tau_B := \inf\{t:Y_t\notin B\}$. In Section 9.4 we described the random measure X_{τ_B} on $(E,\mathcal{E})$ which was used in Dynkin (1991c) t establish a necessary and sufficient condition for R-polarity. In this section we briefly indicate how this random measure can be constructed as a measure functional of the historical process.

To understand this idea we return to the historical particle system. Consider a modified system in which each particle of a branching particle system evolves according to the motion process stopped at τ_B, W_B, and the additive functional is defined by $\kappa_B(dt) := 1_{[0,\tau_B]}(t)dt$. We then consider the (W_B,κ_B,Φ) particle system and superprocess.

The history of the ith particle at time t, $W_i(t)$, consists of its own trajectory pieced together with the trajectories of all its ancestors (i.e. the law of the particle is simply the motion process). The mass distribution at time t is

$$X_t(A) = \varepsilon \sum_i 1_A(W_{B,i}(t)), \quad A\in\mathcal{E},$$

where ε is the particle mass.

Then we define

$$X_{\tau_B}(A) := \varepsilon\cdot \sum_i 1(\tau_B(W_{B,i})<\infty)\cdot 1_A(W_{B,i}(\infty)),$$

$$\mathcal{O}(\tau,A) := \varepsilon \sum_i \int_0^{\tau_B} 1_A(W_{B,i}(s))ds$$

where the sum is over all particles alive at time infinity.

We can then again take the measure-valued limit as $\varepsilon \to 0$, (by analogy with the argument in Chapter 4) yielding $X_{\tau_B},\mathcal{O}_{\tau_B}$. This leads to the following theorem.

<u>Theorem</u> <u>12.3.5.1</u> The joint Laplace functional of $(X_{\tau_B},\mathcal{O}_{\tau_B})$ is given by:

(12.3.5.1) $P_\mu\exp\{-\langle\mathcal{O}_{\tau_B},\psi\rangle-\langle X_{\tau_B},\phi\rangle\} = \exp(-\langle\mu,v\rangle)$,

(12.3.5.2) $v(x) + P_{0,x}\left(\int_0^{\tau_B}\Phi(W_s,v(W_s))ds\right) = P_{0,x}\left[\int_0^{\tau_B}\psi(W_s)ds+\phi(W_{\tau_B})\right]$.

(In fact $X_{\tau_B},\mathcal{O}_{\tau_B}$ can be defined for all coanalytic sets B.)

<u>Proof.</u> Dynkin (1991c).

What is less clear is whether or not the pair $\{X_{\tau_B},\mathcal{O}_{\tau_B}\}$ can be represented as random variables defined in terms of the canonical process $\{X_t:t\geq 0\}$. Note however that if $\{W_t\}$ denotes the Y_t-path process and Δ is an interval in $[0,\infty)$, then formally $1_\Delta(\tau_B)1_A(Y_{\tau_B}) = \int_\Delta 1_A(\pi_t W_t)A_B(W,dt)$ where $A_B(W,[0,t]) = 1(\tau_B(W_t)\leq t)$.

Therefore, formally, $X_{\tau_B}(A) = \int_0^\infty\int_{D_E} 1_A(\pi_t W_t) A_B(W,dt) H_t(dW)$. It is not immediately clear whether the right hand side of this expression is well-defined. However in the case in which A_B is replaced by an additive functional A which has finite support $\{t_i\}$ this is well-defined, namely,

(12.3.5.3)

$$\int_\Delta\int_{D_E} 1_A(\pi_t W_t) A(W,dt) H_t(dW) = \sum_{t_i \in \Delta} \int_{D_E} 1_A(\pi_{t_i} W_{t_i}) A(W,\{t_i\}) H_{t_i}(dW).$$

Dynkin (1991c) showed that this could be extended to the general case by taking limits in an appropriate sense. We briefly state the main tool used to complete this step.

Let $(L,\mathcal{L})$ be a measurable space and $\zeta \in p\mathcal{L}$. *A measure functional* of the path process with values in $M_F(L)$ is given by a measurable mapping η: $(D_E,\mathcal{D},\mathcal{D}_{s,t},W_t,P_{s,y}) \longrightarrow M_F(L)$. We say that $\eta \in \Gamma^o_W(L)$ if it satisfies

(i) $\exists$ $b \in \mathbb{R}_+$ such that $\eta\{\zeta > b\} = 0$ and $\sup_{y \in D_E} \eta\{y,\zeta \le b\} < \infty$,

(ii) $\eta\{B,\zeta<t\}$ is measurable with respect to $\mathcal{D}^{uc}_{s,t}$, $\forall$ $B \in \mathcal{L}$, $t \in \mathbb{R}_+$.

Then we define $\Gamma_W(L) := \{\eta\colon \exists\ \eta_n \in \Gamma^o_W(L),\ \eta_n \uparrow \eta\}$.

Theorem 12.3.5.2 Let $\{H_t\}$ be the (A,c)-historical process. Then to every random measure $\eta \in \Gamma_W(L)$ there corresponds a kernel H^η from $(\Omega,\mathcal{G}^{uc})$ to $(L,\mathcal{L})$ such that: if $\eta \in \Gamma^o_W(L)$, then for every $B \in \mathcal{L}$

$$H^\eta(B\times[0,t))) = \text{medial} \lim_{n\to\infty} \sum_{\substack{i\ge 1\\ i/2^n \le t}} \int_{D_E} \eta(W,B,(i-1)2^{-n} \le \zeta < i2^{-n})\ H_{i/2^n}(dW)$$

a.s.

Proof. Dynkin (1991c, Thm. 1.3).

Example: The expression (12.3.5.3) above corresponds to the special case in which

$L = E\times\mathbb{R}_+$, $\zeta(e,t)=t$, and

$$\eta(W,A,\zeta<t) = \sum_{t_i<t} 1_A(\pi_{t_i} W_{t_i}) A(W,\{t_i\}), \quad A \in \mathcal{E}.$$

Remark: In particular Theorem 12.3.5.2 can be applied to the random measures with values in $M_F(E)$:

$$\eta_1(W,B,\zeta<t) = \int_0^t 1_B(\pi_t W_t) A(W,dt),\ B \in \mathcal{E},\quad A(W,[0,t]) = 1(\tau(W_t) \le t);$$

$$\eta_2(W,B,\zeta<t) = \int_0^t\int_0^s 1_B(\pi_u W_u)du A(W,dt), \qquad B\in \mathcal{E}, \quad A(W,[0,t]) = 1(\tau(W_t)\le t)$$

to yield the random variables X_{τ_B}, $\mathcal{O}_{\tau_B}$, respectively on the canonical probability space of the historical process. This was carried out by Dynkin (1991b) and he showed that the joint Laplace functional satisfies (12.3.5.1), and (12.3.5.2).

12.3.6. Application to Sample Path Behavior.

Since the historical process H_t is an enriched version of the superprocess X_t, there are many potential applications of it. As an illustration the following theorem gives a more complete statement (in a certain sense) of the modulus of continuity result of Theorem 9.3.2.2 for the special case $\beta = 1$.

Theorem 12.3.6.1. Let $Q_{s,\mu}^{H(\Delta/2,1)}$ be the law of the (2,d,1)-historical process. If $\lambda > 2$, then $Q_{s,\mu}^{H(\Delta/2,1)}$-a.s. there is a $\delta(\lambda,\omega)$ such that

$$S(H_t) \subset K(\delta(\lambda),\lambda h_0) \quad \forall\ t > 0$$

where $h_0(u) := [u(\log 1/u)\vee 1]^{1/2}$ and

$K(\delta,h) := \{y\in C([0,\infty),\mathbb{R}^d), |y(t)-y(s)| \le h(t-s)\ \forall\ (s,t),\ |t-s|\le\delta\}$.

Proof. See [DP, Theorem 8.7].

12.4. The Embedding of Branching Particle Systems in the Historical Process

Let H_t denote the H(A,c)-historical process. *We will assume throughout this section that the A-path process satisfies Property S (cf. 12.2.2.6) and that c=2, that is,* $\Phi(\lambda) = -\lambda^2$.

The purpose of this section is to obtain a probabilistic description of H_t for fixed t. We know that H_t consists of a Poisson number of clusters beginning at time zero which are non-extinct at time t. These clusters can be viewed as measure-valued excusions with lifetime greater than t and form a subpopulation with a "common ancestor" at time 0. In any case it suffices to obtain the probabilistic description of the cluster and three complementary descriptions will be derived. The first is an infinite genealogical tree (Theorem 12.4.4), the second is given by an embedded time-inhomogeneous branching particle system which "explodes" at time t, and the third is in terms of an initial atom having a random mass which splits into smaller and smaller atoms (with conservation of mass) as time approaches t (Theorem 12.4.6).

We can obtain a Cox cluster representation of H_t exactly as we did for X_t in Corollary 11.5.3. In particular, conditioned on H_s, H_t can be represented as H_t =

$\sum_j H_{s,t,y_j}$, where $\{y_j\} \subset D_E^s$ are the points of a Poisson random measure with intensity $H(ds)/(t-s)$ and $\{H_{s,t,y_j}\} \subset M_F(D_E^t)$ denote the clusters of age $(t-s)$ associated with the points $\{y_j\}$. The cluster H_{s,t,y_j} can be identified with the subpopulation having a common ancestor with path y_j up to time s. It follows from (4.5.2) (with $\beta=1$) that the masses of the clusters $\langle H_{s,t,y_j},1\rangle$ are independent exponential random variables with mean $(t-s)$. The law of $H_{s,t,y}$, that is, the corresponding *cluster law* is denoted by $\{\Lambda^*_{s,t;y}\}_{0\le s<t;y\in D_E^s}$. For each s,t, and y, $\Lambda^*_{s,t;y}$ is a probability measure on $M_F(D_E^t)$ which describes the distribution of paths up to time t of a subpopulation of individuals alive at time t with a common ancestor at time s whose path up to time s is given by y.

The main objective of this section is to give a probabilistic representation of the *cluster conditioned on its mass*, that is,

$$\Lambda^*_{s,t}(m,y^s;d\nu) := \Lambda^*_{s,t;y}(.\,|\langle H_{s,t,y},1\rangle = m)$$

(this will be done in Theorem 12.4.6).

Most of the results of this section are from [DP, Section 3] and some will be stated without proof. The main objective is to establish the existence of a strong embedding of a hierarchy of historical branching particle systems $\{H_t^{*\varepsilon}:\varepsilon>0\}$ (which were defined in Section 12.3.1) in the historical process H_t. We begin with a proof of the first step in this development.

Let $\mathcal{G}_t^{*\varepsilon} = \sigma\{1(0,\infty)(H_t(A)):A\in\mathcal{D}_{t-\varepsilon}\}$ and let

$$\tilde{H}_t^\varepsilon := Q_{s,\mu}^{H(A,2)}(H_t|\mathcal{G}_t^{*\varepsilon})/\varepsilon \tag{12.4.1}$$

(normalized version of conditional expectation).

Given a measure μ on D^t, let $r_{t-\varepsilon}\mu(A) := \mu(\{w:w^{t-\varepsilon}\in A\})$.

Recall that ε is the expected mass of a cluster of age ε. Therefore $r_{t-\varepsilon}\tilde{H}_t^\varepsilon$ yields a random counting measure on $\mathcal{D}_{t-\varepsilon}$.

<u>Theorem 12.4.1.</u> Assume that the path process satisfies property S, $H_s = \mu \in M_F(D_E)^s$ and that $H_s^{*\varepsilon}$ is a Poisson random measure with intensity μ/ε. Then for $t \ge s+\varepsilon$, both $H_{t-\varepsilon}^{*\varepsilon}$ and $r_{t-\varepsilon}\tilde{H}_t^\varepsilon$ have a Poisson cluster representation of the form

$$L_{\Lambda,\{P_x\}}(\phi) = \exp\left(-\int_{E_1}(1-P_x e^{-\langle .,\phi\rangle})\Lambda(dx)\right), \quad \phi\in bp\mathcal{E}_2 \tag{12.4.2}$$

where $E_1 = D^s$, $E_2 = D^{t-\varepsilon}$, with the same intensity, μ/ε, and with the same cluster measures given by

(12.4.3) $$P_y(A) = Q^{*\varepsilon}_{s,\delta_y}(H^{*\varepsilon}_{t-\varepsilon} \in A), \quad A \in \mathcal{B}(M_F(D^{t-\varepsilon})), \quad y \in D^s.$$

Proof. By the analogue of Corollary 11.5.3 we have for $\phi \in bp\mathcal{D}_{t-\varepsilon}$,

$$Q_{s,\mu}(e^{-\langle \tilde{H}^{\varepsilon}_t,\phi\rangle}) = Q_{s,\mu}\left(\exp(-\langle H_{h-\varepsilon}/\varepsilon,(1-e^{-\phi})\rangle)\right)$$

$$= \exp\left(-\int V_{s,t-\varepsilon}((1-e^{-\phi})/\varepsilon)(y)\mu(dy)\right).$$

We now carry out a formal calculation using the differential equation form of the log-Laplace equation

$$-\frac{\partial V_{s,t-\varepsilon}}{\partial s} = \tilde{A}_s V_{s,t-\varepsilon} - (V_{s,t-\varepsilon})^2, \quad V_{t-\varepsilon,t-\varepsilon} = (1-e^{-\phi})/\varepsilon.$$

If $v_{s,t-\varepsilon} := 1-\varepsilon V_{s,t-\varepsilon}$, $v_{t-\varepsilon,t-\varepsilon} = e^{-\phi}$, then

$$-\frac{\partial v_{s,t-\varepsilon}}{\partial s} = \tilde{A}_s v_{s,t-\varepsilon} + \varepsilon\,((1-v_{s,t-\varepsilon})/\varepsilon))^2$$

$$= (\tilde{A}_s - 2\varepsilon^{-1})v_{s,t-\varepsilon} + 2\varepsilon^{-1}\mathcal{G}(v_{s,t-\varepsilon}).$$

This calculation can be fully justified in its evolution form and yields

(12.4.4) $$Q^{*}_{s,\mu}(e^{-\langle \tilde{H}^{\varepsilon}_t,\phi\rangle}) = \exp\left\{-(\varepsilon)^{-1}\int (1-v_{s,t-\varepsilon})(y)\mu(dy)\right\}.$$

On the other hand by (12.2.4.5)

(12.4.5) $$\int Q^{*\varepsilon}_{s,\mu}\left(e^{-\langle H^{*\varepsilon}_{t-\varepsilon},\phi\rangle}\right)\mathcal{P}ois_{\mu/c\varepsilon}(d\mu)$$

$$= \exp\left(-(\varepsilon)^{-1}\int (1-(G_{s,t-\varepsilon}e^{-\phi})(y))\mu(dy)\right)$$

and $G_{u,t-\varepsilon}e^{-\phi}$ satisfies (12.2.4.3) with $\xi = e^{-\phi}$. Hence

$$(G_{u,t-\varepsilon}e^{-\phi})(y) = v_{u,t-\varepsilon}(y), \quad s \le u \le t-\varepsilon, \text{ by uniqueness.}$$

It follows from (12.4.4) and (12.4.5) that both $r_{t-\varepsilon}\tilde{H}^{\varepsilon}_t$ and $H^{*\varepsilon}_{t-\varepsilon}$ are Poisson cluster random measures with intensity μ/ε and cluster measures $Q^{*\varepsilon}_{s,\delta_y}(H^{*\varepsilon}_{t-\varepsilon} \in .)$.

□

Corollary 12.4.2. (a) For $\varepsilon>0$, the random measure $r_{t-\varepsilon}H_t$ is pure atomic $Q_{s,\mu}$-a.s. with $s \le t-\varepsilon$.

(b) Conditioned on $r_{t-\varepsilon}\tilde{H}^{\varepsilon}_t$, H_t is the sum of independent non-zero clusters with law $\Lambda^{*}_{t-\varepsilon,t;y}$ (cf. Theorem 12.4.6 below) one for each atom of $r_{t-\varepsilon}\tilde{H}^{\varepsilon}_t$, that is,

$$Q^{H(A,2)}_{s,\mu}(e^{-\langle H_t,\phi\rangle}|r_{t-\varepsilon}\tilde{H}^{\varepsilon}_t)$$

$$= \exp\left\{\int \log \Lambda^{*}_{t-\varepsilon,t;y}(e^{-\langle .,\phi\rangle})r_{t-\varepsilon}\tilde{H}^{\varepsilon}_t(dy)\right\}, \quad \phi\in p\mathcal{D}.$$

Proof. (a) This follows from the fact that $r_{t-\varepsilon}H_t(A) = 0$, a.s. if $r_{t-\varepsilon}\tilde{H}^{\varepsilon}_t(A) = 0$.

(b) We apply Theorem 3.4.1.2 to the Cox cluster random measure (cf. 3.3.4) H_t with $I = H_{t-\varepsilon}/\varepsilon$, $E_1=E_2=D_E$, and $X(A\times B) := H_t(\{w:(w^{t-\varepsilon},w)\in A\times B\})$. In the terminology of Theorem 3.4.1.2 we have $\mathcal{G}_0 = \mathcal{G}^{*\varepsilon}_t$ and $\mathcal{G}_1 = \sigma(H_{t-\varepsilon})$.

Corollary 11.5.3 implies that H_t is a Cox cluster random measure with cluster law $P_y = \Lambda^{*}_{t-\varepsilon,t;y}$ and intensity $I = H_{t-\varepsilon}/\varepsilon$ is a.s. nonatomic. Therefore we may use Theorem 3.4.1.2 to conclude that

$$Q_{s,m}(r_{t-\varepsilon}H_t(A)|\sigma(H_{t-\varepsilon})\vee\mathcal{G}^{*\varepsilon}_t)/\varepsilon$$

$$= \int 1_A(y)\int \mu(D_E)\Lambda^{*}_{t-\varepsilon,t;y}(d\mu)\tilde{X}(dy)/\varepsilon$$

$$= \tilde{X}(A).$$

Since $\tilde{X}$ is $\mathcal{G}^{*\varepsilon}_t$-measurable (see the expression for $\tilde{X}$ in Theorem 3.4.1.2) this shows that $r_{t-\varepsilon}\tilde{H}^{\varepsilon}_t(A) = \tilde{X}(A)$ a.s. Then Theorem 3.4.1.2 and the Markov property of H give the first part of (b). The expression for the conditional Laplace functional of H_t now follows from Theorem 3.4.1.2 and the equality $r_{t-\varepsilon}\tilde{H}^{\varepsilon}_t = \tilde{X}$ a.s. □

Corollary 12.4.2 would suggest that this analogy is also valid in the sense of processes. This is established in the next theorem whose proof is too long to be included here.

Theorem 12.4.3 Let $H_0 = \mu\in M_F(D_E)^0$.

(a) (*Ancestral process of age ε.*) $r_0\tilde{H}^{\varepsilon}_{\varepsilon}$ is a Poisson random measure with intensity $\mu/c\varepsilon$ and the process $\{r_t\tilde{H}^{\varepsilon}_{t+\varepsilon}:t\ge 0\}$ is a branching particle system with transition function given by (12.2.4.3), (12.2.4.4).

(b) (*Probabilistic representation of* H_t.)

For $0\le s\le t$, let

(12.4.6) $$r_s\tilde{H}^{t-s,x}_t = r_s\Lambda^{*}_{0,t;x}(H_t|\mathcal{G}^{*t-s}_t)/(t-s)$$

(i.e. a non-zero cluster of $r_s\tilde{H}^{t-s}_t$ starting at $x\in\mathbb{R}^d$).

Then the process $s \longrightarrow r_s\tilde{H}^{t-s,x}_t$, $0\le s\le t$, has the same law as the time inhomogeneous binary branching historical particle system $\{H^{*\nu}_{s,t}:0\le s\le t\}$ with initial measure $H^{*\nu}_{0,t} = \delta_x$, $\nu(ds) = 1/(t-s)$.

Proof. [DP, Theorem 3.9].

We will next formulate the probabilistic description of the historical cluster first (Theorem 12.4.4) in the language of genealogical trees which are associated with the binary branching A-motions via Proposition 12.2.3.1 and then under condition S in the language of the splitting atomic measures (Theorem 12.4.5).

In the former case however we will now consider the infinite binary tree $K := \bigcup_{n=0}^{\infty} \{1,2\}^n$ and $K_\infty := \{1,2\}^{\mathbb{N}\setminus\{0\}}$. $K^* := (E^*)^K$ and $K^*_\infty := (E^*)^{K_\infty}$ where E^* is as in section 12.2.1. For $k_\infty \in K_\infty$ and $n \in \mathbb{N}$, let $[k_\infty]_n := k_1k_2\ldots k_n$. For $k \in K$, we say that $k < k_\infty$ if $k = [k_\infty]_n$ for some n. A typical element of K^* is denoted by $(\tau_k, W_k)_{k\in K}$.

Theorem 12.4.4 *The Infinite Genealogical Tree*

There exists a unique measurable collection $\{\Lambda^a_{t,y}: a > 0,\ t\ge 0,\ y \in D^t_E\}$ of probability measures on K^* such that the following properties hold:

(a) (i) Under $\Lambda^a_{t,y}$, τ_∂ is uniformly distributed over $[t,t+a]$ and conditioned on τ_∂, W_∂ is distributed as an A-path process starting from (t,y) stopped at time $t+\tau_\partial$.

(ii) Under $\Lambda^a_{t,y}$, conditioned on $(\tau_\partial, W_\partial)$ the translated trees (T_1w, T_2w) are independent and follow the law $\Lambda^{a-\tau_\partial}_{t+\tau_\partial, W_\partial(t+\tau_\partial)}$.

(b) (i) $\lim_{n\to\infty} \pi_{t_{[k_\infty]_n}} W_{k_\infty}$ exists for every $k_\infty \in K_\infty$, $\Lambda^a_{t,y}$-a.s.

(ii) With $\Lambda^a_{t,y}$-probability one there exists a unique measure ϑ on K_∞ such that for every $h \in K$

$$\vartheta^a_t(\{k_\infty : h < k_\infty\}) = \lim_{\varepsilon\to 0} 2\varepsilon \Big(\sum_{\substack{k\in K\\ h<k}} 1(s_k \le a-\varepsilon < t_k) \Big). \tag{12.4.7}$$

Proof. Theorem 12.4.4 is proved in Le Gall (1991a, Proposition 8.1 and Theorem 8.2) and is essentially equivalent to the following.

Theorem 12.4.5. *The Branching Particle Picture*

Assume that A satisfies Property S; let c=2, and 0<s<t. Then

(a) Under the law $\Lambda^*_{0,t;x}$, $r_s\tilde{H}^{t-s}_t$ consists of $N^B_t(s)$ atoms of masses $M_1(s),\ldots,M_{N^B_t(s)}(s)$.

(i) $\{N^B_t(s): 0\le s\le t\}$ is a time inhomogeneous Markov process with $N^B_t(0) = 1$, and

$$P(N^B_t(r_2t)=k \mid N^B_t(r_1t)=j) = \binom{k-1}{k-j} \left(\frac{1-r_2}{1-r_1}\right)^j \left(\frac{r_2-r_1}{1-r_1}\right)^{k-j}, \tag{12.4.8}$$

$1\le j\le k,\ k\in\mathbb{Z}_+,\ 0\le r_1<r_2<1,$

$P(N_t^B(0)=k) = e^{-<m,1>/t}(<m,1>/t)^k/k!, \quad k\in\mathbb{Z}_+.$

(ii) Conditioned on $N_t^B(s) = k$, $M_1(s),\dots,M_k(s)$ are i.i.d. exponential random variables with mean $(t-s)$.

(iii) Given $N_t^B(s) = k$, let τ_i $i=1,\dots,k$ denote the splitting times of the particles. Then $\{\tau_i\}$ are i.i.d. and uniformly distributed on $[s,t]$.

(b) We have $\Lambda^*_{0,t,y}$-a.s.

(12.4.9) $\quad \varepsilon\cdot\tilde{H}_t^\varepsilon \longrightarrow H_t$ as $\varepsilon\to 0$ (weak convergence of measures).

<u>Proof.</u> (a) (i) The total mass of the cluster of age t, $M = M_1$, is exponentially distributed with mean t (cf. (4.5.2) with $\beta=1$). From Theorem 12.4.3(b) $N_t^B(s)$, $0\le s\le t$, is equal in law to $\langle H^{*\nu}_{s,t},1\rangle$, $0\le s<t$, and the latter is a time inhomogeneous pure birth process with birth rate at time u equal to $\nu(u)=1/(t-u)$, $s\le u<t$. Then by Theorem 12.2.4.1 the generating function $G(s,t;\theta) = E(\theta^{N_t^B(t-\varepsilon)}\,|\,N_t^B(s)=1)$, $s < t-\varepsilon$, satisfies the differential equation

$$\frac{-\partial G(s,t-\varepsilon;\theta)}{\partial s} = \frac{G^2-G}{t-s}, \qquad G(t-\varepsilon,t-\varepsilon;\theta)=\theta.$$

which can be solved to obtain

$$(12.4.10) \qquad G(s,t-\varepsilon;\theta) = \frac{\varepsilon\theta/(t-s)}{1-\theta(t-s-\varepsilon)/(t-s)},$$

which is the generating function of a geometric random variable with parameter $\varepsilon/(t-s)$. Letting $\varepsilon=t-rt$, $s=0$, we obtain for $0<r<1$

$$P(N_t^B(rt)=k\,|\,N_t^B(0)=1) = r^{k-1}(1-r), \quad k\ge 1.$$

In addition for $1 > r_2 > r_1 > 0$, conditioned on $N_t^B(r_1t)$, $N_t^B(r_2t)$ is distributed as the sum of $N_t^B(r_1t)$ independent geometric random variables with parameter $[1-(r_2-r_1)/(1-r_1)]$, that is, (12.4.8) holds.

The fact that $N_t^B(0)$ is $\mathcal{P}ois(<m,1>/t)$ follows from Theorem 12.4.1.

(ii) $P(\tau_1 > r(t-s)\,|\,N_t^B(s)=1) = P(N_t^B(s+r(t-s))=1\,|\,N_t^B(s)=1) = 1-r$, and similarly

$P(\min_{i\le k}\tau_i > r(t-s)\,|\,N_t^B(s)=k) = (1-r)^k$, both by (12.4.8).

(b) is proved in [DP, Theorem 3.10].

<u>Remarks:</u>

(1) Under $\Lambda^*_{0,t,y}$, the time τ_∂ of the first branch is uniformly distributed ove

$[0,t]$. The life times τ_1 and τ_2 of its immediate descendents are, conditioned on τ_∂, independent and uniformly distributed over $[t-\tau_1,t]$, etc. The atom positions (marks) W_∂ are obtained by running A-path processes along the branches of the tree, starting from (t,y) for the ancestor.

(2) Under Property S we can recover the full tree structure (up to relabelling of the offspring as 1 and 2) at time t from the binary branching particle system $s \longrightarrow r_s\tilde{H}_t^{t-s,x}$, $0\le s\le t$ as in the remark at the end of section 12.2.3.

On the other hand as a consequence of Theorem 12.4.4(b) the measure H_t can be reconstructed as the limit of the (suitably normalized) counting measures on the branches of the tree at $(t-\varepsilon)$. In fact the cluster law $\Lambda^*_{t,t+a;y}(B)$, $B \in \mathcal{B}(M_F(D_E)^{t+a})$ can be represented by $\Lambda^a_{t,y}\left(\int d\vartheta^a_t(k_\infty)\delta_{W_{k_\infty}} \in B\right)$ where ϑ^a_t is defined as in (12.4.7). Thus the measure H_t and the complete genealogical tree at time t can be viewed as two different representations of the same information. Each has its own advantages. For example, the tree representation is more natural when A is a bounded operator. On the other hand the branching particle representation generalizes without difficulty to the general (A,Φ)-superprocess whereas the appropriate tree may no longer be a binary tree and then its description becomes much more complicated.

(3) The infinite genealogical tree associated with super-Brownian motion is an example of a *continuum random tree* with the *leaf-tight* as defined by Aldous (1991a,c, 1992).

Finally, we come to the main result of this section which is the determination of the probabilistic structure of a cluster of given age conditioned on its mass.

Theorem 12.4.6. *The Splitting Atom Process - Conditioned Cluster Law*

Under the assumptions of Theorem 12.4.5, the cluster law $\Lambda^*_{s,t;y}$ on $M_F(D_E)^t$ can be disintegrated as follows

$$(12.4.11) \qquad \Lambda^*_{s,t;y}(d\nu) = \iint \frac{1}{t-s}\, e^{-\eta/(t-s)}\Lambda^*_{s,t}(\eta,y^s;d\nu)d\eta$$

where

$$(12.4.12)\ \Lambda^*_{s,t}(m,y^s;d\nu) := \Lambda^*_{s,t;y}(.\,|\langle H_t,1\rangle = m) \qquad \textit{(conditioned cluster).}$$

$\Lambda^*_{s,t}(m,y^s;d\nu)$ is the law of a *branching atom process* $\sum_{j=1}^{N_t^B(s)} M_j(s)\delta_{y_j(s)}$, $y_j(s)\in D_E^s$, constructed as follows:

starting with one particle of mass m at time s further particles are produced by a process of subdivision. A particle of mass m_i at time $r<t$ divides with Poisson

rate $[m_i t/(t-r)^2]dr$ and on splitting produces two particles with masses m_i' and $m_i - m_i'$ and the ratio m_i'/m_i is uniformly distributed on [0,1]. Between divisions the atoms perform independent A-motions.

<u>Proof.</u> Assume that $N_t^B(0) = 1$ and let $M = M_1(0)$. We first prove that conditioned on $\{M=m\}$, $N_t^B(rt)-1$ (which represents the number of particle splits by time rt) is an inhomogeneous Poisson process with intensity $\frac{m}{t(1-r)^2}\,dr$. From Theorem 12.4.5 M is the sum of $N_t^B(rt)$ (geometric with mean $1/(1-r)$) masses which are independent $\mathcal{E}xp((1-r)t)$. From these facts we can compute

$$P(N_t^B(rt)=k\,|\,M=m)$$

$$= \frac{P(N_t^B(rt)=k)\cdot P(M=m\,|\,N_t^B(rt)=k)}{P(M=m)} = \left(\frac{rm}{(1-r)t}\right)^{k-1} \frac{e^{-mr/(1-r)t}}{(k-1)!}$$

$= \mathcal{P}ois(rm/(1-r)t)$. Similarly for $r_2 > r_1$,

$$P(N_t^B(r_2t)=k\,|\,N_t^B(r_1t)=j, M=m)$$

$$= \frac{(r_2-r_1)^{k-j}\,(m/t)^{k-j}}{[(1-r_2)(1-r_1)]^{k-j}(k-j)!}\exp\left(-\frac{m(r_2-r_1)}{t(1-r_1)(1-r_2)}\right),$$

that is, the conditional distribution is $\mathcal{P}ois\left(\frac{(r_2-r_1)m}{(1-r_1)(1-r_2)t}\right)$.

By Theorem 12.4.5 the joint density of the time of splitting, τ, of a particle and the masses of the two resulting particles, M_1 and M_2, starting at time $r < t$ is

$$\frac{1}{t-r}\ \frac{2\ e^{-(m_1+m_2)/(t-s)}}{(t-s)^2} \qquad r \le s < t,\quad m_i > 0.$$

From this we easily obtain that the marginal density of M is $\mathcal{E}xp(t-r)$, and

$$P(M_1 \le \eta M) = \eta,\ \eta \in [0,1].$$

Refer to [DP, Theorem 3.11] for a more detailed proof. □

<u>Corollary 12.4.7.</u> Under the law $\Lambda^*_{s,t}(m, y^s; d\nu)$, that is, conditioned that the cluster has mass m, $\{N_t^B(s): s \le r < t\}$ is an inhomogeneous Poisson process with $N_t^B(s)=1$ and rate $mt/(t-r)^2$.

12.5. The Fleming-Viot-Genealogical Process

Genealogical processes have been extensively studied in recent years in the literature on mathematical population genetics (e.g. Kingman (1982), Tavaré (1984), Donnelly and Tavaré (1986, 1987), Donnelly (1991), Donnelly and Joyce 1992)). In this section we will develop the genealogical process by analogy with the historical process and also indicate how the genealogy of the Fleming-Viot process can be represented via an enriched version of the Donnelly-Kurtz infinite particle representation of section 11.3. Since much of the development follows along the same lines as for the historical process we will simply sketch the main ideas.

12.5.1. Martingale Problem Formulation.

Recall that (cf. Sect. 12.1) at the level of the Moran process we can introduce the empirical measures on histories as we did for branching particle systems. Carrying out the same limiting procedure as in section 2.7 we obtain a probability measure-valued process called the (A,γ)-*genealogical process* which is denoted by

$$G = (\tilde{\Omega},\tilde{\mathcal{G}},\tilde{\mathcal{G}}_{s,t},(G_t)_{t\geq 0},\{P^{G(A,\gamma)}_{s,\mu}\}_{s\geq 0,\mu\in M_1(D_E)^s})$$

where $\tilde{\Omega} := D([0,\infty),M_1(D_E))$, $G_t(\tilde{\omega}) = \tilde{\omega}(t)$, $\mathcal{G}_{s,t} = \bigcap_{u>t}\sigma(G_r : s\leq r\leq u)$.

Furthermore, $\{P^{G(A,\gamma)}_{s,\mu}\}_{s\geq 0,\mu\in M_1(D_E)^s}$ is characterized as the unique solution to the following martingale problem.

For each $\phi\in D_0(\tilde{A})$, and for $t\geq s$,

$$(12.5.1.1)\qquad Z_t(\phi) := \langle G_t,\phi(t)\rangle - \langle G_s,\phi(s)\rangle - \int_s^t \langle G_r,(\tilde{A}\phi)(r)\rangle dr$$

is a $Q^{G(A,\gamma)}_{s,G_s}$-martingale with increasing process

$$(12.5.1.2)\qquad \langle Z(\phi)\rangle_t = \gamma\int_0^t\left[\int_{D_E}\phi(r,y)^2 G_r(dy)dr - \left(\int_{D_E}\phi(r,y)G_r(dy)\right)^2\right]dr.$$

The fact that this martingale problem uniquely characterizes the genealogical process can be verified by combining the remarks made in section 12.3.3 and the method of moment measures much as in the proof of Theorem 5.5.1 or by the construction of a time inhomogeneous dual process as in the proof of Theorem 5.7.1.

Remarks: The relationship between the $H(A,c)$-historical process and $G(A,\gamma)$-genealogical process is parallel to that between $B(A,c)$ and $S(A,\gamma)$, and the results of Theorems 8.1.1 and 8.1.2 can be obtained in an analogous manner. In other words heuristically the process G_t can be obtained as H_t/m conditioned on $\{\langle H_s,1\rangle \approx m,\ 0\le s\le t\}$. In fact the arguments leading to the proof of Theorem 8.1.2 can be adapted to establish this.

12.5.2. The Particle Representation of the Fixed Time Genealogy.

The purpose of this section is to obtain a probabilistic description at a fixed time t of the $G(A,\gamma)$-genealogical process, $\{G_t\}$. *Throughout this section we assume that A satisfies property* S *and* $\gamma=2$.

In fact we will obtain a description of the evolution of the genealogy of the population starting from the time at which the entire population had a common ancestor. Two such descriptions are derived. The first (Theorem 12.5.2.4) is a splitting atom model analogous to that derived for the branching cluster in Theorem 12.4.6. The second (Theorem 12.5.2.5) is in terms of a hierarchy of Polya urns which in turn is closely related to the infinite genealogical tree.

The de Finetti representation is the key to a probabilistic description of G_t for fixed t in much the same way as the cluster representation was the key ingredient in obtaining the probabilistic representation of H_t. In this section we describe the infinite particle representation of G_t following Donnelly and Kurtz (1992).

G_t can be viewed as the Fleming-Viot process in which the mutation process is given by the A-path process. Next consider the associated $D_E^{\mathbb{N}}$-valued look-down process $\{Y_k(t): t\ge 0,\ k\in\mathbb{N}\}$ constructed exactly as in Section 11.3 except that the A-motion in E is replaced by the A-path process in D_E. The N-particle genealogical look-down process has the following MP-generator: for $\phi\in D_0(\tilde{A}^N)$

(12.5.2.1)
$$(\tilde{K}_N\phi)(s,y_1,\dots,y_N) = (\tilde{A}^N\phi)(s,y_1,\dots,y_N) + 2\sum_{i<j}[(\Theta_{ij}\phi)(s,y_1,\dots,y_N)-\phi(s,y_1,\dots,y_N)]$$

where $\tilde{A}^N$ is the N-particle path process generator and for $y_1,\dots,y_N\in D_E$,

$$(\Theta_{ij}\phi)(s,y_1,\dots,y_N) = \phi(s,y_1,\dots,y_i,\dots,y_{j-1},y_i,y_{j+1},\dots,y_N).$$

In other words at the time of a jump the particle assumes the entire past history of the particle onto which it jumps.

We can also give the following alternative description of the genealogical look-down process. At each fixed time t we consider the $(D_E)^{\mathbb{N}}$-valued random variable obtained by assigning to each $j=1,\dots,N$ a path $W_j(t;s)$ defined by

$$W_j(t;s) = W_j(t;t) \text{ for } s\ge t,$$

$$W_j(t;s) = W_{n(j,s,t)}(t;s) \quad \text{for } 0\le s\le t, \text{ where}$$

$n(j,s,t) := \min\{i\le j: j \text{ is a descendent of } i \text{ in } (s,t]\}$

where j is recursively defined to be a descendent of i in (s,t] if j jumped down to i in (s,t] or if j jumped down to k in (r,t], $s<r\le t$ and k is a descendent of i in (s,r]. Under condition S, two particle paths, $W_i(t;.)$ and $W_j(t;.)$, share an identical path *only* on the interval $[0,s_{ij}]\subset[0,t]$ where s_{ij} is the last time that i and j shared a *common ancestor*.

The collection $\{\{W_j(t;.):j=1,\ldots,m,\ t\ge 0\}:m\in\mathbb{N}\}$ yields a consistent family of D_E-valued Markov processes. Furthermore under an exchangeable initial distribution $\{W_j(t;0):j\in\mathbb{N}\}$, for fixed t the D_E-valued random variables $\{W_j(t;.):j\in\mathbb{N}\}$ are exchangeable. Thus we can consider the resulting empirical measure on D_E given by

$$G_m(t,.) := \frac{1}{m}\sum_{j=1}^{m}\delta_{W_j(t;.)}$$

and following the argument at the beginning of the proof of Theorem 11.3.3 it can be verified that $\{G_m(t):t\ge 0\} \Rightarrow \{G_t:t\ge 0\}$ as $m\to\infty$ where $\{G_t\}$ denotes the (A,2)-genealogical process.

We will next discuss the probabilistic structure of the $\{W_j(t,.):j\in\mathbb{N}\}$. Given n distinct particles $\{W_1(t;.),\ldots,W_n(t;.)\}$ at time t, for $0\le s\le t$, let $N^S_{t,n}(s)$ denote the number of distinct individuals (ancestors) at time s that have descendents among $\{W_1(t;.),\ldots,W_n(t;.)\}$ and let $\Gamma_n(s,t)$ denote the collection of indices of these $N^S_{t,n}(s)$ particles. Since $N^S_{t,n}(s)$ is monotone increasing in s, we can associate with $\{W_j(t;.)\}$ a binary branching A-system in a natural way (cf. remark at the end of section 12.2.3). For u>0 let $R^n(u)$ denote the equivalence relation on $\{1,\ldots,n\}$ where i and j are in the same equivalence class if and only if they have the same ancestor at time t-u. Let $\mathfrak{E}^n$ denote the set of equivalence classes on $\{1,\ldots,n\}$. Let $D^S_{t,n}(u) := N^S_{t,n}((t-u)-)$, $0\le u\le t$, the number of equivalence classes in $R^n(u)$ (and for convenience we take right continuous versions of all processes).

Theorem 12.5.2.1.

(a) The $\mathbb{N}$-valued process $\{D^S_{t,n}(u):u\ge 0\}$ is a pure death process with death rates $d_k= k(k-1)$ and $N^S_{t,n}(t-0) = n$.

(b) Let $N^S_t(s) := \lim_{n\to\infty} N^S_{t,n}(s)$ denote the number of distinct ancestors at time s of the infinite set of particles $\{W_1(t;.),W_2(t;.),\ldots\}$. Then for $s<t$, $N^S_t(s) < \infty$, a.s.

(c) Let $D^S_t(u) := N^S_t((t-u)-)$. Then $\{D^S_t(u):u>0\}$ is a Markov death process started from an entrance boundary at ∞ with death rates $d_k= k(k-1)$.

(d) The process $\{R^n(.)\}$ coincides with Kingman's n-coalescent, that is, it is an

$\mathcal{E}^n$-valued continuous time Markov chain with transition rates:

$q_{\xi\eta}$ = 1 if η is obtained by coalescing two of the equivalence classes of ξ

= 0 otherwise.

Proof. (Also see Dawson and Hochberg (1982, Lemma 6.5)), Donnelly (1991), Donnelly and Joyce (1992).)

(a) The times between jumps in the n-particle look-down process are i.i.d. exponential random variables with mean $1/(n(n-1))$. Therefore the time since the last look-down is exponential with mean $1/(n(n-1))$. To obtain the second-to-last look-down time we then consider the resulting (n-1)-particle look-down system and the distribution of its last look-down is exponential with mean $1/((n-1)(n-2))$. Continuing in this way we get that the time between the (k-1)st last and kth last look-down is an exponential random variable with mean $1/((n-k+1)(n-k))$. Since the times between these look-downs are also independent we conclude that $\{D^S_{t,n}(s):s\geq 0\}$ is a pure death process with death rates $d_k = k(k-1)$.

(b) Let $\tau_{n,k} := \inf\{s:D^S_{t,n}(s) = k\}$. Then $\tau_{n,k} = E_1/(n(n-1)) + \tau_{n-1,k}$ where E_1 and $\tau_{n-1,k}$ are independent random variables and E_1 has an exponential distribution with mean 1. From this we obtain the representation $\tau_{n,k} = E_1/(n(n-1)) + \ldots + E_{n-k}/((k+1)k)$ where $\{E_m\}$ are i.i.d. $\mathcal{E}xp(1)$ random variables. Since $\sum_{j=k}^{\infty} 1/((j+1)j) < \infty$, we conclude that $\lim_{n\to\infty} \tau_{n,k} < \infty$ a.s. Consequently $N^S_t(s) < \infty$ a.s. if $s < t$.

(c) This follows from (a), the consistency of the processes $\{N^S_{t,n}(.):n\in\mathbb{N}\}$ and the construction of N^S_t as the projective limit of the $\{N^S_{t,n}(.):n\in\mathbb{N}\}$.

(d) Transitions in $R^n(.)$ correspond to the coalescence of two equivalence classes and the rate is $k(k-1)$ when $D^S_{t,n}(u)=k$. In terms of the original look-down process it corresponds to a jump in a k-particle look-down process (where each of the k-particles corresponds to an equivalence class). Since in this case each pair (i,j) with $k\geq i>j\geq 1$ experiences a jump $i\to j$ (with the resulting coalescence of the ith and jth classes) after an independent $\mathcal{E}xp(1/2)$-distributed time. This implies that $R^n(.)$ is a Markov process with transition rates equal to $\{q_{\xi\eta}\}$. □

Now let $\Gamma(s,t)$ be the collection of indices of particles at time s<t that have a descendent at time t in the infinite look-down process. By Theorem 12.5.2.1(b) $\Gamma(s,t)$ is a.s. finite and is therefore associated with an equivalence relation on $\mathbb{N}$ having a finite number of equivalence classes which we denote by $R(t-s)$. In other words, $R(u)$ is the equivalence relation on $\mathbb{N}$ in which i and j belong to the same

equivalence class if and only if they have the same ancestor at time t-u.

Let $\mathfrak{E} \subset 2^{\mathbb{N}\times\mathbb{N}}$ denote the set of equivalence relations on $\mathbb{N}$ with the subspace topology when $2^{\mathbb{N}\times\mathbb{N}}$ is given the product topology. Then $\mathfrak{E}$ is a compact metrizable space (cf. Kingman (1982a)). A probability measure on $\mathfrak{E}$ is called *exchangeable* if it is invariant under transformations induced by permutations on $\mathbb{N}$.

From the limiting argument above we conclude that R(s) is a $\mathfrak{E}$-valued continuous time Markov chain called the *coalescent* which is characterized by the property that its restriction to $\{1,\dots,n\}$ is the n-coalescent described above. Having identified R(.) as a coalescent we can then obtain detailed information on the probabilistic structure of G_t by using the following results of Kingman.

Theorem 12.5.2.2. Let $\{R(u):u>0\}$ denote the coalescent process, D(u) denote the number of equivalence classes in R(u) and $\mathcal{R}_k := R(\tau_k)$ where $\tau_k := \inf\{u:D(u)=k\}$ (and hence $R(s) = \mathcal{R}_{D(s)}$).

Then

(a) $\{\mathcal{R}_k : k\in\mathbb{N}\}$ is a reverse time Markov chain with state space $\mathfrak{E}$ and transition probabilities

$$P(\mathcal{R}_{k-1}= \eta \mid \mathcal{R}_k=\xi) = 2/k(k-1)$$

for each of the $\binom{k}{2}$ equivalences η which can be obtained by coalescing two of the equivalence classes of ξ.

(b) The law $\{\mathcal{P}_k\}$ of $\{\mathcal{R}_k\}$ is for each k an exchangeable probability measure on $\mathfrak{E}$, such that

$$\mathcal{P}^k = \int_{\Sigma_{k-1}} P^{\mathbf{x}}\lambda(d\mathbf{x})$$

where λ is the uniform distribution on Σ_{k-1} and for $\mathbf{x} = (x_1,..,x_k)$, $P^{\mathbf{x}}$ is the law of the exchangeable equivalence relation on $\mathbb{N}$ induced by i.i.d. random variables with values in $\{1,\dots,k\}$ and distribution $(x_1,\dots,x_k)$.

(c) $\{D(u):u\geq 0\}$ is independent of $\{\mathcal{R}_k : k\in\mathbb{N}\}$.

(d) If C_j is one of the equivalence classes in R(u), then

$$M_{C_j}(u) := \lim_{n\to\infty} \frac{1}{n}\sum_{k=1}^{n} 1_{C_j}(k) > 0, \quad \text{exists a.s.}$$

(e) Let $C_1(u),\dots,C_{D(u)}$ denote the equivalence classes of R(u). Then for $j_1\neq j_2\neq\dots\neq j_m$,

$$P(j_k\in C_{i_k}(u),k=1,\dots,m \mid D(u), M_{C_1(u)},\dots,M_{C_{D(u)}}(u)) = \prod_{k=1}^{m} M_{C_{i_k}}(u).$$

Proof. See Kingman (1982a). Note that the proofs of (d) and (e) involve an analogue of de Finetti's theorem for exchangeable equivalence classes (cf. Kingman (1982a, Theorem 2)).□

It is also of interest to regard the genealogical development in "forward time". Note that $t-\tau_1$ = sup {s: all particles at time t have a common ancestor at time s} (in the original direction of time). Define $\bar{D}(s) = D((\tau_1-s)-)$, $\bar{R}(s) = R((\tau_1-s)-)$. Then $\bar{D}$ is a pure birth process with $\bar{D}(0) = 2$ and birth rates $k(k-1)$ and a.s. finite explosion time $\hat{\tau}_\infty := \lim_{k\to\infty} \hat{\tau}_k$ where $\hat{\tau}_k := \inf\{s:\bar{D}(s)=k\}$. We denote by $\bar{D}^t$, $\bar{R}^t$ the corresponding processes conditioned on $\{\hat{\tau}_\infty=t\}$.

Starting with the conditioned pure birth process $\bar{D}^t$ there are two alternate routes to the construction of the branching atom representation of G_t. The first of these is based on properties of the coalescent contained in the following result of Kingman.

Corollary 12.5.2.3. (a) At the time of a split, one equivalence class of mass M_C split into two equivalence classes of masses $M_{C'} = UM_C$ and $M_{C''}=(1-U)M_C$ where U is an independent uniform (0,1) random variable.

(b) For each C, the probability that C splits is given by M_C.

Proof. Refer to Kingman (1982a, Section 5).

The application of Corollary 12.5.2.3 to G_t yields the following result.

Theorem 12.5.2.4. *Splitting Atom Process*

Assume that A satisfies condition S and let r_s be defined as in section 12.4. Then for s>0, $\{r_sG_t:s<t\}$ is a pure atomic random measure of the form

(12.5.2.2) $$r_sG_t = \sum_{j=1}^{N_t^S(s)} M_j(s)\delta_{y_j(s)}$$

with $y_j(s) \in D_E^S$, and $(M_1(s),\ldots,M_{N_t^S(s)}(s)) \in \Sigma_{N_t^S(s)-1}$.

At the jump times of $N_t^S(s)$ one of the atoms splits into two smaller atoms. If a jump occurs at time s, then for $j=1,\ldots,N_t^S(s-)$ the probability that it is atom j that splits is given by $M_j(s-)$. The masses of the two atoms resulting from the split of an atom of mass M_j are uniformly distributed on $[0,M_j]$, that is, $M' = UM_j$ and $M'' =(1-U)M_j$ where U is a random variable uniformly distributed on [0,1] and independent of everything else. Between jumps the atoms perform independent A-motions.

Sketch of the Proof.

It follows from Theorem 12.5.2.1 and the construction of the process G_t from the look-down process that the number of atoms is non-decreasing. Property S of A guarantees that there exists a one-to-one correspondence between atoms in $r_s G_t$ and equivalence classes as described by the coalescent $\bar{R}^t(s)$. Furthermore by the construction of G_t using de Finetti's theorem (as in Theorem 11.3.3), the mass assigned to an equivalence class C_j, $j=1,\dots,\bar{D}^t(t-s)$ coincides with $M_j(s)$ as defined above. Hence the mechanism of the splitting of the masses is the same as in Corollary 12.5.2.3. Since the mutation process is independent of the sampling process, we can also conclude that the atom locations in E (which correspond to single particle locations in the look-down process) follow A-motions. □

We now verify that the branching atom process can also be obtained from the hierarchy of Polya urns introduced in Dawson and Hochberg (1982) and also use this viewpoint to obtain another characterization of $M_j(s)$.

Corollary 12.5.2.5. *Polya Urn Hierarchy Representation*

(a) The quantities $M_j(.)$ are also given by

$$(12.5.2.3)\qquad M_j(s) = \lim_{u\uparrow t} \frac{N_t^S(s,u,j)}{\sum_k N_t^S(s,u,k)}, \quad j=1,\dots,N_t^S(s)$$

where $N_t^S(s,u;j)$:= number of atoms at time u with common ancestor j at time s.

(b) At each time s, The random vector $(M_{C_1}(s),\dots,M_{C_{D(s)}}(s))$ is uniformly distributed over the simplex $\Sigma_{D(s)-1}$.

Proof. (a) For fixed s>0 we consider an urn model involving $N_t^S(s)$ types. At each jump time, u, of $N_t^S(.)$, s<u<t, a particle of type j is added with probability

$$(12.5.2.4)\qquad \hat{M}_j(u) := \frac{N_t^S(s,u;j)}{\sum_k N_t^S(s,u;k)}, \quad j=1,\dots,N_t^S(s),$$

that is, it is an $N_t^S(s)$-type Polya urn. Then it follows from the theory of Polya urns (cf. Blackwell and Kendall (1964) or Johnson and Kotz (1977)) (and the fact that we have conditioned on $\{\tau_\infty = t\}$) that

$$(12.5.2.5)\qquad \tilde{M}_j(s) := \lim_{u\to t} \frac{N_t^S(s,u;j)}{\sum_j N_t^S(s,u;j)} \text{ exists}$$

and that the vector $(\tilde{M}_1(s),\dots,\tilde{M}_{N_t^S(s)})$ is uniformly distributed over the simplex

$\Sigma_{N_t^S(s)-1}$ (cf. Blackwell and Kendall (1964)). Moreover the same result implies that at the time of a split, the mass $\tilde{M}_j(u)$ is divided exactly as in Corollary 12.5.2.3(a). Recall that if an m-type Polya urn is started with n_j initial particles of type j, then the joint distribution of the limiting proportions $(x_1,\dots,x_m)$ is an absolutely continuous distribution on Σ_{m-1} with Dirichlet density

$$f(x_1,\dots,x_m) = \frac{\Gamma(\Sigma n_i)}{\Pi\Gamma(n_i)} \Pi\, x_i^{n_i-1} \qquad \text{(cf. Athreya (1969)).}$$

Using this we can show that if a split occurs at time s, then

$$P(\text{jth class splits}\,|\,\tilde{M}_1(\tau_k-),\dots,\tilde{M}_{k-1}(\tau_k-)) = \tilde{M}_j(\tau_k-),\ j=1,\dots,k-1,$$

in other words, this is in agreement with Corollary 12.5.2.3b). Taking advantage of the fact that merging types in a Pólya urn yields a new Pólya urn, it suffices to to prove this for a two-type urn in which case we obtain

$$f_{m,n}(x,1-x) = \frac{\Gamma(m+n)}{\Gamma(m)\Gamma(n)}\, x^{m-1}(1-x)^{n-1},\ x\in(0,1).$$

Then $P(\text{1st class splits}\,|\,\tilde{M}_1=x,\tilde{M}_2=1-x)$

$$= \frac{\frac{m}{m+n} f_{m+1,n}(x,1-x)}{\frac{m}{m+n} f_{m+1,n}(x,1-x) + \frac{n}{m+n} f_{m,n+1}(x,1-x)} = x.$$

Finally it is clear from the above construction that the $\{\hat{M}_j(\tau_k-):\ j=1,\dots,k-1;\ k\in\mathbb{N}\}$ is independent of $\{\bar{D}^t(u):u>0\}$. This completes the proof that the Polya urn scheme and coalescent yield the same probabilistic mechanism, that is,

$$\mathcal{L}aw((\tilde{M}_1(s),\dots,\tilde{M}_{N_t^S(s)}):s<t) = \mathcal{L}aw((M_1(s),\dots,M_{N_t^S(s)}):s<t)$$

Then (12.5.2.3) follows since it is equivalent to (12.5.2.5) which has already been established.

(b) follows either from Theorem 12.5.2.2(b) or from the result of Blackwell and Kendall stated above. □

Note that Theorem 12.5.2.4 is an analogue of Theorem 12.4.5. On the other hand from the Polya urn systems of Corollary 12.5.2.5 together with $N_t^S(s)$ from Theorem 12.5.2.1 we can construct an *infinite Fleming-Viot genealogical tree* analogous to Theorem 12.4.4 (see also Aldous (1992, Section 4.1)).

The whole study of the genealogy of genetic models is a rapidly growing subject which is developing in a number of directions which we cannot describe here in detail. The following remarks will indicate just a few of these.

Remarks.

(1) The Polya urn hierarchy and atomic measures with Dirichlet distributions which arise in Corollary 12.5.2.5 are closely related to the family of Dirichlet processes introduced by Ferguson (1973) and studied by Blackwell and MacQueen (1973) and Feigen and Tweedie (1989).

(2) It is possible to extend $\{N_t^S(s):0\le s\le t\}$ to $\{N_t^S(s):s\ge 0\}$ to yield the genealogical stucture for an "infinitely old population" and then to consider for simplicity $\{N_0^S(s):s\ge 0\}$ as well as the corresponding infinite genealogy, $\tilde{G}_0$, which now becomes a random measure on $D((-\infty,0],E)$. If the A-process has a stationary distribution we denote by $W_1(t,.)$ to the corresponding stationary path process. Having done this it is easy to construct a stationary version of the full infinite system $\{W_j(t):t\in\mathbb{R},j\in\mathbb{N}\}$. If the underlying process W_1 is ergodic, so is the resulting system (cf. Donnelly and Kurtz (1992)).

(3) Note that we can also disintegrate the law of G_t with respect to $W_1(.)$ yielding the Palm distribution of the random probability measure G_t (cf. Corollary 11.2.3). Given the infinite particle representation $\{W_k(.)\}$ the Palm measure of G_t at $y\in \mathbf{D}_E^t$ is given by

(12.5.2.6) $\quad (P)_y = P(G_t\in .\,|W_1(t)=y).$

The representation of Corollary 12.5.2.5 and (12.5.2.4) was (implicitly) used in Dawson and Hochberg (1982) to obtain an upper bound on the carrying dimension of the $S(\Delta,\gamma)$-Fleming-Viot process in $\mathbb{R}^d$ and to verify the compact support property.

(4) A more complete development of the genealogical process in the context of Fleming-Viot processes can be found in Dawson and Hochberg (1982) and Donnelly and Kurtz (1992). For applications to population genetics refer to Kingman (1980, 1982a,b,c), Watterson (1984), Donnelly and Tavaré (1987), and Tavaré (1984, 1989).

(5) It turns out that many of the genealogical processes studied in the literature can be naturally embedded into the FV-genealogical process. Of particular relevance to mathematical population genetics is the genealogical process associated to the infinitely many alleles model in which the mutation operator A does not satisfy condition S but is bounded, as in Section 8.2. In this case the object of interest is the *number of lines of descent*, where a line of descent is defined to be the number of ancestral classes without intervening mutation (cf. Griffiths (1980), Tavaré (1984)). This is closely related to the general genetic model (cf. Cannings (1974)) which is given by a discrete time (generation) process in which there is a fixed population size N, fixed mutation probability and in which a mutation always gives rise to a new type. Let $A_N(m)$ denote the number of ancestral classes originating m

generations back into the past (where two individuals are said to be in the same such class if they have the same ancestor m generations back with no intervening mutation along their lines of descen)t. Let $M_{N,j}(m)$, $j=1,\ldots,A_N(m)$ denote the proportion of the population in each of these classes. Donnelly and Joyce (1992) have proved under suitable technical assumptions that as $N \to \infty$, the processes $(A_N([N\sigma^{-2}.]),M_N([N\sigma^{-2}.])) \Longrightarrow (D(.),M(.))$ as $N\to\infty$ (weak convergence on D_E) where $E := \{0,1,\ldots,\infty\}\times\bar{\mathfrak{A}}_\infty$ and σ^2 is the limiting variance of the offspring distribution of an individual. The limiting process $\{D(t),M(t):t\geq 0\}$ can be embedded in the infinite genealogy $\tilde{G}_0$ associated with the infinitely many alleles model. This unifying weak convergence result was the culmination of a series of papers beginning with Kingman (1982a). Here $\{D(t):t>0\}$ is the Markov death process started from an entrance boundary at ∞ with death rates $k(k+\theta-1)/2$ where θ is the mutation rate. If $\theta>0$, then $D(u)$ eventually reaches 0 (cf. Donnelly (1991), and Donnelly and Joyce (1992)). The use of Polya type urns for the study of these questions has been systematically developed in Hoppe (1987). We will now complete our study of the relation between the Fleming-Viot and branching systems with a comparison of the structures of the associated genealogical-historical processes.

12.6. Comparison of the Branching and Sampling Historical Structures.

It was established in Theorems 12.4.6 and 12.5.2.4 that the fixed time distributions of the B(A,c)-historical process $\{H_t\}$ and $S(A,\gamma)$-genealogical procees $\{G_t\}$ can be described in terms of splitting atom processes. We again assume throughout this section that A satisfies property S and $c=\gamma=2$. Then conditioned on $N_t^S(s)$, $N_t^B(s)$, these two splitting atom processes are identical. Thus the difference between the historical and genealogical processes lies in the processes describing the times of the branching, $N_t^S(s)$ and $N_t^B(s)$.

Moreover, in view of the sampling analogue of Theorem 8.1 we expect that the historical process conditioned on the given trajectory of the total mass process coincides with the genealogical process associated to a time inhomogeneous Fleming-Viot process with highly fluctuating sampling rate. However the latter may not have exactly the same genealogical structure as the Fleming-Viot process with *constant* sampling rate.

Let us begin with a closer look at the process $\{N_t^S(.)\}$. By Theorem 12.5.2.1 the sojourn times $\{\max\,\{s:N_t^S(s)=k\} - \min\,\{s:N_t^S(s)=k\} : k\in\mathbb{N}\setminus\{1\}\}$ can be represented by $\{E_k/(k(k-1)):\ k\in\mathbb{N}\setminus\{1\}\}$ where the $\{E_k\}$ are independent mean one exponential random variables and this yields a representation of $\{N_t^S\}$ on $(\mathbb{R}^{\mathbb{N}},\mathbb{P},\{E_k\}_{k\in\mathbb{N}})$ where $\mathbb{P}$ is an infinite product of mean one exponential probability measures. The remaining time after $N_t^S(.)$ first reaches n, $R_S(n)$, is then represented by $\sum_{k=n+1}^{\infty} E_k/(k(k-1))$.

Note that $\int_{k-1}^{k} x^{-2}dx = 1/(k(k-1))$ and therefore

$\sum_{k=n+1}^{\infty} 1/(k(k-1)) = 1/n.$

Then $\lim_{n\to\infty} E(nR_S(n)) = 1$, and

$$\lim_{n\to\infty} n\mathrm{Var}(nR_S(n)) = \lim_{n\to\infty} n^3\cdot \sum_{k=n+1}^{\infty} \mathrm{Var}(E_k)/(k\cdot(k-1))^2 = 3.$$

We therefore get $nR_S(n) \longrightarrow 1$ in probability as $n \to \infty$ and if $N_t^S(s) := \min\{n:R_S(n)\ge s\}$, then $(t-s)N_t^S(s) \longrightarrow 1$ in probability as $s\to 0$.

Consider the B(A,c)-branching process under the assumptions that A satisfies property S and c=2. Consider a single cluster of age t conditioned to have mass m, that is, we consider N_t^B under the law $\Lambda^*_{0,t}(m,y;.)$ defined by (12.4.12). By Corollary 12.4.7, $N_t^B(s)-1$ is an inhomogeneous Poisson process with intensity $\frac{mt}{(t-s)^2}$, that is, it has independent increments and $N_t^B(s)-1$ is Poisson with mean $sm/(t(t-s))$. Hence $\lim_{s\to t} (t-s)N_t^B(s) = m$ (in probability).

We will now compare the H(A,2)-historical process H_t with m=1 to the G(A,2)-genealogical process G_t, both at t=1. From the above we have $\lim_{s\to t} (t-s)N_t^B(s) = \lim_{s\to t} (t-s)N_t^S(s) = 1$.

For $0 \le u < \infty$ define $Z(u) = N_1^B(u/(1+u))$ (and then $N_1^B(s) = Z(s/(1-s))$). Then Z is a standard Poisson process and $\tilde{E}_n := \tilde{T}_B(n+1)-\tilde{T}_B(n)$ are i.i.d. exponential (1) r.v.'s. Then $\tilde{T}_B(n) = \sum_{i=1}^{n} \tilde{E}_i$, $T_B(n) := \inf\{u:N_1^B(u)\ge n\} = \tilde{T}_B(n)/(1+\tilde{T}_B(n))$ and therefore the remaining time is $R_B(n) := 1-T_B(n) = 1/(1+\tilde{T}_B(n))$. Then $\tilde{T}_B(n) = \frac{1}{R_B(n)} -1$. By the law of large numbers, $\lim_{n\to\infty} \tilde{T}_B(n)/n = nR_B(n) = 1$ a.s.

Now consider $\tilde{T}_S(n) = \frac{1}{R_S(n)} -1$. Then

(12.6.1)

$$F_{n+1} := \tilde{T}_S(n+1)-\tilde{T}_S(n) = \frac{1}{R_S(n+1)} - \frac{1}{R_S(n)}$$

$$= \left\{\left(\sum_{k=n+2}^{\infty} E_k/k(k-1)\right)^{-1} - \left(\sum_{k=n+1}^{\infty} E_k/k(k-1)\right)^{-1}\right\}$$

$$= \left\{E_{n+1}\cdot\left[(n+1)\sum_{k=n+2}^{\infty} E_k/k(k-1)\right]^{-1}\cdot\left[E_{n+1}/(n+1) + n\sum_{k=n+2}^{\infty} E_k/k(k-1)\right]^{-1}\right\}.$$

The main result of this section is obtained by an analysis of the two sequences $\{\tilde{E}_n\}$ and $\{F_n\}$. It concerns the marginal probability laws in $M_1(M_1(D([0,1],E)))$, given by

$$P_H := \Lambda^*_{0,t}(1,y;.)\circ(H_1^{-1}) \text{ and } P_G := P^{G(A,1)}_{0,\delta_y}\circ(G_1^{-1})\left(\ .\ \middle|\ N_1^S(0) = 1\right).$$

Theorem 12.6.1. P_H and P_G are mutually singular.

First we proceed with three auxiliary lemmas.

Lemma 12.6.2. $B_n := n\sum_{k=n+1}^{\infty} E_k/(k(k-1)) < \infty$, P_H-a.s, and

(a) For $0 < \nu < 1/2$

$$P(|B_n-1| > n^{-\nu}) \le 2e^{-3(n^{1-2\nu})/4} + c/n^{1+\nu};$$

(b) $|B_n - 1| \le \mathit{const}\ n^{-\nu}$ for all $n \ge n(\nu)$ P_H-a.s.

(c) Set $A_k^N := \sum_{n=k}^{N}(B_n-1)^2 = \sum_{n=k}^{N} n^2\left(\sum_{\ell=n+1}^{\infty}\frac{(E_\ell-1)^2}{\ell^2(\ell-1)^2}\right)$ where $2 \le k \le N$.

Then $A_2^N/(\frac{1}{3}\log N) \longrightarrow 1$ as $N \to \infty$, P_H-a.s. which implies that for any fixed integer $k \ge 2$, A_k^N diverges to $+\infty$ as $N \to \infty$, P_H-a.s.

Outline of Proof. First note that $P_H(B_n) = 1$ so that $B_n < \infty$, P_H-a.s.
(a) is then proved using a standard exponential Chebyshev estimates (see Lemma 5.1 in Dawson and Vinogradov (1992b) for more details).
(b) follows from (a) by a Borel-Cantelli argument.
(c) is straightforward and relies on a modification of Chebyshev's inequality, estimates for moments, a Borel-Cantelli argument and the verification of conditions of the two-series theorem. □

Lemma 12.6.3. For any integer $n\ge 2$, $\sigma\{F_k:k\ge n\} = \sigma\{E_k:k\ge n\}$.

Outline of Proof. (See Lemma 5.2 in Dawson and Vinogradov (1992b) for the details.) Obviously, $\sigma\{F_k:k\ge n\} \subset \sigma\{E_k:k\ge n\}$. The proof of the reverse inclusion is proved by induction on n (taken in descending order): from the definition of F_n (cf. (12.6.1)) we easily derive that for an arbitrary fixed integer $n\ge 2$

$$F_n = E_n B_n^{-1}\left(E_n/n + \frac{n-1}{n} B_n\right)^{-1}, \text{ and for any integer } 2\le k\le n-1,$$

$$F_n = E_n\left(k \sum_{\ell=k+1}^{n} E_\ell/(\ell(\ell-1)) + \frac{k}{n}B_n\right)^{-1}$$
$$\times\left(E_k/k + (k-1) \sum_{\ell=k+1}^{n} E_\ell/(\ell(\ell-1)) + \frac{k-1}{n} B_n\right)^{-1}.$$

Solving these equations (by induction) with respect to $E_n, E_{n-1},\dots,E_2$ we obtain that for $2\le k\le n$ E_k is measurable with respect to $\sigma(F_2,\dots,F_n,B_n)$. The rest of the proof involves the zero-one law for $\{E_n:n\ge 1\}$ and some properties of B_n established in Lemma 12.6.2. □

Lemma 12.6.4. Set

$$M_k^N := \sum_{m=k}^{N} (E_m-1)(B_m-1) = \sum_{m=k}^{N}(E_m-1)\left(m \sum_{\ell=m+1}^{\infty} \frac{(E_\ell-1)}{\ell(\ell-1)}\right), \; 2 \le k \le N,$$

and $M_{N+1}^N := 0$. Then $\{M_{N+1}^N, M_N^N,\dots,M_3^N,M_2^N\}$ is a square integrable martingale for any integer $N \ge 2$ with the increasing process A_n^N defined in Lemma 12.6.2.c (A_{N+1}^N := 0), and the ratio $M_2^N/A_2^N \to 0$ as $N \to \infty$, P_H-a.s.

Proof. The proof of this lemmas is straightforward and involves Lemma 12.6.2, Borel-Cantelli arguments and the derivation of exponential upper bounds for the probabilities of large deviations for the family of reverse martingales $\{M_n^N\}$ (similar to those obtained in Chapter 4, Section 13 of Liptser and Shiryayev (1989)). The details are given in Lemma 5.3 of Dawson and Vinogradov (1992b). □

Corollary 12.6.5. $\sum_{n=2}^{\infty}\left\{2(1-E_n)(B_n-1) - (1+E_n)(B_n-1)^2/2\right\}$ diverges to $-\infty$, P_H-a.s.

Note that this corollary easily follows from Lemmas 12.6.2 and 12.6.4.

Outline of Proof of Theorem 12.6.1. (See Section 6 of Dawson and Vinogradov (1992b) for a detailed proof.)

It clearly suffices to show that the laws of $\{N_1^B(s):0\le s<1\}$ and $\{N_1^S(s):0\le s<1\}$ are mutually singular. To demonstrate the latter it suffices to show that the laws of the infinite sequences $\{\tilde{E}_n\}$ and $\{F_n\}$ defined above are mutually singular. Consider the basic probability space $(\Omega,\mathcal{F},\mathbb{P})$ where $\Omega = \mathbb{R}^{\mathbb{N}}$ and $\mathbb{P}$ is the product of exponential(1) distributions and let $\{E_n\}$ be the canonical process on $(\Omega,\mathcal{F},\mathbb{P})$. Let $\mathbb{Q}$ denote the law of the $\{F_n\}$. Let $\mathcal{F}_{n,\infty} := \sigma\{E_k:k\geq n\}$ and $\mathcal{F}_\infty := \bigcap_{n\geq 2} \mathcal{F}_{n,\infty}$. Note that $\mathcal{F}_\infty$ is P_H-trivial. Then Lemma 12.6.3 implies that $\mathcal{F}_{n,\infty} = \sigma\{F_k:k\geq n\}$.

Now in order to prove that the two genealogical processes are mutually singular it suffices to prove that $\mathbb{P}$ and $\mathbb{Q}$ are mutually singular. (Actually since we wish to compare genealogies of clusters of age 1 starting with exactly one common ancestor we should actually consider $\mathbb{Q}(.|B_2>0)$ instead of $\mathbb{Q}$ but since $\mathbb{Q}(B_2>0) > 0$ this would not change in any way the following arguments.) Also since the sequences $\{E_n\}$ and $\{\tilde{E}_n\}$ both have law $\mathbb{P}$ we will supress the "$\sim$" in the sequel.

Consider the function $g(y;b) := \frac{ny}{b((n-1)b+y)}$. We then have that

$$F_n = g(E_n,B_n)$$

where $B_n = n\sum_{k=n+1}^{\infty} E_k/(k(k-1))$ and $\mathbb{P}[B_n] = 1$. Furthermore the conditional density of F_n given $\mathcal{F}_{n+1,\infty}$ is given by

$$\text{(12.6.2)}\quad f_{F_n}(y_n|\mathcal{F}_{n+1,\infty}) = \exp\left\{\frac{-(n-1)B_n^2y_n}{n-B_ny_n}\right\}\cdot\frac{n(n-1)\cdot B_n^2}{(n-B_ny_n)^2}, \quad \text{if } 0 \leq y_n < n/B_n$$

$$= 0 \text{ otherwise.}$$

Then the conditional density of the distribution of $(F_2,\ldots,F_n)$ with respect to the product measure $\prod_{j=2}^{n} f_{E_j}(y_j)dy_j$ is given by

(12.6.3)

$$R_n(y_2,\ldots,y_n|\mathcal{F}_{n+1,\infty}) := f_{F_2,\ldots,F_n}(y_2,\ldots,y_n|\mathcal{F}_{n+1,\infty}) \Big/ \prod_{j=2}^{n} f_{E_j}(y_j)$$

$$= \prod_{j=2}^{n} \exp\left\{\frac{-j(B_j^2-1)y_j+B_jy_j(B_j-y_j)}{j-B_jy_j}\right\}\cdot\left(\frac{j(j-1)B_j^2}{(j-B_jy_j)^2}\right)\cdot 1(0\leq y_j<j/B_j).$$

If $\mathbb{Q} \ll \mathbb{P}$ with $d\mathbb{Q}/d\mathbb{P} = R$, then using a conditional expectation argument and Lemma 12.6.3 it can be verified that

$$\text{(12.6.4)}\quad R(y_2,y_3,\ldots) = R_n(y_2,\ldots,y_n|\mathcal{F}_{n+1,\infty})\cdot\mathbb{P}(R|\mathcal{F}_{n+1,\infty}),\ \mathbb{P}\text{-a.e. } (y_2,y_3,\ldots).$$

Since $\mathbb{P}(R) = 1$ and $\mathcal{F}_\infty$ is $\mathbb{P}$-trivial, the martingale convergence theorem implies that $\mathbb{P}(R\,|\,\mathcal{F}_{n+1,\infty}) \longrightarrow 1$ as $n\to\infty$, $\mathbb{P}$-a.s. In turn this implies the $\mathbb{P}$-almost sure convergence of

$$(12.6.5)\quad \prod_{j=2}^{n} \exp\left\{ \frac{-j(B_j^2-1)y_j + B_jy_j(B_j-y_j)}{j-B_jy_j} \right\} \cdot \left(\frac{j(j-1)B_j^2}{(j-B_jy_j)^2} \right) \cdot 1(0\le y_j<j/B_j).$$

on $\{R>0\}$ as $n\to\infty$, and the following representation for R:

$$(12.6.6)\quad R = \prod_{j=2}^{\infty} \exp\left\{ \frac{-j(B_j^2-1)y_j + B_jy_j(B_j-y_j)}{j-B_jy_j} \right\} \cdot \left(\frac{j(j-1)B_j^2}{(j-B_jy_j)^2} \right) \cdot 1(0\le y_j<j/B_j).$$

Since $\mathcal{F}_\infty$ is $\mathbb{P}$-trivial, the usual singularity-equivalence dichotomy holds and $\mathbb{Q} \ll \mathbb{P}$ if and only if $\mathbb{P}(A\cap A_0) = 1$ where

(12.6.7)

$$A := \left\{ 0 < \prod_{j=2}^{\infty} \exp\left\{ \frac{-j(B_j^2-1)y_j + B_jy_j(B_j-y_j)}{j-B_jy_j} \right\} \cdot \left(\frac{j(j-1)B_j^2}{(j-B_jy_j)^2} \right) \cdot 1(0\le y_j<j/B_j) < \infty \right\}$$

and $A_0 := \prod_{n=2}^{\infty} 1(0\le y_n<n/B_n) > 0$ (cf. Shiryayev (1984), p. 496 for a similar criterion). Note that for any $k \ge 1$, $\sum_n \mathbb{P}\{E_n > n/k\} < \infty$ so that $\mathbb{P}(A_0) > 0$. Since $A\cap A_0 \in \mathcal{F}_\infty$, then $\mathbb{P}(A\cap A_0) = 0$ or 1.

Taking logs we wish to show that

$$A\cap A_0 = \left\{ -\infty < \sum_{n=2}^{\infty} \left(\frac{-n(B_n^2-1)y_n + B_ny_n(B_n-y_n)}{n-B_ny_n} + \log\left(\frac{n(n-1)B_n^2}{(n-nB_ny_n)^2} \right) \right) \cdot 1(0\le y_n<n/B_n) < \infty \right\}$$

has $\mathbb{P}$-measure zero.

Therefore in order to prove mutual singularity it suffices to verify the a.s. non-convergence of the series

$$\sum_{n=2}^{\infty} f_n(B_n,E_n)\; 1(0\le E_n<n/B_n)$$

$$= \sum_{n=2}^{\infty} \left(\frac{-n(B_n^2-1)y_n + B_n y_n (B_n - y_n)}{n - B_n y_n} + \log\left(\frac{n(n-1)B_n^2}{(n-nB_n y_n)^2}\right)\right) \cdot 1(0 \le y_n < n/B_n)$$

(recall that $B_n = n \sum_{k=n+1}^{\infty} E_k/(k(k-1)) \to 1$ as $n \to \infty$, $\mathbb{P}$-a.s.).

Using a Taylor expansion for f_n as a function in B_n around $B_n=1$, and some standard probabilistic arguments we can show that this is equivalent to verifying that

$$(12.6.8) \qquad \sum_{n=2}^{N} \left[2\cdot(1-E_n)(B_n-1) - (1+E_n)(B_n-1)^2/2\right]$$

does not converge $\mathbb{P}$-a.s. to a finite limit as $N \to \infty$. But this series in fact diverges to $-\infty$, P_H-a.s. (due to Corollary 12.6.5). □

<u>Remark:</u> In [DP, section 4.2] a 0-1 law was obtained for (the Palm measure) of H_t restricted to an appropriately defined germ σ-algebra which contains the information about motions and branching just immediately before time t. This was used already in these notes to obtain a precise description of the (2,d,1)-superprocess in dimensions d≥3, namely Theorem 9.3.3.3. A similar result is expected for the Fleming-Viot process. However Theorem 12.6.1 shows that we cannot obtain such results directly from the corresponding results for the superprocess as would have been the case if equivalence of P_H and P_G were satisfied.

REFERENCES

R.J. Adler and M. Lewin (1991) An evolution equation for the intersection local times of superdiffusions, Stochastic Analysis, Cambridge Univ. Press, 1-22.

R.J. Adler and M. Lewin (1992). Local time and Tanaka Formulae for super-Brownian motion and super stable processes, Stoch. Proc. Appl., 41, 45-68.

D.J. Aldous (1985). *Exchangeability and related topics*, Lecture Notes in Math. 1117, 2-198.

D.J. Aldous (1991a). Asymptotic fringe distributions for general families of random trees, Ann. Appl. Prob. 1, 228-266.

D.J. Aldous (1991b). The continuum random tree II: an overview, in *Stochastic Analysis,* ed. M.T. Barlow and N.H. Bingham, 23-70, Cambridge Univ. Press.

D.J. Aldous (1992). The continuum random tree III, preprint.

K.B. Athreya (1969). On a characteristic property of Pólya's urn, Stud. Sci. Math. Hung., 4, 31-35.

K.B. Athreya and P.E. Ney (1977). *Branching Processes*, Springer-Verlag.

P. Baras, M. Pierre (1984). Singularités éliminables pour des équations semi-linéaires, Ann. INst. Fourier 34, 185-206.

M.T. Barlow, S.N. Evans and E.A. Perkins (1991) Collision local times and measure-valued processes, Can. J. Math. 43, 897-938.

C. Berg, J.P.R. Christensen and P. Ressel (1984). *Harmonic Analysis on Semigroups*, Springer-Verlag.

P. Billingsley (1968). *Convergence of Probability Measures*, John Wiley.

D. Blackwell and D.G. Kendall (1964). The Martin boundary for Polya's urn scheme and an application to stochastic population growth, J. Appl. Prob. 1, 284-296.

D. Blackwell and J.B. MacQueen (1973). Ferguson distributions via Pólya urn schemes, Ann. Stat. 1, 353-355.

D. Blackwell and L.E. Dubins (1983). An extension of Skorohod's almost sure representation theorem, Proc. A.M.S. 89, 691-692.

R.M. Blumenthal and R.K. Getoor (1968). *Markov Processes and Potential Theory*, Academic Press, New York.

V.S. Borkar (1984). Evolution of interacting particles in a Brownian medium, Stochastics 14, 33-79.

A. Bose and I. Kaj (1991a) Diffusion approximation for an age-structured population, LRSP Tech. Report 148, Carleton Univ.

A. Bose and I. Kaj (1991b). Measure-valued age-structured processes, LRSP Tech. Report 161, Carleton Univ.

L. Breiman (1968). *Probability*, Addison-Wesley.

H. Brezis, L.A. Peletier and D. Terman (1986). A very singular solution of the heat equation with absorption, Arch. Rational Mech. Anal. 95, 185-209.

H. Brezis and L. Veron (1980). Removable singularities of some nonlinear elliptic equations, Arch. Rational Mech. Anal. 75, 1-6.

H. Brezis and A. Friedman (1983) Nonlinear parabolic equations involving measures as initial conditions, J. Math. pures et appl. 62, 73-97.

O.G. Bulycheva and A.D. Vent-tsel' (1989). On the differentiability of expectations of functionals of a Wiener process, Th. Prob. Appl. 34, 509-512.

C. Cannings (1974). The latent roots of certain Markov chains arising in genetics: A new approach 1. Haploid models. Adv. Appl. Probab. 6, 260-290.

B. Chauvin (1986a). Arbres et processus de Bellman-Harris, Ann. Inst. Henri Poincaré 22, 209-232.

B. Chauvin (1986b). Sur la propriéte de branchement, Ann. Inst. Henri Poincaré 22, 233-236.

B. Chauvin, A. Rouault and A. Wakolbinger (1989). Growing conditioned trees, Stoch. Proc. Appl. 39, 117-130.

P.L. Chow (1976). Function space differential equations associated with a stochastic partial differential equation, Indiana Univ. Math. J. 25, 609-627.

P.L. Chow (1978). Stochastic partial differential equations in turbulence related problems. In *Probabilistic Analysis and Related Topics,* Vol. 1, Academic Press.

K.L. Chung, P. Erdos and T. Sirao (1959). On the Lipschitz condition for Brownian motions, J. Math. Soc. Japan 11, 263-274.

Z. Ciesielski and S.J. Taylor (1962). First passage times and sojourn times for Brownian motion and exact Hausdorff measure of the sample path, Trans. Amer. Math. Soc. 103, 434-450.

J.T. Cox and D. Griffeath (1985). Occupation times for critical branching Brownian motions, Ann. Probab. 13, 1108-1132.

J.T. Cox and D. Griffeath (1987). Recent results on the stepping stone model,in Percolation Theory and Ergodic Theory of Infinite Particle Systems, 73-83, IMA Volume 8, ed. H. Kesten, Springer-Verlag.

J.T. Cox and D. Griffeath (1990). Mean-field asymptotics for the planar stepping stone model, Proc. London Math. Soc. 61, 189-208.

J.F. Crow and M. Kimura (1970). *An Introduction to Population Genetics*, Burgess.

C. Cutler (1984a). *Some measure-theoretic and topological results for measure-*valued *and set-valued stochastic processes*, Ph.D. Thesis, Carleton University.

C. Cutler (1984b). A Lebesgue decomposition theorem for random measures and random measure processes, Tech Report 23, LRSP, Carleton University.

Dai Yonglong (1982). On absolute continuity and singularity of random measures (in Chinese), Chinese Annals of Mathematics 3, 241-246.

Yu. Dalecky and S. Fomin (1991). *The measures and differential equations in infinite dimensional spaces*, Kluwer.

D.J. Daley and D. Vere-Jones (1988). *An Introduction to the Theory of Point Processes*, Springer-Verlag.

D.A. Dawson (1975). Stochastic evolution equations and related measure-valued processes, J. Multivariate Analysis 5, 1-52.

D.A. Dawson (1977). The critical measure diffusion, Z. Wahr. verw Geb. 40, 125-145.

D.A. Dawson (1978a). Geostochastic calculus, Canadian Journal of Statistics 6, 143-168.

D.A. Dawson (1978b). Limit theorems for interaction free geostochastic systems, Colloquia Math. Soc. J. Bolyai, 24, 27-47.

D.A. Dawson (1986a). Measure-valued stochastic processes: construction, qualitative behavior and stochastic geometry, Proc. Workshop on Spatial Stochastic Models, Lecture Notes in Mathematics 1212, 69-93, Springer-Verlag.

D.A. Dawson (1986b). Stochastic ensembles and hierarchies, Lecture Notes in Mathematics 1203, 20-37, Springer-Verlag.

D.A. Dawson (1992). Infinitely Divisible Random Measures and Superprocesses, in *Proc. 1990 Workshop on Stochastic Analysis and Related Topics, Silivri, Turkey.*

D.A. Dawson and K. Fleischmann (1991) Critical branching in a highly fluctuating random medium, Probab. Theory Rel. Fields, 90, 241-274.

D.A. Dawson and K. Fleischmann (1992). Diffusion and reaction caused by point catalysts, SIAM J. Appl. Math. 52, 163-180.

D.A. Dawson, K. Fleischmann, R.D. Foley and L.A. Peletier (1986). A critical measure-valued branching process with infinite mean, Stoch. Anal. Appl. 4, 117-129.

D.A. Dawson, K. Fleischmann, and L.G. Gorostiza, (1989). Stable hydrodynamic limit fluctuations of a critical branching particle system, Ann. Probab. 17, 1083-1117.

D.A. Dawson, K. Fleischmann and S. Roelly (1991). Absolute continuity of the measure states in a branching model with catalysts, Seminar on Stochastic processes 1990, Birkhäuser, 117-160.

D.A. Dawson and K.J. Hochberg (1979). The carrying dimmension of a stochastic measure diffusion, Ann. Prob. 7, 693-703.

D.A. Dawson and K.J. Hochberg (1982). Wandering random measures in the Fleming-Viot model, Ann. Prob. 10, 554-580.

D.A. Dawson and K.J. Hochberg (1985). Function-valued duals for measure-valued processes with applications, Contemporary Mathematics 41, 55-69.

D.A. Dawson, K.J. Hochberg and Y. Wu (1990). Multilevel branching systems, in Proc. Bielefeld Encounters in Mathematics and Physics 1989, World Scientific, 93-107.

D.A. Dawson and K.J. Hochberg (1991). A multilevel branching model, Adv. Appl. Prob. 23, 701-715.

D.A. Dawson, I. Iscoe and E.A. Perkins (1989). Super-Brownian motion: path properties and hitting probabilities, Probab. Th. Rel. Fields 83, 135-205.

D.A. Dawson and B.G. Ivanoff (1978). Branching diffusions and random measures. In Stochastic Processes, ed. A. Joffe and P. Ney, 61-104, Dekker, New York.

D.A. Dawson and T.G. Kurtz (1982). Applications of duality to measure-valued processes, Lecture Notes in Control and Inform. Sci. 42, 177-191.

D.A. Dawson and P. March (1992). In preparation.

D.A. Dawson and E.A. Perkins (1991). *Historical processes*, Memoirs of the American Mathematical Society 93, no. 454.

D.A. Dawson and H. Salehi (1980). Spatially homogeneous random evolutions, J. Mult. Anal. 10, 141-180.

D.A. Dawson and V. Vinogradov (1992a). Almost sure path properties of $(2,d,\beta)$ superprocesses, LRSP Tech. Report 195.

D.A. Dawson and V. Vinogradov (1992b). Mutual singularity of genealogical structures of Fleming-Viot and continuous branching processes, LRSP Tech Report 204.

C. Dellacherie and P.A. Meyer (1976). *Probabilités et potentiel*, Hermann, Vol. I 1976, Vol. II 1980, Vol. III 1983, Vol. IV 1987.

A. De Masi and E. Presutti (1991). Mathematical Methods for Hydrodynamic Limits, Lecture Notes in Mathematics 1501, Springer Verlag.

P. Donnelly (1984). The transient behavior of the Moran model in population genetics, Math. Proc. Camb. Phil Soc. 95, 349-358.

P. Donnelly (1985). Dual processes and an invariance result for exchangeable models in population genetics, J. Math. Biol.

P. Donnelly (1986) Partition structures, Polya urns, the Ewens sampling formula and the ages of alleles, Theor. Pop. Biol. 30, 271-288.

P. Donnelly (1991). Weak convergence to a Markov chain with an entrance boundary: ancestral processes in population genetics, Ann. Probab. 19, 1102-1117.

P. Donnelly and P. Joyce (1992). Weak convergence of population genealogical processes to the coalescent with ages, Ann. Prob. 20, 322-341.

P. Donnelly and T.G. Kurtz (1992) The Fleming Viot measure-valued diffusion as an interactive particle system, preprint.

P. Donnelly and S. Tavaré (1986). The ages of alleles and a coalescent, Adv. Appl. Prob. 18, 1-19.

P. Donnelly and S. Tavaré (1987). The population genealogy of the infinitely-many neutral alleles model, J. Math. Biol. 25, 381-391.

J.L. Doob (1984). *Classical Potential Theory and Its Probabilistic Counterpart*, Springer-Verlag.

R. Durrett (1978). The genealogy of critical branching processes, Stoch. Proc. Appl. 8, 101-116.

R. Durrett (1988). *Lecture Notes on Particle Systems and Percolation*, Wadsworth and Brooks/Cole.

E.B. Dynkin (1965). *Markov Processes*, Volumes I and II, Springer-Verlag.

E.B. Dynkin, (1988). Representation for functionals of superprocesses by multiple stochastic integrals, with applications to self intersection local times, Astérisque **157-158**, 147-171.

E.B. Dynkin, (1989a). Superprocesses and their linear additive functionals, Trans. Amer. Math. Soc. , 314, 255-282.

E.B. Dynkin, (1989b). Regular transition functions and regular superprocesses, Trans. Amer. Math. Soc., 316, 623-634.

E.B. Dynkin, (1989c). Three classes of infinite dimensional diffusions, J. Funct. Anal. 86, 75-110.

E.B. Dynkin, (1991a). Branching particle systems and superprocesses, Ann. Probab., 19, 1157-1194.

E.B. Dynkin (1991b), Path processes and historical superprocesses, Probab. Th. Rel. Fields 90, 1-36.

E.B. Dynkin (1991c) A probabilistic approach to one class of nonlinear differential equations, Probab. Th. Rel. Fields 89, 89-115.

E.B. Dynkin (1991d) Additive functionals of superdiffusion processes, in *Random Walks, Brownian Motion and Interacting Particle Systems, A Festschrift in Honor of Frank Spitzer*, 269-282, R. Durrett and H. Kesten, eds., Birkhäuser.

E.B. Dynkin (1992a) Superdiffusions and parabolic nonlinear differential equations. Ann. Probab. 20, 942-962.

E.B. Dynkin (1992b). Superprocesses and partial differential equations, (1991 Wald Memorial Lectures).

E.B. Dynkin, S.E. Kuznetsov and A.V. Skorohod (1992). Branching measure-valued processes, preprint.

N. El Karoui (1985). Non-linear evolution equations and functionals of measure-valued branching processes. In *Stochastic Differential Systems*, ed. M. Metivier and E. Pardoux, Lect. Notes Control and Inf. Sci. 69, 25-34., Springer-Verlag.

N. El Karoui and S. Roelly (1991). Proprietes de martingales, explosion et representation de Lévy-Khinchine d'une classe de processus de branchement à valeurs mesures, Stoch. Proc. Appl. 38, 239-266.

N. El Karoui and S. Méléard (1990) Martingale measures and stochastic calculus,

Prob. Th. Rel Fields. 84, 83-101.

A. Etheridge and P. March (1991) A note on superprocesses, Probab. Theory Rel. Fields, 89, 141-147.

S.N. Ethier (1976). A class of degenerate diffusion processes occurring in population genetics, Comm. Pure Appl. Math. 29, 483-493.

S.N. Ethier (1979). Limit theorems for absorption times of genetic models, Ann. Prob. 7, 622-638.

S.N. Ethier (1981). A class of infinite-dimensional diffusions occurring in population genetics, Indiana Univ. Math. J. 30,925-935.

S.N. Ethier (1988). The infinitely-many-neutral-alleles diffusion model with ages, Adv. Appl. Prob. 22, 1-24.

S.N. Ethier (1990a) On the stationary distribution of the neutral one-locus diffusion model in population genetics, Ann. Appl. Prob. 2, 24-35.

S.N. Ethier (1990b) The distribution of the frequencies of age-ordered alleles in a diffusion model, Adv. Appl. Prob. 22, 519-532.

S.N. Ethier and R.C. Griffiths (1987). The infinitely many sites model as a measure-valued diffusion, Ann. Prob. 15, 515-545.

S.N. Ethier and R.C. Griffiths (1988). The two locus infinitely many neutral alleles diffusion model, preprint.

S.N. Ethier and R.C. Griffiths (1990) The neutral two locus model as a measure-valued diffusion, Adv. Appl. Prob.

S.N. Ethier and R.C. Griffiths (1992) The transition function of a Fleming-Viot process, preprint.

S.N. Ethier and T.G. Kurtz (1981). The infinitely many neutral alleles diffusion model, Adv. Appl. Prob. 13, 429-452.

S.N. Ethier and T.G. Kurtz (1985). *Markov processes: characterization and convergence*, Wiley.

S.N. Ethier and T.G. Kurtz (1987). The infinitely many alleles model with selection as a measure-valued diffusion, Lecture Notes in Biomathematics 70, 72-86.

S.N. Ethier and T.G. Kurtz (1990a) Coupling and ergodic theorems for Fleming-Viot processes, preprint.

S.N. Ethier and T.G. Kurtz (1990b) Convergence to Fleming-Viot processes in the weak atomic topology, Stochatic Proc. Appl. to appear.

S.N. Ethier and T.G. Kurtz (1992a) On the stationary distribution of the neutral diffusion model in population genetics, Ann. Appl. Prob. 2.

S.N. Ethier and T.G. Kurtz (1992b). Fleming-Viot processes in population genetics, preprint.

S.N. Evans (1990). The entrance space of a measure-valued Markov branching process conditioned on non-extinction. Tech. Rept. 230, Dept. of Stat., Univ. of California at Berkeley.

S.N. Evans (1991) Trapping a measure-valued branching process conditioned on non-extinction, Ann. Inst. Henri Poincaré 27, 215-220.

S.N. Evans (1992) The entrance space of a measure-valued Markov branching proces conditioned on non-extinction, Can. Math. Bull., to appear.

S. Evans and E. Perkins (1990). Measure-valued Markov branching processes conditioned on non-extinction, Israel J. Math., 71, 329-337.

S. Evans and E. Perkins (1991). Absolute continuity results for superprocesses with some applications, Trans. Amer. Math. Soc., 325, 661-681.

S. Evans and E.A. Perkins (1992). Measure-valued branching diffusions with singular interaction, preprint.

W.J. Ewens (1979). *Mathematical Population Genetics*, Springer-Verlag.

K.J. Falconer (1985). *The Geometry of Fractal Sets*, Cambridge Univ. Press.

H. Federer (1969). *Geometric measure theory*, Springer-Verlag.

P.D. Feigen and R.L. Tweedie (1989). Linear functionals and Markov chains associated with Dirichlet processes, Math. Proc. Camb. Phil. Soc. 105, 579-585.

W. Feller (1951). Diffusion processes in genetics, Proc. Second Berkeley Symp., Univ. of Calif. Press, Berkeley, 227-246.

T.S. Ferguson (1973). A Bayesian analysis of some nonparametric problems, Ann. Stat. 1, 209-230.

X. Fernique (199.) Fonctions aléatoires à valeurs dans les espaces lusiniens, Expositiones Math.

R.A. Fisher (1958). *The genetic theory of natural selection*, Dover.

P.J. Fitzsimmons (1988). Construction and regularity of measure-valued branching processes, Israel J. Math. 64, 337-361.

P.J. Fitzsimmons (1991). Correction to Construction and regularity of measure-valued branching processes, Israel J. Math. 73, 127.

P.J. Fitzsimmons (1992). On the martingale problem for measure-valued Markov branching processes, in *Seminar on Stochastic Processes, 1991*, E. Cinlar, K.L. Chung and M.J. Sharpe, eds., Birkhäuser.

K. Fleischmann (1988). Critical behavior of some measure-valued processes, Math. Nachr. 135, 131-147.

K. Fleischmann and J. Gärtner (1986). Occupation time process at a critical point, Math. Nachr. 125, 275-290.

K. Fleischmann and U. Prehn (1974). Ein Grenzwertsatz für subkritische Verzweigungsprozesse mit endlich vielen Typen von Teilchen, Math. Nachr. 64, 357-362.

K. Fleischmann and U. Prehn (1975). Subkritische räumlich homogene Verzweigungsprozesse, Math. Nachr. 70, 231-250.

K. Fleischmann and R. Sigmund-Schultze (1977). The structure of reduced critical Galton-Watson processes, Math. Nachr. 74, 233-241.

K. Fleischmann and R. Sigmund-Schultze (1978). An invariance principle for reduced family trees of critically spatially homogeneous branching processes (with discussion), Serdica Bulg. Math. 4, 11-134.

W.H. Fleming and M. Viot (1979). Some measure-valued Markov processes in population genetics theory, Indiana Univ. Math. J. 28, 817-843.

J. Gärtner (1988). On the McKean-Vlasov limit for interacting diffusions, Math. Nachr. 137, 197-248.

R.K. Getoor (1974). *Markov processes: Ray processes and right processes*, Lecture Notes in Math. 440, Springer-Verlag.

R.K. Getoor (1975). On the construction of kernels, Sem. de Prob. IX., Lecture Notes in Mathematics 465, 441-463, Springer-Verlag.

A. Gmira, L. Veron (1984). Large time behavior of the solutions of a semilinear parabolic equation in $\mathbb{R}^N$, J. Diff. Equations 53, 258-276.

D.E. Goldberg (1989). *Genetic Algorithms in Search, Optimization, and Machine Learning*, Addison-Wesley.

L.G. Gorostiza (1981). Limites gaussiennes pour les champs aléatoires ramifiés supercritiques, Colloque CNRS *Aspects statistiques et aspects physiques des processus gaussiens*, 385-398.

L.G. Gorostiza and J.A. López-Mimbela (1990). The multitype measure branching process, Adv. Appl. Prob. 22, 49-67.

L.G. Gorostiza and J. A. López-Mimbela (1992). A convergence criterion for measure-valued processes, and application to continuous superprocesses, Prog. in Probab., Birkhäuser, to appear.

L.G. Gorostiza and S. Roelly-Coppoletta (1990) Some properties of the multitype measure branching process, Stoch. Proc. Appl. 37, 259-274.

L.G. Gorostiza, S. Roelly-Coppoletta and A. Wakolbinger (1990). Sur la persistence du processus de Dawson-Watanabe stable; intervention del la limite en temps et de la renormalization, Sém. Probab. XXIV, Lecture Notes in Math. 1426, 275-281.

L.G. Gorostiza, S. Roelly and A. Wakolbinger (1992) Persistence of critical multitype particle and measure branching processes, Prob. Th. Rel. Fields.

L.G. Gorostiza and A. Wakolbinger (1991). Persistence criteria for a class of critical branching particle systems in continuous time, Ann. Probab. 19, 266-288.

L.G. Gorostiza and A. Wakolbinger (1992). Convergence to equilibrium of critical branching particle systems and superprocesses, and related nonlinear partial differential equations, Acta Appl. Math., to appear.

R.C. Griffiths (1979) A transition density expansion for a multi-allele diffusion model, Adv. Appl. Prob. 11, 310-325.

I. Gyöngy and E. Pardoux (1991). On quasi-linear stochastic partial differential equations, Probab. Th. Rel. Fields.

K. Handa (1990) A measure-valued diffusion process describing the stepping stone model with infinitely many alleles, Stoch. Proc. Appl. 36, 269-296.

T.E. Harris (1963). *The Theory of Branching Processes*, Springer-Verlag.

K.J. Hochberg (1991) Measure-valued processes: techniques and applications. In *Selected Proc. Sheffield Symp. Appl. Probab.* IMS Lecture Notes - Monograph Series 18, 212-235.

K.J. Hochberg (1986). Stochastic population theory: Mathematical evolution of a genetical model, in *New Directions in Applied and Computational Mathematics*, 101-115, Springer.

R.A. Holley and D.W. Stroock (1978). Generalized Ornstein-Uhlenbeck processes and infinite particle branching Brownian motion, Publ. R.I.M.S. Kyoto Univ. 14, 741-788.

R.A. Holley and D.W. Stroock (1979). Central limit phenomena of various interacting systems, Ann. Math. 110, 333-393.

R. Holley and T. Liggett (1975). Ergodic theorems for weakly interacting systems and the voter model, Ann. Prob. 3, 643-663.

F.M. Hoppe (1987). The sampling theory of neutral alleles and an urn model in population genetics, J. Math. Biol. 25, 123-159.

N. Ikeda, M. Nagasawa and S. Watanabe (1968), (1969). Branching Markov processes I,II,III, J. Math. Kyoto Univ. 8, 233-278, 9, 95-160.

N. Ikeda and S. Watanabe (1981). *Stochastic differential equations and diffusion processes*, North Holland.

I. Iscoe (1980). *The man-hour process associated with measure-valued branching random motions in* $\mathbb{R}^d$, Ph.D. thesis, Carleton University.

I. Iscoe (1986a). A weighted occupation time for a class of measure-valued critical branching Brownian motion, Probab. Th. Rel. Fields 71, 85-116.

I. Iscoe (1986b). Ergodic theory and a local occupation time for measure-valued branching processes, Stochastics 18, 197-143.

I. Iscoe (1988). On the supports of measure-valued critical branching Brownian motion, Ann. Prob. 16, 200-221.

S. Itatsu (1981). Equilibrium measures of the stepping stone model in population genetics, Nagoya Math. J. 83, 37-51.

K. Itô and H.P. McKean (1965). *Diffusion processes and their sample paths*, Springer-Verlag.

K. Itô (1984). *Foundations of stochastic differential equations in infinite dimensional space*, SIAM.

B.G. Ivanoff (1981). The multitype branching diffusion, J. Mult. Anal. 11, 289-318.

B.G. Ivanoff (1989). The multitype branching random walk: temporal and spatial limit theorems, preprint.

K. Iwata (1987). An infinite dimensional stochastic differential equation with state space C(ℝ), Prob. Th. Rel. Fields 74, 141-159.

J. Jacod (1979). *Calcul Stochastiques et Problèmes de Martingales*, LNM 714, Springer-Verlag.

J. Jacod and A.N. Shiryaev (1987). *Limit theorems for stochastic processes*, Springer-Verlag.

P. Jagers (1974). Aspects of random measures and point processes. *In Advances in Probability*, P. Ney and S. Port, eds., M. Dekker, 179-238.

P. Jagers (1975). *Branching processes with biological applications*, Wiley.

P. Jagers and O. Nerman (1984). The growth and composition of branching processes, Adv. Appl. Prob. 16, 221-259.

A. Jakubowski (1986). On the Skorohod topology, Ann. Inst. H. Poincaré B22, 263-285.

M. Jirina (1958). Stochastic branching processes with continuous state space, Czechoslovak Math. J. 8., 292-313.

M. Jirina (1964). Branching processes with measure-valued states, In. Trans. Third Prague Conf. on Inf. Th., 333-357.

A. Joffe and M. Métivier (1986). Weak convergence of sequences of semimartingales with applications to multitype branching processes, Adv. Appl. Prob. 18, 20-65.

N.L. Johnson and S. Kotz (1977). *Urn Models and Their Applications*, Wiley.

O. Kallenberg (1977). Stability of critical cluster fields, Math. Nachr. 77, 7-43.

O. Kallenberg (1983). *Random measures*, 3rd ed., Akademie Verlag and Academic Press.

N.L. Kaplan, T. Darden and R.R. Hudson (1988) The coalescent process in models with selction, Genetics 120, 819-829.

K. Kawazu and S. Watanabe (1971). Branching processes with immigration and related limit theorems, Th. Prob. Appl. 26, 36-54.

M. Kimura (1983a). *The neutral theory of molecular evolution*, Cambridge Univ. Press.

M. Kimura (1983b). Diffusion model of intergroup selection, with special reference to evolution of an altruistic character, Proc. Nat. Acad. Sci. USA 80, 6317-6321.

J.F.C. Kingman (1975). Random discrete distributions, J.R. Statist. Soc. B37, 1-22.

J.F.C. Kingman (1978). Uses of exchangeability, Ann. Probab. 6, 183-197.

J.F.C. Kingman (1980) *The mathematics of Genetic Diversity*, CBMS Regional Conf. Series in Appl. Math. 34, SIAM.

J.F.C. Kingman (1982a). The coalescent, Stoch. Proc. Appl. 13, 235-248.

J.F.C. Kingman (1982b) On the genealogy of large populations, J. Appl. Prob. 19A, 27-43.

J.F.C. Kingman (1982c). Exchangeability and the evolution of large populations, in *Exchangeability in Probability and Statistics*, eds. G. Koch and F. Spizzichino, 97-112, North Holland.

F. Knight (1981). *Essentials of Brownian Motion and Diffusion*, Amer. Math. Soc., Providence.

N. Konno and T. Shiga (1988). Stochastic differential equations for some measure-valued diffusions, Prob. Th. Rel Fields 79, 201-225.

P. Kotelenez (1988). High density limit theorems for nonlinear chemical reactions with diffusion, Probab. Th. Rel. Fields 78, 11-37.

P. Kotelenez (1989). A class of function and density valued stochastic partial differential equations driven by space-time white noise, preprint.

S. Krone (1990) Local times for superdiffusions (Abstract), Stoch. Proc. Appl. 35, 199-200.

N.V. Krylov and B.L. Rozovskii (1981). Stochastic evolution equations, J. Soviet Math. (Itogi Nauki i Techniki 14), 1233-1277.

H. Kunita (1986). Stochastic flows and applications, Tata Institute and Springer-Verlag.

H. Kunita (1990). Stochastic flows and stochastic differential equations, Cambridge Univ. Press.

T.G. Kurtz and D. Ocone (1988). A martingale problem for conditional distributions and uniqueness for the nonlinear filtering equations, Ann. Probab.

T.G. Kurtz (1981). *Approximation of Population Processes*, SIAM.

S.E. Kuznetsov (1984). Nonhomogeneous Markov processes, J. Soviet Math. 25, 1380-1498.

J. Lamperti (1967). Continuous state branching processes, Bull. Amer. Math. Soc. 73, 382-386.

T.-Y. Lee (1990).Some limit theorems for critical branching Bessel processes and related semilinear differential equations, Probab. Th. Rel. Fields 84, 505-520.

J.F. Le Gall (1987). Exact Hausdorff measure of Brownian multiple points, in *Seminar on Stochastic Processes, 1986,* E. Cinlar, K.L. Chung and R.K. Getoor, eds., Birkhäuser.

J.F. Le Gall (1989a). Marches aléatoires, mouvement brownien et processes de branchement, L.N.M. 1372, 258-274.

J.F. Le Gall (1989b). Une construction de certains processus de Markov à valeurs mesures, C.R. Acad. Sci. Paris 308, Série I, 533-538.

J.F. Le Gall (1991a). Brownian excursions, trees and measure-valued branching processes, Ann. Probab. 19., 1399-1439.

J.F. Le Gall (1991b). A class of path-valued Markov processes and its applications to superprocesses, preprint.

Y. Le Jan (1989). Limites projectives de processus de branchement markoviens, C.R. Acad. Sci. Paris 309 Série 1, 377-381.

Y. Le Jan (1991). Superprocesses and projective limits of branching Markov processes, Ann. Inst. H. Poincaré 27, 91-106.

C. Léonard (1986). Une loi des grands nombres pour des systèmes de diffusions avec interaction à coefficients non bornés, Ann. Inst. Henri Poincaré 22, 237-262.

Z.-H. Li (1992). A note on the multitype measure branching process, Adv. Appl. Prob. 24, 496-498.

A. Liemant, K. Matthes and A. Wakolbinger (1988). *Equilibrium Distributions of Branching Processes*, Akademie-Verlag, Berlin, and Kluwer Academic Publ., Dordrecht.

T.M. Liggett (1985). *Interacting Particle Systems*, Springer-Verlag.

R. Sh. Liptser and A.N. Shiryayev (1989). *Theory of Martingales*, Kluwer.

R.A. Littler and A.J. Good (1978). Ages, extinction times and first passage probabilities for a multiallele diffusion model with irreversible mutation, Theor. Pop. Biol. 13, 214-225.

L. Liu and C. Mueller (1989). On the extinction of measure valued critical branching Brownian motion, Ann. Probab. 17, 1463-1465.

R. Marcus (1979). Stochastic diffusion on an unbounded domain, Pacific J. Math. 84, 143-153.

G. Matheron (1975). *Random sets and integral geometry*, Wiley.

K. Matthes, J. Kerstan and J. Mecke (1978). *Infinitely Divisible Point Processes*, Wiley.

H.P. McKean (1969). *Stochastic Integrals*, Academic Press.

S. Méléard and S. Roelly-Coppoletta (1990). A generalized equation for a continuous measure branching process, L.N. Math. 1390, 171-186.

S. Méléard and S. Roelly (1991). Discontinuous measure-valued branching processes and generalized stochastic equations, Math. Nachr. 154, 141-156.

M. Métivier (1982). *Semimartingales*, W. de Gruyter.

M. Métivier and J. Pellaumail (1980). *Stochastic integration*, Academic Press.

M. Métivier (1984). Convergence faible et principe d'invariance pour des martingales à valeurs dans des espaces de Sobolev, Ann. Inst. Henri Poincaré 20, 329-348.

M. Métivier (1985). Weak convergence of measure-valued processes using Sobolevimbedding techniques, L.N. Math. 1236, 172-183.

M. Métivier (1986). Quelques problemes liés aux systèmes infini de particules et leur limites, Springer L.N.M., 426-446.

M. Métivier and M. Viot (1987). On weak solutions of stochastic partial differential equations, Springer L.N.M. 1322, 139-150.

N.G. Meyers (1970). A theory of capacities for potentials of functions in Lebesgue classes, Math. Scand. 26, 255-292.

C. Mueller (1991a). Limit results for two stochastic partial differential equations, Stochastics 37, 175-199.

C. Mueller (1991b) On the supports of solutions to the heat equation with noise, Stochastics, 37, 225-246.

C. Mueller (1991). Long time existence for the heat equation with noise, Probab. Th. Rel. Fields 90, 505-518.

C. Mueller and E.A. Perkins (1991). The compact support property for solutions to the heat equation with noise, preprint.

J. Neveu (1964). *Bases Mathématiques du Calcul des Probabilités*, Masson et. Cie, Paris.

J. Neveu (1975). *Discrete-Parameter Martingales*, North-Holland.

J. Neveu (1986). Arbres et processus de Galton-Watson, Ann. Inst. H. Poincaré 22, 199-207.

J. Neveu and J.W. Pitman (1980). The branching process in a Brownian excursion, LNM 1372, 248-257, Springer-Verlag.

J.M. Noble (1992). Evolution equations with random potential, private communication.

M. Notohara and T. Shiga (1980). Convergence to genetically uniform state in stepping stone models of population genetics, J. Math. Biol. 10, 281-294.

K. Oelschläger (1989). On the derivation of reaction-diffusion equations as limit dynamics of systems of moderately interacting stochastic processes, Probab. Th. Rel. Fields 82, 565-586.

K. Oelschläger (1990) Limit theorems for age-structured populations, Ann. Probab. 18, 290-318.

T. Ohta and M. Kimura (1973). A model of mutation appropriate to estimate the number of electrophoretically detectable alleles in a finite population, Genet. Res. 22, 201-204.

E. Pardoux (1975). Equations aux dérivées partielles stochastiques non lineaires monotone. Etude des solutions forte de type Ito, Thèse, Univ. de Paris Sud, Orsay.

K.R. Parthasarathy (1967). *Probability Measures on Metric Spaces*, Academic Press.

A. Pazy (1983). *Semigroups of linear operators and applications to partial differential equations*, Springer-Verlag.

E.A. Perkins (1988). A space-time property of a class of measure-valued branching diffusions, Trans. Amer. Math. Soc., 305, 743-795.

E.A. Perkins (1989). The Hausdorff measure of the closed support of super-Brownian motion, Ann. Inst. Henri Poincaré 25, 205-224.

E.A. Perkins (1990). Polar sets and multiple points for super-Brownian motion, Ann. Probab. 18, 453-491.

E.A. Perkins (1991a) On the continuity of measure-valued processes, Seminar on Stochastic Processes 1990, Birkhauser, 261-268.

E.A. Perkins (1991b) Conditional Dawson-Watanabe processes and Fleming-Viot processes, Seminar in Stochastic Processes, 1991, Birkhauser, 142-155.

E.A. Perkins (1992). Measure-valued branching diffusions with spatial interactions, Probab. Th. Rel. Fields, to appear.

P. Priouret (1974). Processus de diffusion et equations differentielles stochastiques, Lecture Notes in Math. 390, 38-111, Springer-Verlag.

P. Protter (1990). *Stochastic Integration and Differential Equations*, Springer-Verlag.

M. Reimers (1986). *Hyper-finite methods for multi-dimensional stochastic processes*, Ph.D. thesis, U.B.C.

M. Reimers (1987). Hyperfinite methods applied to the critical branching diffusion, Probab. Th. Rel. Fields 81, 11-27.

M. Reimers (1989). One dimenional stochastic partial differential equations and the branching measure diffusion, Probab. Th. Rel. Fields 81, 319-340.

M. Reimers (1992) A new result on the support of the Fleming-Viot process proved by non-standard construction, preprint.

P. Ressel and W. Schmidtechen (1991). A new characterization of Laplace functionals and probability generating functionals, Prob. Th. Rel. Fields 88, 195-213.

D. Revuz and M. Yor (1991). *Continuous Martingales and Brownian Motion*, Springer-Verlag.

S. Roelly-Coppoletta (1986). A criterion of convergence of measure-valued processes: application to measure branching processes, Stochastics 17, 43-65.

S. Roelly and S. Méléard (1990) Interacting branching measure processes, *Proceedings: Stochastic Partial Differential Equations and Applications III*, Trento, Italy, Springer-Verlag.

S. Roelly-Coppoletta and A. Rouault (1989). Processus de Dawson-Watanabe conditioné par le futur lointain, C.R. Acad. Sci. Paris 309, 867-872.

S. Roelly and A. Rouault (1990). Construction et propriétés de martingales des branchements spatiaux interactifs, Int. Stat. Rev. 58, 173-189.

C.A. Rogers (1970). *Hausdorff measures*, Cambridge Univ. Press.

L.C.G. Rogers and D. Williams (1987). *Diffusions, Markov processes and Martingales, Vol. 2, It[Calculus*, Wiley.

J. Rosen (1990). Renormalization and limit theorems for self-intersections of super-processes, preprint.

B.L. Rozovskii (1990) *Stochastic Evolution Equations*, D. Reidel.

S.M. Sagitov (1990). Multi-dimensional critical branching processes generated by large numbers of identical particles, Th. Prob. Appl. 35.

K. Sato (1976a) Diffusion processes and a class of Markov chains related to population genetics, Osaka J. Math. 13, 631-659.

K. Sato (1976b). A class of Markov chains related to selection in population genetics, J. Math. Soc. Japan 28, 621-636.

K. Sato (1978) Convergence to a diffusion of a multi-allelic model in population genetics, Adv. Appl. Prob. 10, 538-562.

K.I. Sato (1983). Limit diffusion of some stepping stone models, J. Appl. Prob. 20, 460-471.

S. Sawyer (1976). Results for the stepping stone model for migration in population genetics, Ann. Prob. 4, 699-728.

S. Sawyer (1979). A limit theorem for patch size in a selectively neutral migration model, J. Appl. Prob. 16, 482-495.

M.J. Sharpe (1988). *General theory of Markov processes*, Academic Press.

B. Schmuland (1991). A result on the infinitely many neutral alleles diffusion model, J. Appl. Prob.

T. Shiga (1980) An interacting system in population genetics, J. Math. Kyoto Univ. 20, 213-242.

T. Shiga (1981) Diffusion processes in population genetics, J. Math. Kyoto Univ. 21, 133-151.

T. Shiga (1982) Wandering phenomena in infinite allelic diffusion models, Adv. Appl. Prob. 14, 457-483.

T. Shiga (1982), Continuous time multi-allelic stepping stone models in population genetics, J. Math. Kyoto Univ. 22, 1-40.

T. Shiga (1985) Mathematical results on the stepping stone model in population genetics, in *Population Genetics and Molecular evolution*, T. Ohta and K. Aoki, eds., Springer-Verlag.

T. Shiga (1987a). Existence and uniqueness of solutions for a class of non-linear diffusion equations, J. Math. Kyoto Univ. 27-2, 195-215.

T. Shiga (1987b). A certain class of infinite dimensional diffusion processes arising in population genetics, J. Math. Soc. Japan 30, 17-25.

T. Shiga (1988) Stepping stone models in population genetics and population dynamics, in S. *Albeverio et al (eds.) Stochastic Processes in Physics and Engineering*, 345-355.

T. Shiga (1990a) A stochastic equation based on a Poisson system for a class of measure-valued diffusions, J. Math. Kyoto Univ. 30(1990), 245-279.

T. Shiga (1990b) Two contrastive properties of solutions for one-dimensional stochastic partial differential equations, preprint.

T. Shiga and A. Shimizu (1980) Infinite dimensional stochastic differential equations and their applications, J. Math. Kyoto Univ. 20, 395-416.

T. Shiga and K. Uchiyama (1986). Stationary states and the stability of the stepping stone model involving mutation and selection, Prob. Th. Rel. Fields 73, 87-117.

N. Shimakura (1985). Existence and uniqueness of solutions for a diffusion model of intergroup selection, J. Math. Kyoto Univ. 25, 775-788.

A. Shimizu (1985). Diffusion approximation of an infinite allele model incorporating gene conversion, in *Population genetics and molecular evolution*, eds. T. Ohta and K. Aoki. Japan Sci. Soc. Press and Springer-Verlag.

A. Shimizu (1987). Stationary distribution of a diffusion process taking values in probability distributions on the partitions, Lecture Notes in Biomath. 70, 100-114.

A. Shimizu (1990). A measure valued diffusion process describing an n locus model incorporating gene conversion, Nagoya Math. J. 119, 81-92.

A.N. Shiryayev (1984). *Probability*, Springer-Verlag.

M.L. Silverstein (1969). Continuous state branching semigroups, Z. Wahr. verw. Geb. 14, 96-112.

D.W. Stroock and S.R.S. Varadhan (1979). *Multidimensional diffusion processes*, Springer-Verlag.

S. Sugitani (1987). Some properties for the measure-valued branching diffusion processes, J. Math. Soc. Japan 41, 437-462.

A.S. Sznitman (1991). Topics in Propagation of Chaos, *Ecole d'été de Probabilités* de *Saint Flour*, L.N.M. 1464, 165-251.

S.J. Taylor (1966). Multiple points for the sample paths of the symmetric stable process, Z. Wahr. verw. Geb. 5, 247-258.

S. Tavaré (1984). Line of descent and genealogical processes, and their applications in population genetics models, Theor. Pop. Biol. 26, 119-164.

S. Tavaré (1989). The genealogy of the birth, death and immigration process, in *Mathematical Evolutionary Theory*, ed. M.W. Feldman, 41-56.

R. Tribe (1989). *Path properties of superprocesses*, Ph.D. thesis, U.B.C.

R. Tribe (1991). The connected components of the closed support of super Brownian motion, Probab. Th. Rel. Fields 89, 75-87.

R. Tribe (1992). The behavior of superprocesses near extinction, Ann. Probab. 20, 286-311.

J. Vaillancourt (1987). *Interacting Fleming-Viot processes and related measure-valued processes*, Ph.D. thesis, Carleton University.

J. Vaillancourt (1988). On the existence of random McKean-Vlasov limits for triangular arrays of exchangeable diffusions, Stoch. Anal.

J. Vaillancourt (1990a). Interacting Fleming-Viot processes, Stoch. Proc. Appl. 36, 45-57.

J. Vaillancourt (1990b). On the scaling theorem for interacting Fleming-Viot processes, Stoch. Proc. Appl. 36, 263-267.

S.R.S. Varadhan (1984). *Large Deviations and Applications*, CBMS-NSF Regional Conf. 46, SIAM.

A.D. Venttsel' (1985). Infinitesimal characteristics of Markov processes in a function space which describes the past, Th. Prob. Appl. 30, 661-676.

A.D. Vent-tsel (1989). Refinement of the functional central limit theorem for stationary processes, Th. Prob. Appl. 34, 402-415.

L. Véron (1981). Singular solutions of some nonlinear elliptic equations, Nonlinear Anal. Theory, Math. Appl. 5, 225-242.

M. Viot (1976). *Solutions faibles d'equations aux dwrivwes partielles non lineaires*, Thèse, Univ. Pierre et Marie Curie, Paris.

J.B. Walsh (1986). An introduction to stochastic partial differential equations, in P.L. Hennequin (ed.), *Ecole d'été de Probabilités de Saint-Flour XIV- 1984*, L.N.M. 1180, 265-439.

F.S. Wang (1982a). Diffusion approximations of age-and-position dependent branching processes, Stoch. Proc. Appl. 13, 59-74.

F.S. Wang (1982b). Probabilities of extinction of multiplicative measure diffusion processes with absorbing boundary, Indiana Univ. Math J. 31, 97-107.

H. Watanabe (1988). Averaging and fluctuations for parabolic equations with rapidly oscillating random coefficients, Probab. Th. Rel. Fields 77, 359-378.

H. Watanabe (1989). On the convergence of partial differential equations of parabolic type with rapidly oscillating coefficients, Appl. Math. Optim. 20, 81-96.

S. Watanabe (1968). A limit theorem of branching processes and continuous state branching, J. Math. Kyoto Univ. 8, 141-167.

S. Watanabe (1969). On two dimensional Markov processes with branching property, Trans. Amer. Math. Soc. 136, 447-466.

G.A. Watterson (1976a) Reversibility and the age of an allele I. Moran's infinitely many neutral alleles model, Theor. Pop. Biol. 10, 239-253.

G.A. Watterson (1976b). The stationary distribution of the infinitely many neutral alleles model, J. Appl. Prob. 13, 639-651.

G.A. Watterson (1984) Lines of descent and the coalescent, Theor. Pop. Biol. 10, 239-253.

A.D. Wentzell (1992). On differentiability of the expectation of functionals of a Markov process, Stochastics and Stochastic Reports 39, 53-65.

S. Wright (1943). Isolation by distance, Genetics 28, 114-138.

S. Wright (1949) Adaptation and selection. In *Genetics, Paleontology and Evolution*, ed. G.L. Jepson et al, 365-389, Princeton Univ. Press.

Y. Wu (1991). Asymptotic behavior of two level branching processes, LRSP Tech. Report 179, Carleton Univ.

Y. Wu (1991). Multilevel birth and death particle system and its continuous diffusion, LRSP Tech. Report 186, Carleton Univ.

Y. Wu (1992). *Dynamic particle systems and multilevel measure branching processes*. Ph.D. thesis, Carleton University.

T. Yamada and S. Watanabe (1971). On the uniqueness of solutions of stochastic differential equations, J. Math. Kyoto Univ. 11, 155-167, 553-563.

M. Yor (1974). Existence et unicité de diffusions à valeurs dans un espace de Hilbert, Ann. Inst. Henri Poincaré 10, 55-88.

U. Zähle (1988a). Self-similar random measures I. Notion,carrying Hausdorff dimension and hyperbolic distribution, Probab. Th. Rel. Fields 80, 79-100.

U. Zähle (1988b). The fractal character of localizable measure-valued processes I - random measures on product spaces, Math. Nachr. 136, 149-155.

U. Zähle (1988c). The fractal character of localizable measure-valued processes II, Localizable processes and backward trees, Math. Nachr. 137, 35-48.

U. Zähle (1988d). The fractal character of localizable measure-valued processes III. Fractal carrying sets of branching diffusions, Math. Nachr. 138, 293-311.

H. Zessin (1983). The method of moments for random measures, Z. Wahr. verw. Geb. 62, 395-409.

V.M. Zolotarev (1957). More exact statements of several theorems in the theory of branching processes, Th. Prob. Appl. 2, 245-253.

Subject Index

PROCESSUS DE MARKOV :

NAISSANCE, RETOURNEMENT, REGENERATION

Bernard MAISONNEUVE

PROCESSUS DE MARKOV : NAISSANCE, RETOURNEMENT, REGENERATION

Bernard MAISONNEUVE
Institut Fourier, Université de Grenoble I
B.P. 74

F-38402 SAINT MARTIN D'HERES Cedex

Introduction

Ce cours a été annoncé avec le titre "Processus de Markov : Naissance et Régénération". Mon projet initial était de présenter une partie des chapitres markoviens (XVII à XX) du dernier volume de *Probabilités et Potentiel* (Dellacherie et Meyer m'ont demandé de collaborer à cet ouvrage, que je désignerai par DMM dans la suite). En préparant les exposés oraux, je me suis décidé à centrer le cours sur les processus de Markov à naissance aléatoire (PMNA) et à les utiliser de manière plus systématique que dans DMM, pour l'étude des mesures et fonctions excessives, pour le retournement du temps, le théorème de Revuz et une généralisation du théorème d'Azéma (de représentation des mesures). Ceci m'a amené à mettre l'accent sur le problème du prolongement en 0 d'un processus de Markov $(X_t)_{t>0}$ (chapitre I). De ce fait la partie "Naissance" a pris beaucoup d'importance et les auditeurs de Saint-Flour ont pu constater que la partie "Régénération" s'est trouvée réduite à un exposé et demi, consacré à la solution de Bertoin et Le Jan du problème de Skorohod (lorsqu'il existe un point régulier et récurrent).

Tout ceci pour dire que je me suis décidé à rédiger le cours, mais que la partie "Régénération" est passée presque entièrement à la trappe (au profit du chapitre intitulé retournement du temps). Je ne désespère pas de réparer ultérieurement cette lacune en rassemblant quelques résultats de régénération stricte (ensembles régénératifs, subordinateurs), un peu de théorie générale des excursions, des décompositions de dernière sortie et des résultats nouveaux de conditionnement des excursions. En attendant, le lecteur peut déjà regarder le chapitre XX (Ensembles aléatoires, excursions) de DMM.

Nous avons supposé dans tout le cours que le semi-groupe (P_t) régissant la propriété de Markov est *droit*, mais pour les idées essentielles le lecteur pourra supposer qu'il est *fellérien*. Pour les généralités concernant les processus de Markov (fellérien ou droits), nous lui conseillons de se reporter au fur et à mesure des besoins au volume *D* de *Probabilités et Potentiel*, ainsi qu'au chapitre XVII de DMM.

Notations. — Nous utiliserons fréquemment les notations suivantes, où $(\Omega, \mathcal{F}, P)$ est un espace mesuré, F une fonction $\mathcal{F}$-mesurable $\geqslant 0$, A un élément de $\mathcal{F}$:

$$P(F) = \int F dP, \quad P_A(F) = \int_A F dP .$$

Lorsque l'espace $(\Omega, \mathcal{F}, P)$ joue un rôle privilégié, nous utiliserons aussi les notations probabilistes $E(F)$, $E(F, A)$, p.s., même si la mesure P est infinie.

Références. — Nos références *principales* seront

[DM] – DELLACHERIE C. et MEYER P.A., *Probabilités et Potentiel*, chapitres XII à XVI, Hermann, 1987.

[DMM] – Dernier volume de *Probabilités et Potentiel*, à paraître.

[FM] – FITZSIMMONS P.J. et MAISONNEUVE B., *Excessive measures and Markov processes with random birth and death*, Prob. Th. Rel. Fields, 72, 1986, 319–336.

La liste des autres références utilisées dans le cours figure à la fin. Le lecteur trouvera dans DMM une bibliographie beaucoup plus complète, ainsi que des commentaires historiques, qui font totalement défaut à la présente rédaction.

Ce cours peut apparaître comme un complément (traité de manière autonome) à DMM. Nous avons donc utilisé le même système de numérotation que dans DMM (et DM).

Je tiens à remercier Arlette Guttin-Lombard pour la belle présentation qu'elle a su donner à mon texte en tenant compte des exigences de Springer.

I. Prolongement en 0 d'un processus markovien $(X_t)_{t>0}$

1. — Les processus de Markov considérés seront, sauf avis contraire, relatifs à un semi-groupe de transition *fixé* (P_t), supposé *droit* (la définition d'un tel s.g. est rappelée au n°4). Nous parlerons souvent de propriété ou de processus de Markov sans mentionner (P_t). Pour $p \in \mathbb{R}_+$ le noyau p-potentiel $\int_0^\infty e^{-pt} P_t dt$ est noté U_p et on pose $U = U_0$.

Étant donné un processus markovien $(X_t)_{t>0}$ continu à droite (càd), défini sur un espace probabilisé $(\Omega, \mathcal{F}, P)$, est-il possible de le prolonger en 0 de manière à obtenir un processus $(X_t)_{t \geqslant 0}$ càd p.s. et markovien, avec le même espace d'états et le même semi-groupe ?

Un tel prolongement n'est évidemment pas toujours possible. Par exemple ($X_t = t,\ t > 0$), considéré comme processus markovien à valeurs dans $]0,\infty[$ par rapport au semi-groupe de la translation uniforme sur $]0,\infty[$ n'est pas prolongeable au sens indiqué. Si on considère ce même processus comme prenant ses valeurs dans $\mathbb{R}_+$ et si on rend le point 0 absorbant, le processus prolongé par continuité n'est plus markovien, faute de régularité suffisante du semi-groupe (qui est droit, mais non fellérien).

LE CAS FELLÉRIEN.

2. — Le problème de prolongement posé au n°1 admet une solution si le semi-groupe (P_t) est *fellérien* sur un espace compact métrisable $(E,\mathcal{B})$ et si $P_t 1 = 1$. En effet, dans ce cas, (P_t) transforme les fonctions continues en fonctions continues et $t \mapsto P_t f$ est càd sur $\mathbb{R}_+$ pour toute fonction f continue sur E. Pour f continue $\geqslant 0$ sur E et $p > 0$ la fonction $U_p f$ est elle-même continue et $(e^{-pt}U_p f \circ X_t)_{t>0}$ est une surmartingale positive, qui converge p.s. lorsque $t \to 0$. Il existe alors une famille dénombrable assez riche de fonctions continues φ telles que $\varphi \circ X_t$ converge p.s. lorsque $t \to 0$. D'où l'existence p.s. d'une limite $X_0 = \lim_{t\to 0} X_t$ dans E. Si $(\mathcal{F}_t)_{t>0}$ est une filtration par rapport à laquelle $(X_t)_{t>0}$ est markovien, le processus prolongé $(X_t)_{t\geqslant 0}$ est encore markovien$/(\mathcal{F}_t)_{t\geqslant 0}$, avec $\mathcal{F}_0 = \mathcal{F}_{0+}$. En effet pour $A \in \mathcal{F}_0$, $f \geqslant 0$ continue sur E et $t > 0$ on a

$$E\Big(\int_t^\infty e^{-ps} f \circ X_s, A\Big) = E\Big(e^{-pt} U_p f \circ X_t, A\Big) ,$$

d'où par passage à la limite

$$E\Big(\int_0^\infty e^{-ps} f \circ X_s, A) = E(U_p f \circ X_0, A\Big) ;$$

les expressions $E(f \circ X_s, A)$ et $E(P_s f \circ X_0, A)$, qui sont càd en $s > 0$, sont donc égales.

Remarque. — Si (P_t) est sous-markovien et fellérien sur un espace localement compact à base dénombrable $(E,\mathcal{B})$, on peut étendre (P_t) au compactifié d'Alexandrof E_δ de E en rendant δ absorbant. Cette extension est alors fellérienne sur E_δ (DM XIII 21) et l'on peut encore appliquer ce qui précède.

SEMI-GROUPES DROITS.

3. — Le caractère fellérien du semi-groupe (P_t) est trop restrictif à cause des opérations que l'on fait parfois subir aux processus de Markov (meurtres, subordination, h-transformées). Dans toute la suite nous supposerons seulement que (P_t) satisfait aux *hypothèses droites de Meyer*, que nous allons maintenant rappeler et qui sont préservées par les transformations précédentes (DM XVI, 7-8-9).

L'espace d'états $(E,\mathcal{B})$ est supposé radonien : E est une partie universellement mesurable d'un compact métrisable $\overline{E}$. Souvent E est même lusinien, c'est-à-dire borélien dans $\overline{E}$. Soit E_δ l'espace agrandi d'un point $\delta \notin E$. On peut supposer E_δ lui-même radonien et muni de sa tribu borélienne $\mathcal{B}_\delta = \mathcal{B} \vee \{\delta\}$. Le semi-groupe (P_t), sous-markovien sur E ($P_t 1 \leqslant 1$), est prolongé en un semi-groupe markovien sur E_δ rendant le point δ absorbant, encore noté (P_t).

4. — Nous dirons qu'un processus de Markov $(X_t)_{t\geqslant 0}$ (défini sur un espace $(\Omega, \mathcal{F}, P)$) est *régulier* s'il est p.s. càd, à valeurs dans une partie lusinienne de E_δ et si de plus toute fonction excessive est presque borélienne et p.s. càd le long des trajectoires de (X_t). Cette notion est très souple, comme le montre le théorème suivant, que nous ne démontrerons pas : les arguments sont ceux de DMM XVII 6.

THÉORÈME. — *Supposons qu'il existe un processus de Markov régulier de loi initiale μ.*

1) Si $(Y_t)_{t\in D}$ est un processus markovien (sur un espace probabilisé arbitraire), indexé par une partie dénombrable dense D de $\mathbb{R}_+$ contenant 0, de loi initiale μ, alors (Y_t) se prolonge p.s. en un processus régulier.

2) Tout processus markovien $(Y_t)_{t\in\mathbb{R}_+}$ de loi initiale μ admet une modification régulière. En particulier, s'il est càd, il est lui-même régulier, et s'il est càg, il admet p.s. des limites à droite et le processus des limites à droite est markovien régulier.

Avec cette terminologie nous pouvons exprimer les *hypothèses droites de Meyer* de la manière suivante.

DÉFINITION. — *Le semi-groupe (P_t) est droit si pour toute loi initiale μ sur E, il existe un processus markovien $/(P_t)$, de loi initiale μ et régulier.*

PROLONGEMENT DE $(X_t)_{t>0}$ (CAS GÉNÉRAL).

5. — Au n° 2 nous avons examiné le problème du prolongement de $(X_t)_{t>0}$ dans le cas particulier d'un semi-groupe de Feller. Dans le cas général où (P_t) est un semi-groupe *droit*, considérons un compactifié de Ray $\widetilde{E}$ de E_δ et l'extension de Ray $(\widetilde{P}_t)$ de (P_t) à $\widetilde{E}$ (DM XVI 6, DMM XVII 12). Noter que dans le cas fellérien on peut prendre $\widetilde{E} = E$, $\widetilde{P}_t = P_t$ et le prolongement obtenu dans le théorème qui suit se réduit à celui du n° 2.

THÉORÈME. — *Soit $(X_t)_{t>0}$ un processus markovien càd relatif à une filtration $(\mathcal{F}_t)_{t>0}$ et soit Ω_0 l'ensemble sur lequel X_t converge au sens de Ray vers une limite dans E $(t \to 0)$. En notant X_0 cette limite ($X_0 = \delta$ sur Ω_0^c) et $\mathcal{F}_0 = \mathcal{F}_{0+}$, le processus $(X_t)_{t\geqslant 0}$ est alors markovien $/(\mathcal{F}_t)_{t\geqslant 0}$ sur Ω_0.*

L'ensemble Ω_0 est essentiellement maximal parmi les $A \in \mathcal{F}_0$ sur lesquels un tel prolongement markovien est possible, car un processus markovien càd est aussi càd pour la topologie de Ray (DMM XVII 12).

Démonstration. — Par le raisonnement du cas fellérien on montre l'existence d'un prolongement dans $\widetilde{E}$ markovien par rapport à $(\widetilde{P}_t)$. Les détails sont laissés au lecteur.

MESURES INFINIES.

6. — Les processus markoviens $(X_t)_{t>0}$ rencontrés dans la suite sont souvent définis sur un espace de mesure *infinie* (mesures de Kuznetsov, mesures d'excursions) et c'est dans ce cadre que nous allons étendre le théorème 5. Voici d'abord quelques définitions. Soit $(X_t)_{t>0}$

un processus càd à valeurs dans E_δ, défini sur un espace mesuré $(\Omega, \mathcal{F}, P)$, adapté à une filtration $(\mathcal{F}_t)_{t>0}$. On suppose que $\zeta = \inf\{t : X_t = \delta\}$ est P-p.s. > 0.

DÉFINITION.

1) Le processus $(X_t)_{t>0}$ est dit markovien par rapport à $(\mathcal{F}_t)$ si pour tout $t > 0$:

a) la mesure restreinte $P_{\{t<\zeta\}}$ est σ-finie sur $\mathcal{F}_t$,

b) $P_{\{t<\zeta\}}(f \circ X_{t+s} | \mathcal{F}_t) = P_s f \circ X_t, \quad (s \geqslant 0,\ f \geqslant 0)$.

2) Une partie $A \in \mathcal{F}_0 = \mathcal{F}_{0+}$ est dite admissible s'il existe une mesure μ sur E telle que $(X_t)_{t>0}$ ait pour loi $P^\mu = \int \mu(dx) P^x$ (μ peut être infinie).

Comme d'habitude P^x désigne la loi d'un processus markovien càd issu de x, considérée comme une mesure sur l'espace canonique usuel ; par restriction du temps à $]0, \infty[$ on obtient un processus dont la loi est encore notée P^x.

Remarques.

1) Comme $\zeta > 0$ p.s., la condition *a)* implique que la mesure sous-jacente P est σ-finie.

2) On exige souvent une forme renforcée de *a)*, en demandant que la loi $\mu_t = X_t(P)$ sur E soit σ-finie pour tout $t > 0$.

3) Si A est admissible, $X_0 = X_{0+}$ existe p.s. sur A pour la topologie de Ray et pour la topologie initiale ; en effet le processus canonique possède une propriété analogue sous chaque P^x, donc aussi sous P^μ.

7. EXEMPLES.

1) Soit P^0 la mesure de Wiener sur l'espace Ω des applications continues de $\mathbb{R}_+$ dans $\mathbb{R}$ et soit $\widehat{P}$ la mesure sur Ω définie par

$$\widehat{P}(f) = P^0\Big(\sum_{s \in G} e^{-s} f \circ \theta_s\Big) ,$$

où (θ_s) est le processus des translations et G est l'ensemble des débuts d'excursions relatives à 0. Sous $\widehat{P}$ le processus des coordonnées $(X_t)_{t>0}$ est alors markovien par rapport au semi-groupe de Wiener (P_t). On a $\widehat{P}(1 - e^{-R}) < \infty$, où $R = \inf\{t > 0 \ : X_t = 0\}$, donc $\widehat{P}\{R > s\} < \infty$ pour tout $s > 0$ et la condition *a)* est bien satisfaite, puisque $\widehat{P}$ est portée par $\{R > 0\}$.

2) Soient k_t les opérateurs de meurtre de l'espace de Wiener et soit (Q_t) le semi-groupe de Wiener tué au temps R. La mesure $\widehat{Q}(f) = \widehat{P}(f \circ k_R)$ n'est autre que la mesure des excursions d'Itô. Sous $\widehat{Q}$ le processus des coordonnées est markovien $/(Q_t)$, avec la condition renforcée de la remarque 6.2).

Dans ces deux exemples, il n'existe pas d'ensemble admissible non négligeable pour $\widehat{P}$ (resp. $\widehat{Q}$).

3) Dans le cas général supposons qu'il existe un point $a \in E$ régulier pour lui-même : on a $R = 0$ p^a-p.s., où $R = \inf\{t > 0 \, ; X_t = a\}$ (X désigne ici le processus canonique). La mesure $\widehat{P}$ (resp. $\widehat{Q}$) peut être définie comme précédemment et les propriétés markoviennes de $(X_t)_{t>0}$ subsistent relativement à (P_t) (resp. (Q_t)). Dans ce cas l'ensemble $\{X_0 \neq a\}$ est l'ensemble maximal sur lequel $(X_t)_{t\geqslant 0}$ soit markovien $/(P_t)$ sous $\widehat{P}$ (resp. $/(Q_t)$ sous $\widehat{Q}$).

8. THÉORÈME. — *Soit $(X_t)_{t>0}$ un processus markovien $/(\mathcal{F}_t)_{t>0}$ au sens de la définition 6 et supposons qu'il existe $g > 0$ sur E, mesurable, et $p \in \mathbb{R}_+$ tels que $a = P(\int_0^\infty e^{-ps} g \circ X_s \, ds) < \infty$ $\quad (g(\delta) = 0)$. Alors l'ensemble*

$$\Omega_0 = \left\{ X_t \xrightarrow[t\to 0]{\text{Ray}} \cdot \in E, \quad \underline{\lim}_{t\to 0} U_p g \circ X_t > 0 \right\}$$

est admissible et essentiellement maximal parmi les ensembles admissibles. De plus la mesure restreinte P_{Ω_0} est σ-finie sur $\mathcal{F}_0 = \mathcal{F}_{0+}$ et le processus prolongé $(X_t)_{t\geqslant 0}$ est markovien $/(\mathcal{F}_t)_{t\geqslant 0}$ sous P_{Ω_0}.

Démonstration. — Soit A admissible. Alors X_t converge p.s. sur A ($t \to 0$) au sens de Ray, vers une limite dans E et $\underline{\lim} U_p g \circ X_t > 0$. En effet, on a une propriété analogue pour le processus canonique sous chaque mesure P^x. Par suite $A \subset \Omega_0$ p.s.. Par ailleurs $\underline{\lim} U_p g \circ X_t = \underline{\lim} e^{-pt} U_p g \circ X_t$ est intégrable, car $P(e^{-pt} U_p g \circ X_t) = P\left(\int_t^\infty e^{-ps} g \circ X_s \, ds \right) \leqslant a$. Donc la mesure P restreinte à $\left\{ \underline{\lim} U_p g \circ X_t > 0 \right\}$ est σ-finie sur $\mathcal{F}_0$ et il résulte du cas où P est finie (n° 5) que Ω_0 est admissible et même que le prolongement est markovien $/(\mathcal{F}_t)_{t\geqslant 0}$ (ce qui signifie que les conditions *6a)* et *6b)* sont exigées pour $t \geqslant 0$).

Remarque. — On peut donner à l'ensemble Ω_0 une forme ne faisant pas intervenir explicitement la topologie de Ray (*cf.* l'ensemble Ω_q de FM, section 3).

FERMETURE D'UNE LOI D'ENTRÉE.

9. — Rappelons qu'une loi d'entrée $(\mu_t)_{t>0}$ est une famille de mesures sur E σ-finies et telles que $\mu_{t+s} = \mu_t P_s$, $(s, t > 0)$. Elle est dite *fermable* s'il existe une mesure σ-finie μ sur E telle que $\mu_t = \mu P_t$. On pourra consulter DMM XVII 13 pour une étude directe de la fermabilité de (μ_t). Nous allons regarder brièvement cette question à la lumière du théorème de prolongement du n° 8. Sous les hypothèses de ce théorème, les mesures $\mu_t = X_t(P)$ sur E forment une loi d'entrée (μ_t est σ-finie, car $\mu_t(U_p g) < \infty$). Si μ est la loi de X_0 sous P_{Ω_0}, on a $\mu_t \geqslant \mu P_t$ et $(\mu P_t)_{t>0}$ est la plus grande loi d'entrée fermable majorée par (μ_t) (μ est σ-finie car $\mu(U_p g) \leqslant a$). De plus *(μ_t) est fermable si et seulement si P est portée par Ω_0.*

Inversement, partons d'une loi d'entrée (μ_t) telle que $\eta = \int_0^\infty e^{-pt} \mu_t \, dt$ soit σ-finie. Elle peut être associée comme précédemment à un processus markovien $(X_t)_{t>0}$ (*cf.* Remarque 9) satisfaisant aux hypothèses du théorème 8 et l'on dispose encore du critère de fermabilité précédent.

Remarque. — La construction d'un processus markovien (X_t) admettant (μ_t) comme loi d'entrée est facile lorsque $\sup_t \mu_t(E) < \infty$. En effet, en posant

$$\nu_t = \mu_t + a_t \varepsilon_\delta, \quad \text{où } a_t = \lim_{\varepsilon \downarrow 0} \uparrow \mu_\varepsilon(1 - P_{t-\varepsilon} 1_E) ,$$

on obtient une loi d'entrée pour le s.g. *markovien* (P_t) sur E_δ et cette loi d'entrée est bornée. La construction est également facile si (P_t) est déjà *markovien* sur E, car le théorème de Kolmogorov reste valable pour des mesures infinies, lorsque l'une des mesures du système projectif est σ-finie. Dans le cas général, la construction fait appel au théorème de Kuznetsov (voir II.9).

II. Processus de Markov à naissance aléatoire (PMNA)

1. — Les processus que nous allons construire et commencer à étudier dans ce chapitre évoluent dans l'espace d'états E de manière markovienne, sur un intervalle stochastique (α, β) de $\mathbb{R}$ (en dehors duquel nous lui attribuons la valeur δ). Leur étude sera surtout développée dans le cas stationnaire, aux chapitres III et IV. Mais dans ce cas la mesure sous-jacente est typiquement *infinie* : elle ne pourra être finie que si $(\alpha, \beta) = \mathbb{R}$ p.s. et si la loi unidimensionnelle du processus (qui est invariante par le semi-groupe) est finie. Cela fait partie du charme de ces processus. Notre méthode de construction repose essentiellement sur celle de Dynkin [4] et Kuznetsov [9], combinée avec une idée de J. Mitro [10] (la définition de Q comme $\bigvee_r Q^r$ dans la démonstration du théorème 6).

2. — Considérons un espace mesuré $(W, \mathcal{G}, Q)$, muni d'une filtration $(\mathcal{G}_u)_{u \in \mathbb{R}}$ et un processus $(Y_u)_{u \in \mathbb{R}}$ à valeurs dans E_δ, adapté à $(\mathcal{G}_u)$. *En général, nous utiliserons les lettres u, v pour noter des instants de $\mathbb{R}$, et s, t pour des instants de $\mathbb{R}_+$.*

Dans la suite p.s. signifie : Q-presque partout. Nous adopterons la définition suivante, un peu différente de celle de DMM.

DÉFINITION. — *Le processus (Y_u) est dit markovien à naissance aléatoire relativement à $(\mathcal{G}_u)$ (et à (P_t)) si*

a) l'ensemble $\{u : Y_u \in E\}$ est p.s. un intervalle stochastique (α, β) de longueur > 0 (sa forme n'est pas précisée pour le moment),

b) la mesure $Q^u = Q_{\{Y_u \in E\}}$ est σ-finie sur $\mathcal{G}_u$ pour tout $u \in \mathbb{R}$,

c) pour tout u, le processus $(Y_{u+t})_{t \geq 0}$ est markovien $/(\mathcal{G}_{u+t})_{t \geq 0}$ sous la mesure Q^u.

Le temps de naissance α est une v.a., car $\alpha = \inf\{r \text{ rationnel} : Y_r \in E\}$. En général, on exigera une propriété *b) renforcée*, en demandant que *la loi η_u de Y_u sur E soit σ-finie.* Lorsque $(\mathcal{G}_u)$ est la tribu naturelle de (Y_u), la propriété *c)* équivaut alors à la propriété de Markov pour tout u de $(Y_{u+t})_{t \geq 0}$ sous Q^u.

3. Exemples.

1) Si $(\eta_u)_{u\in\mathbb{R}}$ est une loi d'entrée sur E pour (P_t) (les mesures η_u sont σ-finies et telles que $\eta_{u+t} = \eta_u P_t$) et si le semi-groupe (P_t) sur E est *markovien*, le théorème de Kolmogorov (avec une extension facile) permet de construire un PMNA (Y_u) tel que $Y_u(Q) = \eta_u$. Dans ce cas $(\alpha, \beta) = \mathbb{R}$. Ce processus est stationnaire ssi $\eta_u = \eta$ fixe.

2) Soit μ une mesure sur E et soit P^μ la mesure correspondante sur l'espace canonique usuel Ω. Sur $E_\delta^{\mathbb{R}}$ on construit la mesure $Q = \int \varphi_u(P^\mu)\, du$, où $\varphi_u : \Omega \to E_\delta^{\mathbb{R}}$ est définie par : $\varphi_u(\omega)(v) = \omega(v-u)$ si $v \geqslant u$, δ sinon. Si la mesure μU est σ-finie (on dit que c'est un *potentiel*) le processus (Y_u) des coordonnées est un PMNA stationnaire : sur E on a

$$Y_u(Q) = \int_u^\infty \mu P_{v-u}\, dv = \mu U .$$

La "loi" de α est ici $\mu(E)\lambda$, où λ est la mesure de Lebesgue.

Analyse de la loi d'un PMNA.

4. — Pour un PMNA, tel qu'il est défini au n° 3, la mesure Q est *σ-finie*, car $W = \bigcup_{r\in D} \{Y_r \in E\}$ p.s. d'après *a)*, D désignant une partie dénombrable dense de $\mathbb{R}$, et $\{Y_r \in E\}$ est une union dénombrable d'éléments de $\mathcal{G}_r$ de Q-mesure finie d'après *b)*. On a $Q = \bigvee_{r\in D} Q^r = \bigvee_{u\in\mathbb{R}} Q^u$.

Dans la suite, nous supposerons que

(4.1) *les mesures* $\eta_u = Q(Y_u \in \cdot)$ *sur* $(E, \mathcal{E})$ *sont σ-finies.*

Ceci constitue un renforcement de la condition *b)*. Il résulte de la propriété de Markov de $(Y_{u+t})_{t\geqslant 0}$ sur $\{Y_u \in E\}$ que $\eta_u P_{v-u} \leqslant \eta_v$ si $u \leqslant v$. Une famille (η_u) de mesures σ-finies sur E satisfaisant à cette propriété s'appelle une *règle d'entrée* (pour le s.g. (P_t)). Pour la règle d'entrée (η_u) associée à (Y_u), la différence $\eta_v^u = \eta_v - \eta_u P_{v-u}$ est égale à la mesure $Q\{Y_u = \delta,\ Y_v \in \cdot\}$ sur E.

5. — Précisons la loi de $(Y_u)_{u\leqslant r}$ sous Q^r pour $r \in \mathbb{R}$. Pour f mesurable $\geqslant 0$ sur E_δ^n et $u_1 < \cdots < u_n = r$ nous pouvons écrire $Q^u(f(Y_{u_1}, \ldots, Y_{u_n}))$ sous la forme :

$$\begin{aligned}
&\int_E \eta_{u_1}(dx_1) \int_E P_{u_2}^{u_1}(x_1, dx_2) \int_E P_{u_3}^{u_2}(x_2, dx_3) \cdots \int_E P_{u_n}^{u_{n-1}}(x_{n-1}, dx_n) f(x_1, \ldots, x_n) \\
&\quad + \int_E \eta_{u_2}^{u_1}(dx_2) \int_E P_{u_3}^{u_2}(x_2, dx_3) \cdots \int_E P_{u_n}^{u_{n-1}}(x_{n-1}, dx_n) f(\delta, x_2, \ldots, x_n) \\
&\quad + \ldots\ldots\ldots\ldots\ldots\ldots \\
&\quad + \int \eta_{u_n}^{u_{n-1}}(dx_n) f(\delta, \ldots, \delta, x_n) ,
\end{aligned} \tag{5.1}$$

où $P_v^u = P_{v-u}$ $(u \leqslant v)$ et où nous avons supposé $n \geqslant 3$.

CONSTRUCTION DES PMNA. LE THÉORÈME DE KUZNETSOV.

6. — Pour le théorème suivant, nous n'utiliserons que l'hypothèse

$$P_t 1_E \longrightarrow 1_E \text{ lorsque } t \to 0 , \tag{6.1}$$

qui est une conséquence des hypothèses droites.

THÉORÈME. — *Soit* (η_u) *une règle d'entrée pour* (P_t). *Il existe une mesure unique* Q *sur l'espace canonique* W *des PMNA sous laquelle le processus des coordonnées soit un PMNA de règle d'entrée* (η_u).

(W est l'ensemble des applications w de $\mathbb{R}$ dans E_δ telles que $\{u : w(u) \in E\}$ est un intervalle de longueur > 0, noté $(\alpha(w), \beta(w))$; sa forme n'est pas précisée, mais on voit facilement qu'on peut le choisir ouvert en $\beta(w)$, à cause de (6.1)).

Démonstration. — L'analyse faite au n° 5 établit l'unicité. Pour l'existence, considérons pour $r \in \mathbb{R}$ la mesure Q^r sur $E_\delta^{\mathbb{R}}$ telle que le processus des coordonnées (Y_u) soit markovien à la droite de r avec η_r comme loi initiale et satisfasse (5.1) à la gauche de r (les mesures définies en (5.1) sont σ-finies et forment un système projectif par rapport aux indices $< r$). Compte tenu de (6.1) et de la forme des mesures (5.1), Q^r est portée par W. Nous la restreignons à W sans changer de notations.

Soit D une partie dénombrable dense de $\mathbb{R}$. Grâce à la σ-finitude des mesures Q^r, on peut définir la mesure $Q = \bigvee_{r \in D} Q^r$ sur W. Pour montrer que c'est la mesure cherchée, il suffit de vérifier que $Q = Q^u$ sur $W_u = \{Y_u \in D\}$. Ceci est vrai pour $u \in D$, car sur W_u on a $Q^r \leqslant Q^u$ pour tout $r \in D$ (d'après la forme des lois marginales). Ensuite, pour $u \in \mathbb{R}$, $r \in D$, $r > u$ on a $Q^u = Q^r = Q$ sur $W_u \cap W_r$, d'où $Q^u = Q$ sur W_u.

7. — Considérons les mesures $\hat{\eta}_u = \sup_{r<u} \eta_r P_{u-r}$. On a $\hat{\eta}_u \leqslant \eta_u$ et la différence $\eta_u - \hat{\eta}_u$ peut être évaluée à l'aide du processus (Y_u) qui vient d'être construit : elle vaut $Q\{\alpha = u,\ Y_u \in \cdot\}$ sur E.

Nous venons ainsi d'établir que $\{u : \hat{\eta}_u \neq \eta_u\}$ *est dénombrable*, à cause de la σ-finitude de Q. Lorsque $\hat{\eta}_u = \eta_u$ pour tout u, on dit que la règle d'entrée est *régulière*. Dans ce cas, on peut modifier comme on veut la valeur en α du processus Y sans changer sa loi temporelle. Nous avons ainsi établi le *théorème de Kuznetsov*, que voici :

THÉORÈME. — *Si* (η_u) *est une règle d'entrée régulière il existe un PMNA* (Y_u) *de règle d'entrée* (η_u), *ayant un intervalle de vie ouvert non vide. Sa loi est uniquement déterminée.*

En utilisant maintenant les hypothèses *droites* complètes, nous pouvons même supposer que le processus Y précédent est càd sur son intervalle de vie (on applique le théorème 3.2) aux processus $(Y_{u+t})_{t \geqslant 0}$ sur $\{Y_u \in E\}$, pour u rationnel).

NOTATIONS.

8. — Il est temps de fixer les notations que nous utiliserons *dans toute la suite du cours*. W désigne l'ensemble des applications w de $\mathbb{R}$ dans E_δ telles que $\{u : w(u) \in E\}$ soit un intervalle ouvert, sur lequel w est càd. On note (Y_u) le processus des coordonnées, $(\mathcal{G}^0_u)$ sa filtration naturelle, $\mathcal{G}^0 = \bigvee_u \mathcal{G}^0_u$, $\alpha = \inf\{u : Y_u \in E\}$, $\beta = \sup\{u : Y_u \in E\}$. Le théorème de Kuznetsov montre que si (η_u) est une règle d'entrée régulière, il existe une mesure unique Q sur $(W, \mathcal{G}^0)$ faisant de (Y_u) un PMNA de règle d'entrée (η_u). Cette mesure Q sera appelée *mesure de Kuznetsov associée à* (η_u).

9. — Si $m = (\mu_t)_{t>0}$ est une loi d'entrée sur E, la famille $\eta_u = \mu_u$ si $u > 0$, 0 si $u \leqslant 0$ est une règle d'entrée, la mesure de Kuznetsov associée sera notée P^m. Elle est portée par $\Omega^+ = \{\alpha = 0\}$, que l'on peut identifier à l'ensemble des applications de $]0, \infty[$ dans E_δ càd et à durée de vie > 0. Lorsque $\mu_t = \mu P_t$ pour une mesure μ, la mesure P^m est notée P^μ ; elle est alors portée par l'ensemble $\Omega = \{\alpha = 0,\ Y_t \underset{t\to 0+}{\longrightarrow} \cdot \in E\}$, qui s'identifie à l'espace canonique habituel des processus de Markov càd indexés par $\mathbb{R}_+$. Sur les espaces Ω et Ω^+, les applications coordonnées d'indices > 0, seront notées X_t, $t > 0$, et $\mathcal{F}^0$ désigne la tribu engendrée par les X_t.

LE CAS STATIONNAIRE.

10. — Soit $\eta_u = \eta$ une règle d'entrée constante : cela revient à se donner une mesure σ-finie η telle que $\eta P_t \leqslant \eta$. Sa régularisée $(\hat{\eta}_u)$ est également constante, et égale à $\hat{\eta} = \sup_{t>0} \eta P_t$, donc $\hat{\eta} = \eta$ d'après 7. La mesure η est donc *excessive* (on pourra consulter DM XII 37 pour une démonstration directe de ce résultat bien connu).

Soit Q_η la mesure de Kuznetsov associée à la règle constante $\eta_u = \eta$; on dit aussi qu'elle est associée à la mesure excessive η. Pour tout $r \in \mathbb{R}$, le processus $(Y_{r+u})_{u\in\mathbb{R}}$ est un PMNA de règle d'entrée $\eta_u = \eta$. D'après l'unicité de la mesure Q_η, on a $\sigma_r(Q_\eta) = Q_\eta$, où σ_r est l'opérateur de *translation* défini par $Y_u \circ \sigma_r = Y_{r+u}$ $(r, u \in \mathbb{R})$. $(W, \mathcal{G}^0, \sigma_u, Q_\eta)$ constitue un *flot mesurable* (le processus (σ_u) est mesurable). Cette propriété importante sera exploitée à fond dans les chapitres suivants. On se reportera au fur et à mesure des besoins à l'appendice A, qui rassemble divers résultats généraux concernant les flots.

PROPRIÉTÉ DE MARKOV FORTE.

11. THÉORÈME. — *Soit Q une mesure de Kuznetsov générale et soit $(\mathcal{G}_u)$ la filtration $(\mathcal{G}^0_{u+})$ Q-complétée. Pour tout t.d'a.T de $(\mathcal{G}_u)$, la mesure $Q_{\{Y_T\in E\}}$ est σ-finie sur $\mathcal{G}_T$ et l'on a, pour f mesurable $\geqslant 0$ sur $(\Omega, \mathcal{F}^0)$:*

(11.1) $$Q_{\{Y_T\in E\}}(f \circ \theta_T | \mathcal{G}_T) = P^{Y_T}(f) \quad (Y_{\pm\infty} = \delta)\,,$$

les opérateurs θ_r étant définis par

(11.2) $$Y_u \circ \theta_r = Y_{r+u} \ \text{ si } \ u > 0,\ \delta \ \text{ si } \ u \leqslant 0\,.$$

On vérifie facilement que les processus (Y_u) et (θ_u) sont mesurables ((Y_u) est même optionnel grâce à sa continuité à droite sur $]\alpha,\infty[$). Noter que $\theta_T \in \Omega^+$ sur $\{T \geqslant \alpha\}$ et $\theta_T \in \Omega$ p.s. sur $\{Y_T \in E\}$.

Démonstration. — Les lois $\eta_r = Y_r(Q)$ sur E sont σ-finies ; la σ-finitude de $Q_{\{Y_T\in E\}}$ sur $\mathcal{G}_T$ résulte de l'égalité $\{Y_T \in E\} = \bigcup_{r \text{ rationnel}} \{r < T < \beta,\ Y_r \in E\}$. La propriété de Markov du processus $(Y_{r+t})_{t\geqslant 0}$ relativement à $(\mathcal{G}_{r+t})_{t\geqslant 0}$ sur $\{Y_r \in E\}$, appliquée au temps $(T-r)^+$, montre que l'on a $Q_A(f\circ\theta_T) = Q_A(P^{Y_T}(f))$ pour $A \in \mathcal{G}_T$ tel que $Q(A) < \infty$, $A \subset \{\alpha < r < T < \beta\}$. La formule (11.1) en résulte.

12. — La propriété de Markov forte peut être améliorée dans le cas *stationnaire* grâce au théorème I.8, en modifiant le processus Y à l'instant α. Soit donc $Q = Q_\eta$ la mesure de Kuznetsov associée à la mesure excessive η. Sur l'espace $\Omega^+ = \{\alpha = 0\}$ considérons la mesure $P = Q_{\{0<\alpha\leqslant 1\}}(\sigma_\alpha \in \cdot)$ (on peut remplacer σ_α par θ_α dans cette formule). La propriété de Markov forte de Y aux instants $\alpha+t$, $t>0$ montre que le processus $(X_t)_{t>0}$ est markovien sous P. De plus, d'après A (3.1) appliqué au temps $S=\alpha$

$$Q(f) \geqslant P\Big(\int f\circ\sigma_u\, du\Big)\ . \tag{12.1}$$

En particulier pour $f = g\circ Y_0$ (noter que $\alpha = 0$ P-p.s.)

$$\eta(g) \geqslant P\int_0^\infty g\circ X_t\, dt\ . \tag{12.2}$$

Nous pouvons donc appliquer le théorème I.8 en prenant $p = 0$, $g > 0$ η-intégrable. Posons $X_0 = \lim_{t\to 0+} X_t$ sur Ω_0, δ ailleurs. La loi $\mu = X_0(P)$ sur E est σ-finie.

THÉORÈME. — *Dans le cas stationnaire l'énoncé du théorème précédent reste valable pour le processus Y modifié au point α par la formule*

$$Y_\alpha = X_0\circ\sigma_\alpha = X_0\circ\theta_\alpha \quad \text{sur} \quad \{\alpha\in\mathbb{R}\}\ .$$

Démonstration. — Contentons nous de traiter le cas où $T=\alpha$. D'après A (2.2) la loi de (θ_α,α) sous $Q_{\{Y_\alpha\in E\}}$ est $P\otimes\lambda$; en particulier celle de (Y_α,α) est $\mu\otimes\lambda$. La mesure $Q_{\{Y_\alpha\in E\}}$ est donc σ-finie sur $\mathcal{G}_\alpha$. Pour établir (11.1) avec la valeur modifiée de Y_α il suffit de montrer que pour des fonctions mesurables $\geqslant 0$ h sur $(\Omega,\mathcal{F}^0)$ et f sur $(\Omega\times\mathbb{R}, \bigcap_{t>0}\mathcal{F}_t^0\otimes\mathcal{B}_{\mathbb{R}})$ (où $\mathcal{F}_t^0 = \mathcal{T}(X_s,\ s\leqslant t)$) on a

$$Q_{\{Y_\alpha\in E\}}\big(f(\theta_\alpha,\alpha)h(\theta_\alpha)\big) = Q_{\{Y_\alpha\in E\}}(f(\theta_\alpha,\alpha)P^{Y_\alpha}(h))\ .$$

Cette égalité s'écrit aussi, en posant $g = \int_{\mathbb{R}} f(\cdot,u)\,du$,

$$P_{\{X_0\in E\}}(gh) = P_{\{X_0\in E\}}(gP^{X_0}(h))\ .$$

Mais g est $\mathcal{F}_t^0$ mesurable pour tout $t>0$ et il suffit d'appliquer la propriété de Markov en 0 sous $P_{\{X_0\in E\}}$.

III. Mesures excessives. Décompositions

Dans ce chapitre η désigne une mesure excessive fixée et $Q = Q_\eta$ est la mesure de Kuznetsov associée (II 10). Sous Q le processus (Y_u) est un PMNA stationnaire. On le suppose *modifié au point* α comme dans le théorème II.12.

LE PRINCIPE DES DÉCOMPOSITIONS.

1. — Étant donné un ensemble $A \in \mathcal{G}_\alpha$ invariant ($\sigma_u^{-1}(A) = A$ pour tout $u \in \mathbb{R}$) la mesure définie sur E par

$$\eta^A = Q_A(Y_u \in \cdot) \tag{1.1}$$

ne dépend pas de u et est excessive (car (Y_u) est encore un PMNA sous Q_A), d'où une décomposition $\eta = \eta^A + \eta^{A^c}$ en deux mesures excessives.

Nous allons montrer, en suivant FM, que les décompositions de Riesz $\eta = \eta_p + \eta_h$ et $\eta = \eta_{\text{pur}} + \eta_i$ (cf. n° 3-4) peuvent être obtenues par ce procédé, en prenant successivement $A = \{Y_\alpha \in E\}$ et $A = \{\alpha \in \mathbb{R}\}$ (noter que $\{Y_\alpha \in E\} \subset \{\alpha \in \mathbb{R}\}$, car $Y_{\pm\infty} = \delta$ par convention). Nous montrerons aussi que la décomposition $\eta = \eta_d + \eta_c$ en partie dissipative et partie conservative s'obtient de manière très simple par ce procédé en prenant $A = \bigcap_{u \in \mathbb{R}} \{\int_{-\infty}^u g \circ Y_r dr < \infty\}$, pour $g > 0$ η-intégrable choisie arbitrairement. Ainsi la partie potentiel, la partie purement excessive et la partie dissipative de η sont liées à la finitude de α et au comportement du processus Y au voisinage de α.

POTENTIELS, UNICITÉ DES MASSES.

2. — La *mesure de Palm* associée au temps stationnaire α est la mesure P sur W définie par

$$P(f) = Q_{\{0<\alpha\leqslant 1\}}(f \circ \sigma_\alpha) .$$

D'après A.3, P est l'unique mesure sur W portée par $\Omega^+ = \{\alpha = 0\}$ et telle que

$$Q_{\{\alpha \in \mathbb{R}\}}(f) = P\Big(\int_{\mathbb{R}} f \circ \sigma_u \, du\Big) . \tag{2.1}$$

Lorsque η s'écrit μU pour une mesure μ sur E, on dit que η est un *potentiel* et dans ce cas $P = P^\mu$ d'après II.3.2).

Une mesure excessive ξ est dite *fortement majorée* par η ($\xi \prec \eta$) si $\xi \leqslant \eta$ et si $\eta - \xi$ est excessive. Cela équivaut à $Q_\xi \leqslant Q$ ou, lorsque ξ est un potentiel νU, à $P^\nu \leqslant P$. Pour $\xi = \nu U$ et $\eta = \mu U$, on a donc $\xi \prec \eta$ ssi $\nu \leqslant \mu$. En particulier une mesure potentiel s'écrit μU pour une mesure μ unique : c'est le principe d'*unicité des masses* (*cf.* DMM XVII 18 pour une démonstration directe de ce fait). On notera aussi que μ est σ-finie, car la fonction Ug est > 0 et μ-intégrable si g est > 0 et μU-intégrable.

PREMIÈRE DÉCOMPOSITION DE RIESZ.

3. THÉORÈME. — *La mesure $\eta^{\{Y_\alpha \in E\}}$ s'écrit μU, où $\mu = P(X_0 \in \cdot)$ sur E, et c'est le plus grand potentiel fortement majoré par η.*

Cette mesure sera notée η_p. D'après ce théorème la mesure $\eta_h = \eta - \eta_p$ est excessive et ne majore fortement aucun potentiel $\neq 0$: on dit qu'elle est *harmonique*.

Démonstration. — Pour f mesurable $\geqslant 0$ sur E ($f(\delta) = 0$)

$$Q_{\{Y_\alpha \in E\}}(f \circ Y_0) = P_{\{X_0 \in E\}}\Big(\int_0^\infty f \circ X_t \, dt\Big) = \mu U f$$

d'après (2.1). Donc $\eta^{\{Y_\alpha \in E\}} = \mu U$. Si $\nu U \prec \eta$, on a $P^\nu \leqslant P$; comme P^ν est portée par $\{X_0 \in E\}$, il vient $P^\nu \leqslant P_{\{X_0 \in E\}} = P^\mu$, d'où $\nu \leqslant \mu$, donc $\nu U \leqslant \eta_p$ (et même $\nu U \prec \eta_p$).

Remarques.

a) Il résulte du théorème précédent que η est un potentiel (resp. une mesure harmonique) ssi Q est portée par $\{Y_\alpha \in E\}$ (resp $\{Y_\alpha = \delta\}$) et que la décomposition $\eta = \eta_p + \eta_h$ en un potentiel et une mesure harmonique est unique.

b) Soit S un t.d'a. de $(\mathcal{G}_u) \geqslant \alpha$ et stationnaire ($S = u + S \circ \sigma_u$, $\forall u \in \mathbb{R}$). Le processus $Z_u = Y_u$ si $u > S$, δ si $u \leqslant S$ est un PMNA stationnaire sous $Q_{\{S<\beta\}}$, donc la mesure sur E

$$\eta^S = Q(S < u,\ Y_u \in \cdot)$$

est excessive. Elle sera appelée *réduite* de η par S (elle a été introduite dans FM sous le nom de S-balayée de η). Lorsque $S > \alpha$ p.s. on a $X_0 \circ \theta_S \in E$ p.s. sur $\{S < \beta\}$, donc la réduit η^S est un potentiel d'après le critère de la remarque a).

DEUXIÈME DÉCOMPOSITION DE RIESZ.

4. — Rappelons qu'une mesure excessive ξ est dite *purement excessive* si $\lim\limits_{t\to\infty} \xi P_t f = 0$ pour $f \geqslant 0$ et ξ-intégrable et *invariante* si $\xi P_t = \xi$ pour tout $t \geqslant 0$. La mesure excessive η s'écrit de manière unique $\eta = \eta_{\mathrm{pur}} + \eta_i$, où η_{pur} est purement excessive et η_i est invariante. On a $\eta_i = \lim\limits_{t\to\infty} \eta P_t$.

THÉORÈME. — *Soit μ_t la loi de X_t sur E sous la mesure P. On a $\eta_{\mathrm{pur}} = \eta^{\{\alpha \in \mathbb{R}\}} = \int_0^\infty \mu_t \, dt$, $\eta_i = \eta^{\{\alpha = -\infty\}}$.*

Démonstration. — D'après (2.1) la mesure $\eta^{\{\alpha \in \mathbb{R}\}}$ s'écrit $\int_0^\infty \mu_t \, dt$, donc elle est purement excessive. D'autre part $\eta^{\{\alpha=-\infty\}} P_t f = Q_{\{\alpha=-\infty\}}(P_t f \circ Y_0) = Q_{\{\alpha=-\infty\}}(f \circ Y_t) = Q_{\{\alpha=-\infty\}}(f \circ Y_0) = \eta^{\{\alpha=-\infty\}}(f)$, donc $\eta^{\{\alpha=-\infty\}}$ est invariante et le théorème en résulte.

MESURES DISSIPATIVES, MESURES CONSERVATIVES.

5. — Il existe de nombreuses manières de définir la dissipativité d'une mesure excessive. Nous adopterons la définition de FM.

DÉFINITION. — *Une mesure excessive ξ est dite dissipative si elle est la borne supérieure des potentiels (simplement) majorés par ξ et conservative si tout potentiel $\leqslant \xi$ est nul.*

Dans la suite g désigne une fonction > 0 sur E, η-intégrable et A désigne l'ensemble $\bigcap_{u\in\mathbb{R}} \left\{ \int_{-\infty}^{u} g\circ Y_r \, dr < \infty \right\}$. Pour $u \in \mathbb{R}$ choisi arbitrairement on a $A = \left\{ \int_{-\infty}^{u} g\circ Y_r \, dr < \infty \right\}$ p.s., puisque $Q\left(\int_{u}^{v} g\circ Y_r \, dr \right) = (v-u)\eta(g) < \infty$ $(u \leqslant v)$. L'ensemble A est $\mathcal{G}_\alpha$-mesurable, et de plus il est invariant.Les mesures η^A et η^{A^c} sont donc excessives. Noter que $A \supset \{\alpha \in \mathbb{R}\}$, $A^c \subset \{\alpha = -\infty\}$ p.s..

THÉORÈME. — *La mesure $\eta_d = \eta^A$ est dissipative et $\eta_c = \eta^{A^c}$ est conservative. De plus*

$$\eta_d = \sup\{\mu U : \mu U \leqslant \eta\} = \eta_{\{Ug<\infty\}}, \quad \eta_c = \eta_{\{Ug=\infty\}} \ .$$

Démonstration. — Les t.d'a. stationnaires $S_n = \inf\left\{ u : \int_{-\infty}^{u} g \circ Y_r \, dr > \frac{1}{n} \right\}$ sont $> \alpha$ sur A . D'après la remarque 3b) la réduite η^{A,S_n} s'écrit $\mu_n U$ ($\mu_n = Q(Y_{S_n} \in \cdot,\ 0 < S_n \leqslant 1)$). Par ailleurs $S_n \downarrow\downarrow \alpha$ sur A, donc $\mu_n U \nearrow \eta^A$. Ceci montre que η^A est dissipative et portée par $\{Ug < \infty\}$, comme chacune des mesures $\mu_n U$ (noter que μ_n est portée par $\{Ug < \infty\}$ et que cet ensemble est absorbant). Pour terminer la démonstration il suffit d'établir

$$\int_{u}^{\infty} g \circ Y_r \, dr = \infty \text{ p.s. sous } Q_{A^c}, \quad (u \in \mathbb{R}) \ . \tag{5.1}$$

En effet (5.1) entraîne que $Ug(Y_u) = \infty$ p.s. (Q_{A^c}) et que η^{A^c} est portée par $\{Ug = \infty\}$, donc est conservative (si $\mu U \leqslant \eta^{A^c}$, μU est portée par $\{Ug = \infty\}$ et par $\{Ug < \infty\}$).

Pour établir (5.1), montrons que le temps stationnaire $S = \inf\left\{ u : \int_{u}^{\infty} g\circ Y_s \, ds < \infty \right\}$ est p.s. infini sous Q_{A^c}. En effet, la mesure $Q' = Q_{A^c \cap \{S<+\infty\}}$ est invariante par les σ_u et est portée par $\{S \in \mathbb{R}\}$. Comme $Q'(g \circ Y_0) < \infty$, on a $\int_{-\infty}^{+\infty} g \circ Y_s \, ds < \infty$ Q p.s., d'après A (3.1), donc Q', portée à la fois par A et A^c, est nulle.

6. — Voici un résultat complémentaire fournissant des critères de dissipativité ou de conservativité. La fonction g est comme précédemment et u est un réel arbitraire.

THÉORÈME.

1) Les conditions suivantes sont toutes équivalentes à la dissipativité de η (p.s. fait référence à la mesure $Q = Q_\eta$) :

a) $\int_{-\infty}^{u} g \circ Y_r \, dr < \infty$ p.s.,

b) il existe une suite décroissante (S_n) de t.d'a. stationnaires telle que $S_n \downarrow\downarrow \alpha$ p.s.,

c) il existe une suite croissante $(\mu_n U)$ de limite η,

d) η est portée par $\{Ug < \infty\}$,

e) $\int_u^\infty g \circ Y_r \, dr < \infty$ *p.s.,*

f) il existe une suite croissante (L_n) *de temps stationnaires telle que* $L_n \uparrow\uparrow \beta$ *p.s.,*

g) il existe un temps stationnaire p.s. fini S*.*

2) Chacune des conditions suivantes équivaut à la conservativité de η *:*

i) $\int_{-\infty}^{+\infty} g \circ Y_r \, dr = \infty$ *p.s.,*

ii) $\int_{-\infty}^{u} g \circ Y_r \, dr = \infty$ *p.s.,*

iii) $\int_u^\infty g \circ Y_r \, dr = \infty$ *p.s.,*

iv) η *est portée par* $\{Ug = \infty\}$*.*

Démonstration.

1) Les implications *a)* $\Longrightarrow$ *b)* $\Longrightarrow$ *c)* $\Longrightarrow$ *d)* ont déjà été établis dans la démonstration du théorème 5 ; *d)* $\Longrightarrow$ *e)* résulte de la propriété de Markov sur $\{Y_u \in E\}$. Pour *e)* $\Longrightarrow$ *f)* il suffit de poser

$$L_n = \sup \left\{ u : \int_u^\infty g \circ Y_r \, dr \geqslant \frac{1}{n} \right\} .$$

Pour *f)* $\Longrightarrow$ *g)* on pose $S = L_N$, où $N = \inf\{n : L_n > -\infty\}$. Enfin *g)* $\Longrightarrow$ $\int_{-\infty}^{+\infty} g \circ Y_r \, dr < \infty$ p.s. (*cf.* démonstration de (5.1)).

2) On a $\int_u^\infty g \circ Y_r \, dr < \infty$ p.s. sur $\{\int_{-\infty}^u g \circ Y_u \, dr < \infty\}$ d'après 1) appliqué à η^A, donc *i)* $\Longleftrightarrow$ *ii)* ; *ii)* $\Longrightarrow$ *iii)* d'après (5.1) ; *iii)* $\Longrightarrow$ $Ug(Y_u) = \infty$ p.s. sur $\{Y_u \in E\}$ $\Longrightarrow$ *iv)*. Enfin *iv)* équivaut à *ii)* ou à $\eta = \eta_c$ d'après le théorème 5.

Remarques.

a) Les inclusions $\{Y_\alpha \in E\} \subset \{\alpha \in \mathbb{R}\} \subset A$ montrent que Pot $\subset$ Pur $\subset$ Dis et Har $\supset$ Inv $\supset$ Csv avec des notations évidentes pour les diverses classes de mesures excessives.

b) Une mesure invariante finie η est conservative ($\int_{-\infty}^{+\infty} g \circ Y_r \, dr = +\infty$ p.s. pour $g \equiv 1$), mais ce n'est plus vrai si η est infinie : la mesure de Lebesgue de $\mathbb{R}^n$ est conservative (resp. dissipative) pour le semi-groupe de Wiener de dimension n si $n \leqslant 2$ (resp. $\geqslant 3$).

c) Soit $(W, \mathcal{G}, (\sigma_u), Q)$ un flot mesurable général. La mesure Q est invariante pour le semi-groupe $(\sigma_t)_{t \geqslant 0}$ sur $(W, \mathcal{G})$ et le processus (σ_u) est markovien par rapport à ce semi-groupe. D'après le théorème 6 les conditions $\int_0^\infty g \circ \sigma_r \, dr < \infty$ p.s. et $\int_{-\infty}^0 g \circ \sigma_r \, dr < \infty$ p.s. sont équivalentes et caractérisent la dissipativité de Q (ici les conditions *d)* et *e)* sont les mêmes). Les conditions $\int_0^\infty g \circ \sigma_r \, dr = \infty$ p.s., $\int_{-\infty}^0 g \circ \sigma_r \, dr = \infty$ p.s., $\int_{-\infty}^{+\infty} g \circ \sigma_r \, dr = \infty$ p.s. sont équivalentes et caractérisent la conservativité de η (g est > 0 et Q intégrable). Dans le cas du flot de Kuznetsov, on voit donc que η est dissipative (conservative) ssi Q l'est. Ce résultat est dû à Fitzsimmons [7].

d) Dans FM et dans DMM XIX, le lecteur pourra trouver des décompositions de dernière sortie de η (ou plutôt de sa réduite) relatives à un ensemble aléatoire homogène.

IV. Retournement du temps

Ce chapitre est consacré à divers résultats liés à la dualité. Nous commencerons par le retournement d'un PMNA, qui s'effectue simplement en remplaçant l'ordre sur $\mathbb{R}$ par l'ordre renversé et en déduirons, selon Dynkin [5], le théorème de Nagasawa [11] de retournement à un temps de retour fini, pour un processus de Markov $(X_t)_{t>0}$.

RETOURNEMENT SUR W.

1. — Soit η une mesure excessive fixée. On suppose qu'il existe un semi-groupe droit $(\widehat{P}_t)$ sur E, en dualité avec (P_t) relativement à η :

$$\eta(fP_tg) = \eta(g\widehat{P}_tf) \tag{1.1}$$

pour f, g mesurables $\geqslant 0$ sur E.

Sous la mesure de Kuznetsov $Q = Q_\eta$ le processus $(Y_u)_{u\in\mathbb{R}}$ est alors markovien $/(\widehat{P}_t)$ pour l'ordre *renversé* d'après le calcul suivant :

$$\begin{aligned} Q(f(Y_{u-t})h(\theta_u), Y_u \in E) &= Q(f(Y_{u-t})g(Y_u),\ Y_u \in E) \\ &= \eta(fP_tg) = \eta(g\widehat{P}_tf) \\ &= Q(\widehat{P}_tf(Y_u)g\circ\theta_u,\ Y_u \in E) \end{aligned}$$

($u \in \mathbb{R}$, $t \in \mathbb{R}^+$, $f \geqslant 0$ sur E, $h \geqslant 0$ sur Ω, $g = P^{\cdot}(h)$).

Comme le semi-groupe $(\widehat{P}_t)$ est droit, le processus (Y_u) admet p.s. des limites à gauche sur $]-\infty,\beta[$ pour la topologie de E, d'après le théorème I.4.2). Le processus (Y_u^-) des limites à gauche sur $]-\infty,\beta[$, valant δ sur $[\beta,\infty[$, est le processus de Kuznetsov associé à $(\widehat{P}_t)$, η (qui est excessive $/(\widehat{P}_t)$) et à l'ordre *renversé*. Pour maximiser la propriété de Markov forte de ce processus, on le *modifie en* β à la manière de II.12, sans changer de notation. On peut alors énoncer le résultat suivant, où $(\overline{\mathcal{G}}_u)$ désigne la filtration pour l'ordre renversé obtenue par complétion de $(\overline{\mathcal{G}}^0_{u-})$, avec $\overline{\mathcal{G}}^0_u = \mathcal{T}(Y_s,\ s \geqslant u)$.

THÉORÈME. — *Pour tout t.d'a. T de $(\overline{\mathcal{G}}_u)$*

a) le processus $(Y^-_{T-t})_{t>0}$ est markovien $/(\widehat{P}_t)$ sous la mesure $Q_{\{\alpha<T\leqslant\beta\}}$,

b) le processus $(Y^-_{T-t})_{t\geqslant 0}$ est markovien $/(\widehat{P}_t)$ sous la mesure $Q_{\{Y^-_T\in E\}}$.

RETOURNEMENT SUR Ω^+.

2. — Soit $m = (\mu_t)_{t>0}$ une loi d'entrée sur E et soit $P = P^m$ la mesure de Markov associée sur Ω^+. Nous supposerons que la mesure $\eta = \int_0^\infty \mu_t\, dt$ est σ-finie, donc excessive, que $(\widehat{P}_t)$ est un semi-groupe droit satisfaisant à (1.1) et que L est un *temps de retour* sur

Ω^+ ($L \circ \theta_t = (L-t)^+$ pour $t \geqslant 0$) universellement mesurable et P-p.s. fini et $\leqslant \zeta$. Lorsque $\zeta < \infty$ p.s. on peut prendre $L = \zeta$. Voici le théorème de Nagasawa (initialement énoncé pour une loi d'entrée fermée $\mu_t = \mu P_t$ et $P = P^\mu$).

THÉORÈME. — *Le processus (X_t) admet P-p.s. des limites à gauche sur $]0, L[$ et le processus retourné $X_t^L = X_{(L-t)-}$ si $0 < t < L$, δ si $t \geqslant L$ est markovien $/(\widehat{P}_t)$.*

Démonstration. — Considérons la mesure $Q = Q_\eta$ et étendons la définition de L à W en posant $L = \alpha + L \circ \theta_\alpha$ sur $\{\alpha > -\infty\}$, $L = -\infty$ sur $\{\alpha = -\infty\}$. L est un t.d'a. $\leqslant \beta$ relatif à $(\overline{\mathcal{G}}_u)$, car $\{L > u,\ \alpha \leqslant u\} = \{L \circ \theta_u > 0,\ \alpha \leqslant u\}$. Sous la mesure $Q_{\{0<L\leqslant 1\}}$ le processus $(Y_{L-t})_{t>0}$ admet p.s. des limites à droite et le processus des limites à droite correspondant est markovien $/(\widehat{P}_t)$, d'après le *a)* du théorème précédent. Par ailleurs le processus (Y_{L-t}) est formé de v.a. invariantes, à cause de la stationnarité de L. Les propriétés énoncées sous $Q_{\{0<L\leqslant 1\}}$ restent valables sous $\Pi = Q_{\{0<\alpha\leqslant 1\}}$, grâce au lemme qui suit. Il reste à remarquer que $P = \theta_\alpha(\Pi)$ et que $Y_{L-t} = X_{L-t} \circ \theta_\alpha$ sur $\{\alpha > -\infty\}$.

LEMME. — *Si S et T sont deux temps stationnaires p.s. finis, on a $Q_{\{0<S\leqslant 1\}} = Q_{\{0<T\leqslant 1\}}$ sur la tribu invariante.*

Démonstration. — Soit $f \geqslant 0$ invariante. On a

$$\begin{aligned} Q(f,\ 0 < S \leqslant 1) &= Q \int \left(f I_{\{0<S+u\leqslant 1,\ 0<T\leqslant 1\}} \right) \circ \sigma_u \, du \\ &= \int Q(f,\ 0 < S + u \leqslant 1,\ 0 < T \leqslant 1)\, du \\ &= Q(f,\ 0 < T \leqslant 1)\,. \end{aligned}$$

ÉTUDE DES FONCTIONS EXCESSIVES.

3. — Supposons toujours que $(\widehat{P}_t)$ soit un semi-groupe droit en dualité avec (P_t) relativement à une mesure excessive η. Soit a une fonction excessive finie. La mesure $\xi = a \cdot \eta$ est alors σ-finie. Elle est même excessive $/(\widehat{P}_t)$ (on dit qu'elle est coexcessive) d'après un calcul évident, donc elle admet une décomposition de Riesz

$$\xi = \nu \widehat{U} + \xi_h\,, \tag{3.1}$$

où $\widehat{U} = \int_0^\infty \widehat{P}_t\, dt$ et où ξ_h est harmonique $/(\widehat{P}_t)$.

La dualité (1.1) est souvent renforcée par les hypothèses d'absolue continuité (*cf.* [2])

$$\begin{aligned} U(x, dy) &= u(x,y)\eta(dy)\,, \\ \widehat{U}(x, dy) &= u(y,x)\eta(dy)\,. \end{aligned} \tag{3.2}$$

La fonction $U\nu(x) = \int u(x,y)\nu(dy)$ est alors le plus grand "potentiel de mesure" fortement dominé par a au sens des fonctions excessives, d'où la décomposition de Riesz de la fonction excessive a.

4. — Revenons à la situation générale du n° 3 et considérons une fonctionnelle additive A portée par $]0,\zeta]$ telle que $a = E^{\cdot}(A_\infty)$ (elle est unique si l'on exige qu'elle soit prévisible). La f.a. A induit sur W une mesure aléatoire homogène (MAH) encore notée A, portée par $]\alpha,\beta]$ et telle que

$$(4.1) \qquad A]u,u+t] = A_t \circ \theta_u \text{ sur } \{u > \alpha\}(u \in \mathbb{R},\ t \in \mathbb{R}_+) .$$

On consultera A4 pour la définition des MAH.

Remarque. — La mesure A est de Radon sur $]\alpha,\infty[$, mais il peut se faire que $A(]\alpha,\alpha+1]) = \infty$. Par exemple pour le semi-groupe de la translation uniforme sur $]0,\infty[$ nous pouvons restreindre Ω à l'ensemble $\{X_t = X_0+t,\ \forall t \in \mathbb{R}_+\}$ et définir la f.a. $A_t = \rho]X_0,X_t]$, où ρ est la mesure $\frac{1}{s}ds$ sur $]0,\infty[$. La MAH associée explose en α sur $\{\alpha > -\infty\}$.

THÉORÈME. — *Soit Q^ξ la mesure sur W faisant de (Y_u) un PMNA stationnaire $/(\widehat{P}_t)$ pour l'ordre renversé et de règle d'entrée ξ.*

1) Pour f mesurable $\geqslant 0$ sur W on a

$$(4.2) \qquad Q^\xi(f) = Q\Big(\int f \circ k_s A(ds)\Big) ,$$

où k_s est l'opérateur de meurtre à s.

2) La mesure ν intervenant dans la décomposition (3.1) s'exprime à l'aide de A par les formules :

$$(4.3) \qquad \nu(h) = Q \int_{]0,1]} h \circ Y_s^- A(ds) ,$$

$$(4.4) \qquad \nu(h) = P^m\Big(\int_0^\infty h \circ X_s^- A(ds)\Big) + P^{\eta_i}\Big(\int_0^\infty e^{-s} h \circ X_s^- dA_s\Big) ,$$

avec les notations de III.4 et $m = (\mu_t)$, $X_s^- = Y_s^-$ sur Ω^+.

On notera que dans ces formules Y^- est le processus modifié en β comme au n° 1.

Démonstration.

1) Soit Q' la mesure définie au second membre de (4.2) et soit $\overline{\theta}_u = (Y_{u+t})_{t<0}$. On a

$$\begin{aligned} Q'(g(\overline{\theta}_u),\ Y_u \in E) &= Q(A]u,\infty[g(\overline{\theta}_u),\ Y_u \in E) \\ &= Q(A]u,\infty[\overline{P}^{Y_u}(g),\ Y_u \in E) \\ &= Q'(\overline{P}^{Y_u}(g),\ Y_u \in E) , \end{aligned}$$

où $\overline{P}^x$ désigne la loi de $(X_{-t})_{t<0}$ sous la mesure $\widehat{P}^x$ associée au semi-groupe $(\widehat{P}_t)$. Pour établir le 1) il reste à vérifier que $Q'(Y_u \in \cdot) = \xi$ sur E, ce qui est immédiat.

2) La formule (4.3) résulte de (4.2) et du théorème III.3 appliqué au semi-groupe $(\widehat{P}_t)$ et au processus (Y_u^-), pour l'ordre renversé. Il reste à établir (4.4).

Soit φ une fonction $\geqslant 0$ sur $\mathbb{R}$ d'intégrale de Lebesgue 1. D'après le théorème A4, on a

$$\nu(h) = Q \int \varphi(s) h(Y_s^-) A(ds)$$

d'où il vient d'après III (2.1)

$$\nu(h) = P \int \Big(\int \varphi(s) h(Y_s^-) A(ds) \Big) \circ \sigma_u \, du$$
$$+ Q_{\eta_i} \Big(\int \varphi(s) h(Y_s^-) A(ds) \Big) .$$

Le second terme s'écrit aussi $P^{\eta_i}\Big(\int_0^\infty e^{-s} h(X_s^-) dA_s \Big)$ et le premier s'écrit

$$P \int \int \varphi(s) h(Y_{s+u}^-) A(\sigma_u, ds) du = P \int \int \varphi(s-u) h(Y_s^-) A(ds) du$$
$$= P \Big(\int_0^\infty h(X_s^-) A(ds) \Big) .$$

Remarque. — La mesure ξ s'écrit $\nu \widehat{U}$ si et seulement si $Q(A\{\beta\}, Y_\beta^- = \delta) = 0$. Cette condition est liée aux hypothèses (exprimées sur Ω^+) faites pour le théorème de Revuz (DMM XVIII 33).

Le théorème d'Azéma de représentation des mesures.

5. — Une mesure ν sur E peut-elle être représentée par une formule du type (4.3) ou (4.4) à l'aide d'une f.a. prévisible A ? Pour cela une condition nécessaire est que ν ne charge pas les ensembles η-polaires à gauche, c'est-à-dire ceux que le processus Y^- ne visite pas. Cette condition est également suffisante (pour ν bornée) avec quelques réserves que nous ne préciserons pas.

Nous concentrerons plutôt notre attention sur un problème de représentation voisin, sans hypothèse de dualité particulière. Nous dirons qu'un ensemble presque borélien est *η-polaire* s'il est Q (ou $P^m + P^{\eta_i}$) p.s. non rencontré par le processus Y modifié selon II.12. Rappelons qu'une fonctionnelle additive gauche (FAG) est un processus croissant (càd à valeurs dans $[0,\infty]$) défini sur Ω tel que $A_{t+s} = A_{t-} + A_s \circ \theta_t$. La v.a. A_0 est $\geqslant 0$, non nécessairement nulle, les v.a. A_t sont supposées universellement mesurables. A la FAG A on associe la mesure aléatoire, encore notée A, portée par $[\alpha, \infty[$ définie sur W par

$$A[u,\; u+t[= A_t \circ \theta_u \quad \text{sur} \quad \{u > \alpha\} ,$$

$$A\{\alpha\} = A_0 \circ \theta_\alpha \text{ sur } \{\alpha \in \mathbb{R},\; Y_\alpha \in E\}, 0 \text{ ailleurs.}$$

THÉORÈME. — *Si ν est une mesure finie sur E ne chargeant pas les ensembles η-polaires, il existe une FAG adaptée A portée par $[0,\zeta[$ telle que*

$$\nu(h) = Q\int_{]0,1]} h(Y_s)A(ds)$$
$$= P^m\int_{[0,\infty[} h(X_s)A(ds) + P^{\eta_i}\int_{[0,\infty[} e^{-s}h(X_s)dA_s\ .$$

(la v.a. X_0 étant définie sur Ω^+ comme en II.12). *De plus la mesure aléatoire associée A est l'unique mesure aléatoire homogène sur W, optionnelle $/(\mathcal{G}_u)$ et prévisible $/(\overline{\mathcal{G}}_u)$, portée par $\{Y \in E\}$ permettant d'écrire l'une des égalités ci-dessus* (l'unicité s'entend à l'indistinguabilité $/Q$ ou $P^m + P^{\eta_i}$ près).

Le cas où η est un potentiel est dû à Azéma [1] et le cas général est dû à Fitzsimmons [6] (5.22), à ceci près que la classe des ensembles η-évanescents de [6] est plus grande que celle des ensembles η-polaires (notre énoncé permet donc de représenter davantage de mesures).

Démonstration. — Si A existe on doit avoir, d'après A.4

$$Q\int f(Y_s,s)A(ds) = \nu\otimes\lambda(f)\ .$$

D'autre part, d'après le lemme qui suit, la projection optionnelle $/(\mathcal{G}_u)$ et la projection prévisible $/(\overline{\mathcal{G}}_u)$ commutent et pour un processus Z mesurable $\geqslant 0$ sur W la projection double $^{\overline{p}0}Z$ est de la forme $f_Z(Y_u,u)$, donc

$$Q\Big(\int Z_sA(ds)\Big) = \nu\otimes\lambda(f_Z)\ . \tag{5.2}$$

La mesure $Q\otimes A$ est donc déterminée par (5.2), d'où l'unicité de A.

Pour l'existence de A on pose

$$\mu(Z) = \nu\otimes\lambda(f_Z)\ , \tag{5.3}$$

ce qui, grâce aux hypothèses faites sur ν, définit bien une mesure sur $W\times\mathbb{R}$, σ-finie, ayant une première projection absolument continue par rapport à Q. En effet, si $Z = 0$, le processus $Z_{s-r}(\sigma_r)$ est également nul pour tout r fixé, donc $(f_Z(Y_{s,s-r}))_{s\in\mathbb{R}}$ est indistinguable de 0 et l'ensemble $\{\int f_Z(\cdot,r)\,dr > 0\}$, p.s. ignoré par Y, est ν-négligeable ; la σ-finitude provient de ce que $\mu(Z) = \nu(h)\lambda(\varphi)$ lorsque Z_u est de la forme $h(Y_u)\varphi(u)$ ($h,\varphi \geqslant 0$). On peut donc écrire μ sous la forme $Q\otimes A$, où A est une mesure aléatoire. Cette mesure A est homogène au sens faible suivant : pour tout $r\in\mathbb{R}$ la mesure $A(\sigma_r,\cdot - r)$ est p.s. égale à A. Par des méthodes un peu fastidieuses (*cf.* [6], [8]) on en déduit une version "parfaite", universellement mesurable de A et il reste à poser

$$A_t = A[0,t] \text{ sur } \Omega \tag{5.4}$$

pour obtenir la FAG cherchée. La deuxième expression de $\nu(h)$ s'obtient à la manière de (4.4).

LEMME.

1) Il existe un noyau $(\overline{P}^x)_{x\in E}$ *formé de mesures sur* $\overline{\Omega} = \{\beta = 0\}$ *tel que, pour tout temps S prévisible* $/(\overline{\mathcal{G}}_u)$

(5.5) $$E(f \circ \overline{\theta}_S | \overline{\mathcal{G}}_S) = \overline{P}^{Y_S}(f) \text{ sur } \{Y_S \in E\} ,$$

où $\overline{\theta}_s = k_0 \circ \sigma_s = (Y_{s+t})_{t<0}$.

2) Tout processus mesurable positif Z porté par $\{Y \in E\}$ *admet une projection double optionnelle* $/(\mathcal{G}_u)$ *et prévisible* $/(\overline{\mathcal{G}}_u)$ *de la forme* $f_Z(Y_u, u)$, *où*

$$f_Z(y,u) = \int \overline{P}^y(d\overline{\omega}) \int P^y(d\omega) Z_u(\sigma_{-u}(\overline{\omega}|0|\omega)) 1_E(y) .$$

(Pour $\overline{\omega} \in \overline{\Omega}$, $\omega \in \Omega$ on note $(\overline{\omega}|0|\omega)$ l'élément de W qui coïncide avec $\overline{\omega}$ et ω respectivement sur $]-\infty, 0[$ et $]0, \infty[$.)

Démonstration. — Le *1)* se démontre par un argument classique de régularisation de pseudo-noyaux (voir par exemple Fitzsimmons [6], mais il faut prendre garde que dans [6] le prolongement utilisé pour Y en α n'est pas le prolongement markovien maximal de II.12, que nous utilisons ici). Cette propriété *1)* est la propriété de Markov modérée de Chung et Walsh [3] pour le processus Y (voir aussi Azéma [1]).

Le *2)* se déduit immédiatement des formules II (11.1) et IV (5.5) en remarquant que sur $\{Y_u \in E\}$

$$Z_u = Z_u(\sigma_{-u}(\sigma_u)) = Z_u(\sigma_{-u}(\overline{\theta}_u|0|\theta_u)) .$$

L'ordre dans lequel on effectue les deux projections n'a pas d'importance, à cause du théorème de Fubini.

Remarque. — On peut choisir les mesures $\overline{P}^x$ de telle sorte qu'elles proviennent d'un semi-groupe $(\overline{P}_t)$ en dualité avec (P_t), mais nous n'avons pas utilisé ce raffinement. Ce semi-groupe ne peut en général être choisi droit, mais seulement gauche (DMM XVIII).

V. Régénération et problème de Skorohod

Après quelques rappels sur le temps local d'un point régulier et la mesure de sortie associée, nous allons présenter dans ce chapitre une approche du problème de Skorohod due à Bertoin et Le Jan [12]. $(\Omega, \mathcal{F}, \mathcal{F}_t, X_t, \theta_t, P^x)$ désigne la réalisation canonique du semi-groupe droit (P_t). On suppose que a est un état fixé de l'espace d'états E et que a est ***régulier*** au sens où le temps d'entrée

$$R = \inf\{t > 0 : X_t = a\}$$

est p.s. nul sous P^a.

TEMPS LOCAL ET MESURE DE SORTIE DE a.

1. — Le *temps local* (standard) de a est la fonctionnelle additive continue L telle que

$$E^{\cdot}\Big(\int_0^\infty e^{-s}dL_s\Big) = E^{\cdot}(e^{-R}) . \tag{1.1}$$

On sait que le support de la mesure dL est p.s. (sous chaque P^x) égal à la fermeture, notée M, de l'ensemble $\{t : X_t = a\}$.

La *mesure de sortie* de a est la mesure $\widehat{P}$ sur Ω définie par

$$\widehat{P}(f) = E^a\Big(\sum_{s\in G} e^{-s} f\circ\theta_s\Big) , \tag{1.2}$$

où G désigne l'ensemble des extrêmités gauches > 0 d'intervalles contigus à M.

THÉORÈME.

1) Soit $m = E^a\Big(\int_0^\infty e^{-s}1_M(s)\,ds\Big)$. *On a*

$$m + \widehat{P}(1 - e^{-R}) = 1 . \tag{1.3}$$

La mesure $\widehat{P}$ *est* σ*-finie.*

2) $1_M(s)\,ds = m\,dL_s$ *p.s..*

3) Soit $D_t = t + R\circ\theta_t = \inf\{s > t : X_s = a\}$. *Pour tout processus* $(\mathcal{F}_{D_t})$ *prévisible positif* Z *et toute fonction mesurable* $\geqslant 0$ f *sur* Ω*, on a la formule de sortie*

$$E^{\cdot}\Big(\sum_{s\in G} Z_s f\circ\theta_s\Big) = \widehat{P}(f) E^{\cdot}\Big(\int_0^\infty Z_s\,dL_s\Big) . \tag{1.4}$$

Démonstration.

1) L'égalité (1.3) s'obtient en intégrant sous P^a l'égalité

$$\int_0^\infty e^{-s}1_M(s)\,ds + \sum_{s\in G} e^{-s}(1 - e^{-R\circ\theta_s}) = \int_R^\infty e^{-u}\,du = e^{-R} .$$

La mesure $\widehat{P}$ est portée par $\{R > 0\}$ et $\widehat{P}(1 - e^{-R}) < \infty$, donc $\widehat{P}$ est σ-finie.

2)

$$E^x\Big(\int_t^\infty e^{-s}1_M(s)\,ds|\mathcal{F}_{D_t}\Big) = E^x\Big(\int_{D_t}^\infty e^{-s}1_M(s)\,ds|\mathcal{F}_{D_t}\Big) = me^{-D_t} ,$$

$$E^x\Big(\int_t^\infty e^{-s}\,dL_s|\mathcal{F}_{D_t}\Big) = E^x\Big(\int_{D_t}^\infty e^{-s}\,dL_s|\mathcal{F}_{D_t}\Big) = e^{-D_t} ,$$

(les deuxièmes égalités de chaque ligne résultent de la propriété de Markov en D_t). Les processus croissants $\int_0^t e^{-s} 1_M(s)\, ds$ et $m \int_0^t e^{-s}\, dL_s$ ont donc même potentiel $/(\mathcal{F}_{D_t})$. Comme ils sont $(\mathcal{F}_{D_t})$ prévisibles, ils sont égaux (DM VI 69).

3) Il suffit d'établir (1.4) pour $f \geqslant 0$ $\widehat{P}$-intégrable. En utilisant le théorème 75 de DM VI, avec la filtration $(\mathcal{F}_{D_t})$, il suffit alors de montrer que pour $t \geqslant 0$

(1.5) $$E^{\cdot}\Big(\sum_{s \in G, s>t} e^{-s} f \circ \theta_s | \mathcal{F}_{D_t} \Big) = \widehat{P}(f) E^{\cdot}\Big(\int_t^{\infty} e^{-s}\, dL_s | \mathcal{F}_{D_t} \Big) .$$

Dans les sommations on peut remplacer la borne t par D_t et la formule (1.5) apparaît alors comme une conséquence de la propriété de Markov en D_t.

Mesure de sortie et théorème d'Azéma.

2. Théorème. — *Soit ν une mesure finie sur $E \setminus \{a\}$ ne chargeant pas les ensembles P^a-polaires. Il existe alors une fonctionnelle additive gauche universellement mesurable A^* telle que dA^* soit portée par $\{X \neq a\} \cap [0, R[$ et telle que, pour f mesurable $\geqslant 0$ sur E*

(2.1) $$\nu(f) = \widehat{P}\Big(\int_{[0,R[} f \circ X_s dA_s^* \Big) .$$

De plus A^* est unique à la $\widehat{P}$-indistinguabilité près.

Démonstration. — Sous la mesure $\widehat{Q} = k_R(\widehat{P})$, le processus $(X_t)_{t>0}$ est markovien relativement au semi-groupe (Q_t) tué à R. On peut alors appliquer le théorème I.8 avec $p = 1$, $g = \mathbf{1}_E$, l'ensemble Ω_0 pouvant ici être remplacé par $\{X \neq a\}$ d'après I.7.3). L'existence de A^* résulte du théorème d'Azéma généralisé IV.5 (la f.a. gauche intervenant dans ce théorème peut être choisie universellement mesurable).

A partir de A^* on construit la mesure aléatoire A sur $\mathbf{R}_+$ portée par $M^c \cup G \cup \{0\}$ telle que

(2.2) $$A(du) = A^*(\theta_s, du - s) \text{ sur } [s, D_s[\text{ pour } s \in G \cup \{0\} .$$

Le processus $(A_t = A[0,t])$ est p.s. fini sous P^a, car on a p.s.

$$\int_{\mathbf{R}_+} e^{-s} A(ds) = \sum_{s \in G} e^{-s} \Big(\int_{[0,R[} e^{-u} A(du) \Big) \circ \theta_s \leqslant \sum_{s \in G} e^{-s} A_R^* \circ \theta_s$$

et $\widehat{P}(A_R^*) = \nu(1) < \infty$.

Le problème de Bertoin et Le Jan.

3. — Supposons maintenant que l'état a soit régulier et *récurrent*. La mesure $\xi = m\varepsilon_a + \widehat{V}$, où

(3.1) $$\widehat{V} f = \widehat{P}\Big(\int_0^R f \circ X_s\, ds \Big) \quad (f \geqslant 0 \text{ sur } E) ,$$

est alors invariante d'après un résultat de Silverstein (voir DMM XIX 46, où est aussi établi un résultat d'unicité). Étant donnée une mesure ν sur E de masse $\leqslant 1$, ne chargeant pas $\{a\}$ et admettant une représentation

$$\nu(f) = E^{\xi}\Big(\int_{[0,1[} f \circ X_s\, dA_s\Big) \tag{3.2}$$

à l'aide d'une f.a. *continue* A, Bertoin et Le Jan [12] considèrent le temps d'arrêt

$$T = \inf\{s : A_s > L_s\} \tag{3.3}$$

et montrent que $\nu = X_T(P^a)$ sur E. Une des clés de la démonstration réside dans la formule

$$\nu(f) = \widehat{P}\Big(\int_0^R f \circ X_s\, dA_s\Big)\ , \tag{3.4}$$

que nous allons déduire de (3.2). En posant $F_t = \int_{[0,t]} e^{-s} f \circ X_s dA_s$, on a $\nu(f) = E^{\xi}(F_\infty)$, donc

$$\begin{aligned}\nu(f) &= E^{\xi}(F_R) + E^{\xi}\Big(\sum_{s\in G} e^{-s} F_R \circ \theta_s\Big)\ ,\\ &= \widehat{V}(E^{\cdot}(F_R)) + \widehat{P}(F_R) E^{\xi}\Big(\int_0^\infty e^{-s} dL_s\Big)\ ,\\ &= \widehat{V}(V_A^1 f) = \widehat{P}(F_R)(m + \widehat{V}\varphi)\end{aligned}$$

où $V_A^1 f = E^{\cdot}\Big(\int_0^R e^{-s} f \circ X_s dA_s\Big)$, $\varphi = E^{\cdot}(e^{-R})$. D'après un calcul classique (démonstration de l'équation résolvante), qui utilise la continuité de A, on a

$$\widehat{V} V_A^1 f = \widehat{P}\Big(\int_0^R (1 - e^{-s}) f \circ X_s\, dA_s\Big)\ ,$$

et par ailleurs $m + \widehat{V}\varphi = m + \widehat{P}(1 - e^{-R}) = 1$, d'où la formule cherchée.

4. — Les mesures ν considérées par Bertoin et Le Jan ne chargent pas les ensembles P^a-semi-polaires. Pour une mesure finie sur $E \setminus \{a\}$, ne chargeant pas les ensembles P^a-polaires, nous disposons de la représentation donnée par le théorème 2 à l'aide d'une fonctionnelle additive gauche. La solution du problème de Shorohod fournie par Bertoin et Le Jan reste valable pour une telle mesure :

THÉORÈME. — *Soit ν une mesure sur $E \setminus \{a\}$ de masse $\leqslant 1$ ne chargeant pas les ensembles P^a-polaires et soit (A_t) le processus croissant qui lui est associé par le théorème 2 (et les indications suivant sa démonstration). On a alors $\nu = X_T(P^a)$ sur E, où T est le temps d'arrêt* $\inf\{s : A_s > L_s\}$.

Remarque. — Lorsque ν ne charge pas les semi-polaires sous P^a, le processus (A_t) est continu p.s.-P^a et l'on retrouve le résultat de [12].

Démonstration. — Sous la mesure P^a, le processus $S_t = A_{\tau_t}$, où $\tau_t = \inf\{s : L_s > t\}$, est un subordinateur purement discontinu (rappelons que $1_{\{X=a\}} dA = 0$ p.s.), de mesure de

Lévy $\pi = .A_R(\widehat{P})$. Comme $\widehat{P}(A_R) = \nu(1)$, on a $\int x\pi(dx) \leqslant 1$ et, d'après le lemme qui suit, le temps $\sigma = \inf\{t > 0 : S_t > t\}$ est p.s. > 0. Or $\sigma = L_T$, d'où $T > 0$ p.s.. En utilisant la propriété de Markov en T et le fait que A ne charge pas $\{X = a\}$, il en résulte que $X_T \neq a$.

Sur $\{T < \infty\}$, G_T est l'unique $s \in G$ tel que $s \leqslant T$, $A_R \circ \theta_s > L_s - A_{s-} = \Delta_s$; de plus $T = s + A^{-1}_{\Delta_s}(\theta_s)$, où A^{-1} désigne l'inverse càd de A. Nous sommes prêts à appliquer la formule de sortie (1.4). En posant

$$\Phi_t = f(X(A_t^{-1}))\mathbf{1}_{\{A_R > t\}},\ g(t) = \widehat{P}(\Phi_t) ,$$

et d'après une extension facile de (1.4) on peut écrire

$$\begin{aligned} E^a(f \circ X_T,\ T < \infty) &= E^a\Big(\sum_{s \in G, s \leqslant T} \Phi_{\Delta_s}(\theta_s)\Big) \\ &= E^a\Big(\int_0^T g(\Delta_s) dL_s\Big) = E^a\Big(\int_0^\sigma g(t - S_t)dt\Big) . \end{aligned}$$

Il reste à appliquer le *2)* du lemme qui suit, puis (3.1) :

$$\begin{aligned} E^a(f(X_T),\ T < \infty) &= \int_0^\infty g(u)\, du = \widehat{P}\Big(\int_0^{A_R} f(X(A_u^{-1})\, du\Big) \\ &= \widehat{P}\Big(\int_{[0,R]} f(X_t)\, dA_t\Big) = \nu(f) . \end{aligned}$$

5. LEMME. — *Soit S un subordinateur purement discontinu, nul en 0, de mesure de Lévy π telle que $\int x\pi(dx) \leqslant 1$. On pose $Y_t = t - S_t$, $\sigma = \inf\{t > 0 : Y_t < 0\}$. Alors*

1) $\sigma > 0$ p.s. (et même $E(\sigma) = \infty$),

2) $E\Big(\int_0^\sigma g(Y_t)\, dt\Big) = \int_0^\infty g(u)\, du$ pour toute fonction $g \geqslant 0$ sur $\mathbb{R}_+$.

Démonstration. — La fonction

$$\psi(p) = p + \int (e^{-py} - 1)\pi(dy) \tag{5.1}$$

est convexe sur $\mathbb{R}_+$; la condition $\int x\pi(dx) \leqslant 1$ entraîne qu'elle est strictement croissante. Nous allons établir que pour $q > 0$

$$E\Big(\int_0^\sigma e^{-qt} g \circ Y_t\, dt\Big) = \int_0^\infty e^{-y\psi^{-1}(q)} g(y)dy . \tag{5.2}$$

En faisant tendre q vers 0, on obtient *2)* et pour $g = 1$ on obtient $E(\sigma) = \lim_{q \to 0} \dfrac{1}{\psi^{-1}(q)} = +\infty$.

Il suffit de vérifier (5.2) dans le cas où la mesure π est finie (le cas général en résulte par approximation). Le graphe de Y a alors l'allure d'une fonction affine par morceaux, de pente 1. Posons $T_y = \inf\{t > 0 : Y_t = y\}$. On a p.s. $Y_{T_y} = y$ sur $\{T_y < \infty\}$, $0 < \sigma < T_0$ et

$Y_t < 0$ sur $\{\sigma \leqslant t < T_0\}$. Dans le premier membre de (5.2) on peut remplacer σ par T_0 (on pose $g(u) = 0$ si $u < 0$). D'après la propriété de Markov en T_0, le premier membre s'écrit

$$aE\Big(\int_0^\infty e^{-qt} g(Y_t)dt\Big), \quad \text{où} \quad a = 1 - E(e^{-qT_0}) .$$

En coupant l'intégrale à l'aide des instants de saut successifs du processus Y, on justifie facilement l'égalité suivante :

$$E\Big(\int_0^\infty e^{-qt} g(Y_t)\, dt\Big) = \int_0^\infty E\Big(\sum_{t \geqslant 0} e^{-qt} I_{\{Y_t = y\}}\Big) g(y) dy .$$

Par la propriété de Markov en T_y, l'espérance au second membre s'écrit pour $y > 0$.

$$E(e^{-qT_y}) E\Big(\sum_{t \geqslant 0} e^{-qt} I_{\{Y_t = 0\}}\Big) = \frac{1}{a} E(e^{-qT_y}) = \frac{1}{a} e^{-\psi^{-1}(q)y}$$

(la dernière égalité s'obtient par arrêt à T_y de la martingale $e^{pY_t - t\psi(p)}$, pour $p = \psi^{-1}(q)$). D'où (6.2).

A. Appendice - Théorie des flots, mesures de Palm

1. — Dans cet appendice nous avons rassemblé quelques éléments de théorie des flots utilisés dans ce cours. Nous gardons les notations $(W, \mathcal{G}^0, \sigma_u, Q)$ des flots de Kuznetsov (II.8,10), mais nous avons ici affaire à un flot général. La mesure Q est supposée σ-finie, le processus $(\sigma_u)_{u \in \mathbb{R}}$, à valeurs dans $(W, \mathcal{G}^0)$, est supposé mesurable et tel que

$$\sigma_{u+v} = \sigma_v \circ \sigma_u, \quad \sigma_0 = \mathrm{id} \quad (u, v \in \mathbb{R}) ,$$

$$\sigma_u(Q) = Q \quad (u \in \mathbb{R}) .$$

La complétée de $\mathcal{G}^0$ par Q est notée $\mathcal{G}$.

TEMPS STATIONNAIRES.

2. — Un temps aléatoire (c'est-à-dire une v.a. $\mathcal{G}$-mesurable à valeurs dans $\overline{\mathbb{R}}$) S est dit *stationnaire* si

$$S = u + S \circ \sigma_u, \quad u \in \mathbb{R} . \tag{2.1}$$

THÉORÈME. — *Étant donné un temps stationnaire S, il existe une mesure unique P_S sur $(W, \mathcal{G}^0)$, appelée mesure de Palm de S telle que*

$$Q_{\{S \in \mathbb{R}\}}(f(\sigma_S, S)) = P_S\Big(\int_{\mathbb{R}} f(\cdot, u) du\Big) \tag{2.2}$$

pour f mesurable $\geqslant 0$ sur $(W \times \mathbb{R}, \mathcal{G}^0 \otimes \mathcal{B}_{\mathbb{R}})$.

Démonstration. — S'il existe une mesure P_S permettant d'écrire (2.2) on doit avoir

$$P_S = Q_{\{0<S\leqslant 1\}}(\sigma_S \in \cdot)\ , \tag{2.3}$$

d'où l'unicité de P_S. Considérons la mesure (2.3). Le premier membre de (2.2) s'écrit

$$Q\int (f(\sigma_S, S-u)I_{\{0<S\leqslant 1\}})\circ\sigma_u\ du \quad \text{(d'après (2.1))}$$

$$= \int Q_{\{0<S\leqslant 1\}} f(\sigma_S, S-u))\ du \quad \text{(Fubini et } \sigma_u(Q)=Q),$$

et ceci est égal au second membre de (2.2) (Fubini).

3. — Voici un résultat complémentaire utile. S est un temps stationnaire fixé.

THÉORÈME.

1) Pour toute fonction mesurable positive g sur $(W, \mathcal{G}^0)$ on a

$$Q_{\{S\in\mathbb{R}\}}(g) = P_S\Big(\int_{-\infty}^{+\infty} g\circ\sigma_u\ du\Big)\ ; \tag{3.1}$$

en particulier la mesure P_S est σ-finie.

2) Si S est $\mathcal{G}^0$-mesurable (ou plus généralement universellement mesurable) *la mesure P_S est l'unique mesure sur $(W, \mathcal{G}^0)$ portée par $\{S=0\}$ satisfaisant à (3.1).*

Démonstration. — En prenant $f(w,u) = g(\sigma_{-u}(\sigma_u w))$ dans (2.2), on trouve (3.1). Lorsque S est universellement mesurable, la mesure P_S est portée par $\{S=0\}$, comme on le voit en prenant $f(w,u) = I_{\{S\neq 0\}}(w)$ dans (2.2) et en remarquant que $S\circ\sigma_S = 0$ sur $\{S\in\mathbb{R}\}$.

Enfin, si P_S est une mesure sur $(W, \mathcal{G}^0)$ satisfaisant à (3.1), cette mesure est nécessairement σ-finie (pour $g>0$ et Q-intégrable, $\int g\circ\sigma_u\ du$ est >0 et P_S-intégrable) et l'on a

$$Q_{\{S\in\mathbb{R}\}}(f(\sigma_S, S)) = P_S\Big(\int f(\sigma_S, S-u)du\Big)\ .$$

Si de plus P_S est portée par $\{S=0\}$, cette expression est égale au second membre de (2.2), d'où l'unité de P_S.

MESURES ALÉATOIRES HOMOGÈNES.

4. — Soit A une mesure aléatoire homogène (MAH), c'est-à-dire une somme dénombrable de noyaux bornés de $(W, \mathcal{G})$ dans $(\mathbb{R}, \mathcal{B}_{\mathbb{R}})$ telle que

$$A(\sigma_u, f(u+\cdot)) = Af \qquad (u\in\mathbb{R},\ f\geqslant 0 \text{ sur } \mathbb{R})\ . \tag{4.1}$$

On note $Q*A$ la mesure sur $(W\times\mathbb{R}, \mathcal{G}^0\otimes\mathcal{B}_{\mathbb{R}})$ définie par

$$Q*A(f) = Q\Big(\int f(\sigma_s, s)A(ds)\Big)\ . \tag{4.2}$$

THÉORÈME. — *La mesure $Q * A$ est Σ-finie (somme dénombrable de mesures finies) ; elle s'écrit $P_A \times \lambda$ pour une mesure unique P_A sur $(W, \mathcal{G}^0)$ et la mesure de Lebesque λ. P_A est appelée mesure de Palm de A :*

La mesure $P_A \times \lambda$ est ici *par définition* la mesure $f \mapsto P_A\big(\int f(\cdot, s)ds\big)$, ce qui est compatible avec la définition usuelle du produit de deux mesures (généralement supposées σ-finies).

Démonstration. — La mesure $\mu = Q * A$ est invariante par translation de la seconde coordonnée, au sens suivant :

$$\int \mu(dw, ds) f(w, s + r) = \mu(f) \quad (r \in \mathbb{R}) . \tag{4.3}$$

En effet

$$Q \int f(\sigma_s, s + r) A(ds) = Q \int f(\sigma_{s+r}, s + r) A(\sigma_r, ds) \quad (\sigma_r Q = Q)$$
$$= Q \int f(\sigma_s, s) A(ds) \quad \text{(d'après (4.1))}.$$

Le théorème résulte alors du lemme suivant, dû à Getoor.

LEMME. — *Si μ est une mesure Σ-finie (resp. σ-finie) sur $W \times \mathbb{R}$ satisfaisant à (4.3), alors il existe une mesure unique π sur W telle que $\mu = \pi \times \lambda$. La mesure π est Σ-finie (resp. σ-finie).*

La démonstration que nous allons en donner repose simplement sur le théorème d'échange des intégrations (Fubini) valable pour les mesures Σ-finies.

Démonstration. — L'unicité est claire : si φ est $\geqslant 0$ et telle que $\lambda(\varphi) = 1$, on a nécessairement

$$\pi(g) = \int \mu(dw, ds) g(w) \varphi(s) ds .$$

Soit donc π la mesure définie par cette formule. On a

$$\mu(f) = \int \mu(dw, ds) \int f(w, s) \varphi(s - r) dr$$
$$= \int dr \int \mu(dw, ds) f(w, s) \varphi(s - r)$$
$$= \int dr \int \mu(dw, ds) f(w, s + r) \varphi(s) \quad \text{(d'après (4.3))}$$
$$= \int \mu(dw, ds) \varphi(s) \int f(w, s + r) dr = \pi \int f(\cdot, u)\, du .$$

La mesure π est Σ-finie. Si μ est σ-finie, il existe $h > 0$ telle que $\pi\big(\int h(\cdot, u)\, du\big) < \infty$, donc π est elle-même σ-finie.

5. *Remarques.*

1) Nous dirons que la mesure aléatoire A est *σ-finie* s'il existe une fonction $f > 0$ mesurable sur $(W \times \mathbb{R}, \mathcal{G}^0 \otimes \mathcal{B}_{\mathbb{R}})$ telle que la fonction $Af = \int A(\cdot, ds) f(\cdot, s)$ soit finie p.s.. La mesure $\mu = Q * A$ est alors elle-même σ-finie. En effet, d'après la σ-finitude de $Af \cdot Q$, il existe $g > 0$ sur $(W, \mathcal{G}^0)$ telle que $Q(gAf) < \infty$. Or $Q(gAf) = \mu(h)$, où $h(w, s) = g(\sigma_{-s} w) f(\sigma_{-s} w, s) > 0$.

2) La condition précédente sur A est satisfaite si A est de Radon sur un intervalle stochastique I portant A (cela signifie que pour tout w la mesure A_w est de Radon sur $I(w)$). Par exemple, si $I =]\alpha, \beta[$, posons $B_{n,k} = \left\{(w,s) : \alpha(w) < \frac{k}{2^n} \leqslant s < \frac{k+1}{2^n} < \beta(w)\right\}$ pour $n \in \mathbb{N}^*$, $k \in \mathbb{R}$. On a alors $A(B_{n,k}) < \infty$ et la σ-finitude de A en résulte. Les MAH considérées dans DMM XIX sont associées à des hélices : elles sont de Radon sur $\mathbb{R}$ tout entier.

3) La formule

$$Q \int A(dr) g(\sigma_r, r, \cdot) = P_A \int \lambda(dr) g(\cdot, r, \sigma_{-r}) \tag{5.1}$$

est valable pour g mesurable $\geqslant 0$ sur $(W \times \mathbb{R} \times W, \mathcal{G}^0 \otimes \mathcal{B}_{\mathbb{R}} \otimes \mathcal{G}^0)$. Lorsque $Q * A$ est σ-finie, (5.1) s'étend aux fonctions $\mathcal{G}^0 \otimes \mathcal{B}_{\mathbb{R}} \otimes \mathcal{G}$-mesurables $\geqslant 0$, par classes monotones et par encadrement. En particulier, si B est une deuxième MAH, on peut écrire pour f $\mathcal{G}^0 \otimes \mathcal{B}_{\mathbb{R}} \otimes \mathcal{G}^0 \otimes \mathcal{B}_{\mathbb{R}}$ mesurable $\geqslant 0$

$$\begin{aligned} Q \int A(dr) \int B(ds) f(\sigma_r, r, \sigma_s, s) &= P_A \int \lambda(dr) \int B_{\sigma_{-r}}(ds) f(\cdot, r, \sigma_{s-r}, s) \\ &= P_A \int \lambda(dr) \int B(ds) f(\cdot, r, \sigma_s, s+r) \end{aligned} \tag{5.2}$$

Si la mesure $Q * B$ est également σ-finie, on peut échanger les rôles de A et B et le second membre de (5.2) s'écrit aussi

$$P_B \int \lambda(dr) \int A(ds) f(\sigma_s, s+r, \cdot, r) . \tag{5.3}$$

Ces identités précisent la formule (9.22) de Getoor [8]. En prenant $f(w, u, w', v) = \varphi(u) g(w', v-u)$ ($\varphi \geqslant 0, \lambda(\varphi) = 1$) on trouve *l'identité de Neveu*

$$P_A \int B(ds) g(\sigma_s, s) = P_B \int A(ds) g(\cdot, -s) , \tag{5.4}$$

qui peut aussi se démontrer directement (DMM XIX 32).

Signalons pour terminer que les identités (5.2) et (5.4) sont également valables pour des MAH universement mesurables (cela assure l'existence des intégrales intervenant dans ces formules).

Références bibliographiques

Les références DM, DMM, FM sont données dans l'introduction

[1] AZEMA J. — *Théorie générale des processus et retournement du temps*, Ann. Sci. ENS **6** (1973), 459–519.

[2] BLUMENTHAL R.M. et GETOOR R. K. — *Markov processes and potential theory*, Academic Press, 1968.

[3] CHUNG K.L. et WALSH J.B. — *To reverse a Markov process*, Acta Math **123** (1970), 225–251.

[4] DYNKIN E.B. — *Integral representation of excessive measures and excessive functions*, Russian Math. Surveys (Uspehi) **27** (1972), 43–84.

[5] DYNKIN E.B. — *An application of flows to time shift and time reversal in stochastic processes*, TAMS **287** (1985), 613–619.

[6] FITZSIMMONS P.J. — *Homogeneous random measures and a weak order on excessive measures of a Markov process*, TAMS **303** (1987), 431–478.

[7] FITZSIMMONS P.J. — *On a connection between Kuznetsov processes and quasi-processes*, Sem. Stoch. proc. (1988), 123–134.

[8] GETOOR R.K. — *Excessive measures*, Birkhäuser, 1990.

[9] KUZNETSOV S.E. — *Construction of Markov processes with random birth and death*, Th. Prob. Appl. **18** (1974), 571–574.

[10] MITRO J.B. — *Dual Markov processes : construction of a useful auxiliary process*, ZW **47** (1979), 139–156.

[11] NAGASAWA M. — *Time reversion of Markov processes*, Nagoya Math. J. **24** (1964), 117–204.

[12] BERTOIN J. et LE JAN Y. — *Representation of measures by balayage from a regular recurrent point*, The Annals of Probability **20** , n°1 (1992), 538–548.

NINE LECTURES ON RANDOM GRAPHS

Joel SPENCER

NINE LECTURES ON RANDOM GRAPHS

Joel Spencer
Courant Institute of Mathematical Sciences
251 Mercer Street

NEW YORK, NY 10012, U.S.A.

Graph Theory Preliminaries A graph G, formally speaking, is a pair $(V(G), E(G))$ where the elements $v \in V(G)$ are called vertices and the elements of $E(G)$, called edges, are two element subsets $\{v, w\}$ of $V(G)$. When $\{v, w\} \in E(G)$ we say v, w are adjacent. (In standard graph theory terminology our graphs are undirected and have no loops and no multiple edges.) Pictorially, we often display the $v \in V(G)$ as points and draw an arc between v and w when they are adjacent. We call $V(G)$ the vertex set of G and $E(G)$ the edge set of G. (When G is understood we shall write simply V and E respectively. We also often write $v \in G$ or $\{v, w\} \in G$ instead of the formally correct $v \in V(G)$ and $\{v, w\} \in E(G)$ respectively.) A set $S \subseteq V$ is called a *clique* if all pairs $x, y \in S$ are adjacent. The clique number, denoted by $\omega(G)$, is the largest size of a clique in G. An independent set S is one for which no pairs $x, y, \in S$ are adjacent, the largest size of an independent set is called the independence number and is denoted $\alpha(G)$. A k-coloring of G is a map $f : V \to \{1, \ldots, k\}$ such that if x, y are adjacent then $f(x) \neq f(y)$. The minimal k for which a k-coloring exists is called the chromatic number of G and is denoted $\chi(G)$. Note $\omega(G) \leq \chi(G)$ since all vertices of a clique much receive distinct colors. The complement of G, denoted $\overline{G}$, has the same vertex set as G and x, y are adjacent in $\overline{G}$ if and only if x, y are not adjacent in G. The complete graph on k vertices, denoted by K_k, consists of a vertex set of size k with all pairs x, y adjacent. The empty graph on k vertices, denoted by I_k, consists of a vertex set of size k with no pairs x, y adjacent.

References. Theorem 2.3.4 refers to Lecture 2, Section 3, theorem 4. Double indexing, such as Theorem 3.4, refers to Section 3, theorem 4 in the current lecture.

Lecture 1: Basics

1 What is a Random Graph

Let n be a positive integer, $0 \leq p \leq 1$. The random graph $G(n, p)$ is a probability space over the set of graphs on the vertex set $\{1, \ldots, n\}$ determined by

$$\Pr[\{i, j\} \in G] = p$$

with these events mutually independent.

Random Graphs is an active area of research which combines probability theory and graph theory. The subject began in 1960 with the monumental paper *On the Evolution of Random Graphs* by Paul Erdős and Alfred Rényi. The book *Random Graphs* by Béla Bollobás is the standard source for the field.

There is a compelling dynamic model for random graphs. For all pairs i, j let $x_{i,j}$ be selected uniformly from $[0, 1]$, the choices mutually independent. Imagine p going from 0 to 1. Originally, all potential edges are "off". The edge from i to j (which we may imagine as a neon light) is turned on when p reaches $x_{i,j}$ and then stays on. At $p = 1$ all edges are "on". At time p the graph of all "on" edges has distribution $G(n,p)$. As p increases $G(n,p)$ *evolves* from empty to full.

In their original paper Erdős and Rényi let $G(n,e)$ be the random graph with n vertices and precisely e edges. Again there is a dynamic model: Begin with no edges and add edges randomly one by one until the graph becomes full. Generally $G(n,e)$ will have very similar properties as $G(n,p)$ with $p \sim \frac{e}{\binom{n}{2}}$. We will work on the probability model exclusively.

2 Threshold Functions

The term "the random graph" is, strictly speaking, a misnomer. $G(n,p)$ is a probability space over graphs. Given any graph theoretic property A there will be a probability that $G(n,p)$ satisfies A, which we write $\Pr[G(n,p) \models A]$. When A is monotone $\Pr[G(n,p) \models A]$ is a monotone function of p. As an instructive example, let A be the event "G is triangle free". Let X be the number of triangles contained in $G(n,p)$. Linearity of expectation gives

$$E[X] = \binom{n}{3} p^3$$

This suggests the parametrization $p = c/n$. Then

$$\lim_{n\to\infty} E[X] = \lim_{n\to\infty} \binom{n}{3} p^3 = c^3/6$$

We shall see that the distribution of X is asymptotically Poisson. In particular

$$\lim_{n\to\infty} \Pr[G(n,p) \models A] = \lim_{n\to\infty} \Pr[X=0] = e^{-c^3/6}$$

Note that

$$\lim_{c\to 0} e^{-c^3/6} = 1$$

$$\lim_{c\to\infty} e^{-c^3/6} = 0$$

When $p = 10^{-6}/n$, $G(n,p)$ is very unlikely to have triangles and when $p = 10^6/n$, $G(n,p)$ is very likely to have triangles. In the dynamic view the first triangles almost always appear at $p = \Theta(1/n)$. If we take a function such as $p(n) = n^{-.9}$ with $p(n) \gg n^{-1}$ then $G(n,p)$ will almost always have triangles. Occasionally we will abuse notation and say, for example, that $G(n, n^{-.9})$ contains a triangle - this meaning that the probability that it contains a triangle approaches 1 as n approaches infinity. Similarly, when $p(n) \ll n^{-1}$, for example, $p(n) = 1/(n \ln n)$, then $G(n,p)$ will almost always not contain a triangle and we abuse notation and say that $G(n, 1/(n \ln n))$ is trianglefree. It was a central observation of Erdös and Rényi that many natural graph theoretic properties become true in a very narrow range of p. They made the following key definition.
Definition. $r(n)$ is called a *threshold function* for a graph theoretic property A if
(i) When $p(n) \ll r(n), \lim_{n\to\infty} \Pr[G(n,p) \models A] = 0$
(ii) When $p(n) \gg r(n), \lim_{n\to\infty} \Pr[G(n,p) \models A] = 1$
or visa versa.

In our example, $1/n$ is a threshold function for A. Note that the threshold function, when one exists, is not unique. We could equally have said that $10/n$ is a threshold function for A.

Lets approach the problem of $G(n,c/n)$ being trianglefree once more. For every set S of three vertices let B_S be the event that S is a triangle. Then $\Pr[B_S] = p^3$. Then "trianglefreeness" is precisely the conjunction $\wedge\overline{B}_S$ over all S. If the B_S were mutually independent then we *would* have

$$\Pr[\wedge\overline{B}_S] = \prod[\overline{B}_S] = (1-p^3)^{\binom{n}{3}} \sim e^{-\binom{n}{3}p^3} \to e^{-c^3/6}$$

The reality is that the B_S are not mutually independent though when $|S \cap T| \leq 1$, B_S and B_T are mutually independent. This is quite a typical situation in the study of random graphs in which we must deal with events that are "almost", but not precisely, mutual independent.

3 Variance

Here we introduce the Variance in a form that is particularly suited to the study of random graphs. The expressions Δ and Δ^* defined in this section will appear often in these notes.

Let X be a nonnegative integral valued random variable and suppose we want to bound $\Pr[X=0]$ given the value $\mu = E[X]$. If $\mu < 1$ we may use the inequality

$$\Pr[X > 0] \leq E[X]$$

so that if $E[X] \to 0$ then $X = 0$ almost always. (Here we are imagining an infinite sequence of X dependent on some parameter n going to infinity.) But now suppose $E[X] \to \infty$. It does *not* necessarily follow that $X > 0$ almost always. For example, let X be the number of deaths due to nuclear war in the twelve months after reading this paragraph. Calculation of $E[X]$ can make for lively debate but few would deny that it is quite large. Yet we may believe - or hope - that $Pr[X \neq 0]$ is very close to zero. We can sometimes deduce $X > 0$ almost always if we have further information about $Var[X]$.

Theorem 3.1

$$\Pr[X=0] \leq \frac{Var[X]}{E[X]^2}$$

Proof. Set $\lambda = \mu/\sigma$ in Chebyschev's Inequality. Then

$$\Pr[X=0] \leq \Pr[|X-\mu| \geq \lambda\sigma] \leq \frac{1}{\lambda^2} = \frac{\sigma^2}{\mu^2} \quad \square$$

We generally apply this result in asymptotic terms.

Corollary 3.2

$$\text{If } Var[X] = o(E[X]^2) \text{ then } X > 0 \text{ a.a.}$$

The proof of the Theorem actually gives that for any $\epsilon > 0$

$$\Pr[|X - E[X]| \geq \epsilon E[X]] \leq \frac{Var[X]}{\epsilon^2 E[X]^2}$$

and thus in asymptotic terms we actually have the following stronger assertion:

Corollary 3.3

$$\text{If } Var[X] = o(E[X]^2) \text{ then } X \sim E[X] \text{ a.a.}$$

Suppose again $X = X_1 + \ldots + X_m$ where X_i is the indicator random variable for event A_i. For indices i,j write $i \sim j$ if $i \neq j$ and the events A_i, A_j are not independent. We set (the sum over ordered pairs)

$$\Delta = \sum_{i \sim j} \Pr[A_i \wedge A_j]$$

Note that when $i \sim j$

$$Cov[X_i, X_j] = E[X_iX_j] - E[X_i]E[X_j] \leq E[X_iX_j] = \Pr[A_i \wedge A_j]$$

and that when $i \neq j$ and not $i \sim j$ then $Cov[X_i, X_j] = 0$. Thus

$$Var[X] \leq E[X] + \Delta$$

Corollary 3.4. If $E[X] \to \infty$ and $\Delta = o(E[X]^2)$ then $X > 0$ almost always. Furthermore $X \sim E[X]$ almost always.

Let us say $X_1, \ldots, X_m$ are *symmetric* if for every $i \neq j$ there is an automorphism of the underlying probability space that sends event A_i to event A_j. Examples will appear in the next section. In this instance we write

$$\Delta = \sum_{i \sim j} \Pr[A_i \wedge A_j] = \sum_i \Pr[A_i] \sum_{j \sim i} \Pr[A_j | A_i]$$

and note that the inner summation is independent of i. We set

$$\Delta^* = \sum_{j \sim i} \Pr[A_j | A_i]$$

where i is any fixed index. Then

$$\Delta = \sum_i \Pr[A_i] \Delta^* = \Delta^* \sum_i \Pr[A_i] = \Delta^* E[X]$$

Corollary 3.5. If $E[X] \to \infty$ and $\Delta^* = o(E[X])$ then $X > 0$ almost always. Furthermore $X \sim E[X]$ almost always.

The condition of Corollary 3.5 has the intuitive sense that conditioning on any specific A_i holding does not substantially increase the expected number $E[X]$ of events holding.

4 Appearance of Small Subgraphs

What is the threshold function for the appearance of a given graph H. This problem was solved in the original papers of Erdős and Rényi. We begin with an instructive special case.
Theorem 4.1 The property $\omega(G) \geq 4$ has threshold function $n^{-2/3}$.
Proof. For every 4-set S of vertices in $G(n,p)$ let A_S be the event "S is a clique" and X_S its indicator random variable. Then

$$E[X_S] = \Pr[A_S] = p^6$$

as six different edges must all lie in $G(n,p)$. Set

$$X = \sum_{|S|=4} X_S$$

so that X is the number of 4-cliques in G and $\omega(G) \geq 4$ if and only if $X > 0$. Linearity of Expectation gives

$$E[X] = \sum_{|S|=4} E[X_S] = \binom{n}{4} p^6 \sim \frac{n^4 p^6}{24}$$

When $p(n) \ll n^{-2/3}$, $E[X] = o(1)$ and so $X = 0$ almost surely.

Now suppose $p(n) \gg n^{-2/3}$ so that $E[X] \to \infty$ and consider the Δ^* of Corollary 3.5. (All 4-sets "look the same" so that the X_S are symmetric.) Here $S \sim T$ if and only if $S \neq T$ and S, T have common edges - i.e., if and only if $|S \cap T| = 2$ or 3. Fix S. There are $O(n^2)$ sets T with $|S \cap T| = 2$ and for each of these $\Pr[A_T|A_S] = p^5$. There are $O(n)$ sets T with $|S \cap T| = 3$ and for each of these $\Pr[A_T|A_S] = p^3$. Thus

$$\Delta^* = O(n^2p^5) + O(np^3) = o(n^4p^6) = o(E[X])$$

since $p \gg n^{-2/3}$. Corollary 3.5 therefore applies and $X > 0$, i.e., there *does* exist a clique of size 4, almost always. □

The proof of Theorem 4.1 appears to require a fortuitous calculation of Δ^*. The following definitions will allow for a description of when these calculations work out.
Definitions. Let H be a graph with v vertices and e edges. We call $\rho(H) = e/v$ the *density* of H. We call H *balanced* if every subgraph H' has $\rho(H') \leq \rho(H)$. We call H *strictly balanced* if every proper subgraph H' has $\rho(H') < \rho(H)$.
Examples. K_4 and, in general, K_k are strictly balanced. The graph

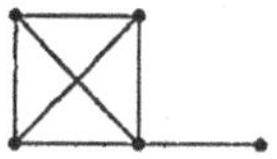

is not balanced as it has density $7/5$ while the subgraph K_4 has density $3/2$. The graph

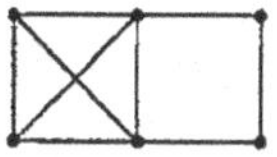

is balanced but not strictly balanced as it and its subgraph K_4 have density $3/2$.
Theorem 4.2 Let H be a balanced graph with v vertices and e edges. Let $A(G)$ be the event that H is a subgraph (not necessarily induced) of G. Then $p = n^{-v/e}$ is the threshold function for A.
Proof. We follow the argument of Theorem 4.1 For each v-set S let A_S be the event that $G|_S$ contains H as a subgraph. Then

$$p^e \leq \Pr[A_S] \leq v!p^e$$

(Any particular placement of H has probability p^e of occuring and there are at most $v!$ possible placements. The precise calculation of $\Pr[A_S]$ is, in general, complicated due to the overlapping of potential copies of H.) Let X_S be the indicator random variable for A_S and

$$X = \sum_{|S|=v} X_S$$

so that A holds if and only if $X > 0$. Linearity of Expectation gives

$$E[X] = \sum_{|S|=v} E[X_S] = \binom{n}{v} \Pr[A_S] = \Theta(n^v p^e)$$

If $p \ll n^{-v/e}$ then $E[X] = o(1)$ so $X = 0$ almost always.

Now assume $p \gg n^{-v/e}$ so that $E[X] \to \infty$ and consider the Δ^* of Corollary 3.5 (All v-sets look the same so the X_S are symmetric.) Here $S \sim T$ if and only if $S \neq T$ and S, T have common edges - i.e., if and only if $|S \cap T| = i$ with $2 \leq i \leq v-1$. Let S be fixed. We split

$$\Delta^* = \sum_{T \sim S} \Pr[A_T|A_S] = \sum_{i=2}^{v-1} \sum_{|T\cap S|=i} \Pr[A_T|A_S]$$

For each i there are $O(n^{v-i})$ choices of T. Fix S, T and consider $\Pr[A_T|A_S]$. There are $O(1)$ possible copies of H on T. Each has - since, critically, H is balanced - at most $\frac{ie}{v}$ edges with both vertices in S and thus at least $e - \frac{ie}{v}$ other edges. Hence

$$\Pr[A_T|A_S] = O(p^{e-\frac{ie}{v}})$$

and

$$\Delta^* = \sum_{i=2}^{v-1} O(n^{v-i}p^{e-\frac{ie}{v}}) = \sum_{i=2}^{v-1} O((n^v p^e)^{1-\frac{i}{v}})$$

$$= \sum_{i=2}^{v-1} o(n^v p^e) = o(E[X])$$

since $n^v p^e \to \infty$. Hence Corollary 3.5 applies. □

Theorem 4.3 In the notation of Theorem 4.2 if H is *not* balanced then $p = n^{-v/e}$ is *not* the threshold function for A.

Proof. Let H_1 be a subgraph of H with v_1 vertices, e_1 edges and $e_1/v_1 > e/v$. Let α satisfy $v/e < \alpha < v_1/e_1$ and set $p = n^{-\alpha}$. The expected number of copies of H_1 is then $o(1)$ so almost always $G(n,p)$ contains no copy of H_1. But if it contains no copy of H_1 then it surely can contain no copy of H. □

The threshold function for the property of containing a copy of H, for general H, was examined in the original papers of Erdős and Rényi. Let H_1 be that subgraph with maximal density $\rho(H_1) = e_1/v_1$. (When H is balanced we may take $H_1 = H$.) They showed that $p = n^{-v_1/e_1}$ is the threshold function. This will follow fairly quickly from the methods of theorem 4.5.

We finish this section with two strengthenings of Theorem 4.2.

Theorem 4.4 Let H be strictly balanced with v vertices, e edges and a automorphisms. Let X be the number of copies of H in $G(n,p)$. Assume $p \gg n^{-v/e}$. Then almost always

$$X \sim \frac{n^v p^e}{a}$$

Proof. Label the vertices of H by $1, \ldots, v$. For each ordered $x_1, \ldots, x_v$ let $A_{x_1,\ldots,x_v}$ be the event that $x_1, \ldots, x_v$ provides a copy of H in that order. Specifically we define

$$A_{x_1,\ldots,x_v} : \{i,j\} \in E(H) \Rightarrow \{x_i, x_j\} \in E(G)$$

We let $I_{x_1,\ldots,x_v}$ be the corresponding indicator random variable. We define an equivalence class on v-tuples by setting $(x_1, \ldots, x_v) \equiv (y_1, \ldots, y_v)$ if there is an automorphism σ of $V(H)$ so that $y_{\sigma(i)} = x_i$ for $1 \leq i \leq v$.Then

$$X = \sum I_{x_1,\ldots,x_v}$$

gives the number of copies of H in G where the sum is taken over one entry from each equivalence class. As there are $(n)_v/a$ terms

$$E[X] = \frac{(n)_v}{a} E[I_{x_1,\ldots,x_v}] = \frac{(n)_v p^e}{a} \sim \frac{n^v p^e}{a}$$

Our assumption $p \gg n^{-v/e}$ implies $E[X] \to \infty$. It suffices therefore to show $\Delta^* = o(E[X])$. Fixing $x_1, \ldots, x_v$,

$$\Delta^* = \sum_{(y_1,\ldots,y_v)\sim(x_1,\ldots,x_v)} \Pr[A_{(y_1,\ldots,y_v)} | A_{(x_1,\ldots,x_v)}]$$

There are $v!/a = O(1)$ terms with $\{y_1, \ldots, y_v\} = \{x_1, \ldots, x_v\}$ and for each the conditional probability is at most one (actually, at most p), thus contributing $O(1) = o(E[X])$ to Δ^*. When $\{y_1, \ldots, y_v\} \cap \{x_1, \ldots, x_v\}$ has i elements, $2 \le i \le v-1$ the argument of Theorem 4.2 gives that the contribution to Δ^* is $o(E[X])$. Altogether $\Delta^* = o(E[X])$ and we apply Corollary 3.5 □

Theorem 4.5 Let H be *any* fixed graph. For every subgraph H' of H (including H itself) let $X_{H'}$ denote the number of copies of H' in $G(n,p)$. Assume p is such that $E[X_{H'}] \to \infty$ for every H'. Then

$$X_H \sim E[X_H]$$

almost always.

Proof. Let H have v vertices and e edges. As in Theorem 4.4 it suffices to show $\Delta^* = o(E[X])$. We split Δ^* into a finite number of terms. For each H' with w vertice and f edges we have those $(y_1, \ldots, y_v)$ that overlap with the fixed $(x_1, \ldots, x_v)$ in a copy of H'. These terms contribute, up to constants,

$$n^{v-w}p^{e-f} = \Theta\left(\frac{E[X_H]}{E[X_{H'}]}\right) = o(E[X_H])$$

to Δ^*. Hence Corollary 3.5 does apply. □

Lecture 2: More Random Graphs

1 Clique Number

Now we fix edge probability $p = \frac{1}{2}$ and consider the clique number $\omega(G)$. We set

$$f(k) = \binom{n}{k} 2^{-\binom{k}{2}},$$

the expected number of k-cliques. The function $f(k)$ drops under one at $k \sim 2\log_2 n$. (Very roughly, $f(k)$ is like $n^k 2^{-k^2/2}$.)

Theorem 1.1 Let $k = k(n)$ satisfy $k \sim 2\log_2 n$ and $f(k) \to \infty$. Then almost always $\omega(G) \ge k$.

Proof. For each k-set S let A_S be the event "S is a clique" and X_S the corresponding indicator random variable. We set

$$X = \sum_{|S|=k} X_S$$

so that $\omega(G) \ge k$ if and only if $X > 0$. Then $E[X] = f(k) \to \infty$ and we examine Δ^*. Fix S and note that $T \sim S$ if and only if $|T \cap S| = i$ where $2 \le i \le k-1$. Hence

$$\Delta^* = \sum_{i=2}^{k-1} \binom{k}{i}\binom{n-k}{k-i} 2^{\binom{i}{2}-\binom{k}{2}}$$

and so

$$\frac{\Delta^*}{E[X]} = \sum_{i=2}^{k-1} g(i)$$

where we set

$$g(i) = \frac{\binom{k}{i}\binom{n-k}{k-i}}{\binom{n}{k}} 2^{\binom{i}{2}}$$

Observe that $g(i)$ may be thought of as the probability that a randomly chosen T will intersect a fixed S in i points times the factor increase in $\Pr[A_T]$ when it does. Setting $i = 2$,

$$g(2) = 2\frac{\binom{k}{2}\binom{n-k}{k-2}}{\binom{n}{k}} \sim \frac{k^4}{n^2} = o(1)$$

At the other extreme $i = k-1$

$$g(k-1) = \frac{k(n-k)2^{-(k-1)}}{\binom{n}{k}2^{-\binom{k}{2}}} \sim \frac{2kn2^{-k}}{E[X]}$$

As $k \sim 2\log_2 n$ the numerator is $n^{-1+o(1)}$. The denominator approaches infinity and so $g(k-1) = o(1)$. Some detailed calculation (which we omit) gives that the remaining $g(i)$ are also negligible so that Corollary 1.3.5 applies. □

Theorem 1.1 leads to a strong concentration result for $\omega(G)$. For $k \sim 2\log_2 n$

$$\frac{f(k+1)}{f(k)} = \frac{n-k+1}{k+1}2^{-k} = n^{-1+o(1)} = o(1)$$

Let $k_0 = k_0(n)$ be that value with $f(k_0) \geq 1 > f(k_0+1)$. For "most" n the function $f(k)$ will jump from a large $f(k_0)$ to a small $f(k_0+1)$. The probability that G contains a clique of size k_0+1 is at most $f(k_0+1)$ which will be very small. When $f(k_0)$ is large Theorem 1.1 implies that G contains a clique of size k_0 with probability nearly one. Together, with very high probability $\omega(G) = k_0$. For some n one of the values $f(k_0), f(k_0+1)$ may be of moderate size so this argument does not apply. Still one may show a strong concentration result found independently by Bollobás, Erdős [1976] and Matula [1976].

Corollary 1.2 There exists $k = k(n)$ so that

$$\Pr[\omega(G) = k \text{ or } k+1] \to 1$$

2 Chromatic Number

Again let us fix $p = \frac{1}{2}$ and this time we consider the chromatic number $\chi(G)$ with $G \sim G(n,p)$. Our results in this section will be improved in Lecture 4.

Theorem 2.1. Almost surely

$$\frac{n}{2\log_2 n}(1+o(1)) \leq \chi(G) \leq \frac{n}{log_2 n}(1+o(1))$$

For the lower bound we use the general bound

$$\chi(G) \geq n/\omega(\overline{G})$$

which is true since each color class must be a clique in $\overline{G}$ and so can be used at most $\omega(\overline{G})$ times. But $\overline{G}$ has the same distribution as G so almost surely $\omega(\overline{G}) \leq (2\log_2 n)(1+o(1))$. This will turn out to be the right asymptotic answer.

For the lower bound (which is not best possible) we outline an analysis of the following "greedy algorithm". We find an independent set C on G as follows. Set $S_0 = V(G)$, $a_1 = 1$ and S_1 equal the set of vertices not adjacent to a_1. Having determined $a_1, \ldots, a_i$ and S_i let a_{i+1} be the least vertex of S_i and let S_{i+1} be those $x \in S_i - \{a_i\}$ not adjacent to a_{i+1}. Continue until $S_t = \emptyset$ and set $C = \{a_1, \ldots a_t\}$. A fairly straightforward analysis gives that $|C| \sim \log_2 n$ almost surely, and moreover that the probability (for any given $\epsilon > 0$ that $|C| < (log_2 n)(1-\epsilon)$ is $o(n^{-1})$. Call this one pass of the algorithm. Now we give all points of C color "one", remove vertices C from G and iterate. Let G^1 be G with C removed. Critically, it finding C we only "exposed" edges involving C so that we can consider G^1 to have distribution $G(n_1, \frac{1}{2})$, where $n_1 = n - |C|$ is the number of vertices. Letting n_j be the number of vertices remaining after the j-th pass, almost surely we have $n_{j+1} < n_j - (1-\epsilon)\log_2 n_j$ so that the algorithm is completed using less than $\frac{n}{\log_2 n}(1+\epsilon')$ colors. (Actually, to avoid end effects we can stop the algorithm when there are $o(n/\log n)$ vertices remaining and simply give each such vertex a separate color.)

It is tempting to improve the lower bound as follows. We know that almost surely G contains an independent set of size $\sim 2\log_2 n$. Let C be that set, remove C from G giving G^1 and iterate. The problem is, of course, that G^1 no longer has distribution $G(n_1, \frac{1}{2})$ and no proof has been found along these lines of the true result that $\chi(G) \sim \frac{n}{2\log_2 n}$ almost surely.

3 Connectivity

In this section we give a relatively simple example of what we call the Poisson Paradigm: the rough notion that if there are many rare and nearly independent events then the number of events that hold has approximately a Poisson distribution. This will yield one of the most beautiful of the Erdős- Rényi results, a quite precise description of the threshold behavior for connectivity. A vertex $v \in G$ is *isolated* if it is adjacent to no $w \in V$. In $G(n,p)$ let X be the number of isolated vertices.

Theorem 3.1. Let $p = p(n)$ satisfy $n(1-p)^{n-1} = \mu$. Then

$$\lim_{n\to\infty} \Pr[X = 0] = e^{-\mu}$$

We let X_i be the indicator random variable for vertex i being isolated so that $X = X_1 + \cdots + X_n$. Then $E[X_i] = (1-p)^{n-1}$ so by linearity of expectation $E[X] = \mu$. Now consider the r-th factorial moment $E[(X)_r]$ for any fixed r. By the symmetry $E[(X)_r] = (n)_r E[X_1 \cdots X_r]$. For vertices $1, \ldots, r$ to all be isolated the $r(n-1) - \binom{r}{2}$ pairs $\{i, x\}$ overlapping $1, \ldots, r$ must all not be edges. Thus

$$E[(X)_r] = (n)_r (1-p)^{r(n-1)-\binom{r}{2}} \sim n^r (1-p)^{r(n-1)} \sim \mu^r$$

(That is, the dependence among the X_i was asymptotically negligible.) As all the moments of X approach those of $P(\mu)$, X approaches $P(\mu)$ in distribution and in particular the theorem holds. □

Now we give the Erdős-Rényi famous "double exponential" result.

Theorem 3.2. Let

$$p = p(n) = \frac{\log n}{n} + \frac{c}{n} + o(\frac{1}{n})$$

Then

$$\lim_{n\to\infty} \Pr[G(n,p) \text{ is connected}] = e^{-e^{-c}}$$

For such p, $n(1-p)^{n-1} \sim \mu = e^{-c}$ and by the above argument the probability that X has no isolated vertices approaches $e^{-\mu}$. If G has no isolated vertices but is not connected there is a component of k vertices for some $2 \leq k \leq \frac{n}{2}$. Letting B be this event

$$\Pr[B] \leq \sum_{k=2}^{n/2} \binom{n}{k} k^{k-2} p^{k-1} (1-p)^{k(n-1)-\binom{k}{2}}$$

The first factor is the choice of a component set $S \subset V(G)$. The second factor is a choice of tree on S. The third factor is the probability that those tree pairs are in $E(G)$. The final factor is that there be no edge from S to $V(G) - S$. Some calculation (which we omit but note that $k = 2$ provides the main term) gives that $\Pr[B] = o(1)$ so that $X \neq 0$ and connectivity have the same limiting probability. □

4 The Probabilistic Method

In 1947 Paul Erdős started what is now called the Probabilistic Method with a three page paper in the Bulletin of the American Mathematical Society. The Ramsey function $R(k,l)$ is defined as the least n such that if the edges of K_n are colored Red and Blue then there is either a Red K_k or a Blue K_l. The existence of such an n is a consequence of Ramsey's Theorem and will not concern us here. Rather, we are interested in lower bounds on the Ramsey function. To unravel the definition $R(k,l) > n$ means that there *exists* a Red-Blue coloring of K_n with neither Red K_k nor Blue K_l. In his 1947 paper Erdős considered the case $k = l$.
Theorem 4.1. If

$$\binom{n}{k} 2^{1-\binom{k}{2}} < 1$$

then $R(k,l) > n$.
Proof. Let $G \sim G(n, \frac{1}{2})$ and consider the random two-coloring given by coloring the edges of G red and the other edges of K_n blue. Let X be the number of monochromatic K_k. Then the left hand side above is simply $E[X]$. With $E[X] < 1$, $\Pr[X = 0] > 0$. Hence there is a point in the probability space - i.e., a graph G, whose coloring has $X = 0$ monochromatic K_k. □

Note here a subtle (for some) point. With positive probability $G(n, \frac{1}{2})$ has the desired property and therefore there must - absolutely, positively - exist a G with the desired property. Random Graphs and the Probabilistic Method are closely related. In Random Graphs we study the probability of $G(n,p)$ having certain properties. In the Probabilistic Method our goal is to prove the existence of a G having certain properties. We create a probability space in which the probability of the random G having these properties is positive, and from that it follows that some such G must exist.

Applying some simple asymptotics to the theorem yields that $R(k,k) > \sqrt{2}^{n(1+o(1))}$. In 1935 Erdős and George Szekeres found the upper bound $R(k,k) < 4^{n(1+o(1))}$ by nonrandom means. While there have been improvements in lower order terms, these bounds remain the best known up to $(1+o(1))^n$ terms. It is also interesting that no exponential lower bound is known by constructive means.

A general lower bound is the following.
Theorem 4.2. If there exists $p \in [0,1]$ with

$$\binom{n}{k} p^{\binom{k}{2}} + \binom{n}{l} (1-p)^{\binom{l}{2}} < 1$$

then $R(k,l) > n$.

Proof. Let $G \sim G(n,p)$ and color the edges of G red and the other edges of K_n blue. Then the left hand side above is simply the expectation of the number of red K_k plus the number of blue K_l. For some G this is zero and that G gives the desired coloring. □

Dealing with the asymptotics of this result can be quite tricky. For example, what does this imply about $R(k, 2k)$?

5 High Girth and High Chromatic Number

Many consider the following one of the most pleasing uses of the probabilistic method, as the result is surprising and does not appear to call for nonconstructive techniques. The *girth* of a graph G is the size of its smallest circuit.

Theorem 5.1(Erdős [1959]). For all k, l there exists a graph G with $girth(G) > l$ and $\chi(G) > k$.

Proof. Fix $\theta < 1/l$ and let $G \sim G(n,p)$ with $p = n^{\theta-1}$. Let X be the number of circuits of size at most l. Then

$$E[X] = \sum_{i=3}^{l} \frac{(n)_i}{2i} p^i \le \sum_{i=3}^{l} \frac{n^{\theta i}}{2i} = o(n)$$

as $\theta l < 1$. In particular

$$\Pr[X \ge \frac{n}{2}] = o(1)$$

Set $x = \lceil \frac{3}{p} \ln n \rceil$ so that

$$\Pr[\alpha(G) \ge x] \le \binom{n}{x} (1-p)^{\binom{x}{2}} < \left[ne^{-p(x-1)/2}\right]^x = o(1)$$

Let n be sufficiently large so that both these events have probability less than .5. Then there is a specific G with less than $n/2$ cycles of length less than l and with $\alpha(G) < 3n^{1-\theta} \ln n$. Remove from G a vertex from each cycle of length at most l. This gives a graph G^* with at least $n/2$ vertices. G^* has girth greater than l and $\alpha(G^*) \le \alpha(G)$. Thus

$$\chi(G^*) \ge \frac{|G^*|}{\alpha(G^*)} \ge \frac{n/2}{3n^{1-\theta} \ln n} = \frac{n^\theta}{6 \ln n}$$

To complete the proof, let n be sufficiently large so that this is greater than k.

Lecture 3: The Poisson Paradigm

When X is the sum of many rare indicator "mostly independent" random variables and $\mu = E[X]$ we would like to say that X is close to a Poisson distribution with mean μ and, in particular, that $\Pr[X = 0]$ is nearly $e^{-\mu}$. We call this rough statement the Poisson Paradigm. We give a number of situations in which this Paradigm may be rigorously proven.

1 The Janson Inequalities

In many instances we would like to bound the probability that none of a set of bad events $B_i, i \in I$ occur. If the events are mutually independent then

$$\Pr[\wedge_{i \in I} \overline{B_i}] = \prod_{i \in I} \Pr[\overline{B_i}]$$

When the B_i are "mostly" independent the Janson Inequalities allow us, sometimes, to say that these two quantities are "nearly" equal.

Let Ω be a finite universal set and let R be a random subset of Ω given by

$$\Pr[r \in R] = p_r,$$

these events mutually independent over $r \in \Omega$. (In application to $G(n,p)$, Ω is the set of pairs $\{i,j\}$, $i,j \in V(G)$ and all $p_r = p$ so that R is the edge set of $G(n,p)$.) Let $A_i, i \in I$, be subsets of Ω, I a finite index set. Let B_i be the *event* $A_i \subseteq R$. (That is, each point $r \in \Omega$ " flips a coin" to determine if it is in R. B_i is the event that the coins for all $r \in A_i$ came up "heads".) Let X_i be the indicator random variable for B_i and $X = \sum_{i \in I} X_i$ the number of $A_i \subseteq R$. The event $\wedge_{i\in I}\overline{B_i}$ and $X = 0$ are then identical. For $i,j \in I$ we write $i \sim j$ if $i \neq j$ and $A_i \cap A_j \neq \emptyset$. Note that when $i \neq j$ and not $i \sim j$ then B_i, B_j are independent events since they involve separate coin flips. Furthermore, and this plays a crucial role in the proofs, if $i \notin J \subset I$ and not $i \sim j$ for all $j \in J$ then B_i is mutually independent of $\{B_j | j \in J\}$, i.e., independent of any Boolean function of those B_j. This is because the coin flips on A_i and on $\cup_{j\in J} A_j$ are independent. We define

$$\Delta = \sum_{i \sim j} \Pr[B_i \wedge B_j]$$

Here the sum is over ordered pairs so that $\Delta/2$ gives the same sum over unordered pairs. (This will be the same Δ as in Lecture 1. We set

$$M = \prod_{i\in I} \Pr[\overline{B_i}],$$

the value of $\Pr[\wedge_{i\in I}\overline{B_i}]$ if the B_i were independent.

Theorem 1.1 (The Janson Inequality). Let $B_i, i \in I, \Delta, M$ be as above and assume all $\Pr[B_i] \leq \epsilon$. Then

$$M \leq \Pr[\wedge_{i\in I}\overline{B_i}] \leq M e^{\frac{1}{1-\epsilon}\frac{\Delta}{2}}$$

Now set

$$\mu = E[X] = \sum_{i\in I} \Pr[B_i]$$

For each $i \in I$

$$\Pr[\overline{B_i}] = 1 - \Pr[B_i] \leq e^{-\Pr[B_i]}$$

so, multiplying over $i \in I$,

$$M \leq e^{-\mu}$$

It is often more convenient to replace the upper bound of Theorem 1.1 with

$$\Pr[\wedge_{i\in I}\overline{B_i}] \leq e^{-\mu + \frac{1}{1-\epsilon}\frac{\Delta}{2}}$$

As an example, set $p = cn^{-2/3}$ and consider the probability that $G(n,p)$ contains no K_4. The B_i then range over the $\binom{n}{4}$ potential K_4 - each being a 6-element subset of Ω. Here, as is often the case, $\epsilon = o(1)$, $\Delta = o(1)$ (as calculated previously) and μ approaches a constant, here $k = c^6/24$. In these instances $\Pr[\wedge_{i\in I}\overline{B_i}] \to e^{-k}$. Thus we have the fine structure of the threshold function of $\omega(G) = 4$.

As Δ becomes large the Janson Inequality becomes less precise. Indeed, when $\Delta \geq 2\mu(1-\epsilon)$ it gives an upper bound for the probability which is larger than one. At that point (and even somewhat before) the following result kicks in.

Theorem 1.2 (The Generalized Janson Inequality). Under the assumptions of Theorem 1.1 and the further assumption that $\Delta \geq \mu(1-\epsilon)$

$$\Pr[\wedge_{i\in I}\overline{B_i}] \leq e^{-\frac{\mu^2(1-\epsilon)}{2\Delta}}$$

Theorem 1.2 (when it applies) often gives a much stronger result than Chebyschev's Inequality as used earlier. We can bound $Var[X] \leq \mu + \Delta$ so that

$$\Pr[\wedge_{i\in I}\overline{B_i}] = \Pr[X=0] \leq \frac{Var[X]}{E[X]^2} \leq \frac{\mu+\Delta}{\mu^2}$$

Suppose $\epsilon = o(1)$, $\mu \to \infty$, $\mu \ll \Delta$, and $\gamma = \frac{\mu^2}{\Delta} \to \infty$. Chebyschev's upper bound on $\Pr[X=0]$ is then roughly γ^{-1} while Janson's upper bound is roughly $e^{-\gamma}$.

2 The Proofs

The original proofs of Janson are based on estimates of the Laplace transform of an appropriate random variable. The proof presented here follows that of Boppana and Spencer [1989]. We shall use the inequalities

$$\Pr[B_i| \wedge_{j\in J} \overline{B_j}] \leq \Pr[B_i]$$

valid for all index sets $J \subset I, i \notin J$ and

$$\Pr[B_i|B_k \wedge \bigwedge_{j\in J} \overline{B_j}] \leq \Pr[B_i|B_k]$$

valid for all index sets $J \subset I, i, k \notin J$. The first follows from general Correlation Inequalities. The second is equivalent to the first since conditioning on B_k is the same as assuming $p_r = \Pr[r \in R] = 1$ for all $r \in A_k$.

Proof of Theorem 1.1 The lower bound follows immediately. Order the index set $I = \{1,\ldots,m\}$ for convenience. For $1 \leq i \leq m$

$$Pr[B_i| \wedge_{1\leq j<i} \overline{B_j}] \leq \Pr[B_i]$$

so

$$Pr[\overline{B_i}| \wedge_{1\leq j<i} \overline{B_j}] \geq Pr[\overline{B_i}]$$

and

$$\Pr[\wedge_{i\in I}\overline{B_i}] = \prod_{i=1}^{m} Pr[\overline{B_i}| \wedge_{1\leq j<i} \overline{B_j}] \geq \prod_{i=1}^{m} Pr[\overline{B_i}]$$

Now the upper bound. For a given i renumber, for convenience, so that $i \sim j$ for $1 \leq j \leq d$ and not for $d+1 \leq j < i$. We use the inequality $\Pr[A|B \wedge C] \geq \Pr[A \wedge B|C]$, valid for any A, B, C. With $A = B_i$, $B = \overline{B_1} \wedge \ldots \wedge \overline{B_d}$, $C = \overline{B_{d+1}} \wedge \ldots \wedge \overline{B_{i-1}}$

$$Pr[B_i| \wedge_{1\leq j<i} \overline{B_j}] = \Pr[A|B \wedge C] \geq Pr[A \wedge B|C] = Pr[A|C]Pr[B|A \wedge C]$$

From the mutual independence $Pr[A|C] = Pr[A]$. We bound

$$\Pr[B|A \wedge C] \geq 1 - \sum_{j=1}^{d} Pr[B_j|B_i \wedge C] \geq 1 - \sum_{j=1}^{d} Pr[B_j|B_i]$$

from the Correlation Inequality. Thus

$$Pr[B_i | \wedge_{1\leq j<i} \overline{B_j}] \geq \Pr[B_i] - \sum_{j=1}^{d} Pr[B_j \wedge B_i]$$

Reversing

$$Pr[\overline{B_i} | \wedge_{1\leq j<i} \overline{B_j}] \leq \Pr[\overline{B_i}] + \sum_{j=1}^{d} Pr[B_j \wedge B_i]$$

$$\leq \Pr[\overline{B_i}] \left(1 + \frac{1}{1-\epsilon} \sum_{j=1}^{d} Pr[B_j \wedge B_i]\right)$$

since $\Pr[\overline{B_i}] \geq 1-\epsilon$. Employing the inequality $1 + x \leq e^x$,

$$Pr[\overline{B_i} | \wedge_{1\leq j<i} \overline{B_j}] \leq \Pr[\overline{B_i}] e^{\frac{1}{1-\epsilon}\sum_{j=1}^{d} Pr[B_j \wedge B_i]}$$

For each $1 \leq i \leq m$ we plug this inequality into

$$\Pr[\wedge_{i\in I} \overline{B_i}] = \prod_{i=1}^{m} Pr[\overline{B_i} | \wedge_{1\leq j<i} \overline{B_j}]$$

The terms $\Pr[\overline{B_i}]$ multiply to M. The exponents add: for each $i, j \in I$ with $j < i$ and $j \sim i$ the term $Pr[B_j \wedge B_i]$ appears once so they add to $\Delta/2$. □
Proof of Theorem 1.2 As discussed earlier, the proof of Theorem 1.1 gives

$$\Pr[\wedge_{i\in I} \overline{B_i}] \leq e^{-\mu + \frac{1}{1-\epsilon}\frac{\Delta}{2}}$$

which we rewrite as

$$-\ln[\Pr[\wedge_{i\in I} \overline{B_i}]] \geq \sum_{i\in I} \Pr[B_i] - \frac{1}{2(1-\epsilon)} \sum_{i\sim j} \Pr[B_i \wedge B_j]$$

For any set of indices $S \subset I$ the same inequality applied only to the $B_i, i \in S$ gives

$$-\ln[\Pr[\wedge_{i\in S} \overline{B_i}]] \geq \sum_{i\in S} \Pr[B_i] - \frac{1}{2(1-\epsilon)} \sum_{i,j\in S, i\sim j} \Pr[B_i \wedge B_j]$$

Let now S be a random subset of I given by

$$\Pr[i \in S] = p$$

with p a constant to be determined, the events mutually independent. (Here we are using probabilistic methods to prove a probability theorem!) Each term $\Pr[B_i]$ then appears with probability p and each term $\Pr[B_i \wedge B_j]$ with probability p^2 so that

$$E\left[-\ln[\Pr[\wedge_{i\in S} \overline{B_i}]\right] \geq E\left[\sum_{i\in S} \Pr[B_i]\right] - \frac{1}{2(1-\epsilon)} E\left[\sum_{i,j\in S, i\sim j} \Pr[B_i \wedge B_j]\right]$$

$$= p\mu - \frac{1}{1-\epsilon} p^2 \frac{\Delta}{2}$$

We set

$$p = \frac{\mu(1-\epsilon)}{\Delta}$$

so as to maximize this quantity. The added assumption of Theorem 1.2 assures us that the probability p is at most one. Then

$$E\left[-\ln[\Pr[\wedge_{i\in S}\overline{B_i}]\right] \geq \frac{\mu^2(1-\epsilon)}{2\Delta}$$

Therefore there is a specific $S \subset I$ for which

$$-\ln[\Pr[\wedge_{i\in S}\overline{B_i}] \geq \frac{\mu^2(1-\epsilon)}{2\Delta}$$

That is,

$$\Pr[\wedge_{i\in S}\overline{B_i}] \leq e^{-\frac{\mu^2(1-\epsilon)}{2\Delta}}$$

But

$$\Pr[\wedge_{i\in I}\overline{B_i}] \leq \Pr[\wedge_{i\in S}\overline{B_i}]$$

completing the proof. □

3 Appearance of Small Subgraphs Revisited

Generalizing the fine threshold behavior for the appearance of K_4 we find the fine threshold behavior for the appearance of any strictly balanced graph H.

Theorem 3.1 Let H be a *strictly* balanced graph with v vertices, e edges and a automorphisms. Let $c > 0$ be arbitrary. Let A be the property that G contains no copy of H. Then with $p = cn^{-v/e}$,

$$\lim_{n\to\infty} \Pr[G(n,p) \models A] = exp[-c^e/a]$$

Proof. Let $A_\alpha, 1 \leq \alpha \leq \binom{n}{v}v!/a$, range over the edge sets of possible copies of H and let B_α be the event $G(n,p) \supseteq A_\alpha$. We apply Janson's Inequality. As

$$\lim_{n\to\infty} \mu = lim_{n\to\infty}\binom{n}{v}v!p^e/a = c^e/a$$

we find

$$\lim_{n\to\infty} M = exp[-c^e/a]$$

Now we examine (similar to Theorem 1.4.2)

$$\Delta = \sum_{\alpha\sim\beta} \Pr[B_\alpha \wedge B_\beta]$$

We split the sum according to the number of *vertices* in the intersection of copies α and β. Suppose they intersect in j vertices. If $j = 0$ or $j = 1$ then $A_\alpha \cap A_\beta = \emptyset$ so that $\alpha \sim \beta$ cannot occur. For $2 \leq j \leq v$ let f_j be the maximal $|A_\alpha \cap A_\beta|$ where $\alpha \sim \beta$ and α, β intersect in j vertices. As $\alpha \neq \beta$, $f_v < e$. When $2 \leq j \leq v-1$ the critical observation is that $A_\alpha \cap A_\beta$ is a subgraph of H and hence, as H is strictly balanced,

$$\frac{f_j}{j} < \frac{e}{v}$$

There are $O(n^{2v-j})$ choices of α, β intersecting in j points since α, β are determined, except for order, by $2v - j$ points. For each such α, β

$$\Pr[B_\alpha \wedge B_\beta] = p^{|A_\alpha \cup A_\beta|} = p^{2e-|A_\alpha \cap A_\beta|} \leq p^{2e-f_j}$$

Thus

$$\Delta = \sum_{j=2}^{v} O(n^{2v-j})O(n^{-\frac{v}{e}(2e-f_j)})$$

But

$$2v - j - \frac{v}{e}(2e - f_j) = \frac{vf_j}{e} - j < 0$$

so each term is $o(1)$ and hence $\Delta = o(1)$. By Janson's Inequality

$$\lim_{n\to\infty} \Pr[\wedge \overline{B}_\alpha] = \lim_{n\to\infty} M = exp[-c^e/a]$$

completing the proof. □

The fine threshold behavior for the appearance of an arbitrary graph H has been worked out but it can get quite complicated.

Lecture 4: The Chromatic Number Resolved!

The centerpiece of this lecture is the result of Béla Bollobás that, with $G \sim G(n, \frac{1}{2})$, the chromatic number $\chi(G)$ is asymptotically $n/(2\log_2 n)$ almost surely.

1 Clique Number Revisited

In this section we fix $p = 1/2$, (other values yield similar results), let $G \sim G(n,p)$ and consider the clique number $\omega(G)$. For a fixed $c > 0$ let $n, k \to \infty$ so that

$$\binom{n}{k} 2^{-\binom{k}{2}} \to c$$

As a first approximation

$$n \sim \frac{k}{e\sqrt{2}}\sqrt{2}^k$$

and

$$k \sim \frac{2\ln n}{\ln 2}$$

Here $\mu \to c$ so $M \to e^{-c}$. The Δ term was examined earlier. For this k, $\Delta = o(E[X]^2)$ and so $\Delta = o(1)$. Therefore

$$\lim_{n,k\to\infty} \Pr[\omega(G(n,p)) < k] = e^{-c}$$

Being more careful, let $n_0(k)$ be the minimum n for which

$$\binom{n}{k} 2^{-\binom{k}{2}} \geq 1.$$

Observe that for this n the left hand side is $1+o(1)$. Note that $\binom{n}{k}$ grows, in n, like n^k. For any $\lambda \in (-\infty, +\infty)$ if

$$n = n_0(k)[1+\frac{\lambda+o(1)}{k}]$$

then

$$\binom{n}{k}2^{-\binom{k}{2}} = [1+\frac{\lambda+o(1)}{k}]^k = e^{\lambda}+o(1)$$

and so

$$\Pr[\omega(G(n,p)) < k] = e^{-e^{\lambda}} + o(1)$$

As λ ranges from $-\infty$ to $+\infty$, $e^{-e^{\lambda}}$ ranges from 1 to 0. As $n_0(k+1) \sim \sqrt{2}n_0(k)$ the ranges will not "overlap" for different k. More precisely, let K be arbitrarily large and set

$$I_k = [n_0(k)[1-\frac{K}{k}], n_0(k)[1+\frac{K}{k}]]$$

For $k \geq k_0(K)$, $I_{k-1} \cap I_k = \emptyset$. Suppose $n \geq n_0(k_0(K))$. If n lies between the intervals (which occurs for "most" n), which we denote by $I_k < n < I_{k+1}$, then

$$\Pr[\omega(G(n,p)) < k] \leq e^{-e^{K}} + o(1),$$

nearly zero, and

$$\Pr[\omega(G(n,p)) < k+1] \geq e^{-e^{-K}} + o(1),$$

nearly one, so that

$$\Pr[\omega(G(n,p)) = k] \geq e^{-e^{-K}} - e^{-e^{K}} + o(1),$$

nearly one. When $n \in I_k$ we still have $I_{k-1} < n < I_{k+1}$ so that

$$\Pr[\omega(G(n,p)) = k \text{ or } k-1] \geq e^{-e^{-K}} - e^{-e^{K}} + o(1),$$

nearly one. As K may be made arbitrarily large this yields the celebrated two point concentration theorem on clique number given as Corollary 2.1.2. Note, however, that for most n the concentration of $\omega(G(n,1/2))$ is actually on a single value!

2 Chromatic Number

Again fix $p = 1/2$ (there are similar results for other p) and let $G \sim G(n,\frac{1}{2})$. We shall find bounds on the chromatic number $\chi(G)$. The original proof of Bollobás used martingales and will be discussed later. Set

$$f(k) = \binom{n}{k}2^{-\binom{k}{2}}$$

Let $k_0 = k_0(n)$ be that value for which

$$f(k_0-1) > 1 > f(k_0)$$

Then $n = \sqrt{2}^{k(1+o(1))}$ so for $k \sim k_0$,

$$f(k+1)/f(k) = \frac{n}{k}2^{-k}(1+o(1)) = n^{-1+o(1)}$$

Set

$$k = k(n) = k_0(n) - 4$$

so that

$$f(k) > n^{3+o(1)}$$

Now we use the Generalized Janson Inequality to estimate $\Pr[\omega(G) < k]$. Here $\mu = f(k)$. (Note that Janson's Inequality gives a lower bound of $2^{-f(k)} = 2^{-n^{3+o(1)}}$ to this probability but this is way off the mark since with probability $2^{-\binom{n}{2}}$ the random G is empty!) The value Δ was examined in Lecture 2 and we showed

$$\frac{\Delta}{\mu^2} = \frac{\Delta^*}{\mu} = \sum_{i=2}^{k-1} g(i)$$

There $g(2) \sim k^4/n^2$ and $g(k-1) \sim 2kn2^{-k}/\mu$ were the dominating terms. In our instance $\mu > n^{3+o(1)}$ and $2^{-k} = n^{-2+o(1)}$ so $g(2)$ dominates and

$$\Delta \sim \frac{\mu^2 k^4}{n^2}$$

Hence we bound the *clique* number probability

$$\Pr[\omega(G) < k] < e^{-\mu^2(1+o(1))/2\Delta} = e^{-(n^2/k^4)(1+o(1))} = e^{-n^{2+o(1)}}$$

as $k = \Theta(\ln n)$. (The possibility that G is empty gives a lower bound so that we may say the probability is $e^{-n^{2+o(1)}}$, though a $o(1)$ in the hyperexponent leaves lots of room.)
Theorem 2.1. (Bollobás [1988]) Almost always

$$\chi(G)) \sim \frac{n}{2\log_2 n}$$

Proof. The argument that

$$\chi(G) \geq \frac{n}{\alpha(G)} \geq \frac{n}{2\log_2 n}(1+o(1))$$

almost always was given in Lecture 2.

The reverse inequality was an open question for a full quarter century! Set $m = \lfloor n/\ln^2 n \rfloor$. For any set S of m vertices the restriction $G|_S$ has the distribution of $G(m, 1/2)$. Let $k = k(m) = k_0(m) - 4$ as above. Note

$$k \sim 2\log_2 m \sim 2\log_2 n$$

Then

$$\Pr[\alpha[G|_S] < k] < e^{-m^{2+o(1)}}$$

There are $\binom{n}{m} < 2^n = 2^{m^{1+o(1)}}$ such sets S. Hence

$$\Pr[\alpha[G|_S] < k \text{ for some } m\text{-set } S] < 2^{m^{1+o(1)}} e^{-m^{2+o(1)}} = o(1)$$

That is, almost always *every* m vertices contain a k-element independent set.

Now suppose G has this property. We pull out k-element independent sets and give each a distinct color until there are less than m vertices left. Then we give each point a distinct color. By this procedure

$$\chi(G) \leq \lceil\frac{n-m}{k}\rceil + m \leq \frac{n}{k} + m = \frac{n}{2\log_2 n}(1+o(1)) + o(\frac{n}{\log_2 n})$$

$$= \frac{n}{2\log_2 n}(1+o(1))$$

and this occurs for almost all G. □

3 Some Very Low Probabilities

Let A be the property that G does not contain K_4 and consider $\Pr[G(n,p) \models A]$ as p varies. (Results with K_4 replaced by an arbitrary H are discussed at the end of this section.) We know that $p = n^{-2/3}$ is a threshold function so that for $p \gg n^{-2/3}$ this probability is $o(1)$. Here we want to estimate that probability. Our estimates here will be quite rough, only up to a $o(1)$ additive factor in the hyperexponent, though with more care the bounds differ by "only" a constant factor in the exponent. If we were to consider all potential K_4 as giving mutually independent events then we would be led to the estimate $(1-p^6)^{\binom{n}{4}} = e^{-n^{4+o(1)}p^6}$. For p appropriately small this turns out to be correct. But for, say, $p = \frac{1}{2}$ it would give the estimate $e^{-n^{4+o(1)}}$. This must, however, be way off the mark since with probability $2^{-\binom{n}{2}} = e^{-n^{2+o(1)}}$ the graph G could be empty and hence trivially satisfy A.

Rather than giving the full generality we assume $p = n^{-\alpha}$ with $\frac{2}{3} > \alpha > 0$. The result is:

$$\Pr[G(n,p) \models A] = e^{-n^{4-6\alpha+o(1)}}$$

for $\frac{2}{3} > \alpha \geq \frac{2}{5}$ and

$$\Pr[G(n,p) \models A] = e^{-n^{2-\alpha+o(1)}}$$

for $\frac{2}{5} \geq \alpha > 0$.

The upper bound follows from the inequality

$$\Pr[G(n,p) \models A] \geq \max\left[(1-p^6)^{\binom{n}{4}}, (1-p)^{\binom{n}{2}}\right]$$

This is actually two inequalities. The first comes from the probability of G not containing a K_4 being at most the probability as if all the potential K_4 were independent. The second is the same bound on the probability that G doesn't contain a K_2 - i.e., that G has no edges. Calculation shows that the "turnover" point for the two inequalities occurs when $p = n^{-2/5+o(1)}$.

The upper bound follows from the Janson inequalities. For each four set α of vertices B_α is that that 4-set gives a K_4 and we want $\Pr[\wedge \overline{B_\alpha}]$. We have $\mu = \Theta(n^4p^6)$ and $-\ln M \sim \mu$ and (as shown in Lecture 1) $\Delta = \Theta(\mu\Delta^*)$ with $\Delta^* = \Theta(n^2p^5 + np^3)$. With $p = n^{-\alpha}$ and $\frac{2}{3} > \alpha > \frac{2}{5}$ we have $\Delta^* = o(1)$ so that

$$\Pr[\wedge \overline{B_\alpha}] \leq e^{-\mu(1+o(1))} = e^{-n^{4-6\alpha+o(1)}}$$

When $\frac{2}{5} > \alpha > 0$ then $\Delta^* = \Theta(n^2p^5)$ (somewhat surprisingly the np^3 never is significant in these calculations) and the extended Janson inequality gives

$$\Pr[\wedge \overline{B_\alpha}] \leq e^{-\Theta(\mu^2/\Delta)} = e^{-\Theta(\mu/\Delta^*)} = e^{-n^{2-\alpha}}$$

The general result has been found by T. Luczak, A. Rucinski and S. Janson. Let H be any fixed graph and let A be the property of not containing a copy of H. For any subgraph H' of H the correlation inequality gives

$$\Pr[G(n,p) \models A] \leq e^{-E[X_{H'}]}$$

where $X_{H'}$ is the number of copies of H' in G. Now let $p = n^{-\alpha}$ where we restrict to those α for which p is past the threshold function for the appearance of H. Then

$$\Pr[G(n,p) \models A] = e^{n^{o(1)}} \min_{H'} e^{-E[X_{H'}]}$$

Lecture 5: Counting Extensions and Zero-One Laws

The threshold behavior for the existence of a copy of H in $G(n,p)$ is well understood. Now we turn to what, in a logical sense, is the next level which we call *extension statements.* We want $G(n,p)$ to have the property that every $x_1,\ldots,x_r$ belong to a copy of H. For example ($r=1$), every vertex lies in a triangle. We find the fine threshold behavior for this property and further show - continuing this example - that for p a bit larger almost surely every vertex lies in about the same number of triangles.

1 Every Vertex in a Triangle

Let A be the property that every vertex lies in a triangle.
Theorem 1.1. Let $c>0$ be fixed and let $p=p(n)$, $\mu=\mu(n)$ satisfy

$$\binom{n-1}{2}p^3=\mu$$

$$e^{-\mu}=\frac{c}{n}$$

Then

$$\lim_{n\to\infty}\Pr[G(n,p)\models A]=e^{-c}$$

Proof. First fix $x\in V(G)$. For each unordered $y,z\in V(G)-\{x\}$ let B_{xyz} be the event that $\{x,y,z\}$ is a triangle of G. Let C_x be the event $\wedge\overline{B_{xyz}}$ and X_x the corresponding indicator random variable. We use Janson's Inequality to bound $E[X_x]=\Pr[C_x]$. Here $p=o(1)$ so $\epsilon=o(1)$. $\sum Pr[B_{xyz}]=\mu$ as defined above. Dependency $xyz\sim xuv$ occurs if and only if the sets overlap (other than in x). Hence

$$\Delta=\sum_{y,z,z'}\Pr[B_{xyz}\wedge B_{xyz'}]=O(n^3)p^5=o(1)$$

since $p=n^{-2/3+o(1)}$. Thus

$$E[X_x]\sim e^{-\mu}=\frac{c}{n}$$

Now define

$$X=\sum_{x\in V(G)}X_x,$$

the number of vertices x not lying in a triangle. Then from Linearity of Expectation

$$E[X]=\sum_{x\in V(G)}E[X_x]\to c$$

We need show that the Poisson Paradigm applies to X. To do this we show that all moments of X are the same as for the Poisson distribution. Fix r. Then

$$E[X^{(r)}/r!]=S^{(r)}=\sum\Pr[C_{x_1}\wedge\ldots\wedge C_{x_r}],$$

the sum over all sets of vertices $\{x_1,\ldots,x_r\}$. All r-sets look alike so

$$E[X^{(r)}/r!]=\binom{n}{r}\Pr[C_{x_1}\wedge\ldots\wedge C_{x_r}]\sim\frac{n^r}{r!}\Pr[C_{x_1}\wedge\ldots\wedge C_{x_r}]$$

where $x_1, \ldots, x_r$ are some particular vertices. But

$$C_{x_1} \wedge \ldots \wedge C_{x_r} = \wedge \overline{B_{x_i yz}},$$

the conjunction over $1 \leq i \leq r$ and all y, z. We apply Janson's Inequality to this conjunction. Again $\epsilon = p^3 = o(1)$. The number of $\{x_i, y, z\}$ is $r\binom{n-1}{2} - O(n)$, the overcount coming from those triangles containing two (or three) of the x_i. (Here it is crucial that r is fixed.) Thus

$$\sum \Pr[B_{x_i yz}] = p^3 \left(r\binom{n-1}{2} - O(n) \right) = r\mu + O(n^{-1+o(1)})$$

As before Δ is p^5 times the number of pairs $x_i yz \sim x_j y'z'$. There are $O(rn^3) = O(n^3)$ terms with $i = j$ and $O(r^2 n^2) = O(n^2)$ terms with $i \neq j$ so again $\Delta = o(1)$. Therefore

$$\Pr[C_{x_1} \wedge \ldots \wedge C_{x_r}] \sim e^{-r\mu}$$

and

$$E[X^{(r)}/r!] \sim \frac{(ne^{-\mu})^r}{r!} = \frac{c^r}{r!}$$

Hence X has limiting Poisson distribution, in particular $\Pr[X = 0] \to e^{-\mu}$. □

2 Rooted Graphs

The above result was only a special case of a general result of Spencer[1990] which we now state. By a *rooted graph* is meant a pair (R, H) consisting of a graph $H = (V(H), E(H))$ and a specified proper subset $R \subset V(H)$ of vertices called the roots. For convenience let the vertices of H be labelled $a_1, \ldots, a_r, b_1, \ldots, b_v$ with $R = \{a_1, \ldots, a_r\}$. In a graph G we say that vertices $y_1, \ldots, y_v$ make an (R, H)-extension of vertices $x_1 \ldots, x_r$ if all these vertices are distinct; y_i, y_j are adjacent in G whenever b_i, b_j are adjacent in H; and x_i, y_j are adjacent in G whenever a_i, b_j are adjacent in H. So G on $x_1, \ldots, x_r, y_1, \ldots, y_v$ gives a copy of H which may have additional edges – except that edges between the x's are not examined. We let $Ext(R, H)$ be the property the for all $x_1, \ldots, x_r$ there exist $y_1, \ldots, y_v$ giving an (R, H) extension. For example, when H is a triangle and R one vertex $Ext(R, H)$ is the statement that every vertex lies in a triangle. When H is a path of length t and R the endpoints $Ext(R, H)$ is the statement that every pair of vertices lie on a path of length t. When $R = \emptyset$ $Ext(\emptyset, H)$ is the already examined statement that there exists a copy of H. As in that situation we have a notion of balanced and strictly balanced. We say (R, H) has type (v, e) where v is the number of nonroot vertices and e is the number of edges of H, not counting edges with both vertices in R. For every S with $R \subset S \subseteq V(H)$ let (v_S, e_S) be the type of $(R, H|_S)$. We call (R, H) balanced if $e_S/v_S \leq e/v$ for all such S and we call (R, H) strictly balanced if $e_S/v_S < e/v$ for all proper $S \subset V(H)$. We call (R, H) nontrivial if every root is adjacent to at least one nonroot.

Theorem 2.1. Let (R, H) be a nontrivial strictly balanced rooted graph with type (v, e) and $r = |R|$. Let c_1 be the number of graph automorphism $\sigma : V(H) \to V(H)$ with $\sigma(x) = x$ for all roots x. Let c_2 be the number of bijections $\sigma : R \to R$ which are extendable to some graph automorphism $\lambda : V(H) \to V(H)$. Let $\mu > 0$ be arbitrary and fixed. Let $p = p(n)$ satisfy

$$\frac{n^v p^e}{c_1} = \ln\left(\frac{n^r}{c_2 \mu}\right)$$

Then

$$\lim_{n \to \infty} \Pr[G(n, p) \models Ext(R, H)] = e^{-\mu}$$

While the counting of automorphisms leads to some technical complexities the proof is essentially that of the "every vertex in a triangle" case.

3 All Vertices in nearly the same number of Triangles

Returning to the example of §1, let $N(x)$ denote the number of triangles containing vertex x. Set $\mu = \binom{n-1}{2}p^3$ as before.

Theorem 3.1. For every $\epsilon > 0$ there exists K so that if $p = p(n)$ is such that $\mu = K \log n$ then almost surely

$$(1-\epsilon)\mu < N(x) < (1+\epsilon)\mu$$

for *all* vertices x.

We shall actually show that for a given vertex x

$$\Pr[|N(x) - \mu| > \epsilon\mu] = o(n^{-1})$$

If the distribution of $N(x)$ were Poisson with mean μ then this would follow by Large Deviation results and indeed our approach will show that $N(x)$ is closely approximated by the Poisson distribution.

We call F a maximal disjoint family of extensions if F consists of pairs $\{x_i, y_i\}$ such that all x, x_i, y_i are triangles in $G(n,p)$, the x_i, y_i are all distinct, and there is no $\{x', y'\}$ with x, x', y' a triangle and x', y' both distinct from all the x_i, y_i. Let $Z^{(s)}$ denote the number of maximal disjoint families of size s. Lets restrict $0 \le s \le \log^2 n$ (a technical convenience) and bound $E[Z^{(s)}]$. There are $\sim \binom{n-1}{2}^s / s!$ choices for F. Each has probability $(p^3)^s$ that all x_i, y_i do indeed give extensions. We further need that the $n-1-2s \sim n$ other vertices contain no extension. The calculation of §1 may be carried out here to show that this probability is $\sim e^{-\mu}$. All together

$$E[Z^{(s)}] \le (1+o(1))\frac{\mu^s}{s!}e^{-\mu}$$

But now the right hand side is asymptotically the Poisson distribution so that we can choose K so that

$$\sum *E[Z^{(s)}] = o(n^{-1}) \qquad (*)$$

where $\sum^*$ is over $s < \log^2 n$ with $|s - \mu| > \epsilon\mu$.

When $s > \log^2 n$ we ignore the condition that F be maximal so that $E[Z^{(s)}] < \mu^s/s! = o(n^{-10})$, say. Thus $(*)$ holds with $\sum^*$ over all s with $|s - \mu| > \epsilon\mu$. Thus with probability $1 - o(n^{-1})$ all maximal disjoint families of extensions F have $|s - \mu| < \epsilon\mu$. But there *must* be some maximal disjoint family of extensions. Thus with probability $1 - o(n^{-1})$ there is a maximal disjoint family of extensions F with $|s - \mu| < \epsilon\mu$. As F consists of extensions

$$\Pr[N(x) < (1-\epsilon)\mu] = o(n^{-1})$$

To complete the upper bound we need show that $N(x)$ will not be much larger than $|F|$. Here we use only that $p = n^{-2/3+o(1)}$. There is $o(n^{-1})$ probability that $G(n,p)$ has an edge $\{x, x'\}$ lying in ten triangles. There is a $o(n^{-1})$ that $G(n,p)$ has a vertex x with $u_i, v_i, w_i, 1 \le i \le 7$ all distinct and all x, u_i, v_i and x, v_i, w_i triangles. When these do not occur $N(x) \le |F| + 70$ for any maximal disjoint family of extensions $|F|$ and so for any $\epsilon' > \epsilon$

$$\Pr[N(x) > (1+\epsilon')\mu] < o(n^{-1}) + \Pr[\text{some } |F| > (1+\epsilon)\mu] = o(n^{-1})$$

With some additional work one can find K so that the conclusions of the theorem hold for any $p = p(n)$ with $\mu > K \log n$. The general result is stated in terms of rooted graphs. For a given rooted graph (R, H) let $N(x_1, \ldots, x_r)$ denote the number of $(y_1, \ldots, y_v)$ giving an (R, H)

extension. Set $\mu = \binom{n-r}{v} p^e$, the expected value of N in $G(n,p)$.
Theorem 3.2. Let (R, H) be strictly balanced. Then for all $\epsilon > 0$ there exists K so that if $p = p(n)$ is such that $\mu > K \log n$ then almost surely

$$|N(x_1, \ldots, x_r) - \mu| < \epsilon\mu$$

for all $x_1, \ldots, x_r$.

In particular if $\mu \gg \log n$ then almost surely all $N(x_1, \ldots, x_r) \sim \mu$.

4 Zero-One Laws

In this section we restrict our attention to graph theoretic properties expressible in the First Order theory of graphs. The language of this theory consists of variables $(x, y, z, \ldots)$, which always represent vertices of a graph, equality and adjacency $(x = y, x \sim y)$, the usual Boolean connectives $(\wedge, \neg, \ldots)$ and universal and existential quanfication $(\forall_x, \exists_y)$. Sentences must be finite. As examples, one can express the property of containing a triangle

$$\exists_x \exists_y \exists_z [x \sim y \wedge x \sim z \wedge y \sim z]$$

having no isolated point

$$\forall_x \exists_y [x \sim y]$$

and having radius at most two

$$\exists_x \forall_y [\neg(y = x) \wedge \neg(y \sim x) \longrightarrow \exists_z [z \sim y \wedge y \sim x]]$$

For any property A and any n, p we consider the probability that the random graph $G(n,p)$ satisfies A, denoted

$$\Pr[G(n,p) \models A]$$

Our objects in this section will be the theorem of Glebskii et.al. [1969] and independently Fagin[1976]
Theorem 4.1 For any fixed p, $0 < p < 1$ and any First Order A

$$\lim_{n \to \infty} \Pr[G(n,p) \models A] = 0 \text{ or } 1$$

and that of Shelah and Spencer[1988]
Theorem 4.2 For any *irrational* α, $0 < \alpha < 1$, setting $p = p(n) = n^{-\alpha}$

$$\lim_{n \to \infty} \Pr[G(n,p) \models A] = 0 \text{ or } 1$$

Both proofs are only outlined.

We shall say that a function $p = p(n)$ satisfies the Zero-One Law if the above equality holds for every First Order A.

The Glebskii/Fagin Theorem has a natural interpretation when $p = .5$ as then $G(n,p)$ gives equal weight to every (labelled) graph. It then says that any First Order property A holds for either almost all graphs or for almost no graphs. The Shelah/Spencer Theorem may be interpreted in terms of threshold functions. For example, $p = n^{-2/3}$ is a threshold function for containment of a K_4. That is, when $p \ll n^{-2/3}$, $G(n,p)$ almost surely does not contain a K_4 whereas when $p \gg n^{-2/3}$ it almost surely does contain a K_4. In between, say at $p = n^{-2/3}$, the probability is between 0 and 1, in this case $1 - e^{-1/24}$. The (admittedly rough) notion is that *at* a threshold function the Zero-One Law will not hold and so to say that $p(n)$ satisfies the Zero-One Law is to say that $p(n)$ is not a threshold function - that it is a boring place in the evolution of the random graph, at least through the spectacles of the First Order language. In stark terms: What happens in the evolution of $G(n,p)$ at $p = n^{-\pi/7}$? The answer: Nothing!

Our approach to Zero-One Laws will be through a variant of the Ehrenfeucht Game, which we now define. Let G, H be two vertex disjoint graphs and t a positive integer. We define a perfect information game, denoted $EHR[G,H,t]$, with two players, denoted Spoiler and Duplicator. The game has t rounds. Each round has two parts. First the Spoiler selects either a vertex $x \in V(G)$ or a vertex $y \in V(H)$. He chooses which graph to select the vertex from. Then the Duplicator must select a vertex in the other graph. At the end of the t rounds t vertices have been selected from each graph. Let $x_1, \ldots, x_t$ be the vertices selected from $V(G)$ and $y_1, \ldots, y_t$ be the vertices selected from $V(H)$ where x_i, y_i are the vertices selected in the i-th round. Then Duplicator wins if and only if the induced graphs on the selected vertices are order-isomorphic: i.e., if for all $1 \leq i < j \leq t$

$$\{x_i, x_j\} \in E(G) \longleftrightarrow \{y_i, y_j\} \in E(H)$$

As there are no hidden moves and no draws one of the players must have a winning strategy and we will say that that player wins $EHR[G,H,t]$.
Lemma 4.3 For every First Order A there is a $t = t(A)$ so that if G, H are any graphs with $G \models A$ and $H \models \neg A$ then Spoiler wins $EHR[G,H,t]$.

A detailed proof would require a formal analysis of the First Order language so we give only an example. Let A be the property $\forall_x \exists_y [x \sim y]$ of not containing an isolated point and set $t = 2$. Spoiler begins by selecting an isolated point $y_1 \in V(H)$ which he can do as $H \models \neg A$. Duplicator must pick $x_1 \in V(G)$. As $G \models A$, x_1 is not isolated so Spoiler may pick $x_2 \in V(G)$ with $x_1 \sim x_2$ and now Duplicator cannot pick a "duplicating" y_2.
Theorem 4.4 A function $p = p(n)$ satisfies the Zero-One Law if and only if for every t, letting $G(n,p(n)), H(m,p(m))$ be independently chosen random graphs on disjoint vertex sets

$$\lim_{m,n \to \infty} \Pr[\text{ Duplicator wins} EHR[G(n,p(n)), H(m,p(m)), t]] = 1$$

Remark. For any given choice of G, H somebody must win $EHR[G,H,t]$. (That is, there is no random play, the play is perfect.) Given this probability distribution over (G,H) there will be a probability that $EHR[G,H,t]$ will be a win for Duplicator, and this must approach one.
Proof. We prove only the "if" part. Suppose $p = p(n)$ did not satisfy the Zero-One Law. Let A satisfy

$$\lim_{n \to \infty} \Pr[G(n,p(n)) \models A] = c$$

with $0 < c < 1$. Let $t = t(A)$ be as given by the Lemma. With limiting probability $2c(1-c) > 0$ exactly one of $G(n,p(n)), H(n,p(n))$ would satisfy A and thus Spoiler would win, contradicting the assumption. This is not a full proof since when the Zero-one Law is not satisfied $\lim_{n\to\infty} \Pr[G(n,p(n)) \models A]$ might not exist. If there is a subsequence n_i on which the limit is $c \in (0,1)$ we may use the same argument. Otherwise there will be two subsequences n_i, m_i on which the limit is zero and one respectively. Then letting $n, m \to \infty$ through n_i, m_i respectively, Spoiler will win $EHR[G,H,t]$ with probability approaching one. □

Theorem 4.4 provides a bridge from Logic to Random Graphs. To prove that $p = p(n)$ satisfies the Zero-One Law we now no longer need to know anything about Logic - we just have to find a good strategy for the Duplicator.

We say that a graph G has the full level s extension property if for every distinct $u_1, \ldots, u_a$ and $v_1, \ldots, v_b$ i G with $a+b \leq s$ there is an $x \in V(G)$ with $\{x, u_i\} \in E(G)$, $1 \leq i \leq a$ and $\{x, v_j\} \notin V(G)$, $1 \leq j \leq b$. Suppose that G, H both have the full level $s-1$ extension property. Then Duplicator wins $EHR[G, H, s]$ by the following simple strategy. On the i-th round, with $x_1, \ldots, x_{i-1}, y_1, \ldots, y_{i-1}$ already selected, and Spoiler picking, say, x_i, Duplicator simply picks y_i having the same adjacencies to the $y_j, j < i$ as x_i has to the $x_j, j < i$. The full extension property says that such a y_i will surely exist.

Theorem 4.5 For any fixed p, $0 < p < 1$, and any s, $G(n,p)$ almost always has the full level s extension property.

Proof. For every distinct $u_1, \ldots, u_a, v_1, \ldots, v_b, x \in G$ with $a+b \leq s$ let $E_{u_1,\ldots,u_a,v_1,\ldots,v_b,x}$ be the event that $\{x, u_i\} \in E(G)$, $1 \leq i \leq a$ and $\{x, v_j\} \notin V(G)$, $1 \leq j \leq b$. Then

$$\Pr[E_{u_1,\ldots,u_a,v_1,\ldots,v_b,x}] = p^a(1-p)^b$$

Now define

$$E_{u_1,\ldots,u_a,v_1,\ldots,v_b} = \wedge_x \overline{E_{u_1,\ldots,u_a,v_1,\ldots,v_b,x}}$$

the conjunction over $x \neq u_1, \ldots, u_a, v_1, \ldots, v_b$. But these events are mutually independent over x since they involve different edges. Thus

$$\Pr[\wedge_x \overline{E_{u_1,\ldots,u_a,v_1,\ldots,v_b,x}}] = [1 - p^a(1-p)^b]^{n-a-b}$$

Set $\epsilon = \min(p, 1-p)^s$ so that

$$\Pr[\wedge_x \overline{E_{u_1,\ldots,u_a,v_1,\ldots,v_b,x}}] \leq (1-\epsilon)^{n-s}$$

The key here is that ϵ is a fixed (dependent on p, s) positive number. Set

$$E = \vee E_{u_1,\ldots,u_a,v_1,\ldots,v_b}$$

the disjunction over all distinct $u_1, \ldots, u_a, v_1, \ldots, v_b \in G$ with $a+b \leq s$. There are less than $s^2 n^s$ such choices as we can choose a, b and then the vertices. Thus

$$\Pr[E] \leq s^2 n^s (1-\epsilon)^{n-s}$$

But

$$\lim_{n \to \infty} s^2 n^s (1-\epsilon)^{n-s} = 0$$

and so E holds almost never. Thus $\neg E$, which is precisely the statement that $G(n,p)$ has the full level s extension property, holds almost always. □

But now we have proven Theorem 4.1. For any $p \in (0,1)$ and any fixed s as $m, n \to \infty$ with probability approaching one both $G(n,p)$ and $H(m,p)$ will have the full level s extension property and so Duplicator will win $EHR[G(n,p), H(m,p), s]$.

Why can't Duplicator use this strategy when $p = n^{-\alpha}$? We illustrate the difficulty with a simple example. Let $.5 < \alpha < 1$ and let Spoiler and Duplicator play a three move game on G, H. Spoiler thinks of a point $z \in G$ but doesn't tell Duplicator about it. Instead he picks $x_1, x_2 \in G$, both adjacent to z. Duplicator simply picks $y_1, y_2 \in H$, either adjacent or not adjacent dependent on whether $x_1 \sim x_2$. But now wily Spoiler picks $x_3 = z$. $H \sim H(m, m^{-\alpha})$ does not have the full level 2 extension property. In particular, most pairs y_1, y_2 do not have a common neighbor. Unless Duplicator was lucky, or shrewd, he then cannot find $y_3 \sim y_1, y_2$ and so he loses. This example does not say that Duplicator will lose with perfect play - indeed, we will show that he almost always wins with perfect play - it only indicates that the strategy used need be more complex. Now let us fix $\alpha \in (0,1)$, α irrational.

Now recall our notion of rooted graphs (R, H) but this time from the perspective of a particular $p = n^{-\alpha}$. We say (R, H) is *dense* if $v - e\alpha < 0$ and *sparse* if $v - e\alpha > 0$. The irrationality of α assures us that all (R, H) are in one of these categories. We call (R, H) *rigid* if for all S with $R \subseteq S \subset V(H)$, (S, H) is dense.

For any r, t there is a finite list (up to isomorphism) of rigid rooted graphs (R, H) containing r roots and with $v(R, H) \leq t$. In any graph G we define the t-closure $cl_t(x_1, \ldots, x_r)$ to be the union of all $y_1, \ldots, y_v$ with (crucially) $v \leq t$ which form an (R, H) extension where (R, H) is rigid. If there are no such sets we define the default value $cl_t(x_1, \ldots, x_r) = \{x_1, \ldots, x_r\}$. We say two sets $x_1, \ldots, x_r$ and $x'_1, \ldots, x'_r$ have the same t-type if their t-closures are isomorphic. (To be precise, these are ordered r-tuples and the isomorphism must send x_i into x'_i.)

Example. Taking $\alpha \sim .51$ (but irrational, of course), $cl_1(x_1, x_2)$ consists of x_1, x_2 and all y adjacent to both of them. $cl_3(x_1, x_2)$ has those points and all y_1, y_2, y_3 which together with x_1 form a K_4 (note that this gives an (R, H) with $v = 3, e = 6$) and a finite number of other possibilities.

We can already describe the nature of Duplicator's strategy. At the end of the r-th move, with $x_1, \ldots, x_r$ and $y_1, \ldots, y_r$ having been selected from the two graphs, Duplicator will assure that these sets have the same a_r – *type*. We shall call this the $(a_1, \ldots, a_t)$ *lookahead strategy.* Here a_r must depend only on t, the total number of moves in the game and α. We shall set $a_t = 0$ so that at the end of the game, if Duplicator can stick to the $(a_1, \ldots, a_t)$ lookahead strategy then he has won. If, however, Spoiler picks, say, x_r so that there is no corresponding y_r with $x_1, \ldots, x_r$ and $y_1, \ldots, y_r$ having the same a_r-type then the strategy fails and we say that Spoiler wins. The values a_r give the "lookahead" that Duplicator uses but before defining them we need some preliminary results.

Lemma 4.6 Let α, $r, t > 0$ be fixed. Then there exists $K = K(\alpha, r, t)$ so that in $G(n, n^{-\alpha})$ a.s.

$$|cl_t(x_1, \ldots, x_r)| \leq K$$

for *all* $x_1, \ldots, x_r \in G$.

Proof. Set $K = r + t(L-1)$. If $X = \{x_1, \ldots, x_r\}$ has t-closure with more than K points then there will be L sets $Y^1, \ldots, Y^L$ disjoint from X, all $|Y^j| \leq t$ so that each $(X, X \cup Y^j)$ forms a rigid extension and with each Y^j having at least one point not in $Y^1 \cup \ldots Y^{j-1}$. Begin with X and add the Y^j in order. Adding Y^j will add, say, v_j vertices and e_j edges. Since $(X, X \cup Y^j)$ was *rigid*, $(X \cup Y^1 \cup \ldots \cup Y^{j-1}, X \cup Y^1 \cup \ldots \cup Y^j)$ is dense and so $v_j - e_j\alpha < 0$. As $v_j \leq t$ there are only a finite number of possible values of $v_j - e_j\alpha$ and so there is an $\epsilon = \epsilon(\alpha, r, t)$ so that all $v_j - e_j\alpha \leq -\epsilon$. Pick L (and therefore K) so that $r - L\epsilon < 0$. The existence of a t-closure of size greater than K would imply the existence in $G(n, n^{-\alpha})$ of one of a finite number of graphs that would have some $r + v_1 + \ldots + v_L$ vertices and at least $e_1 + \ldots + e_L$ edges. But the probability of G containing such a graph is bounded by

$$\begin{aligned} n^{r+v_1+\ldots+v_L} p^{e_1+\ldots+e_L} &= n^{r+v_1+\ldots+v_L-\alpha(e_1+\ldots+e_L)} \\ n^{r+(v_1-\alpha e_1)+\ldots+(v_L-\alpha e_L)} &\leq n^{r-L\epsilon} \\ = o(1) & \end{aligned}$$

so a.s. no such t-closures exist. □

Remark. The value of K given by the above proof depends strongly on how close α may be approximated by rationals of denominator at most t. This is often the case. If, for example, $\frac{1}{2} + \frac{1}{s+1} < \alpha < \frac{1}{2} + \frac{1}{s}$ then a.s. there will be two points $x_1, x_2 \in G(n, n^{-\alpha})$ having s common neighbors so that $|cl_1(x_1, x_2)| = s + 2$.

Now we define the $a_1, \ldots, a_t$ of the lookahead strategy by reverse induction. We set $a_t = 0$. If at the end of the game Duplicator can assure that the 0-types of $x_1, \ldots, x_t$ and $y_1, \ldots, y_t$ are the same then they have the same induced subgraphs and he has won. Suppose, inductively, that $b = a_{r+1}$ has been defined. Let, applying the Lemma, K be a.s. an upper bound on all $cl_b(z_1, \ldots, z_{r+1})$. We then define $a = a_r$ by $a = K + b$.

Now we need show that a.s. this strategy works. Let $G_1 \sim G(n, n^{-\alpha})$, $G_2 \sim G(m, m^{-\alpha})$ and suppose Duplicator tries to play the $(a_1, \ldots, a_t)$ lookahead strategy on $EHR(G_1, G_2, t)$.

Set $a = a_1$ and consider the first move. Spoiler will select, say, $y = y_1 \in G_2$. Duplicator then must play $x = x_1 \in G_1$ with $cl_a(x) \cong cl_a(y)$. Can he always do so - that is, do a.s. G_1 and G_2 have the same values of $cl_a(x)$? The size of $cl_a(x)$ is a.s. bounded so it suffices to show for any potential H that either there almost surely is an x with $cl_a(x) \cong H$ or there almost surely is no x with $cl_a(x) \cong H$.

Let H have v vertices and e edges. Suppose H has a subgraph H' (possibly H itself) with v' vertices, e' edges and $v' - \alpha e' < 0$. The expected number of copies of H' in G_1 is

$$\Theta(n^{v'} p^{e'}) = \Theta(n^{v' - \alpha e'}) = o(1)$$

so a.s. G_1 contains no copy of H', hence no copy of H, hence no x with $cl_a(x) \cong H$. If this does not occur then (since, critically, α is irrational) all $v' - \alpha e' > 0$ so the expected number of copies of all such H' approaches infinity. From Theorem 1.4.5 a.s. G_1 has $\Theta(n^{v-\alpha e})$ copies of H. For x in appropriate position in such a copy of H we cannot deduce $cl_a(x) \cong H$ but only that $cl_a(x)$ contains H as a subgraph. (Essentially, x may have additional extension properties.) For each such x as $cl_a(x)$ is bounded, $cl_a(x)$ contains only a bounded number of copies of H. Hence there are $\Theta(n^{v-\alpha e})$ *different* $x \in G_1$ so that $cl_a(x)$ contains H as a subgraph.

Let H' be a possible value for $cl_a(x)$ that contains H as a subgraph. Let H' have v' vertices and e' edges. As (x, H') is rigid, (H, H') is dense and so

$$(v' - v) - \alpha(e' - e) < 0$$

There are $\Theta(n^{v' - \alpha e'})$ different x with $cl_a(x)$ containingH' but since $v' - \alpha e' < v - \alpha e$ this is $o(n^{v-\alpha e})$. Subtracting off such x for all the boundedly many such potential H' there a.s. remain $\Theta(n^{v-\alpha e})$, hence at least one, x with $cl_a(x) \cong H$.

Now, in general, consider the $(r+1)$-st move. We set $b = a_{r+1}$, $a = a_r$ for notational convenience and recall $a = K + b$ where K is an upper bound on $cl_b(z_1, \ldots, z_{r+1})$. Points $x_1, \ldots, x_r \in G_1$, $y_1, \ldots, y_r \in G_2$ have been selected with

$$cl_a(x_1, \ldots, x_r) \cong cl_a(y_1, \ldots, y_r)$$

Spoiler picks, say, $x_{r+1} \in G_1$. We distinguish two cases. We say Spoiler has moved Inside if

$$x_{r+1} \in cl_K(x_1, \ldots, x_r)$$

Otherwise we say Spoiler has moved Outside.

Suppose Spoiler moves Inside. Then

$$cl_b(x_1, \ldots, x_r, x_{r+1}) \subseteq cl_{K+b}(x_1, \ldots, x_r) = cl_a(x_1, \ldots, x_r)$$

The isomorphism from $cl_a(x_1, \ldots, x_r)$ to $cl_a(y_1, \ldots, y_r)$ sends x_{r+1} to some y_{r+1} which Duplicator selects.

Suppose Spoiler moves Outside. Set $H = cl_b(x_1, \ldots, x_r, x_{r+1})$. Let H_0 be the union of all rigid extensions of any size of $x_1, \ldots, x_r$ in H. If $x_{r+1} \in H_0$ then, as $|H| \leq K$, $x_{r+1} \in$

$cl_K(x_1,\ldots,x_r)$ and Spoiler moved Inside. Hence $x_{r+1} \notin H_0$. Since $|H| \le K \le a$, H_0 lies inside $cl_a(x_1,\ldots,x_r)$. The isomorphism between $cl_a(x_1,\ldots,x_r)$ and $cl_a(y_1,\ldots,y_r)$ maps H_0 into a copy of itself in the graph G_2.

For any copy of H_0 in G_2, let $N(H_0)$ denote the number of extensions of H_0 to H. From Theorem 3.2 one can show that a.s all $N(H_0) = \Theta(n^{v-\alpha e})$, with $v = v(H_0, H)$, $e = e(H_0, H)$ and $v-\alpha e > 0$. For a given H_0 each y_{r+1} is in only a bounded number of copies of H since all copies of H lie in $cl_b(y_1,\ldots,y_r,y_{r+1})$. Hence there are $\Theta(n^{v-\alpha e})$ vertices y_{r+1} so that $cl_b(y_1,\ldots,y_r,y_{r+1})$ contains H. Arguing as with the first move there a.s. are $\Theta(n^{v-\alpha e})$, hence at least one, y_{r+1} with $cl_b(y_1,\ldots,y_r,y_{r+1}) \cong H$. Duplicator selects such a y_{r+1}.

Lecture 6: A Number Theory Interlude

We take a break from Graph Theory and explore applications of these methods to Number Theory.

1 Prime Factors

The second moment method is an effective tool in number theory. Let $v(n)$ denote the number of primes p dividing n. (We do not count multiplicity though it would make little difference.) The following result says, roughly, that "almost all" n have "very close to" $\ln\ln n$ prime factors. This was first shown by Hardy and Ramanujan in 1920 by a quite complicated argument. We give the proof of Paul Turan [1934] a proof that played a key role in the development of probabilistic methods in number theory.
Theorem 1.1 Let $\omega(n) \to \infty$ arbitrarily slowly. Then the number of x in $\{1,\ldots,n\}$ such that

$$|v(x) - \ln\ln n| > \omega(n)\sqrt{\ln\ln n}$$

is $o(n)$.
Proof. Let x be randomly chosen from $\{1,\ldots,n\}$. For p prime set

$$X_p = \begin{cases} 1 & \text{if } p|x \\ 0 & \text{otherwise} \end{cases}$$

and set $X = \sum X_p$, the summation over all primes $p \le n$, so that $X(x) = v(x)$. Now

$$E[X_p] = \frac{\lfloor n/p \rfloor}{n}$$

As $y - 1 < \lfloor y \rfloor \le y$

$$E[X_p] = 1/p + O(1/n)$$

By linearity of expectation

$$E[X] = \sum_{p \le n} \frac{1}{p} + O(\frac{1}{n}) \sim \ln\ln n$$

Now we bound the variance

$$Var[X] \le (1 + o(1)) \ln\ln n + \sum_{p \ne q} Cov[X_p, X_q]$$

With p, q distinct primes, $X_pX_q = 1$ if and only if $p|x$ and $q|x$ which occurs if and only if $pq|x$. Hence

$$Cov[X_p, X_q] = E[X_p]E[X_q] - E[X_pX_q] = \frac{\lfloor n/pq \rfloor}{n} - \frac{\lfloor n/p \rfloor}{n}\frac{\lfloor n/q \rfloor}{n}$$

$$\leq \frac{1}{pq} - \left(\frac{1}{p} - \frac{1}{n}\right)\left(\frac{1}{q} - \frac{1}{n}\right) \leq \frac{1}{n}\left(\frac{1}{p} + \frac{1}{q}\right)$$

Thus

$$\sum_{p \neq q} Cov[X_p, X_q] \leq \frac{1}{n}\sum_{p \neq q} \frac{1}{p} + \frac{1}{q} = \frac{\pi(n) - 1}{n}\sum_{p} \frac{2}{p}$$

where $\pi(n) \sim \frac{n}{\ln n}$ is the number of primes $p \leq n$. So

$$\sum_{p \neq q} Cov[X_p, X_q] < \frac{(n/\ln n)}{n}(2 \ln\ln n) = o(1)$$

That is, the covariances do not affect the variance, $Var[X] \sim \ln\ln n$ and Chebyschev's Inequality actually gives

$$\Pr[|v(n) - \ln\ln n| > \lambda\sqrt{\ln\ln n}] < \lambda^{-2} + o(1)$$

for any constant λ. □

In a classic paper Paul Erdős and Marc Kac [1940] showed, essentially, that X does behave like a normal distribution with mean and variance $\ln\ln n + o(1)$. Here is their precise result.
The Erdős-Kac Theorem. Let λ be fixed, positive, negative or zero. Then

$$\lim_{n \to \infty} \frac{1}{n}|\{x : 1 \leq x \leq n, v(x) \geq \ln\ln n + \lambda\sqrt{\ln\ln n}\}| = \int_{\lambda}^{\infty} \frac{1}{\sqrt{2\pi}} e^{-t^2/2} dt$$

We do not prove this result here.

2 Four Squares with Few Squares

The classic theorem of Lagrange states that every nonnegative integer n is the sum of four squares. How "sparse" can a set of squares be and still retain the four square property. For any set X of nonnegative integers set $N_X(x) = |\{i \in X, i \leq x\}|$. Let $S = \{0, 1, 4, 9, \ldots\}$ denote the squares. If $X \subseteq S$ and every $n \geq 0$ can be expressed as the sum of four elements of X then how slow can be the growth rate of $N_X(x)$? Clearly we must have $N_X(x) = \Omega(x^{1/4})$. Our object here is to give a quick proof of the following result of Wirsing.
Theorem. There is a set $X \subseteq S$ such that every $n \geq 0$ can be expressed as the sum of four elements of X and

$$N_X(x) = O(x^{1/4}(\ln x)^{1/4})$$

In 1828 Jacobi showed that the number $r_4(n)$ of solutions in integers to $n = a^2 + b^2 + c^2 + d^2$ is given by eight times the sum of those $d|n$ with $d \not\equiv 0 (\text{mod} 4)$. In 1801 Gauss found an exact expression for the number $r_2(n)$ of solutions in integers to $n = a^2 + b^2$. We will need only $r_2(n) = n^{o(1)}$ which follows easily from his results. From this the number $r_3(n)$ of solutions to $n = a^2 + b^2 + c^2$ is $O(n^{1/2+o(1)})$. Now suppose $n \not\equiv 0(\text{mod}4)$. Then $r_4(n) > 8n$ so, excluding order there are at least $n/48$ different solutions to $n = a^2 + b^2 + c^2 + d^2$ in nonnegative integers. From $r_2(n) = n^{o(1)}$ it follows that there are $O(n^{1/2+o(1)})$ solutions with $a = b$. Hence there are at least $(1 + o(1))n/48$ *sets* F of four squares adding to n .

Define a random subset $X \subseteq S$ by

$$\Pr[y \in X] = p_y = 10(\ln y)^{1/4} y^{-1/4}$$

for $y \in S$, $y \geq 10^8$. For definiteness say $\Pr[y \in X] = p_y = 1$ for $y \in S$, $y < 10^8$. Then

$$E[N_X(x)] = \sum_{i=0}^{x^{1/2}} \Pr[i^2 \in X] = O(x^{1/4}(\ln x)^{1/4})$$

and large deviation results give $N_X(x) = O(x^{1/4}(\ln x)^{1/4})$ almost always.

For any given $n \not\equiv 0(\text{mod}4)$, $n \geq 10^8$, let $\mathcal{F}_n$ denote the family of sets F of four squares adding to n. For each $F \in \mathcal{F}_n$ let A_F be the event $F \subseteq X$. We apply Janson's Inequality to give an upper bound to $\Pr[\wedge_{F \in \mathcal{F}_n} \overline{A_F}]$. Observe that this probability increases when the p_y decrease so, as the function p_y is decreasing in y, we may make the simplifying assumption

$$p_y = p = 10(\ln n)^{1/4} n^{-1/4}$$

for all $y \in S$, $y \leq n$. Then

$$\Pr[A_F] = p^4 = 10^4(\ln n)/n$$

and

$$\mu \geq (1 + o(1))(n/48)10^4(\ln n)/n \geq (100 + o(1))(\ln n)$$

Thus $e^{-\mu} < n^{-100+o(1)}$. The addends of Δ break into two parts, those $\Pr[A_F \wedge A_{F'}]$ with $|F \cap F'| = 1$ and those with $|F \cap F'| = 2$. The bounds on $r_3(n)$ give that there are at most $n^{3/2+o(1)}$ pairs F, F' of the first type and each has

$$\Pr[F \cap F'] = p^7 = n^{-7/4+o(1)}$$

The bounds on $r_2(n)$ give that there are at most $n^{1+o(1)}$ pairs F, F' of the second type and each has

$$\Pr[F \cap F'] = p^6 = n^{-3/2+o(1)}$$

Hence

$$\Delta \leq n^{3/2+o(1)-7/4+o(1)} + n^{1+o(1)-3/2+o(1)} = o(1)$$

Thus

$$\Pr[\wedge_{F \in \mathcal{F}_n} \overline{A_F}] \leq (1 + o(1))e^{-\mu} \leq n^{-100+o(1)}$$

As $\sum n^{-100+o(1)}$ converges the Borel-Cantelli lemma gives that almost always all sufficiently large $n \not\equiv 0(\text{mod}4)$ will be the sum of four elements of X.

Remark The constant "10" could be made smaller as long as the exponent of n here is less than -1.

Let X be a particular set having the above properties. (As customary, the probabilistic method does not actually "construct" X.) Suppose all $n \geq n_0$, $n \not\equiv 0(\text{mod}4)$ are the sum of four elements of X. Add to X all squares up to n_0. This does not affect the asymptotics of $N_X(x)$ and now all $n \not\equiv 0(\text{mod}4)$ are the sum of four elements of X. Finally, replace X by $X \cup 4X \cup 4^2X \cup 4^3X \cup \ldots$. This affects the asyptotics of $N_X(x)$ only by a constant and now *all* integers are the sum of four elements of X.

3 Counting Representations

For a given set S of natural numbers let (for every $n \in N$) $f(n) = f_S(n)$ denote the number of representations $n = x + y$, $x, y \in S, x \neq y$. For many years it was an open question whether there existed an S with $f(n) \geq 1$ for all sufficiently large n and yet $f(n) \leq n^{o(1)}$.
Theorem 3.1. (Erdős (1956)) There is a set S for which $f(n) = \Theta(\ln n)$. That is, there is a set S and constants c_1, c_2 so that for all sufficiently large n

$$c_1 \ln n \leq f(n) \leq c_2 \ln n$$

Proof. Define S randomly by

$$\Pr[x \in S] = p_x = \min\left[10\sqrt{\frac{\ln x}{x}}, 1\right]$$

Fix n. Now $f(n)$ is a random variable with mean

$$\mu = E[f(n)] = \sum_{x+y=n} p_x p_y$$

Roughly there are n addends with $p_x p_y > p_n^2 = 100\frac{\ln n}{n}$. We have $p_x p_x = \Theta(\frac{\ln n}{n})$ except in the regions $x = o(n), y = o(n)$ and care must be taken that those terms don't contribute significantly to μ. Careful asymptotics (and first year Calculus!) yield

$$\mu \sim (100 \ln n) \int_0^1 \frac{dx}{\sqrt{x(1-x)}} = 100\pi \ln n$$

The negligible effect of the $x = o(n), y = o(n)$ terms reflects the finiteness of the indefinite integral at poles $x = 0$ and $x = 1$. The possible representations $x + y = n$ are mutually independent events so that from basic Large Deviation results

$$\Pr[|f(n) - \mu| > \epsilon\mu] < 2(1-\delta)^{\mu}$$

for constants ϵ, δ. To be specific we take $\epsilon = .9, \delta = .1$ and

$$\Pr[|f(n) - \mu| > .9\mu] < .9^{314 \ln n} < n^{-1.1}$$

for n sufficiently large. Take $c_1 < .1(100\pi)$ and $c_2 > 1.9(100\pi)$.

Let A_n be the event that $c_1 \ln n \leq f(n) \leq c_2 \ln n$ does *not* hold. We have $\Pr[A_n] < n^{-1.1}$ for n sufficiently large. The Borel Cantelli Lemma applies, almost always all A_n fail for n sufficiently large. Therefore there exists a specific point in the probability space, i.e., a specific set S, for which $c_1 \ln n \leq f(n) \leq c_2 \ln n$ for all sufficiently large n. □

Now for a given set S of natural numbers let $g(n) = g_S(n)$ denote the number of representations $n = x + y + z$, $x, y, z \in S$, all unequal.
Theorem 3.2. (Erdős, Tetali[1990]) There is a set S and a positive constants c_1, c_2 so that

$$c_1 \log n \leq g(n) \leq c_2 \log n$$

for all sufficiently large n.

The full result of Erdős and Tetali was that for each k there is a set S and constants c_1, c_2 so that the number of representations of n as the sum of k terms of S lies between $c_1 \log n$ and

$c_2 \log n$ for all sufficiently large n.
Proof. Define S randomly by

$$\Pr[x \in S] = p_x = \min\left[10\left(\frac{\ln x}{x^2}\right)^{1/3}, \frac{1}{2}\right]$$

Fix n. Now $g(n)$ is a random variable and

$$\mu = E[g(n)] = \sum_{x+y+z=n} p_x p_y p_z$$

Careful asymptotics give

$$\mu \sim 10^3 \ln n \int_{x=0}^{1} \int_{y=0}^{1-x} \frac{dxdy}{[xy(1-x-y)]^{2/3}} = K \ln n$$

where K is large. (We may make K arbitrarily large by increasing "10".) We apply the Janson inequality. Here $\epsilon = 1/8$ as all $p_x \leq 1/2$. Also

$$\Delta = \sum p_x p_y p_z p_{y'} p_{z'},$$

the sum over all five-tuples with $x + y + z = x + y' + z' = n$. Roughly there are n^3 terms, each $\sim p_n^5 = n^{-10/3+o(1)}$ so that the sum is $o(1)$. Care must be taken that those terms with one (or more) small variables don't contribute much to the sum.

Now we emulate the argument of Theorem 5.3.1. Call F a maximal disjoint family of solutions if F is a family of sets $\{x_i, y_i, z_i\}$ with all x_i, y_i, z_i distinct, all $x_i + y_i + z_i = n$, all $x_i, y_i, z_i \in S$ and so that there is no $x, y, z \in S$ with $x + y + z = n$ and x, y, z distinct from all x_i, y_i, z_i. Let $Z^{(s)}$ denote the number of maximal disjoint families of solutions of size s. As in Theorem 5.3.1 when $s < \log^2 n$

$$E[Z^{(s)}] < \frac{\mu^s}{s!} e^{-\mu(1+o(1))}$$

while for $s \geq \log^2 n$

$$E[Z^{(s)}] < \mu^s / s!$$

so that $\sum^* E[Z^{(s)}] = o(n^{-10})$, say, where $\sum^*$ is over those s with $|s - \mu| > \epsilon\mu$. (Here ϵ is fixed and K must be sufficiently large.) With probability $1 - o(n^{-10})$ there is an F with $|s - \mu| < \epsilon\mu$.

When this occurs $g(n) \geq |F| \geq (1 - \epsilon)\mu$ but again we must worry about $g(n)$ being considerably larger than $|F|$. Here we use only that $p = n^{-2/3+o(1)}$. Note that the number of representations of $n = x + y + z$ with a given x is the number of representations $m = y + z$ of $m = n - x$.

Lemma 3.3. Almost surely no sufficiently large m has four (or more) representations as $m = y + z$, $y, z \in S$.

Proof. Here $\mu = \Theta(m^{-1/3})$ so the expected number of 4-tuples of representatives is $O(m^{-4/3})$ and so the probability of having four representatives is $O(m^{-4/3})$. Apply Borel-Cantelli. □

Now almost surely there is a C so that *no* m has more than C representations $m = y + z$. Let S be such that this holds and that all maximal disjoint families of solutions F have

$$K(1 - \epsilon) \log n < |F| < K(1 + \epsilon) \log n$$

Each triple $x, y, z \in S$ with $x+y+z = n$ must include one of the at most $3K(1+\epsilon)\log n$ elements of sets of F and each such element is in less than C such triples so that $g(n) < 3CK(1+\epsilon)\log n$. Take $c_1 = K(1 - \epsilon)$ and $c_2 = 3KC(1 + \epsilon)$.

With additional work one can prove Theorem 3.2 with $c_1 = K(1-\epsilon'), c_2 = K(1+\epsilon')$ for arbitrarily small ϵ' and K dependent only on ϵ'.

Lecture 7: The Phase Transition

1 Branching Processes

Paul Erdős and Alfred Rényi, in their original 1960 paper, discovered that the random graph $G(n,p)$ undergoes a remarkable change at $p = 1/n$. Speaking roughly, let first $p = c/n$ with $c < 1$. Then $G(n,p)$ will consist of small components, the largest of which is of size $\Theta(\ln n)$. But now suppose $p = c/n$ with $c > 1$. In that short amount of "time" many of the components will have joined together to form a "giant component" of size $\Theta(n)$. The remaining vertices are still in small components, the largest of which has size $\Theta(\ln n)$. They dubbed this phenomenon the *Double Jump.* We prefer the descriptive term Phase Transition because of the connections to percolation (e.g., freezing) in mathematical physics.

To better understand the Phase Transition we make a lengthy detour into the subject of Branching Processes. Imagine that we are in a unisexual universe and we start with a single organism. Imagine that this organism has a number of children given by a given random variable Z. (For us, Z will be Poisson with mean c.) These children then themselves have children, the number again being determined by Z. These grandchildren then have children, etc. As $Z = 0$ will have nonzero probability there will be some chance that the line dies out entirely. We want to study the total number of organisms in this process, with particular eye to whether or not the process continues forever. (The original application of this model was to a study of the -gasp!- male line of British peerage.)

Now lets be more precise. Let $Z_1, Z_2, \ldots$ be independent random variables, each with distribution Z. Define $Y_0, Y_1, \ldots$ by the recursion

$$Y_0 = 1$$

$$Y_i = Y_{i-1} + Z_i - 1$$

and let T be the least t for which $Y_t = 0$. If no such t exists (the line continuing forever) we say $T = +\infty$. The Y_i and Z_i mirror the Branching Process as follows. We view all organisms as living or dead. Initially there is one live organism and no dead ones. At each time unit we select one of the live organisms, it has Z_i children, and then it dies. The number Y_i of live organisms at time i is then given by the recursion. The process stops when $Y_t = 0$ (extinction) but it is a convenient fiction to define the recursion for all t. Note that T is not affected by this fiction since once $Y_t = 0$, T has been defined. T (whether finite or infinite) is the total number of organisms, including the original, in this process. (A natural approach, found in many probability texts, is to have all organisms of a given generation have their children at once and study the number of children of each generation. While we may think of the organisms giving birth by generation it will not affect our model.)

We shall use the major result of Branching Processes that when $E[Z] = c < 1$ with probability one the process dies out ($T < \infty$) but when $E[Z] = c > 1$ then there is a nonzero probability that the process goes on forever ($T = \infty$).

When a branching process dies we call $H = (Z_1, \ldots, Z_T)$ the *history* of the process. A sequence $(z_1, \ldots, z_t)$ is a possible history if and only if the sequence y_i given by $y_0 = 1, y_i =$

$y_{i-1} + z_i - 1$ has $y_i > 0$ for $0 \le i < t$ and $y_t = 0$. When Z is Poisson with mean λ

$$\Pr[H = (z_1, \ldots, z_t)] = \prod_{i-1}^{t} \frac{e^{-\lambda}\lambda^{z_i}}{z_i!} = \frac{e^{-\lambda}(\lambda e^{-\lambda})^{t-1}}{\prod_{i=1}^{t} z_i!}$$

since $z_1 + \ldots + z_t = t - 1$.

We call $d < 1 < c$ a conjugate pair if

$$de^{-d} = ce^{-c}$$

The function $f(x) = xe^{-x}$ increases from 0 to e^{-1} in [0,1) and decreases back to 0 in $(1, \infty)$ so that all $c \neq 1$ have a uniqe conjugate. Let $c > 1$ and $y = \Pr[T < \infty]$ so that $y = e^{c(y-1)}$. Then $(cy)e^{-cy} = ce^{-c}$ so

$$d = cy$$

Duality Principle. Let $d < 1 < c$ be conjugates. The Branching Process with mean c, conditional on extinction, has the same distribution as the Branching Process with mean d.
Proof. It suffices to show that for every history $H = (z_1, \ldots, z_t)$

$$\frac{e^{-c}(ce^{-c})^{t-1}}{y\prod_{i=1}^{t} z_i!} = \frac{e^{-d}(de^{-d})^{t-1}}{\prod_{i=1}^{t} z_i!}$$

This is immediate as $ce^{-c} = de^{-d}$ and $ye^{-d} = ye^{-cy} = e^{-c}$.

2 The Giant Component

Now let's return to random graphs. We define a procedure to find the component $C(v)$ containing a given vertex v in a given graph G. We are motivated by Karp [1990] in which this approach is applied to random digraphs. In this procedure vertices will be live, dead or neutral. Originally v is live and all other vertices are neutral, time $t = 0$ and $Y_0 = 1$. Each time unit t we take a live vertex w and check all pairs $\{w, w'\}$, w' neutral, for membership in G. If $\{w, w'\} \in G$ we make w' live, otherwise it stays neutral. After searching all neutral w' we set w dead and let Y_t equal the new number of live vertices. When there are no live vertices the process terminates and $C(v)$ is the set of dead vertices. Let Z_t be the number of w' with $\{w, w'\} \in G$ so that

$$Y_0 = 1$$

$$Y_t = Y_{t-1} + Z_t - 1$$

With $G = G(n,p)$ each neutral w' has independent probability p of becoming live. Here, critically, no pair $\{w, w'\}$ is ever examined twice so that the conditional probability for $\{w, w'\} \in G$ is always p. As $t - 1$ vertices are dead and Y_{t-1} are live

$$Z_t \sim B[n - (t-1) - Y_{t-1}, p]$$

Let T be the least t for which $Y_t = 0$. Then $T = |C(v)|$. As in Section 1 we continue the recursive definition of Y_t, this time for $0 \le t \le n$.
Claim 2.1 For all t

$$Y_t \sim B[n-1, 1-(1-p)^t] + 1 - t$$

It is more convenient to deal with

$$N_t = n - t - Y_t$$

the number of neutral vertices at time t and show, equivalently,

$$N_t \sim B[n-1,(1-p)^t]$$

This is reasonable since each $w \neq v$ has independent probability $(1-p)^t$ of staying neutral t times. Formally, as $N_0 = n-1$ and

$$\begin{aligned} N_t &= n-t-Y_t &&= n-t-B[n-(t-1)-Y_{t-1},p]-Y_{t-1}+1 \\ & &&= N_{t-1}-B[N_{t-1},p] \\ & &&= B[N_{t-1},1-p] \end{aligned}$$

the result follows by induction. □

We set $p = c/n$. When t and Y_{t-1} are small we may approximate Z_t by $B[n,c/n]$ which is approximately Poisson with mean c. Basically small components will have size distribution as in the Branching Process. The analogy must break down for $c > 1$ as the Branching Process may have an infinite population whereas $|C(v)|$ is surely at most n. Essentially, those v for which the Branching Process for $C(v)$ does not "die early" all join together to form the giant component.

Fix c. Let $Y_0^*, Y_1^*, \ldots, T^*, Z_1^*, Z_2^*, \ldots, H^*$ refer to the Branching Process and let the unstarred $Y_0, Y_1, \ldots, T, Z_1, Z_2, \ldots, H$ refer to the Random Graph process. For any possible history $(z_1, \ldots, z_t)$

$$\Pr[H^* = (z_1,\ldots,z_t)] = \prod_{i=1}^{t} \Pr[Z^* = z_i]$$

where Z^* is Poisson with mean c while

$$\Pr[H = (z_1,\ldots,z_t)] = \prod_{i=1}^{t} \Pr[Z_i = z_i]$$

where Z_i has Binomial Distribution $B[n-1-z_1-\ldots-z_{i-1},c/n]$. The Poisson distribution is the limiting distribution of Binomials. When $m = m(n) \sim n$ and c, i are fixed

$$\lim_{n\to\infty} \Pr[B[m,c/n]=i] = \lim_{n\to\infty} \binom{m}{z} (\frac{c}{n})^z (1-\frac{c}{n})^{m-z} = e^{-c}c^z/z!$$

hence

$$\lim_{n\to\infty} \Pr[H = (z_1,\ldots,z_t)] = \Pr[H^* = (z_1,\ldots,z_t)] .$$

Assume $c < 1$. For any fixed t, $\lim_{n\to\infty} \Pr[T=t] = \Pr[T^*=t]$. We now bound the size of the largest component. For any t

$$\Pr[T > t] \le \Pr[Y_t > 0] = \Pr[B[n-1,1-(1-p)^t] \ge t] \le \Pr[B[n,tc/n] \ge t]$$

as $1-(1-p)^t \le tp$ and $n-1 < n$. By Large Deviation Results

$$\Pr[T > t] < e^{-\alpha t}$$

where $\alpha = \alpha(c) > 0$. Let $\beta = \beta(c)$ satisfy $\alpha\beta > 1$. Then

$$\Pr[T > \beta \ln n] < n^{-\alpha\beta} = o(n^{-1})$$

There are n choices for initial vertex v. Thus almost always *all* components have size $O(\ln n)$.

Now assume $c > 1$. For any fixed t, $\lim_{n\to\infty} \Pr[T = t] = \Pr[T^* = t]$ but what corresponds to $T^* = \infty$? For $t = o(n)$ we may estimate $1 - (1-p)^t \sim pt$ and $n - 1 \sim n$ so that

$$\Pr[Y_t \leq 0] = \Pr[B[n-1, 1-(1-p)^t] \leq t-1] \sim \Pr[B[n, tc/n] \leq t]$$

drops exponentially in t by Large Deviation results. When $t = \alpha n$ we estimate $1 - (1-p)^t$ by $1 - e^{-c\alpha}$. The equation $1 - e^{-c\alpha} = \alpha$ has solution $\alpha = 1 - y$ where y is the extinction probability. For $\alpha < 1 - y$, $1 - e^{-c\alpha} > \alpha$ and

$$\Pr[Y_t \leq 0] \sim \Pr[B[n, 1 - e^{-c\alpha}] \leq \alpha n]$$

is exponentially small while for $\alpha > 1 - y$, $1 - e^{-c\alpha} < \alpha$ and $\Pr[Y_t \leq 0] \sim 1$. Thus almost always $Y_t = 0$ for some $t \sim (1-y)n$. Basically, $T^* = \infty$ corresponds to $T \sim (1-y)n$. Let $\epsilon, \delta > 0$ be arbitrarily small. With somewhat more care to the bounds we may show that there exists t_0 so that for n sufficiently large

$$\Pr[t_0 < T < (1-\delta)n(1-y) \text{ or } T > (1+\delta)n(1-y)] < \epsilon$$

Pick t_0 sufficiently large so that

$$y - \epsilon \leq \Pr[T^* \leq t_0] \leq y$$

Then as $\lim_{n\to\infty} \Pr[T \leq t_0] = \Pr[T^* \leq 0]$ for n sufficiently large

$$y - 2\epsilon \leq \Pr[T \leq t_0] \leq y + \epsilon$$

$$1 - y - 2\epsilon \leq \Pr[(1-\delta)n(1-y) < T < (1+\delta)n(1-y)] < 1 - y + 3\epsilon$$

Now we expand our procedure to find graph components. We start with $G \sim G(n,p)$, select $v = v_1 \in G$ and compute $C(v_1)$ as before. Then we delete $C(v_1)$, pick $v_2 \in G - C(v_1)$ and iterate. At each stage the remaining graph has distribution $G(m,p)$ where m is the number of vertices. (Note, critically, that no pairs $\{w, w'\}$ in the remaining graph have been examined and so it retains its distribution.) Call a component $C(v)$ small if $|C(v)| \leq t_0$, giant if $(1-\delta)n(1-y) < |C(v)| < (1+\delta)n(1-y)$ and otherwise failure. Pick $s = s(\epsilon)$ with $(y+\epsilon)^s < \epsilon$. (For ϵ small $s \sim K \ln \epsilon^{-1}$.) Begin this procedure with the full graph and terminate it when either a giant component or a failure component is found or when s small components are found. At each stage, as only small components have thus far been found, the number of remaining points is $m = n - O(1) \sim n$ so the conditional probabilities of small, giant and failure remain asymptotically the same. The chance of ever hitting a failure component is thus $\leq s\epsilon$ and the chance of hitting all small components is $\leq (y+\epsilon)^s \leq \epsilon$ so that with probability at least $1 - \epsilon'$, where $\epsilon' = (s+1)\epsilon$ may be made arbitrarily small, we find a series of less than s small components followed by a giant component. The remaining graph has $m \sim yn$ points and $pm \sim cy = d$, the conjugate of c as defined earlier. As $d < 1$ the previous analysis gives the maximal components. In summary: almost always $G(n, c/n)$ has a giant component of size $\sim (1-y)n$ and all other components of size $O(\ln n)$. Furthermore, the Duality Principle has a discrete analog.
Discrete Duality Principle. Let $d < 1 < c$ be conjugates. The structure of $G(n, c/n)$ with its giant component removed is basically that of $G(m, d/m)$ where m, the number of vertices not in the giant component, satisfies $m \sim ny$.

The small components of $G(n, c/n)$ can also be examined from a static view. For a fixed k let X be the number of tree components of size k. Then

$$E[X] = \binom{n}{k} k^{k-2} p^{k-1} (1-p)^{k(n-k)+\binom{k}{2}-(k-1)}$$

Here we use the nontrivial fact, due to Cayley, that there are k^{k-2} possible trees on a given k-set. For c, k fixed

$$E[X] \sim n\frac{e^{-ck}k^{k-2}c^{k-1}}{k!}$$

As trees are strictly balanced a second moment method gives $X \sim E[X]$ almost always. Thus $\sim p_k n$ points lie in tree components of size k where

$$p_k = \frac{e^{-ck}(ck)^{k-1}}{k!}$$

It can be shown analytically that $p_k = \Pr[T = k]$ in the Branching Process with mean c. Let Y_k denote the number of cycles of size k and Y the total number of cycles. Then

$$E[Y_k] = \frac{(n)_k}{2k}(\frac{c}{n})^k \sim \frac{c^k}{2k}$$

for fixed k. For $c < 1$

$$E[Y] = \sum E[Y_k] \to \sum_{k=1}^{\infty} \frac{c^k}{2k}$$

has a finite limit whereas for $c > 1$, $E[Y] \to \infty$. Even for $c > 1$ for any fixed k the number of k-cycles has a limiting expectation and so do not asymptotically affect the number of components of a given size.

3 Inside the Phase Transition

In the evolution of the random graph $G(n,p)$ a crucial change takes place in the vicinity of $p = c/n$ with $c = 1$. The small components at that time are rapidly joining together to form a giant component. This corresponds to the Branching Process when births are Poisson with mean 1. There the number T of organisms will be finite almost always and yet have infinite expectation. No wonder that the situation for random graphs is extremely delicate. In recent years there has been much interest in looking "inside" the phase transition at the growth of the largest components. (See, e.g. Luczak [1990].) The appropriate parametrization is, perhaps surprisingly,

$$p = \frac{1}{n} + \frac{\lambda}{n^{4/3}}$$

When $\lambda = \lambda(n) \to -\infty$ the phase transition has not yet started. The largest components are $o(n^{2/3})$ and there are many components of nearly the largest size. When $\lambda = \lambda(n) \to +\infty$ the phase transition is over - a largest component, of size $>> n^{2/3}$ has emerged and all other components are of size $o(n^{2/3})$. Let's fix λ, c and let X be the number of tree components of size $k = cn^{2/3}$. Then

$$E[X] = \binom{n}{k} k^{k-2} p^{k-1} (1-p)^{k(n-k)+\binom{k}{2}-(k-1)}$$

Watch the terms cancel!

$$\binom{n}{k} = \frac{(n)_k}{k!} \sim \frac{n^k e^k}{k^k\sqrt{2\pi k}} \prod_{i=1}^{k-1}(1-\frac{i}{n})$$

For $i < k$

$$-\ln(1-\frac{i}{n}) = \frac{i}{n} + \frac{i^2}{2n^2} + O(\frac{i^3}{n^3})$$

so that

$$\sum_{i=1}^{k-1} -\ln(1-\frac{i}{n}) = \frac{k^2}{2n} + \frac{k^3}{6n^2} + o(1) = \frac{k^2}{2n} + \frac{c^3}{6} + o(1)$$

Also

$$p^{k-1} = n^{1-k}(1+\frac{\lambda}{n^{1/3}})^{k-1}$$

$$(k-1)\ln(1+\frac{\lambda}{n^{1/3}}) = (k-1)(\frac{\lambda}{n^{1/3}} - \frac{\lambda^2}{2n^{2/3}} + O(n^{-1})) = \frac{\lambda k}{n^{1/3}} - \frac{\lambda^2 c}{2} + o(1)$$

Also

$$\ln(1-p) = -p + O(n^{-2}) = -\frac{1}{n} - \frac{\lambda}{n^{4/3}} + O(n^{-2})$$

and

$$k(n-k) + \binom{k}{2} - (k-1) = kn - \frac{k^2}{2} + O(n^{2/3})$$

so that

$$[k(n-k) + \binom{k}{2} - (k-1)]\ln(1-p) = -k + \frac{k^2}{2n} - \frac{\lambda k}{n^{1/3}} + \frac{\lambda c^2}{2} + o(1)$$

and

$$E[X] \sim \frac{n^k k^{k-2}}{k^k\sqrt{2\pi k}n^{k-1}}e^A$$

where

$$A = k - \frac{k^2}{2n} - \frac{c^3}{6} + \frac{\lambda k}{n^{1/3}} - \frac{\lambda^2 c}{2} - k + \frac{k^2}{2n} - \frac{\lambda k}{n^{1/3}} + \frac{\lambda c^2}{2} + o(1)$$

$$= -\frac{c^3}{6} - \frac{\lambda^2 c}{2} + \frac{\lambda c^2}{2} + o(1)$$

so that

$$E[X] \sim n^{-2/3} e^{-\frac{c^3}{6} - \frac{\lambda^2 c}{2} + \frac{\lambda c^2}{2}} c^{-5/2}(2\pi)^{-1/2}$$

For any particular such k $E[X] \to 0$ but if we sum k between $cn^{2/3}$ and $(c+dc)n^{2/3}$ we multiply by $n^{2/3}dc$. Going to the limit gives an integral: For any fixed a, b, λ let X be the number of tree components of size between $an^{2/3}$ and $bn^{2/3}$. Then

$$\lim_{n\to\infty} E[X] = \int_a^b e^{-\frac{c^3}{6} - \frac{\lambda^2 c}{2} + \frac{\lambda c^2}{2}} c^{-5/2}(2\pi)^{-1/2} dc$$

The large components are not all trees. E.M. Wright [1977] proved that for fixed l there are asymptotically $c_l k^{k-2+\frac{3}{2}l}$ connected graphs on k points with $k-1+l$ edges, where c_l was given by a specific recurrence. Asymptotically in l,

$$c_l \sim \left(\frac{e}{12l}(1+o(1))\right)^{l/2}.$$

The calculation for $X^{(l)}$, the number of such components on k vertices, leads to extra factors of $c_l k^{\frac{3}{2}l}$ and n^{-l} which gives $c_l c^{\frac{3}{2}l}$. For fixed a, b, λ, l the number $X^{(l)}$ of components of size between $an^{2/3}$ and $bn^{2/3}$ with $l-1$ more edges than vertices satisfies

$$\lim_{n\to\infty} E[X^{(l)}] = \int_a^b e^{-\frac{c^3}{6} - \frac{\lambda^2 c}{2} + \frac{\lambda c^2}{2}} c^{-5/2}(2\pi)^{-1/2}(c_l c^{\frac{3}{2}l}) dc$$

and letting X^* be the total number of components of size between $an^{2/3}$ and $bn^{2/3}$

$$\lim_{n\to\infty} E[X^*] = \int_a^b e^{-\frac{c^3}{6}-\frac{\lambda^2 c}{2}+\frac{\lambda c^2}{2}} c^{-5/2}(2\pi)^{-1/2} g(c)dc$$

where

$$g(c) = \sum_{l=0}^{\infty} c_l c^{\frac{3}{2}l}$$

a sum convergent for all c, (here $c_0 = 1$). A component of size $\sim cn^{2/3}$ will have probability $c_l c^{\frac{3}{2}l}/g(c)$ of having $l-1$ more edges than vertices, independent of λ. As $\lim_{c\to 0} g(c) = 1$, most components of size $\epsilon n^{2/3}$, $\epsilon << 1$, are trees but as c gets bigger the distribution on l moves inexoribly higher.

An Overview. For any fixed λ the sizes of the largest components are of the form $cn^{2/3}$ with a distribution over the constant. For $\lambda = -10^6$ there is some positive limiting probability that the largest component is bigger than $10^6 n^{2/3}$ and for $\lambda = +10^6$ there is some positive limiting probability that the largest component is smaller than $10^{-6}n^{2/3}$, though both these probabilities are minuscule. The functions integrated have a pole at $c = 0$, reflecting the notion that for any λ there should be many components of size near $\epsilon n^{2/3}$ for $\epsilon = \epsilon(\lambda)$ appropriately small. When λ is large negative (e.g., -10^6) the largest component is likely to be $\epsilon n^{2/3}$, ϵ small, and there will be many components of nearly that size. The nontree components will be a negligible fraction of the tree components.

Now consider the evolution of $G(n,p)$ in terms of λ. Suppose that at a given λ there are components of size $c_1 n^{2/3}$ and $c_2 n^{2/3}$. When we move from λ to $\lambda + d\lambda$ there is a probability $c_1 c_2 d\lambda$ that they will merge. Components have a peculiar gravitation in which the probability of merging is proportional to their sizes. With probability $(c_1^2/2)d\lambda$ there will be a new internal edge in a component of size $c_1 n^{2/3}$ so that large components rarely remain trees. Simultaneously, big components are eating up other vertices.

With $\lambda = -10^6$, say, we have feudalism. Many small components (castles) are each vying to be the largest. As λ increases the components increase in size and a few large components (nations) emerge. An already large France has much better chances of becoming larger than a smaller Andorra. The largest components tend strongly to merge and by $\lambda = +10^6$ it is very likely that a giant component, Roman Empire, has emerged. With high probability this component is nevermore challenged for supremacy but continues absorbing smaller components until full connectivity - One World - is achieved.

An Continuous Model. In discussions at St. Flour it became apparent that there was a continuous model underlying the asymptotic behavior of $G(n,p)$ with $p = n^{-1} + \lambda n^{-4/3}$. The following should be regarded as only tentative steps toward defining of that continuous model. For fixed λ and k arbitrarily large but fixed one can look at the k largest components of $G(n,p)$ and parametrize them $x_1 n^{2/3}, \ldots, x_k n^{2/3}$ in decreasing order. One can give explicitly a limiting distribution function $H(x_1, \ldots, x_k)$ for these values. Now one can go to the limit with k and consider the "state" $P(\lambda)$ at "time" λ to be an infinite sequence $x_1 > x_2 > \ldots$ of decreasing reals. There will be a distribution over the possible sequences. The sequences must be well-behaved; one can show, for example, that the number of x_i bigger than c must be asymptotic to $\frac{2}{3}(2\pi)^{-1/2}c^{-3/2}$ as $c \to 0$. (There is further information concerning the nature of the components - e.g., are they trees, unicyclic,...- that could also be added.) Now the intriguing thing is the "gravity" that defines $P(\lambda + d\lambda)$ in terms of $P(\lambda)$ in an appropriate limiting sense. If $P(\lambda)$ has terms x_i, x_j then with probability $x_i x_j d\lambda$ they will "merge" and form a single term with value $x_i + x_j$. This corresponds to certain coagulation models in physics though in the physical world the probability of coagulation depends on the surface area (and perhaps other invariants) of the objects) while here it depends only on their sizes. So it seems there should be a probability space whose elements are histories - i.e., the value of $P(\lambda)$ for all real λ - where the change from $P(\lambda)$ to $P(\lambda + d\lambda)$ is governed by these coagulation laws and where further there have to be some appropriate entry laws so that each $P(\lambda)$ has the appropriate distribution. Not that any of this has been done - but in theory there is a theory!

Lecture 8: Martingales

1 Definitions

A martingale is a sequence $X_0, \ldots, X_m$ of random variables so that for $0 \leq i < m$,

$$E[X_{i+1}|X_i] = X_i$$

The Edge Exposure Martingale Let the random graph $G(n,p)$ be the underlying probability space. Label the potential edges $\{i,j\} \subseteq [n]$ by $e_1, \ldots, e_m$, setting $m = \binom{n}{2}$ for convenience, in any specific manner. Let f be any graphtheoretic function. We define a martingale $X_0, \ldots, X_m$ by giving the values $X_i(H)$. $X_m(H)$ is simply $f(H)$. $X_0(H)$ is the expected value of $f(G)$ with $G \sim G(n,p)$. Note that X_0 is a constant. In general (including the cases $i = 0$ and $i = m$)

$$X_i(H) = E[f(G)|e_j \in G \longleftrightarrow e_j \in H, 1 \leq j \leq i]$$

In words, to find $X_i(H)$ we first expose the first i pairs $e_1, \ldots, e_i$ and see if they are in H. The remaining edges are not seen and considered to be random. $X_i(H)$ is then the conditional expectation of $f(G)$ with this partial information. When $i = 0$ nothing is exposed and X_0 is a constant. When $i = m$ all is exposed and X_m is the function f. The martingale moves from no information to full information in small steps.

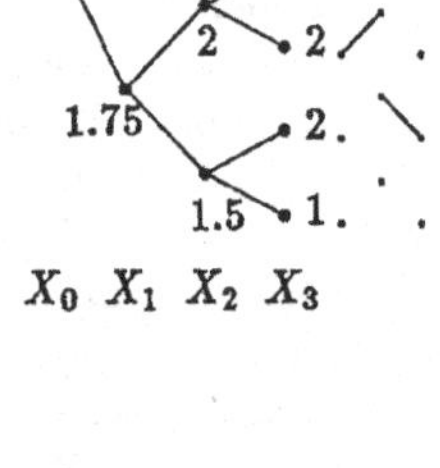

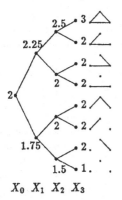

The edge exposure martingale with $n = m = 3, f$ the chromatic number, and the edges exposed in the order "bottom,left,right". The values $X_i(H)$ are given by tracing from the central node to the leaf labelled H.

The figure shows why this is a martingale. The conditional expectation of $f(H)$ knowing the first $i-1$ edges is the weighted average of the conditional expectations of $f(H)$ where the i-th edge has been exposed. More generally - in what is sometimes referred to as a Doob martingale process - X_i may be the conditional expectation of $f(H)$ after certain information is revealed as long as the information known at time i includes the information known at time $i-1$.

The Vertex Exposure Martingale. Again let $G(n,p)$ be the underlying probability space and f any graphtheoretic function. Define $X_1,\ldots,X_n$ by

$$X_i(H) = E[f(G)|\text{for } x,y \leq i, \{x,y\} \in G \longleftrightarrow \{x,y\} \in H]$$

In words, to find $X_i(H)$ we expose the first i vertices and all their internal edges and take the conditional expectation of $f(G)$ with that partial information. By ordering the edges appropriately the vertex exposure martingale may be considered a subsequence of the edge exposure martingale. Note that $X_1(H) = E[f(G)]$ is constant as no edges have been exposed and $X_n(H) = f(H)$ as all edges have been exposed.

2 Large Deviations

Maurey [1979] applied a large deviation inequality for martingales to prove an isoperimetric inequality for the symmetric group S_n. This inequality was useful in the study of normed spaces; see Milman and Schechtman [1986] for many related results. The applications of martingales in Graph Theory also all involve the same underlying martingale results used by Maurey, which are the following.

Theorem 2.1 (Azuma's Ineqality) Let $0 = X_0,\ldots,X_m$ be a martingale with

$$|X_{i+1} - X_i| \leq 1$$

for all $0 \leq i < m$. Let $\lambda > 0$ be arbitrary. Then

$$\Pr[X_m > \lambda\sqrt{m}] < e^{-\lambda^2/2}$$

Corollary 2.2 Let $c = X_0,\ldots,X_m$ be a martingale with

$$|X_{i+1} - X_i| \leq 1$$

for all $0 \leq i < m$. Then

$$\Pr[|X_m - c| > \lambda\sqrt{m}] < 2e^{-\lambda^2/2}.$$

A graph theoretic function f is said to satisfy the *edge Lipschitz condition* if whenever H and H' differ in only one edge then $|f(H) - f(H')| \leq 1$. It satisfies the *vertex Lipschitz condition* if whenever H and H' differ at only one vertex $|f(H) - f(H')| \leq 1$.

Theorem 2.3 When f satisfies the edge Lipschitz condition the corresponding edge exposure martingale satisfies $|X_{i+1} - X_i| \leq 1$. When f satisfies the vertex Lipschitz condition the corresponding vertex exposure martingale satisfies $|X_{i+1} - X_i| \leq 1$.

We prove these results in a more general context later. They have the intuitive sense that if knowledge of a particular vertex or edge cannot change f by more than one then exposing a vertex or edge should not change the expectation of f by more than one. Now we give a simple application of these results.

Theorem 2.4 (Shamir, Spencer[1987]) Let n, p be arbitrary and let $c = E[\chi(G)]$ where $G \sim G(n,p)$. Then

$$\Pr[|\chi(G) - c| > \lambda\sqrt{n-1}] < 2e^{-\lambda^2/2}$$

Proof. Consider the vertex exposure martingale $X_1, \ldots, X_n$ on $G(n,p)$ with $f(G) = \chi(G)$. A single vertex can always be given a new color so the vertex Lipschitz condition applies. Now apply Azuma's Inequality. □

Letting $\lambda \to \infty$ arbitrarily slowly this result shows that the distribution of $\chi(G)$ is "tightly concentrated" around its mean. The proof gives no clue as to where the mean is.

3 Chromatic Number

We have previously shown that $\chi(G) \sim n/2\log_2 n$ almost surely, where $G \sim G(n, 1/2)$. Here we give the original proof of Béla Bollobás using martingales. We follow the earlier notations setting $f(k) = \binom{n}{k}2^{-\binom{k}{2}}$, k_0 so that $f(k_0 - 1) > 1 > f(k_0)$, $k = k_0 - 4$ so that $k \sim 2\log_2 n$ and $f(k) > n^{3+o(1)}$. Our goal is to show

$$\Pr[\omega(G) < k] = e^{-n^{2+o(1)}},$$

where $\omega(G)$ is the size of the maximum clique of G. We shall actually show in Theorem 3.2 a more precise bound. The remainder of the argument is as given earlier.

Let $Y = Y(H)$ be the maximal size of a family of edge disjoint cliques of size k in H. This ingenious and unusual choice of function is key to the martingale proof.

Lemma 3.1. $E[Y] \geq \frac{n^2}{2k^4}(1 + o(1))$

Proof. Let $\mathcal{K}$ denote the family of k-cliques of G so that $f(k) = \mu = E[|\mathcal{K}|]$. Let W denote the number of unordered pairs $\{A, B\}$ of k-cliques of G with $2 \leq |A \cap B| < k$. Then $E[W] = \Delta/2$, with Δ as described earlier, $\Delta \sim \mu^2 k^4 n^{-2}$. Let $\mathcal{C}$ be a random subfamily of $\mathcal{K}$ defined by setting, for each $A \in \mathcal{K}$,

$$\Pr[A \in \mathcal{C}] = q,$$

q to be determined. Let W' be the number of unordered pairs $\{A, B\}$, $A, B \in \mathcal{C}$ with $2 \leq |A \cap B| < k$. Then

$$E[W'] = E[W]q^2 = \Delta q^2/2$$

Delete from $\mathcal{C}$ one set from each such pair $\{A, B\}$. This yields a set $\mathcal{C}^*$ of edge disjoint k-cliques of G and

$$E[Y] \geq E[|\mathcal{C}^*|] \geq E[|\mathcal{C}|] - E[W'] = \mu q - \Delta q^2/2 = \mu^2/2\Delta \sim n^2/2k^4$$

where we choose $q = \mu/\Delta$ (noting that it is less than one!) to minimize the quadratic. □

We conjecture that Lemma 3.1 may be improved to $E[Y] > cn^2/k^2$. That is, with positive probability there is a family of k-cliques which are edge disjoint and cover a positive proportion of the edges.

Theorem 3.2.

$$\Pr[\omega(G) < k] < e^{-(c+o(1))\frac{n^2}{\ln^8 n}}$$

with c a positive constant.

Proof. Let $Y_0, \ldots, Y_m$, $m = \binom{n}{2}$, be the edge exposure martingale on $G(n, 1/2)$ with the function Y just defined. The function Y satisfies the edge Lipschitz condition as adding a single edge can only add at most one clique to a family of edge disjoint cliques. (Note that the Lipschitz condition would not be satisfied for the number of k-cliques as a single edge might yield many new cliques.) G has no k-clique if and only if $Y = 0$. Apply Azuma's Inequality with $m = \binom{n}{2} \sim n^2/2$ and $E[Y] \geq \frac{n^2}{2k^4}(1+o(1))$. Then

$$\begin{aligned} \Pr[\omega(G) < k] &= \Pr[Y=0] & &\leq \Pr[Y - E[Y] \leq -E[Y]] \\ &\leq e^{-E[Y]^2/2\binom{n}{2}} & &\leq e^{-(c'+o(1))n^2/k^8} \\ &= e^{-(c+o(1))n^2/\ln^8 n} \end{aligned}$$

as desired. □

Here is another example where the martingale approach requires an inventive choice of graphtheoretic function.

Theorem 3.3. Let $p = n^{-\alpha}$ where α is fixed, $\alpha > \frac{5}{6}$. Let $G = G(n,p)$. Then there exists $u = u(n,p)$ so that almost always

$$u \leq \chi(G) \leq u+3$$

That is, $\chi(G)$ is concentrated in four values.

We first require a technical lemma that had been well known.

Lemma 3.4. Let α, c be fixed $\alpha > \frac{5}{6}$. Let $p = n^{-\alpha}$. Then almost always every $c\sqrt{n}$ vertices of $G = G(n,p)$ may be 3-colored.

Proof. If not, let T be a minimal set which is not 3-colorable. As $T - \{x\}$ is 3-colorable, x must have internal degree at least 3 in T for all $x \in T$. Thus if T has t vertices it must have at least $\frac{3t}{2}$ edges. The probability of this occuring for some T with at most $c\sqrt{n}$ vertices is bounded from above by

$$\sum_{t=4}^{c\sqrt{n}} \binom{n}{t} \binom{\binom{t}{2}}{\frac{3t}{2}} p^{3t/2}$$

We bound

$$\binom{n}{t} \leq \left(\frac{ne}{t}\right)^t \text{ and } \binom{\binom{t}{2}}{\frac{3t}{2}} \leq \left(\frac{te}{3}\right)^{3t/2}$$

so each term is at most

$$\left[\frac{ne}{t}\frac{t^{3/2}e^{3/2}}{3^{3/2}}n^{-3\alpha/2}\right]^t \leq \left[c_1 n^{1-\frac{3\alpha}{2}} t^{1/2}\right]^t \leq \left[c_2 n^{1-\frac{3\alpha}{2}} n^{1/4}\right]^t = \left[c_2 n^{-\epsilon}\right]^t$$

with $\epsilon = \frac{3\alpha}{2} - \frac{5}{4} > 0$ and the sum is therefore $o(1)$.

Proof of Theorem 3.3. Let $\epsilon > 0$ be arbitrarily small and let $u = u(n,p,\epsilon)$ be the least integer so that

$$\Pr[\chi(G) \leq u] > \epsilon$$

Now define $Y(G)$ to be the minimal size of a set of vertices S for which $G - S$ may be u-colored. This Y satisfies the vertex Lipschitz condition since at worst one could add a vertex to S. Apply the vertex exposure martingale on $G(n,p)$ to Y. Letting $\mu = E[Y]$

$$\Pr[Y \leq \mu - \lambda\sqrt{n-1}] < e^{-\lambda^2/2}$$

$$\Pr[Y \leq \mu + \lambda\sqrt{n-1}] < e^{-\lambda^2/2}$$

Let λ satisfy $e^{-\lambda^2/2} = \epsilon$ so that these tail events each have probability less than ϵ. We defined u so that with probability at least ϵ G would be u-colorable and hence $Y = 0$. That is, $\Pr[Y = 0] > \epsilon$. The first inequality therefore forces $c \leq \lambda\sqrt{n-1}$. Now employing the second inequality

$$\Pr[Y \geq 2\lambda\sqrt{n-1}] \leq \Pr[Y \geq \mu + \lambda\sqrt{n-1}] \leq \epsilon$$

With probability at least $1-\epsilon$ there is a u-coloring of all but at most $c'\sqrt{n}$ vertices. By the Lemma almost always, and so with probability at least $1-\epsilon$, these points may be colored with 3 further colors, giving a $u+3$-coloring of G. The minimality of u guarantees that with probability at least $1-\epsilon$ at least u colors are needed for G. Altogether

$$\Pr[u \leq \chi(G) \leq u+3] \geq 1-3\epsilon$$

and ϵ was arbitrarily small. □

Using the same technique similar results can be achieved for other values of α. For any fixed $\alpha > \frac{1}{2}$ one finds that $\chi(G)$ is concentrated on some fixed number of values.

4 A General Setting

The martingales useful in studying Random Graphs generally can be placed in the following general setting which is essentially the one considered in Maurey [1979] and in Milman and Schechtman [1986]. Let $\Omega = A^B$ denote the set of functions $g : B \to A$. (With B the set of pairs of vertices on n vertices and $A = \{0,1\}$ we may identify $g \in A^B$ with a graph on n vertices.) We define a measure by giving values p_{ab} and setting

$$\Pr[g(b) = a] = p_{ab}$$

with the values $g(b)$ assumed mutually independent. (In $G(n,p)$ all $p_{1b} = p, p_{0b} = 1-p$.) Now fix a gradation

$$\emptyset = B_0 \subset B_1 \subset \ldots \subset B_m = B$$

Let $L : A^B \to R$ be a functional. (E.g., clique number.) We define a martingale $X_0, X_1, \ldots, X_m$ by setting

$$X_i(h) = E[L(g) | g(b) = h(b) \text{ for all } b \in B_i]$$

X_0 is a constant, the expected value of L of the random g. X_m is L itself. The values $X_i(g)$ approach $L(g)$ as the values of $g(b)$ are "exposed". We say the functional L satisfies the Lipschitz condition relative to the gradation if for all $0 \leq i < m$

$$h, h' \text{ differ only on } B_{i+1} - B_i \Rightarrow |L(h') - L(h)| \leq 1$$

Theorem 4.1. Let L satisfy the Lipschitz condition. Then the corresponding martingale satisfies

$$|X_{i+1}(h) - X_i(h)| \leq 1$$

for all $0 \leq i < m$, $h \in A^B$.
Proof. Let H be the family of h' which agree with h on B_{i+1}. Then

$$X_{i+1}(h) = \sum_{h' \in H} L(h') w_{h'}$$

where $w_{h'}$ is the conditional probability that $g = h'$ given that $g = h$ on B_{i+1}. For each $h' \in H$ let $H[h']$ denote the family of h^* which agree with h' on all points except (possibly) $B_{i+1} - B_i$. The $H[h']$ partition the family of h^* agreeing with h on B_i. Thus we may express

$$X_i(h) = \sum_{h' \in H} \sum_{h^* \in H[h']} [L(h^*) q_{h^*}] w_{h'}$$

where q_{h^*} is the conditional probability that g agrees with h^* on B_{i+1} given that it agrees with h on B_i. (This is because for $h^* \in H[h']$ $w_{h'}$ is also the conditional probability that $g = h^*$ given that $g = h^*$ on B_{i+1}.) Thus

$$\begin{aligned}|X_{i+1}(h) - X_i(h)| &= \left|\sum_{h' \in H} w_{h'}[L(h') - \sum_{h^* \in H[h']} L(h^*) q_{h^*}]\right| \\ &\leq \sum_{h' \in H} w_{h'} \sum_{h^* \in H[h']} |q_{h^*}[L(h') - L(h^*)]|\end{aligned}$$

The Lipschitz condition gives $|L(h') - L(h^*)| \leq 1$ so

$$|X_{i+1}(h) - X_i(h)| \leq \sum_{h' \in H} w_{h'} \sum_{h^* \in H[h']} q_{h^*} = \sum_{h' \in H} w_{h'} = 1 \qquad \square$$

Now we can express Azuma's Inequality in a general form.
Theorem 4.2. Let L satisfy the Lipschitz condition relative to a gradation of length m and let $\mu = E[L(g)]$. Then for all $\lambda > 0$

$$\Pr[L(g) > \mu + \lambda\sqrt{m}] < e^{-\lambda^2/2}$$

$$\Pr[L(g) < \mu - \lambda\sqrt{m}] < e^{-\lambda^2/2}$$

5 Three Illustrations

Let g be the random function from $\{1, \ldots, n\}$ to itself, all n^n possible function equally likely. Let $L(g)$ be the number of values not hit, i.e., the number of y for which $g(x) = y$ has no solution. By Linearity of Expectation

$$E[L(g)] = n\left(1 - \frac{1}{n}\right)^n \sim \frac{n}{e}$$

Set $B_i = \{1, \ldots, i\}$. L satisfies the Lipschitz condition relative to this gradation since changing the value of $g(i)$ can change $L(g)$ by at most one. Thus
Theorem 5.1.

$$\Pr[|L(g) - \frac{n}{e}| > \lambda\sqrt{n}] < 2e^{-\lambda^2/2}$$

Deriving these asymptotic bounds from first principles is quite cumbersome.

As a second illustration let B be any normed space and let $v_1, \ldots, v_n \in B$ with all $|v_i| \leq 1$. Let $\epsilon_1, \ldots, \epsilon_n$ be independent with

$$\Pr[\epsilon_i = +1] = \Pr[\epsilon_i = -1] = \frac{1}{2}$$

and set

$$X = |\epsilon_1 v_1 + \ldots + \epsilon_n v_n|$$

Theorem 5.2.

$$\Pr[X - E[X] > \lambda\sqrt{n}] < e^{-\lambda^2/2}$$

$$\Pr[X - E[X] < -\lambda\sqrt{n}] < e^{-\lambda^2/2}$$

Proof. Consider $\{-1, +1\}^n$ as the underlying probability space with all $(\epsilon_1, \ldots, \epsilon_n)$ equally likely. Then X is a random variable and we define a martingale $X_0, \ldots, X_n = X$ by exposing one ϵ_i

at a time. The value of ϵ_i can only change X by two so direct application of Theorem 4.1 gives $|X_{i+1} - X_i| \leq 2$. But let ϵ, ϵ' be two n-tuples differing only in the i-th coordinate.

$$X_i(\epsilon) = \frac{1}{2}\left[X_{i+1}(\epsilon) + X_{i+1}(\epsilon')\right]$$

so that

$$|X_i(\epsilon) - X_{i+1}(\epsilon)| = \frac{1}{2}|X_{i+1}(\epsilon') - X_{i+1}(\epsilon)| \leq 1$$

Now apply Azuma's Inequality. □

For a third illustration let ρ be the Hamming metric on $\{0,1\}^n$. For $A \subseteq \{0,1\}^n$ let $B(A,s)$ denote the set of $y \in \{0,1\}^n$ so that $\rho(x,y) \leq s$ for some $x \in A$. ($A \subseteq B(A,s)$ as we may take $x = y$.)

Theorem 5.3. Let $\epsilon, \lambda > 0$ satisfy $e^{-\lambda^2/2} = \epsilon$. Then

$$|A| \geq \epsilon 2^n \Rightarrow |B(A, 2\lambda\sqrt{n})| \geq (1-\epsilon)2^n$$

Proof. Consider $\{0,1\}^n$ as the underlying probability space, all points equally likely. For $y \in \{0,1\}^n$ set

$$X(y) = \min_{x \in A} \rho(x,y)$$

Let $X_0, X_1, \ldots, X_n = X$ be the martingale given by exposing one coordinate of $\{0,1\}^n$ at a time. The Lipschitz condition holds for X: If y, y' differ in just one coordinate then $X(y) - X(y') \leq 1$. Thus, with $\mu = E[X]$

$$\Pr[X < \mu - \lambda\sqrt{n}] < e^{-\lambda^2/2} = \epsilon$$

$$\Pr[X > \mu + \lambda\sqrt{n}] < e^{-\lambda^2/2} = \epsilon$$

But

$$\Pr[X = 0] = |A|2^{-n} \geq \epsilon$$

so $\mu \leq \lambda\sqrt{n}$. Thus

$$\Pr[X > 2\lambda\sqrt{n}] < \epsilon$$

and

$$|B(A, 2\lambda\sqrt{n})| = 2^n \Pr[X \leq 2\lambda\sqrt{n}] \geq 2^n(1-\epsilon) \qquad \square$$

Actually, a much stronger result is known. Let $B(s)$ denote the ball of radius s about $(0,\ldots,0)$. The Isoperimetric Inequality proved by Harper in 1966 states that

$$|A| \geq |B(r)| \Rightarrow |B(A,s)| \geq |B(r+s)|$$

One may actually use this inequality as a beginning to give an alternate proof that $\chi(G) \sim n/2\log_2 n$ and to prove a number of the other results we have shown using martingales.

Lecture 9: The Lovász Local Lemma

1 The Lemma

In a typical probabilistic proof of a combinatorial result, one usually has to show that the probability of a certain event is positive. However, many of these proofs actually give more and show that the probability of the event considered is not only positive but is large. In fact, most probabilistic proofs deal with events that hold with high probability, i.e., a probability that tends to 1 as the dimensions of the problem grow. On the other hand, there is a trivial case in which one can show that a certain event holds with positive, though very small, probability. Indeed, if we have n mutually independent events and each of them holds with probability at least $p > 0$, then the probability that all events hold simultaneously is at least p^n, which is positive, although it may be exponentially small in n.

It is natural to expect that the case of mutual independence can be generalized to that of rare dependencies, and provide a more general way of proving that certain events hold with positive, though small, proability. Such a generalization is, indeed, possible, and is stated in the following lemma, known as the Lovász Local Lemma. This simple lemma, first proved in [Erdős-Lovász (1975)] is an extremely powerful tool, as it supplies a way for dealing with rare events.

Lemma 1.1 (The Local Lemma; General Case):

Let $A_1, A_2 \ldots A_n$ be events in an arbitrary probability space. A directed graph $D = (V, E)$ on the set of vertices $V = \{1, 2 \ldots n\}$ is called a *dependency digraph* for the events $A_1 \ldots A_n$ if for each i, $1 \le i \le n$, the event A_i is mutually independent of all the events $\{A_j : (i,j) \notin E\}$. Suppose that $D = (V, E)$ is a dependency digraph for the above events and suppose there are real numbers $x_1 \ldots x_n$ such that $0 \le x_i < 1$ and $Pr(A_i) \le x_i \prod_{(i,j)\in E}(1 - x_j)$ for all $1 \le i \le n$. Then $Pr\left(\bigwedge_{i=1}^{n} \overline{A_j}\right) \ge \prod\limits_{i=1}^{n}(1 - x_i)$. In particular, with positive probability no event A_i holds.

We first prove, by induction on s, that for any $S \subset \{1 \ldots n\}$, $|S| = s < n$ and any $i \notin S$

$$\Pr\left(A_i | \bigwedge_{j\in S} \overline{A_j}\right) \le x_i$$

This is certainly true for $s = 0$. Assuming it holds for all $s' < s$, we prove it for S. Put

$$S_1 = \{j \in S; (i,j) \in E\}, S_2 = S - S_1$$

Then

$$\Pr\left(A_i | \bigwedge_{j\in S} \overline{A_j}\right) = \frac{\Pr\left(A_i \wedge (\bigwedge_{j\in S_1} \overline{A_j}) | \bigwedge_{l\in S_2} \overline{A_l}\right)}{\Pr\left(\bigwedge_{j\in S_1} \overline{A_j} | \bigwedge_{l\in S_2} \overline{A_l}\right)}$$

To bound the numerator observe that since A_i is mutually independent of the events $\{A_l : l \in S_2\}$

$$\Pr\left(A_i \wedge (\bigwedge_{j\in S_1} \overline{A_j}) | \bigwedge_{l\in S_2} \overline{A_l}\right) \le \Pr\left(A_i | \bigwedge_{l\in S_2} \overline{A_l}\right) = Pr(A_i) \le x_i \prod_{(i,j)\in E}(1 - x_j)$$

The denominator, on the other hand, can be bounded by the induction hypothesis. Indeed, suppose $S_1 = \{j_1, j_2 \ldots j_r\}$. If $r = 0$ then the denominator is 1, and (1.1) follows. Otherwise, setting $B = \bigwedge_{l\in S_2} \overline{A_l}$,

$$\Pr\left(\overline{A_{j_1}} \wedge \overline{A_{j_2}} \ldots \overline{A_{j_r}} | B\right) = (1 - \Pr(A_{j_1}|B)) \cdot$$

$$\cdot\left(1-\Pr\left(A_{j_2}|\overline{A_{j_1}}\wedge B\right)\right)\cdots\left(1-\Pr\left(A_{j_r}|\overline{A_{j_1}}\wedge\ldots\wedge\overline{A_{j_{r-1}}}\wedge B\right)\right)$$
$$\geq(1-x_{j_1})\cdots(1-x_{j_r})\geq\prod_{(i,j)\in E}(1-x_j)$$

Substituting we conclude that $\Pr\left(A_i|\bigwedge_{j\in S}\overline{A_j}\right)\leq x_i$, completing the proof of the induction. The assertion of Lemma 1.1 now follows easily, as

$$\Pr\left(\bigwedge_{i=1}^{n}\overline{A_i}\right)=(1-Pr(A_1))\cdot(1-Pr(A_2|\overline{A_1}))\cdot\ldots\cdot(1-Pr(A_n|\bigwedge_{i=1}^{n-1}\overline{A_i})\geq\prod_{i=1}^{n}(1-x_i)$$

completing the proof. □

Corollary 1.2 (Lovász Local Lemma; Symmetric Case): Let $A_1, A_2\ldots A_n$ be events in an arbitrary probability space. Suppose that each event A_i is mutually independent of a set of all the other events A_j but at most d, and that $Pr(A_i)\leq p$ for all $1\leq i\leq n$. If

$$ep(d+1)\leq 1$$

then $\Pr\left(\bigwedge_{i=1}^{n}\overline{A_i}\right)>0$.

If $d=0$ the result is trivial. Otherwise, by the assumption there is a dependency digraph $D=(V,E)$ for the events $A_1\ldots A_n$ in which for each i $|\{j:(i,j)\in E\}|\leq d$. The result now follows from Lemma 1.1 by taking $x_i=1/(d+1)(<1)$ for all i and using the fact that for any $d\geq 2$, $\left(1-\frac{1}{d+1}\right)^d>1/e$.□

It is worth noting that as shown by Shearer in 1985,the constant "e" is the best possible constant in inequality (1.5). Note also that the proof of Lemma 1.1 indicates that the conclusion remains true even when we replace the two assumptions that each A_i is mutually independent of $\{A_j:(i,j)\notin E)$ and that $Pr(A_i)\leq x_i\prod_{(ij)\in E}(1-x_j)$ by the weaker assumption that for each i and each $S_2\subset\{1\ldots n\}-\{j:(i,j)\in E\}$, $Pr\left(x_i|\bigwedge_{j\in S_2}\overline{A_j}\right)\leq x_i\prod_{(i,j)\in E}(1-x_j)$. This turns out to be useful in certain applications.

In the next few sections we present various applications of the Lovász Local Lemma for obtaining combinatorial results. There is no known proof of any of these results, which does not use the this Lemma.

2 Property B and multicolored sets of real numbers

A hypergraph $H=(V,E)$ is said to have property B if there is a coloring of V by two colors so that no edge $f\in E$ is monochromatic.

Theorem 2.1. Let $H=(V,E)$ be a hypergraph in which every edge has at least k elements, and suppose that each edge of H intersects at most d other edges. If $e(d+1)\leq 2^{k-1}$ then H has property B.

Color each vertex v of H, randomly and independently, either blue or red (with equal probability). For each edge $f\in E$, let A_f be the event that f is monochromatic. Clearly $Pr(A_f)=2/2^{|f|}\leq 1/2^{k-1}$. Moreover, each event A_f is clearly mutually independent of all the other events $A_{f'}$ for all edges f' that do not intersect f. The result now follows from Corollary 1.2. □

A special case of Theorem 2.1 is that for any $k\geq 9$, any k-uniform k-regular hypergraph H has property B. Indeed, since any edge f of such an H contains k vertices, each of which

is incident with k edges (including f), it follows that f intersects at most $d = k(k-1)$ other edges. The desired result follows, since $e(k(k-1)+1) < 2^{k-1}$ for each $k \geq 9$. This special case has a different proof (see [Alon-Bregman (1988)]), which works for each $k \geq 8$. It is plausible to conjecture that in fact for each $k \geq 4$ each k-uniform k-regular hypergraph is has Property B. The next result we consider, which appeared in the original paper of Erdős and Lovász, deals with k-colorings of the real numbers. For a k-coloring $c : R \to \{1, 2 \ldots k\}$ of the real numbers by the k colors $1, 2 \ldots k$, and for a subset $T \subset R$, we say that T is *multicolored* (with respect to c) if $c(T) = \{1, 2 \ldots k\}$, i.e., if T contains elements of all colors.
Theorem 2.2. Let m and k be two positive integers satisfying

$$e\left(m(m-1)+1\right)k\left(1-\frac{1}{k}\right)^m \leq 1$$

Then, for any set S of m real numbers there is a k-coloring so that each translation $x + S$ (for $x \in R$) is multicolored.

Notice that the condition holds whenever $m > (3 + o(1))k \log k$. There is no known proof of existence of any $m = m(k)$ with this property without using the local lemma.

We first fix a *finite* subset $X \subset R$ and show the existence of a k-coloring so that each translation $x + S$ (for $x \in X$) is multicolored. This is an easy consequence of the Lovász Local Lemma. Indeed, put $Y = \bigcup_{x \in X}(x + S)$ and let $c : Y \to \{1, 2 \ldots k\}$ be a random k-coloring of Y obtained by choosing, for each $y \in Y$, randomly and independently, $c(y) \in \{1, 2 \ldots, k\}$ according to a uniform distribution on $\{1, 2 \ldots k\}$. For each $x \in X$, let A_x be the event that $x + S$ is not multicolored (with respect to c). Clearly $Pr(A_x) \leq k\left(1 - \frac{1}{k}\right)^m$. Moreover, each event A_x is mutually independent of all the other events $A_{x'}$ but those for which $(x + S) \cap (x' + S) \neq \emptyset$. As there are at most $m(m-1)$ such events the desired result follows from Corollary 1.2.

We can now prove the existence of a coloring of the set of all reals with the desired properties, by a standard compactness argument. Since the discrete space with k points is (trivially) compact, Tychanov's Theorem (which is equivalent to the axiom of choice) implies that an arbitrary product of such spaces is compact. In particular, the space of all functions from the reals to $\{1, 2 \ldots k\}$, with the usual product topology, is compact. In this space for every fixed $x \in R$, the set C_x of all colorings c, such that $x + S$ is multicolored is closed. (In fact, it is both open and closed, since a basis to the open sets is the set of all colorings whose values are prescribed in a finite number of places). As we proved above, the intersection of any finite number of sets C_x is nonempty. It thus follows, by compactness, that the intersection of all sets C_x is nonempty. Any coloring in this intersection has the properties in the conclusion of Theorem 2.2.□

Note that it is impossible, in general, to apply the Lovász Local Lemma to an infinite number of events and conclude that in some point of the probability space none of them holds. In fact, there are trivial examples of countably many mutually independent events A_i, satisfying $Pr(A_i) = 1/2$ and $\bigwedge_{i \geq 1} \overline{A_i} = \emptyset$. Thus the compactness argument is essential in the above proof.

3 Lower bounds for Ramsey numbers

The deriviation of lower bounds for Ramsey numbers by Erdős in 1947 was one of the first applications of the probabilistic method. The Lovász Local Lemma provides a simple way of improving these bounds. Let us obtain, first, a lower bound for the diagonal Ramsey number $R(k, k)$. Consider a random 2-coloring of the edges of K_n. For each set S of k vertices of K_n, let A_S be the event that the complete graph on S is monochromatic. Clearly $\Pr(A_S) = 2^{1-\binom{k}{2}}$.

It is obvious that each event A_S is mutually independent of all the events A_T, but those which satisfy $|S \cap T| \geq 2$, since this is the only case in which the corresponding complete graphs share an edge. We can therefore apply Corollary 1.2 with $p = 2^{1-\binom{k}{2}}$ and $d = \binom{k}{2}\binom{n}{k-2}$ to conclude;

Proposition 3.1. If $e\left(\binom{k}{2}\binom{n}{k-2}+1\right)\cdot 2^{1-\binom{k}{2}} < 1$ then $R(k,k) > n$.

A short computation shows that this gives $R(k,k) > \frac{\sqrt{2}}{e}(1+o(1))k2^{k/2}$, only a factor 2 improvement on the bound obtained by the straightforward probabilistic method. Although this minor improvement is somewhat disappointing it is certainly not surprising; the Local Lemma is most powerful when the dependencies between events are rare, and this is not the case here. Indeed, there is a total number of $K = \binom{n}{k}$ events considered, and the maximum outdegree d in the dependency digraph is roughly $\binom{k}{2}\binom{n}{k-2}$. For large k and much larger n (which is the case of interest for us) we have $d > K^{1-O(1/k)}$, i.e., quite a lot of dependencies. On the other hand, if we consider small sets S, e.g., sets of size 3, we observe that out of the total $K = \binom{n}{3}$ of them each shares an edge with only $3(n-3) \approx K^{1/3}$. This suggests that the Lovász Local Lemma may be much more significant in improving the off-diagonal Ramsey numbers $R(k,l)$, especially if one of the parameters, say l, is small. Let us consider, for example, following Spencer (1977), the Ramsey number $R(k,3)$. Here, of course, we have to apply the nonsymmetric form of the Lovász Local Lemma. Let us 2-color the edges of K_n randomly and independently, where each edge is colored blue with probability p. For each set of 3 vertices T, let A_T be the event that the triangle on T is blue. Similarly, for each set of k vertices S, let B_S be the event that the complete graph on S is red. Clearly $Pr(A_T) = p^3$ and $Pr(B_S) = (1-p)^{\binom{k}{2}}$. Construct a dependency digraph for the events A_T and B_S by joining two vertices by edges (in both directions) iff the corresponding complete graphs share an edge. Clearly, each A_T-node of the dependency graph is adjacent to $3(n-3) < 3n$ $A_{T'}$-nodes and to at most $\binom{n}{k}$ $B_{S'}$-nodes. Similarly, each B_S-node is adjacent to $\binom{k}{2}(n-k) < k^2n/2$ A_T nodes and to at most $\binom{n}{k}$ $B_{S'}$-nodes. It follows from the general case of the Lovász Local Lemma that if we can find a $0 < p < 1$ and two real numbers $0 \leq x < 1$ and $0 \leq y < 1$ such that

$$p^3 \leq x(1-x)^{3n}(1-y)^{\binom{n}{k}}$$

and

$$(1-p)^{\binom{k}{2}} \leq y(1-x)^{k^2n/2}(1-y)^{\binom{n}{k}}$$

then $R(k,3) > n$.

Our objective is to find the largest possible $k = k(n)$ for which there is such a choice of p, x and y. An elementary computation (if you have a spare weekend!) shows that the best choice is when $p = c_1n^{-1/2}$, $k = c_2n^{1/2}\log n$, $x = c_3/n^{3/2}$ and $y = c_4e^{-n^{1/2}\log^2 n}$. This gives that $R(k,3) > c_5k^2/\log^2 k$. A similar argument gives that $R(k,4) > k^{5/2+o(1)}$. In both cases the amount of computation required is considerable. However, the hard work does pay; the bound $R(k,3) > c_5k^2/\log^2 k$ matches a lower bound of Erdős proved in 1961 by a highly complicated probabilistic argument. The bound above for $R(k,4)$ is better than any bound for $R(k,4)$ known to be proven without the Local Lemma.

4 A geometric result

A family of open unit balls F in the 3-dimensional Euclidean space R^3 is called a *k-fold covering* of R^3 if any point $x \in R^3$ belongs to at least k balls. In particular, a 1-fold covering is simply called a *covering*. A k-fold covering F is called *decomposable* if there is a partition of F into two pairwise disjoint families F_1 and F_2, each being a covering of R^3. Mani and Pach [1988] constructed, for any integer $k \geq 1$, a non-decomposable k-fold covering of R^3 by open unit balls. On the other hand they proved that any k-fold covering of R^3 in which no point is covered by more than $c2^{k/3}$ balls is decomposable. This reveals a somewhat surprising phenomenon that it is more difficult to decompose coverings that cover some of the points of R^3 too often, than to decompose coverings that cover every point about the same number of times. The exact statement of the Mani-Pach Theorem is the following.

Theorem 4.1. Let $F = \{B_i\}_{i \in I}$ be a k-fold covering of the 3 dimensional Euclidean space by open unit balls. Suppose, further, than no point of R^3 is contained in more than t members of F. If

$$e \cdot t^3 2^{18} / 2^{k-1} \leq 1$$

then F is decomposable.

Let $\{C_j\}_{j \in J}$ be the connected components of the set obtained from R^3 by deleting all the boundaries of the balls B_i in F. Let $H = (V(H), E(H))$ be the (infinite) hypergraph defined as follows; the set of vertices of H, $V(H)$ is simply $F = \{B_i\}_{i \in I}$. The set of edges of H is $E(H) = \{E_j\}_{j \in J}$, where $E_j = \{B_i : i \in I \text{ and } C_j \subseteq B_i\}$. Since F is a k-fold covering, each edge E_j of H contains at least k vertices. We claim that each edge of H intersects less than $t^3 2^{18}$ other edges of H. To prove this claim, fix an edge E_l, corresponding to the connected component C_l, where $l \in J$. Let E_j be an arbitrary edge of H, corresponding to the component C_j, that intersects E_l. Then there is a ball B_i containing both C_l and C_j. Therefore, any ball that contains C_j intersects B_i. It follows that all the unit balls that contain or touch a C_j, for some j that satisfies $E_j \cap E_l \neq \emptyset$ are contained in a ball B of radius 4. As no point of this ball is covered more than t times we conclude, by a simple volume argument, that the total number of these unit balls is at most $t \cdot 4^3 = t \cdot 2^6$. It is not too difficult to check that m balls in R^3 cut R^3 into less than m^3 connected components, and since each of the above C_j is such a component we have $|\{j : E_j \cap E_l \neq \emptyset\}| < (t \cdot 2^6)^3 = t^3 2^{18}$, as claimed.

Consider, now, any finite subhypergraph L of H. Each edge of L has at least k vertices, and it intersects at most $d < t^3 2^{18}$ other edges of L. Since, by assumption, $e(d+1) \leq 2^{k-1}$, Theorem 2.1 (which is a simple corollary of the local lemma), implies that L is 2-colorable. This means that one can color the vertices of L blue and red so that no edge of L is monochromatic. Since this holds for any finite L, a compactness argument, analogous to the one used in the proof of Theorem 2.2, shows that H is 2-colorable. Given a 2-coloring of H with no monochromatic edges, we simply let F_1 be the set of all blue balls, and F_2 be the set of all red ones. Clearly, each F_i is a covering of R^3, completing the proof of the theorem.□

It is worth noting that Theorem 4.1 can be easily generalized to higher dimensions. We omit the detailed statement of this generalization.

5 Latin Transversals

Following the proof of the Lovász Local Lemma we noted that the mutual independency assumption in this lemma can be replaced by the weaker assumption that the conditional probability of each event, given the mutual non-occurance of an arbitrary set of events, each nonadjacent to it in the dependency digraph, is sufficiently small. In this section we describe an application, from Erdős-Spencer [1991], of this modified version of the lemma. Let $A = (a_{ij})$ be an n of n matrix with, say, integer entries. A permutation π is called a *Latin transversal* (of A) if the entries $a_{i\pi(i)}$ $(1 \leq i \leq n)$ are all distinct.

Theorem 6.1. Suppose $k \leq (n-1)/(4e)$ and suppose that no integer appears in more than k

entries of A. Then A has a Latin Transversal.

Let π be a random permutation of $\{1, 2 \leq n\}$, chosen according to a uniform distribution among all possible $n!$ permutations. Denote by T the set of all ordered fourtuples (i, j, i', j') satisfying $i < i', j \neq j'$ and $a_{ij} = a_{i'j'}$. For each $(i, j, i', j') \in T$, let $A_{iji'j'}$ denote the event that $\pi(i) = j$ and $\pi(i') = j'$. The existence of a Latin transversal is equivalent to the statement that with positive probability none of these events hold. Let us define a symmetric digraph, (i.e., a graph) G on the vertex set T by making (i, j, i', j') adjacent to (p, q, p', q') if and only if $\{i, i'\} \cap \{p, p'\} \neq \emptyset$ or $\{j, j'\} \cap \{q, q'\} \neq \emptyset$. Thus, these two fourtuples are not adjacent iff the four cells $(i, j), (i', j'), (p, q)$ and (p', q') occupy four distinct rows and columns of A. The maximum degree of G is less than $4nk$; indeed, for a given $(i, j, i', j') \in T$ there are $4n$ choices of (p, q) with either $p \in \{i, i'\}$ or $q \in \{j, j'\}$, and for each of these choices of (p, q) there are less than k choices for $(p', q') \neq (p, q)$ with $a_{pq} = a_{p'q'}$. Since $e \cdot 4nk \cdot \frac{1}{n(n-1)} \leq 1$, the desired result follows from the above mentioned strengthening of the symmetric version of the Lovász Local Lemma, if we can show that

$$\Pr(A_{iji'j'} | \bigwedge_S \overline{A_{pqp'q'}}) \leq 1/n(n-1)$$

for any $(i, j, i', j') \in T$ and any set S of members of T which are nonadjacent in G to (i, j, i', j'). By symmetry, we may assume that $i = j = 1, i' = j' = 2$ and that hence none of the p's nor q's are either 1 or 2. Let us call a permutation π *good* if it satisfies $\bigwedge_S \overline{A_{pqp'q'}}$, and let S_{ij} denote the set of all good permutations π satisfying $\pi(1) = i$ and $\pi(2) = j$. We claim that $|S_{12}| \leq |S_{ij}|$ for alll $i \neq j$. Indeed, suppose first that $i, j > 2$. For each good $\pi \in S_{12}$ define a permutation π^* as follows. Suppose $\pi(x) = i$, $\pi(y) = j$. Then define $\pi^*(1) = i, \pi^*(2) = j, \pi^*(x) = 1, \pi^*(y) = 2$ and $\pi^*(t) = \pi(t)$ for all $t \neq 1, 2, x, y$. One can easily check that π^* is good, since the cells $(1, i), (2, j), (x, 1), (y, 2)$ are not part of any $(p, q, p', q') \in S$. Thus $\pi^* \in S_{ij}$, and since the mapping $\pi \rightarrow \pi^*$ is injective $|S_{12}| \leq |S_{ij}|$, as claimed. Similarly one can define injective mappings showing that $|S_{12}| \leq |S_{ij}|$ even when $\{i, j\} \cap \{1, 2\} \neq \emptyset$. It follows that $\Pr(A_{1122} \bigwedge \bigwedge_S \overline{A_{pqp'q'}}) \leq Pr(A_{1i2j} \bigwedge \bigwedge_S \overline{A_{pqp'q'}})$ for all $i \neq j$ and hence that $Pr(A_{1122} | \bigwedge \bigwedge_S \overline{A_{pqp'q'}}) \leq 1/n(n-1)$. By symmetry, this implies (6.1) and completes the proof.□

References

N. Alon and Z. Bregman (1988), Every 8-uniform 8-regular hypergraph is 2-colorable, Graphs and Combinatorics *4*, 303-305.

B. Bollobás (1985), *Random Graphs*, Academic Press.

B. Bollobás (1988), *The chromatic number of random graphs*, Combinatorica *8*, 49-55.

B. Bollobás and P. Erdős (1976), Cliques in Random Graphs, Math. Proc. Camb. Phil. Soc. *80*, 419-427

R. B. Boppana and J. H. Spencer (1989), A useful elementary correlation inequality, J. Combinatorial Theory Ser. A, *50*, 305-307.

P. Erdős (1947), Some remarks on the theory of graphs, Bull. Amer. Math. Soc. *53*, 292-294.

P. Erdős (1956), Problems and results in additive number theory, Colloque sur le Théorie des Nombres (CBRM, Bruselles) 127-137

P. Erdős (1959), Graph theory and probability, Canad. J. Math. *11* (1959), 34-38.

P. Erdős and M. Kac (1940), The Gaussian Law of Errors in the Theory of Additive Number Theoretic Functions, Amer. J. Math. *62*, 738-742

P. Erdős and L. Lovász (1975), Problems and results on 3-chromatic hypergraphs and some related questions, in: Infinite and Finite Sets (A. Hajnal et al., eds.), North-Holland, Amsterdam, pp. 609-628.

P. Erdős and A. Rényi (1960), On the Evolution of Random Graphs, Mat Kutató Int. Közl. *5*, 17-60

P. Erdős and J. Spencer (1991), Lopsided Lovász Local Lemma and Latin Transversals, Disc. Appl. Math. *30*, 151-154

P. Erdős and G. Szekeres (1935), A Combinatorial Problem in Geometry, Compositio Math. *2*, 463-470

P. Erdős and P. Tetali (1990), Representations of integers as the Sum of k Terms, Random Structures and Algorithms *1*, 245-261.

R. Fagin (1976), Probabilities in Finite Models, J. Symbolic Logic *41*, 50-58

Y.V. Glebskii, D.I. Kogan, M.I. Liagonkii and V.A. Talanov (1969), Range and degree of realizability of formulas the restricted predicate calculus, Cybernetics *5*, 142-154 (Russian original: Kibernetica *5*, 1969, 17-27)

L. Harper (1966), Optimal numberings and isoperimetric problems on graphs, J. Combinatorial Theory *1*, 385-394.

S. Janson, T. Łuczak, A. Rucinski (1990), An Exponential Bound for the Probability of Nonexistence of a Specified Subgraph in a Random Graph, *in* Random Graphs '87 (M. Karonski, J. Jaworski, A. Rucinski, eds.), John Wiley, 73-87

R.M. Karp (1990), The transitive closure of a Random Digraph, Random Structures and Algorithms *1*, 73-94

T. Łuczak (1990), Component Behavior near the Critical Point of the Random Graph Process, Random Structures and Algorithms *1*, 287-310.

P. Mani-Levitska and J. Pach (1988), Decomposition problems for multiple coverings with unit balls, to appear.

D.W. Matula (1976), The Largest Clique Size in a Random Graph, Tech. Rep. Dept. Comp. Sci. Southern Methodist University, Dallas

B. Maurey (1979), Construction de suites symétriques, Compt. Rend. Acad. Sci. Paris *288*, 679-681.

V. D. Milman and G. Schechtman (1986), Asymptotic Theory of Finite Dimensional Normed Spaces, Lecture Notes in Mathematics 1200, Springer Verlag, Berlin and New York.

E. Shamir and J. Spencer (1987), Sharp concentration of the chromatic number in random graphs $G_{n,p}$, Combinatorica *7*, 121-130

J. Shearer (1985), On a problem of Spencer, Combinatorica *5*, 241-245.

S. Shelah and J. Spencer (1988), Zero-One Laws for Sparse Random Graphs, J. Amer. Math. Soc. *1*, 97-115

J. Spencer (1977), Asymptotic Lower Bounds for Ramsey Functions, Disc. Math. *20*, 69-76

J. Spencer (1990a), Threshold Functions for Extension Statements, J. Comb. Th. (Ser A) *53*, 286-305

J. Spencer (1990b), Counting Extension, J. Combinatorial Th. (Ser A) *55*, 247-255.

P. Turán (1934), On a theorem of Hardy and Ramanujan, J. London Math Soc. *9*, 274-276

E.M. Wright (1977), The number of connected sparsely edged graphs, Journal of Graph Theory *1*, 317-330.

EXPOSES 1991

ABRAHAM Romain
Mouvement brownien et processus de branchement

AMIDI Ali
Le théorème central limite pour les variables aléatoires dépendantes, sous des conditions différentes de régularité.

BALLY Vlad
On the Malliavin covariance matrix for a class of stochastic processes

BELITSKY Vladimir
Dynamics of interacting particle systems

BLASZCZYSZYN Bartlomiej
Queues in series in light traffic

BRYC Wlodek
Large deviations for weakly dependent stationary random fields on Z^d

BUICULESCU Mioara
Irreducible Markov Processes

CABALLERO Maria Emilia
A chain rule for the derivation operator on the Wiener Space

DEHAY Dominique
Processus presque périodiquement corrélé

EL KAROUI Nicole
Evaluation dans les marchés incomplets et programmation dynamique

FERLAND René
Equations de Boltzmann généralisées : compacité en loi des fluctuations

FERNANDEZ Begona
Stability of a class of linear transformations of distribution-valued process and stochastic evolution equations

FLORCHINGER Patrick
Décroissance non exponentielle du noyau de la chaleur

FOURATI Sonia

GALLARDO Léonard
Le théorème de continuité de Paul Lévy dans un espace produit

GOROSTIZA Luis G.
Equilibrium states of measure branching processes

GRORUD Axel
Calcul stochastique anticipatif d'ordre 2

JEULIN Thierry
Décompositions non canoniques de semi-martingales

LE GLAND François
Approximation particulaire des EDP stochastiques du 1er ordre

LEONARD Christian
Grandes déviations pour un système de particules associé à l'équation de Boltzmann spatialement homogène

LOPEZ-MIMBELA José Alfredo
Some asymptotics of demographic variations in branching random fields

MARTIN-LOF Anders
Treshold Limit Theorems for Epidemic Models

MATHIEU Pierre

MICLO Laurent
Recuit simulé en dimension infinie

NOBLE John

NUALART David
Propriété de Markov pour la solution de certaines équations différentielles stochastiques

RAUTU Gheorghe
On some random sheets and symmetric diffusions

TREBICKI Jerzy
Maximum entropy principle in stochastic dynamics

VAN CASTEREN Johannes
Sur une approche stochastique de la théorie spectrale

VETS Peter
Non-commutative central limits

VONDRACEK Zoran
A characterization of Brownian motion by exit distributions

YCART Bernard
The philosopher's process : an ergodic reversible nearest particle system

ZIBAITIS Bronius
General scheme of approximation of stochastic integral

LISTE DES AUDITEURS

Mr. ABRAHAM Romain	Ecole Normale Supérieure, Paris
Mr. AMIDI Ali	Centre des Presses Universitaires, Iran
Mr. ANSEL Jean-Pascal	Université de Besançon
Mr. ARNAUDON Marc	Université de Strasbourg I
Mr. ASPANDIJAROV Sanjar	Moscow State University (Russie)
Mr. ATTAL Stéphane	Université de Strasbourg I
Mr. AZEMA Jacques	Université de Paris VI
Mr. BADRIKIAN Albert	Université Blaise Pascal (Clermont II)
Mr. BALLY Vlad	Université de Bucarest (Roumanie)
Mr. BELITSKY Vladimir	Technion Institute of Technology (Israël)
Mr. BEN AROUS Gérard	Université de Paris-Sud, Orsay
Mr. BERNARD Pierre	Université Blaise Pascal, (Clermont II)
Mr. BLASZCZYSZYN Bartlomiej	Université de Wroclaw (Pologne)
Mr. BRYC Wlodek	University of Cincinnati (U.S.A.)
Melle BUICULESCU Mioara	Université de Bucarest (Roumanie)
Mme CABALLERO Maria Emilia	Universidad Nacional Autonoma, Mexico (Mexique)
Mme CHALEYAT-MAUREL Mireille	Université de Paris VI
Mr. CIPRIANI Fabio	Ecole Supérieure Internationale, Trieste (Italie)
Mr. DEHAY Dominique	I.U.T. Lannion, Rennes
Mme EL KAROUI Nicole	Université de Paris VI
Mr. FERLAND René	Université de Montréal (Canada)
Melle FERNANDEZ Begona	Universidad Nacional Autonoma , Mexico (Mexique)
Mr. FERRANTE Marco	International School for Advanced Studies Trieste (Italie)
Mr. FLORCHINGER Patrick	Université de Metz
Mme FOURATI Sonia	I.N.S.A. de Rouen
Mr. FRANCOIS Olivier	Faculté de Médecine de Grenoble
Mr. GALLARDO Léonard	Université de Brest
Mr. GIROUX Gaston	Université de Sherbrooke, Canada
Mr. GOROSTIZA Luis G.	Cinvestav del I.P.N., Mexico (Mexique)
Mr. GRADINARU Mihai	Université de Paris-Sud, Orsay
Mr. GRORUD Axel	Université de Provence, Aix-Marseille
Mr. GUIMIER Alain	Institut Supérieur des Sciences Nouakchott (Mauritanie)
Mr. HENNEQUIN Paul-Louis	Université Blaise Pascal (Clermont II)
Mr. HU Ying	Université Claude Bernard, Lyon I
Mr. JEULIN Thierry	Université de Paris VII
Mr. LE GALL Jean-François	Université de Paris VI
Mr. LE GLAND François	I.N.R.I.A. Sophia Antipolis, Valbonne
Mr. LEHMAN Eric	Université de Caen
Mr. LEONARD Christian	Université de Paris-Sud, Orsay
Mr.LOPEZ-MIMBELA José Alfredo	Centro de Investigacion en Matematicas Université de Mexico (Mexique)
Melle MANCINO Maria Elvira	Université de Trente (Italie)
Mr. MARTIN-LÖF Anders	Université de Stockholm (Suède)
Mr. MATHIEU Pierre	Ecole Normale Supérieure, Paris
Mr. MAZLIAK Laurent	Université de Paris VI
Mme MELEARD Sylvie	Université de Paris VI
Mr. MICLO Laurent	Université Louis Pasteur, Strasbourg I
Mr. NOBLE John Masson	University of California, Irvine (USA)

Mr. NUALART David	Universitat de Barcelone (Espagne)
Mr. PARDOUX Etienne	Université de Provence, Aix-Marseille
Mme PERISIC Vesna	Universität des Saarlandes Saarbrücken (R.F.A.)
Mr. PICARD Jean	Université Blaise Pascal (Clermont II)
Mr. RAUTU Gheorghe	Université de Bucarest (Roumanie)
Melle REINERT Gesine	Institut für Angewandte Mathematik University of Zurich (Allemagne)
Mr. ROUAULT Alain	Université Paris-Sud, Orsay
Melle SAVONA Catherine	Université Blaise Pascal (Clermont II)
Mr. SERLET Laurent	E.N.S.A.E., Malakoff
Mr. SOLE i CLIVILLES Josep	Universitat de Barcelone (Espagne)
Mr. TREBICKI Jerzy	Institute of Technological Research Warszawa (Pologne)
Mr. VAN CASTEREN Johannes	Universitaire Instelling Antwerpen Wilrijk (Belgique)
Mr. VAN DER WEIDE Hans	Faculty TWI, Delft (Pays-Bas)
Mr. VETS Peter	Université de Louvain (Belgique)
Mr. VONDRACEK Zoran	Université de Zagres (Yougoslavie)
Mr. WATANABE Hisao	Okayama University of Science (Japon)
Mr. YCART Bernard	Université de Pau
Mr. ZIBAITIS Bronius	Université de Vilnius (Lituanie)

LIST OF PREVIOUS VOLUMES OF THE "Ecole d'Eté de Probabilités"

1971 - J.L. Bretagnolle (LNM 307)
"Processus à accroissements indépendants"
S.D. Chatterji
"Les martingales et leurs applications analytiques"
P.A. MEYER
"Présentation des processus de Markov"

1973 - P.A. MEYER (LNM 390)
"Transformation des processus de Markov"
P. PRIOURET
"Processus de diffusion et équations différentielles stochastiques"
F. SPITZER
"Introduction aux processus de Markov à paramètres dans Z_v"

1974 - X. FERNIQUE (LNM 480)
"Régularité des trajectoires des fonctions aléatoires gaussiennes"
J.P. CONZE
"Systèmes topologiques et métriques en théorie ergodique"
J. GANI
"Processus stochastiques de population"

1975 A. BADRIKIAN (LNM 539)
"Prolégomènes au calcul des probabilités dans les Banach"
J.F.C. KINGMAN
"Subadditive processes"
J. KUELBS
"The law of the iterated logarithm and related strong convergence theorems for Banach space valued random variables"

1976 J. HOFFMANN-JORGENSEN (LNM 598)
"Probability in Banach space"
T.M. LIGGETT
"The stochastic evolution of infinite systems of interacting particles"
J. NEVEU
"Processus ponctuels"

1977 D. DACUNHA-CASTELLE (LNM 678)
"Vitesse de convergence pour certains problèmes statistiques"
H. HEYER
"Semi-groupes de convolution sur un groupe localement compact et applications à la théorie des probabilités"
B. ROYNETTE
"Marches aléatoires sur les groupes de Lie"

1978 R. AZENCOTT (LNM 774)
"Grandes déviations et applications"
Y. GUIVARC'H
"Quelques propriétés asymptotiques des produits de matrices aléatoires"
R.F. GUNDY
"Inégalités pour martingales à un et deux indices : l'espace H^p"

1979 J.P. BICKEL (LNM 876)
"Quelques aspects de la statistique robuste"
N. EL KAROUI
"Les aspects probabilistes du contrôle stochastique"
M. YOR
"Sur la théorie du filtrage"

1980 J.M. BISMUT (LNM 929)
"Mécanique aléatoire"
L. GROSS
"Thermodynamics, statistical mechanics and random fields"
K. KRICKEBERG
"Processus ponctuels en statistique"

1981 X. FERNIQUE (LNM 976)
"Régularité de fonctions aléatoires non gaussiennes"
P.W. MILLAR
"The minimax principle in asymptotic statistical theory"
D.W. STROOCK
"Some application of stochastic calculus to partial differential equations"
M. WEBER
"Analyse infinitésimale de fonctions aléatoires"

1982 R.M. DUDLEY (LNM 1097)
"A course on empirical processes"
H. KUNITA
"Stochastic differential equations and stochastic flow of diffeomorphisms"
F. LEDRAPPIER
"Quelques propriétés des exposants caractéristiques"

1983 D.J. ALDOUS (LNM 1117)
"Exchangeability and related topics"
I.A. IBRAGIMOV
"Théorèmes limites pour les marches aléatoires"
J. JACOD
"Théorèmes limite pour les processus"

1984 R. CARMONA (LNM 1180)
"Random Schrödinger operators"
H. KESTEN
"Aspects of first passage percolation"
J.B. WALSH
"An introduction to stochastic partial differential equations"

1985-87 S.R.S. VARADHAN (LNM 1362)
"Large deviations"
P. DIACONIS
"Applications of non-commutative Fourier analysis to probability theorems
H. FOLLMER
"Random fields and diffusion processes"
G.C. PAPANICOLAOU
"Waves in one-dimensional random media"
D. ELWORTHY
Geometric aspects of diffusions on manifolds"
E. NELSON
"Stochastic mechanics and random fields"

1986 O.E. BARNDORFF-NIELSEN (LNS 50)
"Parametric statistical models and likelihood"

1988 A. ANCONA (LNM 1427)
"Théorie du potentiel sur les graphes et les variétés"
D. GEMAN
"Random fields and inverse problems in imaging"
N. IKEDA
"Probabilistic methods in the study of asymptotics"

1989 D.L. BURKHOLDER (LNM 1464)
"Explorations in martingale theory and its applications"
E. PARDOUX
"Filtrage non linéaire et équations aux dérivées partielles stochastiques associées"
A.S. SZNITMAN
"Topics in propagation of chaos"

1990 M.I. FREIDLIN (LNM 1527)
"Semi-linear PDE's and limit theorems for large deviations"
J.F. LE GALL
"Some properties of planar Brownian motion"

1991 D.A. DAWSON (LNM 1541)
"Measure-valued Markov processes"
B. MAISONNEUVE
"Processus de Markov : Naissance, Retournement, Régénération"
J. SPENCER
"Nine Lectures on Random Graphs"

Printing: Druckhaus Beltz, Hemsbach
Binding: Buchbinderei Schäffer, Grünstadt